Lecture Notes in Computer Science 13764

More information about this series at https://link.springer.com/bookseries/558

Patrizio Angelini · Reinhard von Hanxleden (Eds.)

Graph Drawing and Network Visualization

30th International Symposium, GD 2022
Tokyo, Japan, September 13–16, 2022
Revised Selected Papers

 Springer

Editors
Patrizio Angelini ⓘ
John Cabot University
Rome, Italy

Reinhard von Hanxleden ⓘ
Kiel University
Kiel, Germany

ISSN 0302-9743   ISSN 1611-3349   (electronic)
Lecture Notes in Computer Science
ISBN 978-3-031-22202-3   ISBN 978-3-031-22203-0   (eBook)
https://doi.org/10.1007/978-3-031-22203-0

This Springer imprint is published by the registered company Springer Nature Switzerland AG
The registered company address is: Gewerbestrasse 11, 6330 Cham, Switzerland

Preface

This volume contains the papers presented at GD 2022, the 30th International Symposium on Graph Drawing and Network Visualization, held during September 13–16, 2022 in Tokyo, Japan. Graph drawing is concerned with the geometric representation of graphs and constitutes the algorithmic core of network visualization. Graph drawing and network visualization are motivated by applications where it is crucial to visually analyze and interact with relational datasets. Information about the conference series and past symposia is maintained at http://www.graphdrawing.org.

As with GD 2020 and GD 2021, this 2022 conference was held under extraordinary circumstances. After the GD 2020 conference, which sadly had to be held wholly online due to the COVID-19 pandemic, and GD 2021, which took place as a hybrid conference at Universität Tübingen, we had another hybrid conference at the Tokyo Institute of Technology. The credit for this remarkable achievement in such uncertain times goes wholly to the local organizers.

A total of 29 participants from ten different countries attended the conference in person, with a further 115 registered participants from 17 countries online.

With regards to the program itself, regular papers could be submitted to one of two distinct tracks: Track 1 for papers on combinatorial and algorithmic aspects of graph drawing and Track 2 for papers on experimental, applied, and network visualization aspects. Short papers were given a separate category, which welcomed both theoretical and applied contributions. An additional track was devoted to poster submissions. All the tracks were handled by a single Program Committee. As committed to during the GD 2021 business meeting, particular attention was given to the design of Track 2. Broadly speaking, Track 2 should serve as platform that links the theory with the practice of graph drawing, and is seen as vital component for the overall relevance and further development of the community. Aiming for that goal prompted fairly deep discussions involving the current and former Program Chairs and the Steering Committee and resulted in a slightly revised track description in the Call for Papers (CfP). As in previous editions of GD, the papers in the different tracks did not compete with each other, but all Program Committee members were invited to review papers from either track.

In response to the CfP, the Program Committee received a total of 70 submissions, consisting of 65 papers (32 in Track 1, 16 in Track 2, and 17 in the short paper category) and five posters. As a novelty for GD, the review process switched from a single-blind to a "lightweight double-blind" process, where authors were asked to not disclose their identities but were free to disseminate draft versions of the paper prior to the conference and to give talks on the topic as they normally would.

More than 210 reviews were provided, about a third having been contributed by external sub-reviewers. After extensive electronic discussions by the Program Committee via EasyChair, interspersed with virtual meetings of the Program Chairs producing incremental accept/reject proposals, 25 long papers, seven short papers, and five posters were selected for inclusion in the scientific program of GD 2022. This resulted in an overall paper acceptance rate (not considering posters) of 49% (53% in Track 1, 50% in Track 2, and 41% in the short paper category). As is common in GD, some hard choices had to be made, in particular during the final acceptance/rejection round, where several papers that clearly had merit still did not make it. However, the number of submitted high-quality papers speaks for the community, and we were also pleased that Track 2 ended up well-represented.

In total, 20 of the 34 oral presentations (including two invited talks) were delivered on-site; the remaining 14 were delivered remotely using Zoom. Two posters were displayed on-site, all posters were also presented in Zoom breakout rooms.

Authors published an electronic version of their accepted papers on the arXiv e-print repository; a conference index with links to these contributions was made available before the conference.

There were two invited lectures at GD 2022. Ulrik Brandes from ETH Zürich (Switzerland) discussed "Positions in Social and Other Spaces", while Kazuo Misue from the University of Tsukuba (Japan) talked about "Graph Drawing for Thinking Support." Abstracts of both invited lectures are included in these proceedings.

The conference gave out best paper awards in Track 1 and Track 2, as well as a best presentation award and a best poster award. The award for the best paper in Track 1 was given to "Unavoidable patterns in complete simple topological graphs" by Andrew Suk and Ji Zeng, and the award for the best paper in Track 2 was assigned to "FORBID: Fast Overlap Removal By Stochastic Gradient Descent for Graph Drawing" by Loann Giovannangeli, Frédéric Lalanne, Romain Giot, and Romain Bourqui. Based on a majority vote of conference participants, the best presentation award was given to Philipp Kindermann for his presentation of the paper "Morphing Rectangular Duals." There was a tie for the best poster award, which was given to "The Witness Unit Disk Representability Problem" by Giuseppe Liotta, Maarten Löffler, Fabrizio Montecchiani, Alessandra Tappini, and Soeren Terziadis and to "Edge Bundling by Density-based Pathfinding Approach" by Ryosuke Saga, Tomoki Yoshikawa, and Tomoharu Nakashima. Many thanks to Springer whose sponsorship funded the prize money for these awards.

A PhD School was held on the two days prior to the conference. Three half-day sessions led by six invited lecturers covered both theoretical and practical topics in graph drawing and network visualization.

As is traditional, the 30th Annual Graph Drawing Contest was held during the conference. The contest was divided into two parts, creative topics and the live challenge. The creative topics task featured two graphs, an Opera Network (the data represented a collection of opera performances that took place across Europe between 1775 and 1833) and an Aesthetic Experience Network (the data set represented eight networks that model an aesthetic experience of the viewers when observing artworks). The live challenge focused on minimizing the planar polyline edge-length ratio on a fixed grid, with planar undirected inputs. There were two categories: manual and automatic. We thank the Contest Committee, chaired by Philipp Kindermann, for preparing interesting and challenging contest problems. A report about the contest is included in these proceedings.

Many people and organizations contributed to the success of GD 2022. We would like to thank all members of the Program Committee and the external reviewers for carefully reviewing and discussing the submitted papers and posters; this was crucial for putting together a strong and interesting program. Thanks to all authors who chose GD 2022 as the publication venue for their research.

We are grateful for the support of our "Gold" sponsor Tom Sawyer Software, our "Silver" sponsor yWorks, and our "Bronze" sponsor Springer.

Our special thanks go to all the members of the organizing committee based at Ochanomizu University, Tokyo Institute of Technology, Hokkaido Information University, St. Polten University of Applied Sciences, IBM Research, Japan, and the National Institute of Advanced Industrial Science and Technology (AIST), Japan.

The 31st International Symposium on Graph Drawing and Network Visualization (GD 2023) will take place during September 20–22, 2023, in Palermo, Italy. Michael Bekos and Markus Chimani will co-chair the Program Committee, and Emilio Di Giacomo, Fabrizio Montecchiani, and Alessandra Tappini will co-chair the Organizing Committee.

October 2022   Patrizio Angelini
                                                    Reinhard von Hanxleden

Organization

Steering Committee

Patrizio Angelini	John Cabot University, Italy
Michael A. Bekos	University of Ioannina, Greece
Markus Chimani	Osnabrück University, Germany
Giuseppe Di Battista	Roma Tre University, Italy
Emilio Di Giacomo	University of Perugia, Italy
Reinhard von Hanxleden	University of Kiel, Germany
Stephen G. Kobourov (Chair)	University of Arizona, USA
Anna Lubiw	University of Waterloo, Canada
Helen Purchase	Monash University, Australia
Ignaz Rutter	University of Passau, Germany
Roberto Tamassia	Brown University, USA
Ioannis G. Tollis	ICS-FORTH and University of Crete, Greece
Alexander Wolff	University of Würzburg, Germany

Program Committee

Patrizio Angelini (Co-chair)	John Cabot University, Italy
Therese Biedl	University of Waterloo, Canada
Sabine Cornelsen	University of Konstanz, Germany
Giordano Da Lozzo	Roma Tre University, Italy
Stephan Diehl	University of Trier, Germany
Henry Förster	Universität Tübingen, Germany
Martin Gronemann	TU Wien, Austria
Yasuhiro Hashimoto	University of Aizu, Japan
Michael Hoffmann	ETH Zurich, Switzerland
Hiroshi Hosobe	Hosei University, Japan
Yifan Hu	Yahoo!, USA
Takayuki Itoh	Ochanomizu University, Japan
Philipp Kindermann	Universität Trier, Germany
Karsten Klein	University of Konstanz, Germany
Stephen Kobourov	University of Arizona, USA
Jan Kratochvil	Charles University, Czech Republic
Kim Marriott	Monash University, Australia
Irene Parada	DTU, Denmark
Sergey Pupyrev	University of Arizona, USA
Helen Purchase	Monash University, Australia
Arnaud Sallaberry	LIRMM, Université Paul-Valéry Montpellier 3, Université de Montpellier, CNRS, France

Ingo Scholtes	University of Zurich, Switzerland
Falk Schreiber	University of Konstanz, Germany
André Schulz	FernUniversität in Hagen, Germany
Andrew Suk	University of California, San Diego, USA
Antonios Symvonis	National Technical University of Athens, Greece
Alessandra Tappini	University of Perugia, Italy
Reinhard von Hanxleden (Co-chair)	Christian-Albrechts-Universität zu Kiel, Germany
Tatiana von Landesberger	University of Cologne, Germany
Meirav Zehavi	Ben-Gurion University, Israel

Organizing Committee

Takayuki Itoh (Chair)	Ochanomizu University, Japan
Ken Wakita	Tokyo Institute of Technology, Japan
Masahiko Itoh	Hokkaido Information University, Japan
Hsiang-Yun Wu	St. Pölten University of Applied Sciences, Austria
Rina Nakazawa	IBM Research, Japan
Yuri Miyagi	National Institute of Advanced Industrial Science and Technology (AIST), Japan

Contest Committee

Philipp Kindermann (Chair)	University of Trier, Germany
Fabian Klute	Utrecht University, The Netherlands
Tamara Mchedlidze	Utrecht University, The Netherlands
Wouter Meulemans	TU Eindhoven, The Netherlands

External Reviewers

Ahmed, Abu Reyan
Aichholzer, Oswin
Alegría, Carlos
Archambault, Daniel
Bläsius, Thomas
Chen, Kun-Ting
Crnovrsanin, Tarik
Di Bartolomeo, Sara
Di Giacomo, Emilio
Didimo, Walter
Durocher, Stephane
Eppstein, David
Felsner, Stefan
Fink, Simon D.
Frati, Fabrizio

Giovannangeli, Loann
Grilli, Luca
Grosso, Fabrizio
Karim, Md. Rezaul
Keszegh, Balázs
Kryven, Myroslav
Kuckuk, Axel
Lahiri, Abhiruk
Li, Guangping
Löffler, Maarten
Martins, Rafael M.
Meulemans, Wouter
Miltzow, Till
Montecchiani, Fabrizio
Morin, Pat

Moy, Cameron
Nishat, Rahnuma Islam
Nöllenburg, Martin
Ortali, Giacomo
Pfister, Maximilian
Saffo, David
Schlipf, Lena
Schnider, Patrick

Schröder, Felix
Sonke, Willem
Sorge, Manuel
Spence, Richard
Stumpf, Peter
T. P., Sandhya
Verbeek, Kevin
Wood, David R.

Sponsors

Gold Sponsor

Silver Sponsor

Bronze Sponsor

Obituary

Takao Nishizeki, 1947–2022

Peter Eades[1], Seok-Hee Hong[1], Shin-ichi Nakano[2],
and Md. Saidur Rahman[3]

[1] The University of Sydney, Australia
[2] Gunma University, Japan
[3] Bangladesh University of Engineering and Technology (BUET), Bangladesh

It was with great sadness that we received the news that Takao Nishizeki passed away on January 30, 2022. Takao was a great contributor to Graph Drawing.

Takao Nishizeki studied at Tohoku University, and spent most of his career there, including some years as the Dean of Graduate School of Information Sciences. He retired in 2010, but continued research and teaching at Kwansei Gakuin University.

The Graph Drawing community knows Takao's books on planar graphs and planar graph drawing. His 1985 paper "Drawing plane graphs nicely" introduced us to the "CYN" graph drawing algorithm; this was the first method to construct a planar straight-line drawing of a graph in linear time. Since the resulting drawing has *convex* faces, the CYN algorithm forms the basis for many subsequent planar graph drawing algorithms, such as symmetric drawings and star-shaped drawings, as well as beyond planar graphs, such as straight-line drawings of 1-planar graphs.

Takao's main focus was for fundamental Graph Theory and Algorithms, and he produced many ground-breaking theorems and algorithms: in arboricity, planarity, matchings, Hamiltonicity, and colouring. Nevertheless, his work was broad; for example, his paper on Secret Sharing is very highly cited.

He served as an editor and Program Committee member for many prestigious journals and conferences, such as Algorithmica, Journal of Combinatorial Optimization and JGAA, and was a PC chair of the 15th International Symposium on Graph Drawing 2007, held in Australia.

In particular, the ISAAC (International Symposium on Algorithms and Computation) conference was founded by Takao Nishizeki, and he chaired the Advisory committee for many years. It is no exaggeration to say that the basis for a collaborative and supportive Algorithms community in Asia comes from Takao: his calm manner, his generous yet efficient way of steering ISAAC, as well as love of beer and karaoke. He ensured that ISAAC became the most internationally recognised Algorithms conference that was based in Asia. He was also the key motivating person behind the establishment of WALCOM conference.

We remember Takao fondly as a mentor and as a colleague.

Peter Eades, Seok-Hee Hong, Shin-ichi Nakano, and Md. Saidur Rahman on behalf of the Graph Drawing community, September 2022.

Invited Talks

Graph Drawing for Thinking Support

Kazuo Misue

University of Tsukuba, Tsukuba, Japan
misue@cs.tsukuba.ac.jp

In human intellectual activities, organizing the objects of thought is an important task. When the object of thought is large or complex, it is difficult for many people to organize it in their minds. In such cases, it is effective to take these objects out of our minds and observe them externally. When we externalize our thoughts in this way, the form of representation of the objects also influences the effectiveness and efficiency of our thinking process. When we organize fragments of information, we often establish a relationship between two fragments and group several fragments together. In the case of externalization, these relationships and groups are represented graphically, that is, the words and figures representing the fragments are connected by line segments or surrounded by closed lines. Some of the so-called "thinking support techniques" systematize such diagramming methods. The *KJ method* is one such technique.

In the diagram used in the KJ method, if each fragment of information is represented by a node, links connecting two nodes and closed curves surrounding the nodes are used. Moreover, in the KJ method, we can organize objects of thought by transforming a compound diagram that comprises a node-link diagram and an Euler diagram. The original KJ method was designed to be performed entirely by hand using analog tools such as cards, pens, and cords, and it was not intended for the use with computers. This, however, is also a factor that discourages the active use of the KJ method, although its usefulness as an ideation technique has been recognized. For example, despite the effectiveness of repeated drawing of diagrams on the same topic, we often do so only once because repeating the process requires time and effort.

The transformation operation of a compound diagram can be viewed as the transformation operation of a compound graph; to obtain a visual representation of such a compound graph, automatic drawing techniques are required. If we can separate the transforming operations of a compound diagram from the thinking process and if the transforming operations can be supported by a computer, the hurdles associated with the use of such an ideation technique will be reduced, and the efficiency of the ideation process could increase. Eventually, this increased efficiency will lead to an improvement in the quality of the ideation process and its products.

This talk introduces the research and development that has been carried out with this motivation.

Contents

Properties of Drawings of Complete Graphs

Unavoidable Patterns in Complete Simple Topological Graphs 3
  Andrew Suk and Ji Zeng

Compatible Spanning Trees in Simple Drawings of K_n 16
  Oswin Aichholzer, Kristin Knorr, Wolfgang Mulzer,
  Nicolas El Maalouly, Johannes Obenaus, Rosna Paul,
  Meghana M. Reddy, Birgit Vogtenhuber, and Alexandra Weinberger

Mutual Witness Gabriel Drawings of Complete Bipartite Graphs. 25
  William J. Lenhart and Giuseppe Liotta

Empty Triangles in Generalized Twisted Drawings of K_n 40
  Alfredo García, Javier Tejel, Birgit Vogtenhuber,
  and Alexandra Weinberger

Shooting Stars in Simple Drawings of $K_{m,n}$ 49
  Oswin Aichholzer, Alfredo García, Irene Parada, Birgit Vogtenhuber,
  and Alexandra Weinberger

Stress-based Visualizations of Graphs

FORBID: Fast Overlap Removal by Stochastic Gradient Descent for Graph
Drawing. 61
  Loann Giovannangeli, Frederic Lalanne, Romain Giot,
  and Romain Bourqui

Spherical Graph Drawing by Multi-dimensional Scaling. 77
  Jacob Miller, Vahan Huroyan, and Stephen Kobourov

Shape-Faithful Graph Drawings 93
  Amyra Meidiana, Seok-Hee Hong, and Peter Eades

Planar and Orthogonal Drawings

Planar Confluent Orthogonal Drawings of 4-Modal Digraphs 111
  Sabine Cornelsen and Gregor Diatzko

Unit-length Rectangular Drawings of Graphs 127
  Carlos Alegría, Giordano Da Lozzo, Giuseppe Di Battista,
  Fabrizio Frati, Fabrizio Grosso, and Maurizio Patrignani

Strictly-Convex Drawings of 3-Connected Planar Graphs   144
Michael A. Bekos, Martin Gronemann, Fabrizio Montecchiani,
and Antonios Symvonis

Rectilinear Planarity of Partial 2-Trees   157
Walter Didimo, Michael Kaufmann, Giuseppe Liotta,
and Giacomo Ortali

Drawings and Properties of Directed Graphs

Testing Upward Planarity of Partial 2-Trees   175
Steven Chaplick, Emilio Di Giacomo, Fabrizio Frati, Robert Ganian,
Chrysanthi N. Raftopoulou, and Kirill Simonov

Computing a Feedback Arc Set Using PageRank   188
Vasileios Geladaris, Panagiotis Lionakis, and Ioannis G. Tollis

st-Orientations with Few Transitive Edges   201
Carla Binucci, Walter Didimo, and Maurizio Patrignani

Beyond Planarity

Quasiplanar Graphs, String Graphs, and the Erdős-Gallai Problem   219
Jacob Fox, János Pach, and Andrew Suk

Planarizing Graphs and Their Drawings by Vertex Splitting   232
Martin Nöllenburg, Manuel Sorge, Soeren Terziadis, Anaïs Villedieu,
Hsiang-Yun Wu, and Jules Wulms

The Thickness of Fan-Planar Graphs is At Most Three   247
Otfried Cheong, Maximilian Pfister, and Lena Schlipf

An FPT Algorithm for Bipartite Vertex Splitting   261
Reyan Ahmed, Stephen Kobourov, and Myroslav Kryven

Dynamic Graph Visualization

On Time and Space: An Experimental Study on Graph Structural
and Temporal Encodings   271
Velitchko Filipov, Alessio Arleo, Markus Bögl, and Silvia Miksch

Small Point-Sets Supporting Graph Stories   289
Giuseppe Di Battista, Walter Didimo, Luca Grilli, Fabrizio Grosso,
Giacomo Ortali, Maurizio Patrignani, and Alessandra Tappini

On the Complexity of the Storyplan Problem 304
  Carla Binucci, Emilio Di Giacomo, William J. Lenhart, Giuseppe Liotta,
  Fabrizio Montecchiani, Martin Nöllenburg, and Antonios Symvonis

Visualizing Evolving Trees. 319
  Kathryn Gray, Mingwei Li, Reyan Ahmed, and Stephen Kobourov

Improved Scheduling of Morphing Edge Drawing. 336
  Kazuo Misue

Linear Layouts

Queue Layouts of Two-Dimensional Posets 353
  Sergey Pupyrev

Recognizing DAGs with Page-Number 2 Is NP-complete. 361
  Michael A. Bekos, Giordano Da Lozzo, Fabrizio Frati,
  Martin Gronemann, Tamara Mchedlidze,
  and Chrysanthi N. Raftopoulou

The Rique-Number of Graphs. 371
  Michael A. Bekos, Stefan Felsner, Philipp Kindermann,
  Stephen Kobourov, Jan Kratochvíl, and Ignaz Rutter

Contact and Visibility Graph Representations

Morphing Rectangular Duals 389
  Steven Chaplick, Philipp Kindermann, Jonathan Klawitter, Ignaz Rutter,
  and Alexander Wolff

Visibility Representations of Toroidal and Klein-bottle Graphs 404
  Therese Biedl

Coloring Mixed and Directional Interval Graphs 418
  Grzegorz Gutowski, Florian Mittelstädt, Ignaz Rutter,
  Joachim Spoerhase, Alexander Wolff, and Johannes Zink

Outside-Obstacle Representations with All Vertices on the Outer Face. 432
  Oksana Firman, Philipp Kindermann, Jonathan Klawitter, Boris Klemz,
  Felix Klesen, and Alexander Wolff

Arrangements of Pseudocircles: On Digons and Triangles 441
  Stefan Felsner, Sandro Roch, and Manfred Scheucher

GD Contest Report

Graph Drawing Contest Report........................... 459
  Philipp Kindermann, Fabian Klute, Tamara Mchedlidze,
  and Wouter Meulemans

Posters

Visualizing Node-Specific Hierarchies in Directed Networks 473
  Mykyta Shvets, Ehsan Moradi, and Debajyoti Mondal

Can an NN Model Plainly Learn Planar........................ 476
  Simon van Wageningen and Tamara Mchedlidze

Edge Bundling by Density-Based Pathfinding Approach 480
  Ryosuke Saga, Tomoki Yoshikawa, and Tomoharu Nakashima

The Witness Unit Disk Representability Problem 483
  Giuseppe Liotta, Maarten Löffler, Fabrizio Montecchiani,
  Alessandra Tappini, and Soeren Terziadis

Aggregating Hypergraphs by Node Attributes.................... 487
  David Trye, Mark Apperley, and David Bainbridge

Author Index 491

Properties of Drawings of Complete Graphs

Unavoidable Patterns in Complete Simple Topological Graphs

Andrew Suk and Ji Zeng$^{(\boxtimes)}$

Department of Mathematics, University of California at San Diego,
La Jolla, CA 92093, USA
{asuk,jzeng}@ucsd.edu

Abstract. We show that every complete n-vertex simple topological graph contains a topological subgraph on at least $(\log n)^{1/4-o(1)}$ vertices that is weakly isomorphic to the complete convex geometric graph or the complete twisted graph. This is the first improvement on the bound $\Omega(\log^{1/8} n)$ obtained in 2003 by Pach, Solymosi, and Tóth. We also show that every complete n-vertex simple topological graph contains a plane path of length at least $(\log n)^{1-o(1)}$.

Keywords: Topological graph · Unavoidable patterns · Plane path

1 Introduction

A *topological graph* is a graph drawn in the plane or, equivalently, on the sphere, such that its vertices are represented by points and its edges are represented by non-self-intersecting arcs connecting the corresponding points. The arcs are not allowed to pass through vertices different from their endpoints, and if two edges share an interior point, then they must properly cross at that point in common. A topological graph is *simple* if every pair of its edges intersect at most once, either at a common endpoint or at a proper crossing point. If the edges are drawn as straight-line segments, then the graph is said to be *geometric*. If the vertices of a geometric graph are in convex position, then it is called *convex*.

Simple topological graphs have been extensively studied [11,13,15,17,21], and are sometimes referred to as *good drawings* [1,2], or simply as *topological graphs* [14]. In this paper, we are interested in finding large unavoidable patterns in complete simple topological graphs. Two simple topological graphs G and H are *isomorphic* if there is a homeomorphism of the sphere that transforms G to H. We say that G and H are *weakly isomorphic* if there is an incidence preserving bijection between G and H such that two edges of G cross if and only if the corresponding edges in H cross as well. Clearly, any two complete convex geometric graphs on m vertices are weakly isomorphic. Hence, let C_m denote any complete convex geometric graph with m vertices.

By the famous Erdős-Szekeres convex polygon theorem [6] (see also [20]), every complete n-vertex geometric graph contains a geometric subgraph on $m =$

Supported by NSF CAREER award DMS-1800746 and NSF award DMS-1952786.

P. Angelini and R. von Hanxleden (Eds.): GD 2022, LNCS 13764, pp. 3–15, 2023.
https://doi.org/10.1007/978-3-031-22203-0_1

$\Omega(\log n)$ vertices that is weakly isomorphic to C_m. (Note that no three vertices in a complete geometric graph are collinear.) Interestingly, the same is not true for simple topological graphs. The *complete twisted graph* T_m is a complete simple topological graph on m vertices with the property that there is an ordering on the vertex set $V(T_m) = \{v_1, v_2, \ldots, v_m\}$ such that edges $v_i v_j$ and $v_k v_\ell$ cross if and only if $i < k < \ell < j$ or $k < i < j < \ell$. See Fig. 1. It was first observed by Harborth and Mengerson [10] that T_m does not contain a topological subgraph that is weakly isomorphic to C_5. However, in 2003, Pach, Solymosi, and Tóth [14] showed that it is impossible to avoid both C_m and T_m in a sufficiently large complete simple topological graph.

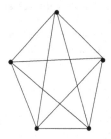 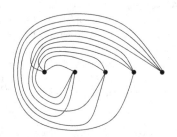

Fig. 1. C_5 and T_5.

Theorem 1 (Pach-Solymosi-Tóth). *Every complete n-vertex simple topological graph contains a topological subgraph on $m \geq \Omega(\log^{1/8} n)$ vertices that is weakly isomorphic to C_m or T_m.*

The main result of this paper is the following improvement.

Theorem 2. *Every complete n-vertex simple topological graph has a topological subgraph on $m \geq (\log n)^{1/4 - o(1)}$ vertices that is weakly isomorphic to C_m or T_m.*

In the other direction, let us consider the following construction. Let $V = \{1, 2, \ldots, n\}$ be n vertices placed on the x-axis, and for each pair $\{i, j\} \in V$, draw a half-circle connecting i and j, with this half-circle either in the upper or lower half of the plane uniformly randomly. By applying the standard probabilistic method [3], one can show that there is a complete n-vertex simple topological graph that does not contain a topological subgraph on $m = \lceil 8 \log n \rceil$ vertices that is weakly isomorphic to C_m or T_m. Another construction, observed by Scheucher [18], is to take n points in the plane with no $2\lceil \log n \rceil$ members in convex position, and then draw straight-line segments between all pairs of points.

It is not hard to see that both C_m and T_m contain a plane (i.e. crossing-free) subgraph isomorphic to any given tree T with at most m vertices (see, e.g., [9]). Thus, as a corollary of Theorem 2, we obtain the following.

Corollary 1. *Every complete n-vertex simple topological graph contains a plane subgraph isomorphic to any given tree T with at most $(\log n)^{1/4 - o(1)}$ vertices.*

In the case when T is a path, we improve this bound with the following result, which is also recently obtained in [2] independently.

Theorem 3. *Every complete n-vertex simple topological graph contains a plane path of length at least $(\log n)^{1-o(1)}$.*

In order to avoid confusion between topological and combinatorial edges, we write uv when referring to a topological edge in the plane, and write $\{u, v\}$ when referring to an edge (pair) in a graph. Likewise, we write $\{u_1, \ldots, u_k\}$ when referring to an edge (k-tuple) in a k-uniform hypergraph. We systemically omit floors and ceilings whenever they are not crucial for the sake of clarity in our presentation. All logarithms are in base 2.

2   Monotone Paths and Online Ramsey Numbers

Before we prove Theorem 2, let us recall the following lemmas. Let H be a k-uniform hypergraph with vertex set $[n] = \{1, 2, \ldots, n\}$. We say that H contains a *monotone k-path* of length m if there are m vertices $v_1 < v_2 < \cdots < v_m$ such that $\{v_i, v_{i+1}, \ldots, v_{i+k-1}\} \in E(H)$ for $1 \leq i \leq m - k + 1$. We say that the edge set $E(H)$ is *transitive* if for any $v_1 < v_2 < \cdots < v_{k+1}$ in $[n]$, the condition $\{v_1, v_2, \ldots, v_k\}, \{v_2, v_3, \ldots, v_{k+1}\} \in E(H)$ implies all k-element subsets of $\{v_1, \ldots, v_{k+1}\}$ are in $E(H)$. We will need the following lemma due to Fox, Pach, Sudakov, and Suk.

Lemma 1 ([7]). *Let $n > k$, and let H be a k-uniform hypergraph with vertex set $[n]$, which contains a monotone path of length n, that is, $\{i, i+1, \ldots, i+k-1\} \in E(H)$ for all $1 \leq i \leq n - k + 1$. If $E(H)$ is transitive, then H is the complete k-uniform hypergraph on $[n]$.*

Next, we need a lemma from Online Ramsey Theory. The *vertex online Ramsey game* is a game played by two players, *builder* and *painter*. Let $t \geq 1$ and suppose vertices $v_1, v_2, \ldots, v_{t-1}$ are present. At the beginning of stage t, a new vertex v_t is added. Then for each $v_i \in \{v_1, \ldots, v_{t-1}\}$, builder decides (in any order) whether to create the edge $\{v_i, v_t\}$. If builder creates the edge, then painter has to immediately color it red or blue. When builder decides not to create any more edges, stage t ends and stage $t + 1$ begins by adding a new vertex. Moreover, builder must create at least one edge at every stage except for the first one. The *vertex online Ramsey number* $r(m)$ is the minimum number of edges builder has to create to guarantee a monochromatic monotone path of length m in a vertex online Ramsey game. Clearly, we have $r(m) \leq O(m^4)$, which is obtained by having builder create all possible edges at each stage and applying Dilworth's theorem [5] on the m^2 vertices. Fox, Pach, Sudakov, and Suk proved the following.

Lemma 2 ([7]). *We have $r(m) = (1 + o(1))m^2 \log_2 m$.*

3   Convex Geometric Graph Versus Twisted Graph

In this section, we prove the following theorem, from which Theorem 2 quickly follows.

Theorem 4. *Let m_1, m_2, n be positive integers such that*

$$9(m_1 m_2)^2 \log(m_1) \log(m_2) < \log n.$$

Then every complete n-vertex simple topological graph contains a topological subgraph that is weakly isomorphic to C_{m_1} or T_{m_2}.

Proof. Let $G = (V, E)$ be a complete n-vertex simple topological graph. Notice that the edges of G divide the plane into several cells (regions), one of which is unbounded. We can assume that there is a vertex $v_0 \in V$ such that v_0 lies on the boundary of the unbounded cell. Indeed, otherwise we can project G onto a sphere, then choose an arbitrary vertex v_0 and then project G back to the plane such that v_0 lies on the boundary of the unbounded cell, moreover, the new drawing is isomorphic to the original one as topological graphs.

Consider the topological edges emanating out from v_0, and label their endpoints $v_1, \ldots, v_{n-1}$ in clockwise order. For convenience, we write $v_i \prec v_j$ if $i < j$. Given subsets $U, W \subset \{v_1, \ldots, v_{n-1}\}$, we write $U \prec W$ if $u \prec w$ for all $u \in U$ and $w \in W$. Following the notation used in [14], we color the triples of $\{v_1, \ldots, v_{n-1}\}$ as follows. For $v_i \prec v_j \prec v_k$, let $\chi(v_i, v_j, v_k) = xyz$, where $x, y, z \in \{0, 1\}$ such that

1. setting $x = 1$ if edges $v_j v_k$ and $v_0 v_i$ cross, and let $x = 0$ otherwise;
2. setting $y = 1$ if edges $v_i v_k$ and $v_0 v_j$ cross, and let $y = 0$ otherwise;
3. setting $z = 1$ if edges $v_i v_j$ and $v_0 v_k$ cross, and let $z = 0$ otherwise.

Pach, Solymosi, and Tóth observed the following.

Observation 1 ([14])**.** *The only colors that appear with respect to χ are 000, 001, 010, and 100.*

Fig. 2. Configurations for 000, 010, 001, 100 respectively.

See Fig. 2 for an illustration. We now make another observation.

Lemma 3. *Colors* 001 *and* 100 *are transitive. That is, for* $v_i \prec v_j \prec v_k \prec v_\ell$,

1. *if* $\chi(v_i, v_j, v_k) = \chi(v_j, v_k, v_\ell) = 001$, *then* $\chi(v_i, v_j, v_\ell) = \chi(v_i, v_k, v_\ell) = 001$;
2. *if* $\chi(v_i, v_j, v_k) = \chi(v_j, v_k, v_\ell) = 100$, *then* $\chi(v_i, v_j, v_\ell) = \chi(v_i, v_k, v_\ell) = 100$.

Proof. Suppose $\chi(v_i, v_j, v_k) = \chi(v_j, v_k, v_\ell) = 001$. Since edges $v_0 v_\ell$ and $v_j v_k$ cross, vertex v_ℓ must lie in the closed region bounded by edges $v_j v_k$, $v_i v_j$, and $v_0 v_k$. See Fig. 3. Hence, edge $v_0 v_\ell$ crosses both $v_i v_j$ and $v_i v_k$. Therefore, we have $\chi(v_i, v_j, v_\ell) = \chi(v_i, v_k, v_\ell) = 001$ as wanted. If $\chi(v_i, v_j, v_k) = \chi(v_j, v_k, v_\ell) = 100$, a similar argument shows that we must have $\chi(v_i, v_j, v_\ell) = \chi(v_i, v_k, v_\ell) = 100$. $\square$

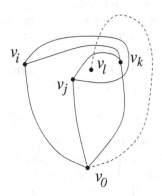

Fig. 3. The closed region bounded by edges $v_j v_k$, $v_i v_j$, and $v_0 v_k$ in Lemma 3.

Based on the coloring χ, we define a coloring ϕ of the pairs of $\{v_1, v_2, \ldots, v_{n-1}\}$ as follows. For $v_i \prec v_j$, let $\phi(v_i, v_j) = (a, b)$ where a is the length of the longest monotone 3-path ending at $\{v_i, v_j\}$ in color 100, and b is the length of the longest monotone 3-path ending at $\{v_i, v_j\}$ in color 001. We can assume that $a, b < m_2$. Otherwise, by Lemmas 3 and 1, we would have a subset $U \subset V$ of size m_2 whose triples are all of the same color, 100 or 001. And it is not hard to argue by induction that such a U corresponds to a topological subgraph that is weakly isomorphic to T_{m_2} as wanted.

Before we continue, let us give a rough outline of the rest of the proof. In what follows, we will construct disjoint vertex subsets $V^{a,b} \subset \{v_1, \ldots, v_{n-1}\}$, where $1 < a, b < m_2$, such that ϕ colors every pair in $V^{a,b}$ with color (a, b). For each $V^{a,b}$, we will play the vertex online Ramsey game by letting the builder create an edge set $E^{a,b}$ and designing a painter's strategy, which gives rise to a coloring ψ on $E^{a,b}$. We then apply Lemma 2 to show that if n is sufficiently large, some vertex set $V^{a,b}$ will contain a monochromatic monotone 2-path of length m_1 with respect to ψ. Finally, we will show that this monochromatic monotone 2-path will correspond to a topological subgraph that is weakly isomorphic to C_{m_1}. The detailed argument follows.

For integers $t \geq 0$ and $1 < a, b < m_2$, we construct a vertex subset $V_t^{a,b} \subset \{v_1, \ldots, v_{n-1}\}$, an edge set $E_t^{a,b}$ of pairs in $V_t^{a,b}$, and a subset $S_t \subset \{v_1, \ldots, v_{n-1}\}$ such that the following holds.

1. We have $\sum\limits_{1 < a,b < m_2} |V_t^{a,b}| = t$.
2. For all $1 < a, b < m_2$, we have $V_t^{a,b} \prec S_t$.
3. For $u_1 \in V_t^{a,b}$, we have $\phi(u_1, u_2) = (a, b)$ for every $u_2 \in V_t^{a,b} \cup S_t$ with $u_1 \prec u_2$.
4. For each edge $\{u_1, u_2\} \in E_t^{a,b}$, where $u_1 \prec u_2$, we have $\chi(u_1, u_2, u_3) = \chi(u_1, u_2, u_4)$ for all $u_3, u_4 \in V_t^{a,b}$ such that $u_1 \prec u_2 \prec u_3 \prec u_4$.

We start by setting $V_0^{a,b} = \emptyset$ for all $1 < a, b < m_2$, and $S_0 = \{v_1, \ldots, v_{n-1}\}$. After stage t, we have $V_t^{a,b}$, $E_t^{a,b}$, for $1 < a, b < m_2$, and S_t as described above.

At the beginning of stage $t + 1$, let w_{t+1} be the smallest element in S_t with respect to $\prec$. By the pigeonhole principle, there exists integers $1 < \alpha, \beta < m_2$ and a subset $S_{t,0} \subset S_t \setminus \{w_{t+1}\}$ of size at least $(|S_t| - 1)/m_2^2$, such that $\phi(w_{t+1}, u) = (\alpha, \beta)$ for all $u \in S_{t,0}$. Then we set $V_{t+1}^{\alpha,\beta} := V_t^{\alpha,\beta} \cup \{w_{t+1}\}$. For all $1 < a, b < m_2$ with $(a, b) \neq (\alpha, \beta)$, we set $V_{t+1}^{a,b} := V_t^{a,b}$ and $E_{t+1}^{a,b} := E_t^{a,b}$.

Claim 1. *For all $u \in V_t^{\alpha,\beta}$ and $v \in S_{t,0}$, we have $\chi(u, w_{t+1}, v) \in \{000, 010\}$.*

Proof. For the sake of contradiction, suppose $\chi(u, w_{t+1}, v) = 100$, where $u \in V_t^{\alpha,\beta}$ and $v \in S_{t,0}$. Since $\phi(u, w_{t+1}) = (\alpha, \beta)$, the longest monotone 3-path in color 100 ending at $\{u, w_{t+1}\}$ has length α. Hence, the longest monotone 3-path in color 100 ending at $\{w_{t+1}, v\}$ has length at least $\alpha + 1$. This contradicts the fact that $\phi(w_{t+1}, v) = (\alpha, \beta)$. A similar argument follows if $\chi(u, w_{t+1}, v) = 001$. $\square$

Now that we have constructed $V_{t+1}^{\alpha,\beta}$ by adding w_{t+1} to $V_t^{\alpha,\beta}$, we play the vertex online Ramsey game so that builder chooses and creates edges of the form $\{u, w_{t+1}\}$, where $u \in V_t^{\alpha,\beta}$, according to his strategy. After each edge $\{u, w_{t+1}\}$ is created, painter immediately colors it $\psi(u, w_{t+1}) \in \{000, 010\}$ as follows. In painter's strategy, after the j-th edge $\{u_j, w_{t+1}\}$ is created and colored, a set $S_{t,j} \subset S_{t,0}$ will be constructed such that all triples $\{u_j, w_{t+1}, v\}$ with $v \in S_{t,j}$ are colored by χ with the same color in $\{000, 010\}$. After the $(j + 1)$-th edge $\{u_{j+1}, w_{t+1}\}$ is created, painter looks at all triples of the form $\{u_{j+1}, w_{t+1}, v\}$ with $v \in S_{t,j}$. Since $\chi(u_{j+1}, w_{t+1}, v) \in \{000, 010\}$ by Claim 1, the pigeonhole principle implies that there exists a subset $S_{t,j+1} \subset S_{t,j}$ with size at least $|S_{t,j}|/2$ such that all triples $\{u_{j+1}, w_{t+1}, v\}$ with $v \in S_{t,j+1}$ are colored by χ with the same color $xyz \in \{000, 010\}$. Then painter sets $\psi(u_{j+1}, w_{t+1}) = xyz$.

If builder decides to stop creating edges from w_{t+1} to $V_t^{\alpha,\beta}$ after j edges are created and colored, the stage ends and we set $S_{t+1} = S_{t,j}$, and we let $E_{t+1}^{\alpha,\beta}$ be the union of $E_t^{\alpha,\beta}$ and all edges built during this stage. Let e_{t+1} denote the total number of edges builder creates in stage $t + 1$. Recall that $e_{t+1} \geq 1$ unless $V_t^{\alpha,\beta} = \emptyset$. As long as $|S_{t+1}| > 0$, we continue this construction process by starting the next stage. Clearly, $V_{t+1}^{a,b}$, $E_{t+1}^{a,b}$, for all $1 < a, b < m_2$, and S_{t+1} have the four properties described above. We now make the following claim.

Claim 2. *For $t \geq 1$, we have*

$$|S_t| \geq \frac{n-1}{m_2^{2t} \cdot 2^{\sum_{i=2}^t e_i}} - \sum_{i=2}^t \frac{1}{m_2^{2(t+1-i)} \cdot 2^{\sum_{j=i}^t e_j}}.$$

Proof. We proceed by induction on t. For the base case $t = 1$, there's no edge for the builder to build in the first stage, so $|S_1| = |S_{0,0}| \geq (n-1)/m_2^2$ as desired. For the inductive step, assume the statement holds for $t \geq 1$. When we start stage $t + 1$ and introduce vertex w_{t+1}, the set S_t shrinks to $S_{t,0}$ whose size is guaranteed to be at least $(|S_t| - 1)/m_2^2$, and each time builder creates an edge from w_{t+1} to $V_t^{\alpha,\beta}$, our set decreases by a factor of two. Since builder creates e_{t+1} edges during stage $t + 1$, we have

$$|S_{t+1}| \geq \frac{|S_t| - 1}{m_2^2 2^{e_{t+1}}} \geq \frac{n-1}{m_2^{2(t+1)} \cdot 2^{\sum_{i=2}^{t+1} e_i}} - \sum_{i=2}^t \frac{1}{m_2^{2((t+1)+1-i)} \cdot 2^{\sum_{j=i}^{t+1} e_j}} - \frac{1}{m_2^2 2^{e_{t+1}}}$$

$$= \frac{n-1}{m_2^{2(t+1)} \cdot 2^{\sum_{i=2}^{t+1} e_i}} - \sum_{i=2}^{t+1} \frac{1}{m_2^{2((t+1)+1-i)} \cdot 2^{\sum_{j=i}^{t+1} e_j}},$$

which is what we want. $\square$

After t stages, builder has created a total of $\sum_{i=1}^t e_i$ edges, such that each edge has color 000 or 010 with respect to ψ. If there is no monochromatic 2-path of length m_1 with respect to ψ on any $(V_t^{a,b}, E_t^{a,b})$, this implies that

$$\sum_{i=1}^t e_i < m_2^2 r(m_1) \leq 2(m_1 m_2)^2 \log m_1.$$

Also, since $e_i \geq 1$ for all but m_2^2 many indices $1 \leq i \leq t$, we have

$$t \leq m_2^2 + \sum_{i=1}^t e_i < 3(m_1 m_2)^2 \log m_1.$$

Since we assumed

$$n > 2^{9(m_1 m_2)^2 \log(m_1) \log(m_2)},$$

we have

$$|S_t| \geq \frac{n-1}{m_2^{2t} \cdot 2^{\sum_{i=2}^t e_i}} - \sum_{i=2}^t \frac{1}{m_2^{2(t+1-i)} \cdot 2^{\sum_{j=i}^t e_j}}$$

$$\geq \frac{n-1}{2^{8(m_1 m_2)^2 \log(m_1) \log(m_2)}} - \sum_{i=2}^t \frac{1}{2^{t-i+1}} > 1.$$

Hence, we can continue to the next stage and introduce vertex w_{t+1}. Therefore, when this process stops, say at stage s, we must have a monochromatic monotone 2-path of length m_1 with respect to ψ on some $(V_s^{a,b}, E_s^{a,b})$.

Now let $W^* = \{w_1^*, \ldots, w_{m_1}^*\}$, where $w_1^* \prec \cdots \prec w_{m_1}^*$, be the vertex set that induces a monochromatic monotone 2-path of length m_1 with respect to ψ on $(V_s^{a,b}, E_s^{a,b})$. Since ϕ colors every pair in W^* with the color (a, b), by following the proof of Claim 1, we have $\chi(w_i^*, w_j^*, w_k^*) \in \{000, 010\}$ for every $i < j < k$. Hence, the following argument due to Pach, Solymosi, and Tóth [14] shows that W^* induces a topological subgraph that is weakly isomorphic to C_{m_1}. For the sake of completeness, we include the proof.

Claim 3. *Let $W^* = \{w_1^*, \ldots, w_{m_1}^*\}$ be as described above. Then W^* induces a topological subgraph that is weakly isomorphic to C_{m_1}.*

Proof. Suppose $\psi(w_i^*, w_{i+1}^*) = 000$ for all i. It suffices to show that every triple in W^* has color 000 with respect to χ. For the sake of contradiction, suppose we have $w_i^* \prec w_j^* \prec w_k^*$ such that $\chi(w_i^*, w_j^*, w_k^*) = 010$, and let us assume that $j - i$ is minimized among all such examples. Since $\{w_i^*, w_{i+1}^*\} \in E_s^{a,b}$, we have $\chi(w_i^*, w_{i+1}^*, w_k^*) = \psi(w_i^*, w_{i+1}^*) = 000$. This implies that $j > i + 1$ and the edge $w_{i+1}^* w_k^*$ crosses $v_0 w_j^*$ (see Fig. 4), which contradicts the minimality condition. A similar argument follows if $\psi(w_i^*, w_{i+1}^*) = 010$ for all i. □

This completes the proof of Theorem 4   □

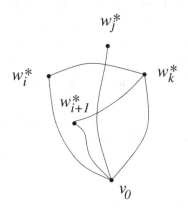

Fig. 4. A figure illustrating Claim 3.

4   Plane Path

In this section, we prove Theorem 3. We will need the following lemma, which was observed by Fulek and Ruiz-Vargas in [8].

Lemma 4. *If a complete simple topological graph G contains a topological subgraph that is isomorphic to a plane K_{2,m^2}, then G contains a plane path of length $\Omega(m)$.*

Let us briefly explain how to establish this lemma, as it is not explicitly stated in [8]. In [22], Tóth proved that every n-vertex geometric graph with more than $2^9 k^2 n$ edges contains k pairwise disjoint edges. His proof easily generalizes to simple topological graphs whose edges are drawn as x-monotone curves, and, in fact, shows the existence of a plane path of length $2k$.

Given a plane topological subgraph K_{2,m^2} inside a complete simple topological graph G, Fulek and Ruiz-Vargas [8] showed that there exists a topological subgraph $G' \subset G$, with m^2 vertices and $\Omega(m^4)$ edges, that is weakly-isomorphic to an x-monotone simple topological graph G''. Hence, we can conclude Lemma 4 by applying Tóth's result stated above with $k = \Omega(m)$.

Proof (of Theorem 3). First, we keep the following notations from the proof of Theorem 2. Let $G = (V, E)$ be a complete n-vertex simple topological graph. We can assume that there is a vertex $v_0 \in V$ such that v_0 lies on the boundary of the unbounded cell. We label the other vertices by $v_1, \ldots, v_{n-1}$ such that the edges $v_0 v_i$, for $1 \le i < n$, emanate out from v_0 in clockwise order. We write $v_i \prec v_j$ if $i < j$, and color every triple $v_i \prec v_j \prec v_k$ by $\chi(v_i, v_j, v_k) \in \{000, 010, 100, 001\}$.

For each v_i, we arrange the vertices $\{v_{i+1}, \ldots, v_{n-1}\}$ into a sequence $\theta(v_i) = (v_{j_1}, \ldots, v_{j_{n-1-i}})$ such that the topological edges $v_i v_0, v_i v_{j_1}, v_i v_{j_2}, \ldots, v_i v_{j_{n-1-i}}$ emanate out from v_i in counterclockwise order. See Fig. 5. We call a sequence of vertices $S = (v_{i_1}, \ldots, v_{i_k})$ increasing (or decreasing) if $v_{i_1} \prec v_{i_2} \prec \cdots \prec v_{i_k}$ (or $v_{i_1} \succ v_{i_2} \succ \cdots \succ v_{i_k}$).

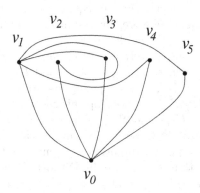

Fig. 5. An example with $\theta(v_1) = (v_4, v_3, v_2, v_5)$.

Lemma 5. *If there exists a vertex u such that $\theta(u)$ contains an increasing subsequence $(u_1, \ldots, u_{m^2})$, then the edges $v_0 u_i$ and $u u_i$, for all $1 \le i \le m^2$, form a plane subgraph K_{2,m^2}.*

Proof. It suffices to show v_0u_i and uu_j do not cross each other for every $1 \le i, j \le k$. When $i = j$, this follows from G being simple. When $j > i$, by the increasing assumption, the edges uv_0, uu_i, and uu_j emanate out from u in counterclockwise order. Observe that this condition forces u_j to be outside the region $\Delta_{v_0uu_i}$ bounded by the topological edges v_0u, uu_i, and u_iv_0. Then the Jordan arc uu_j starting at u, initially outside $\Delta_{v_0uu_i}$, cannot enter $\Delta_{v_0uu_i}$ then leave again to end at u_j. In particular, uu_j doesn't cross v_0u_i. See Fig. 6 for an illustration. A similar argument follows if $j < i$. $\square$

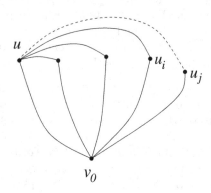

Fig. 6. An increasing subsequence of $\theta(u)$ induce a plane K_{2,m^2}.

We set $m = \left\lfloor \frac{\log n}{2 \log \log n} \right\rfloor$ and prove that G contains a plane path of length $\Omega(m)$. We can assume $m > 1$, otherwise there's nothing to prove. If some sequence $\theta(v_i)$ contains an increasing subsequence of length m^2, then by Lemmas 5 and 4, we are done. Therefore, we assume that $\theta(v_i)$ doesn't contain an increasing subsequence of length m^2 for every i.

For integer $t \ge 1$, we inductively construct subsets $U_t, S_t \subset \{v_1, \ldots, v_{n-1}\}$ with $U_t = \{u_1, \ldots, u_t\}$, where $u_1 \prec \cdots \prec u_t$, and $U_t \prec S_t$. Initially we set $U_1 = \{u_1 := v_1\}$ and $S_1 = \{v_2, \ldots, v_{n-1}\}$. Suppose for some t, we have already constructed U_t and S_t. If $|S_t| \le m^2$, we stop this construction process, otherwise we continue to construct U_{t+1} and S_{t+1} as follows: Let θ' be the subsequence of $\theta(u_t)$ that contains exactly those vertices in S_t. Note that the length of θ' equals to $|S_t|$. According to our assumption, the length of the longest increasing subsequence in θ' is less than m^2. Hence, by Dilworth's theorem [5], θ' contains a decreasing subsequence of length at least $|S_t|/m^2$. Let S'_{t+1} be the set of vertices that appear in this decreasing subsequence of θ'. Next, we take u_{t+1} to be the smallest element of S'_{t+1} with respect to $\prec$ and let $U_{t+1} := U_t \cup \{u_{t+1}\}$. Consider the region $\Delta_{v_0u_tu_{t+1}}$ bounded by the topological edges v_0u_t, u_tu_{t+1}, and $u_{t+1}v_0$. Each vertex in $S'_{t+1} \setminus \{u_{t+1}\}$ is either inside or outside $\Delta_{v_0u_tu_{t+1}}$. So, by the pigeonhole principle, there exists a subset $S_{t+1} \subset S'_{t+1} \setminus \{u_{t+1}\}$ with $|S_{t+1}| \ge |S'_{t+1} \setminus \{u_{t+1}\}|/2$ such that the whole set S_{t+1} is either inside or outside $\Delta_{v_0u_tu_{t+1}}$. Clearly, we have $U_{t+1} \prec S_{t+1}$ and

$$|S_{t+1}| \geq \frac{|S_t|/m^2 - 1}{2} \geq \frac{|S_t|}{(2m)^2}.$$

Using the inequality above and the fact that $|S_1| = n - 2$, we can inductively prove $|S_t| \geq \frac{n}{(2m)^{2t}}$. When $t = m - 1$, this gives us

$$|S_{m-1}| \geq \frac{n}{(2m)^{2(m-1)}} > \frac{n}{(2m)^{\log n/\log\log n - 2}} > m^2 \cdot \frac{n}{(\log n)^{\log n/\log\log n}} \geq m^2.$$

Hence, the construction process ends at a certain t larger than $m - 1$, and we will always construct $U_m = \{u_1, \ldots, u_m\}$.

Now we show that $u_i u_{i+1}$, for $1 \leq i < m$, form a plane path. Our argument is based on the following two claims.

Claim 4. *For any vertices $u_i \prec u_{i+1} \prec u_j \prec u_k$, we have u_j and u_k either both inside or both outside the region $\Delta_{v_0 u_i u_{i+1}}$.*

Claim 4 is obviously guaranteed by the construction process of U_m.

Claim 5. *For any vertices $u_i \prec u_j \prec u_k$, the topological edges $v_0 u_i$ and $u_j u_k$ do not cross each other.*

Proof. Consider the region $\Delta_{v_0 u_i u_j}$ bounded by the topological edges $v_0 u_i$, $u_i u_j$, and $u_j v_0$, then u_k is either inside or outside $\Delta_{v_0 u_i u_j}$. If u_k is inside $\Delta_{v_0 u_i u_j}$, then $v_0 u_k$ must cross $u_i u_j$. By Observation 1, we have $\chi(u_i, u_j, u_k) = 001$, which implies $v_0 u_i$ and $u_j u_k$ do not cross. See the third configuration in Fig. 2.

Suppose u_k is outside $\Delta_{v_0 u_i u_j}$. By the construction process of U_m, the edges $u_i v_0$, $u_i u_k$ and $u_i u_j$ must emanate from u_i in counterclockwise order, this implies that $u_i u_k$ crosses $v_0 u_j$. Then, by Observation 1, $\chi(u_i, u_j, u_k) = 010$ and $u_j u_k$ doesn't cross $v_0 u_i$. See the second configuration in Fig. 2. $\square$

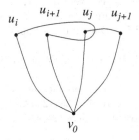
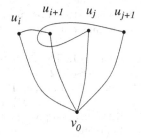

Fig. 7. For $u_i u_{i+1}$ and $u_j u_{j+1}$ with $i+1 < j$ to cross each other, either u_j and u_{j+1} are not both inside or both outside $\Delta_{v_0 u_i u_{i+1}}$ (left graph), or the topological edge $u_j u_{j+1}$ crosses one edge in $\{v_0 u_i, v_0 u_{i+1}\}$ (right graph).

Finally, we argue that the edges $u_i u_{i+1}$ and $u_j u_{j+1}$ do not cross for any $i < j$. When $j = i+1$, this follows from G being simple. When $j > i+1$, by Claim 4, the vertices u_j and u_{j+1} are either both inside or both outside the region $\Delta_{v_0 u_i u_{i+1}}$. So, the edge $u_j u_{j+1}$ crosses the boundary of $\Delta_{v_0 u_i u_{i+1}}$ an even number of times. On the other hand, by Claim 5, $u_j u_{j+1}$ doesn't cross $v_0 u_i$ or $v_0 u_{i+1}$. So $u_j u_{j+1}$ doesn't cross $u_i u_{i+1}$. See Fig. 7 for an illustration. This concludes the proof of Theorem 3.   □

5   Concluding Remarks

Answering a question of Pach and Tóth [15], Suk showed that every complete n-vertex simple topological graph contains $\Omega(n^{1/3})$ pairwise disjoint edges [19] (see also [8]). This bound was later improved to $n^{1/2-o(1)}$ by Ruiz-Vargas in [16]. Hence, for plane paths, we conjecture a similar bound should hold.

Conjecture 1. There is an absolute constant $\varepsilon > 0$, such that every complete n-vertex simple topological graph contains a plane path of length n^ε.

Let $h = h(n)$ be the smallest integer such that every complete n-vertex simple topological graph contains an edge crossing at most h other edges. A construction due to Valtr (see page 398 in [4]) shows that $h(n) \geq \Omega(n^{3/2})$. In the other direction, Kynčl and Valtr [12] used an asymmetric version of Theorem 1 to show that $h(n) = O(n^2/\log^{1/4} n)$. By using Theorem 4 instead, their arguments show that $h(n) \leq n^2/(\log n)^{1/2-o(1)}$. We conjecture the following.

Conjecture 2. There is an absolute constant $\varepsilon > 0$ such that $h(n) \leq n^{2-\varepsilon}$.

References

1. Ábrego, B., et al.: All good drawings of small complete graphs. In: Proceedings 31st European Workshop on Computational Geometry, pp. 57–60 (2015)
2. Aichholzer, O., García, A., Tejel, J., Vogtenhuber, B., Weinberger, A.: Twisted ways to find plane structures in simple drawings of complete graphs, In Proceedings 38th Symposium Computational Geometry, LIPIcs, Dagstuhl, Germany, pp. 1–18 (2022)
3. Alon, N., Spencer, J.: The Probabilistic Method, 4th edn. John Wiley & Sons Inc., Hoboken, New Jersey (2016)
4. Brass, P., Moser, W., Pach, J.: Research Problems in Discrete Geometry. Springer-Verlag, Berlin, Germany (2005). https://doi.org/10.1007/0-387-29929-7
5. Dilworth, R.P.: A decomposition theorem for partially ordered sets. Ann. of Math. **51**, 161–166 (1950)
6. Erdős, P., Szekeres, G.: A combinatorial problem in geometry. Compos. Math. **2**, 463–470 (1935)
7. Fox, J., Pach, J., Sudakov, B., Suk, A.: Erdős-Szekeres-type theorems for monotone paths and convex bodies. Proc. Lond. Math. Soc. **105**, 953–982 (2012)
8. Fulek, R., Ruiz-Vargas, A.: Topological graphs: empty triangles and disjoint matchings. In: Proceedings 29th Symposium Computational Geometry, pp. 259–265. ACM Press, New York (2013)

9. Gritzmann, P., Mohar, B., Pach, J., Pollack, R.: Embedding a planar triangulation with vertices at specified points. Am. Math. Mon. **98**, 165–166 (1991)

10. Harborth, H., Mengersen, I.: Drawings of the complete graph with maximum number of crossings. Congr. Numer. **88**, 225–228 (1992)

11. Kynčl, J.: Improved eumeration of simple topological graphs. Disc. Comput. Geom. **50**, 727–770 (2013)

12. Kynčl, J., Valtr, P.: On edges crossing few other edges in simple topological complete graphs. Discret. Math. **309**, 1917–1923 (2009)

13. Pach, J.: Geometric graph theory. In: Goodman, J., O'Rourke, J., Tóth, C. (eds.) Handbook of Discrete and Computational Geometry, 3rd edition, CRC Press, Boca Raton, Florida, pp. 257–279 (2017)

14. Pach, J., Solymosi, J., Tóth, G.: Unavoidable configurations in complete topological graphs. Disc. Comput. Geom. **30**, 311–320 (2003)

15. Pach, J., Tóth, G.: Disjoint edges in topological graphs. In: Akiyama, J., Baskoro, E.T., Kano, M. (eds.) IJCCGGT 2003. LNCS, vol. 3330, pp. 133–140. Springer, Heidelberg (2005). https://doi.org/10.1007/978-3-540-30540-8_15

16. Ruiz-Vargas, A.: Many disjoint edges in topological graphs. Comput. Geom. **62**, 1–13 (2017)

17. Ruiz-Vargas, A., Suk, A., Tóth, C.: Disjoint edges in topological graphs and the tangled-thrackle conjecture. European J. Combin. **51**, 398–406 (2016)

18. Scheucher, M.: Personal communication

19. Suk, A.: Disjoint edges in complete topological graphs. Disc. Comput. Geom. **49**, 280–286 (2013)

20. Suk, A.: On the Erdos-Szekeres convex polygon problem. J. Amer. Math. Soc. **30**, 1047–1053 (2017)

21. Suk, A., Walczak, B.: New bounds on the maximum number of edges in k-quasi-planar graphs. Comput. Geom. **50**, 24–33 (2015)

22. Tóth, G.: Note on geometric graphs. J. Combin. Theory Ser. A **89**, 126–132 (2000)

Compatible Spanning Trees in Simple Drawings of K_n

Oswin Aichholzer[1], Kristin Knorr[2], Wolfgang Mulzer[2],
Nicolas El Maalouly[3], Johannes Obenaus[2], Rosna Paul[1(✉)],
Meghana M. Reddy[3], Birgit Vogtenhuber[1], and Alexandra Weinberger[1]

[1] Institute of Software Technology, Graz University of Technology, Graz, Austria
{oaich,ropaul,bvogt,weinberger}@ist.tugraz.at
[2] Institut für Informatik, Freie Universität Berlin, Berlin, Germany
{kristin.knorr,wolfgang.mulzer,johannes.obenaus}@fu-berlin.de
[3] Department of Computer Science, ETH Zürich, Zürich, Switzerland
{nicolas.elmaalouly,meghana.mreddy}@inf.ethz.ch

Abstract. For a simple drawing D of the complete graph K_n, two (plane) subdrawings are *compatible* if their union is plane. Let $\mathcal{T}_D$ be the set of all plane spanning trees on D and $\mathcal{F}(\mathcal{T}_D)$ be the *compatibility graph* that has a vertex for each element in $\mathcal{T}_D$ and two vertices are adjacent if and only if the corresponding trees are compatible. We show, on the one hand, that $\mathcal{F}(\mathcal{T}_D)$ is connected if D is a cylindrical, monotone, or strongly c-monotone drawing. On the other hand, we show that the subgraph of $\mathcal{F}(\mathcal{T}_D)$ induced by stars, double stars, and twin stars is also connected. In all cases the diameter of the corresponding compatibility graph is at most linear in n.

Keywords: Compatibility graph · Plane spanning tree · Simple drawing

1 Introduction

A *drawing* D of a graph G is a representation of G in the Euclidean plane such that the vertices of G are distinct points and the edges are Jordan arcs

This work was initiated at the 6th DACH Workshop on Arrangements and Drawings in Stels, August 2021. We thank all participants, especially Nicolas Grelier and Daniel Perz, for fruitful discussions. O.A., R.P. and A.W. are supported by FWF grant W1230. K.K. is supported by the German Science Foundation (DFG) within the research training group 'Facets of Complexity' (GRK 2434). W.M. is partially supported by the German Research Foundation within the collaborative DACH project *Arrangements and Drawings* as DFG Project MU 3501/3–1, and by ERC StG 757609. J.O. is supported by ERC StG 757609. M.M.R. is supported by the Swiss National Science Foundation within the collaborative DACH project *Arrangements and Drawings* as SNSF Project 200021E-171681. (Also note that this author's full last name consists of two words and is *Mallik Reddy*. However, she consistently refers to herself with the first word of her last name being abbreviated.) B.V. was partially supported by the Austrian Science Fund (FWF) within the collaborative DACH project *Arrangements and Drawings* as FWF project I 3340-N35.

P. Angelini and R. von Hanxleden (Eds.): GD 2022, LNCS 13764, pp. 16–24, 2023.
https://doi.org/10.1007/978-3-031-22203-0_2

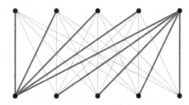

Fig. 1. A simple drawing of the complete bipartite graph with a tree (drawn in red, bold edges) that is an isolated vertex in the corresponding compatibility graph. (Color figure online)

connecting their incident vertices such that no edge passes through any other vertex. A drawing is *simple* if any pair of edges intersect at most once - either in a common vertex or a proper *crossing* in the relative interior of the edges. All drawings considered in this paper are simple and the term simple is mostly omitted. A drawing is *plane* if it does not contain any crossing.

For a fixed integer n let D be a simple drawing of the complete graph K_n and let $\mathcal{T}_D$ be the set of all drawings of plane spanning trees which are subdrawings of D. Note that $\mathcal{T}_D$ is non-empty, as it contains at least the n stars in D (where a *star* contains all edges incident to a single vertex). Unless explicitly stated otherwise, the word *tree* always refers to a plane spanning tree in $\mathcal{T}_D$, where the drawing D is either clear from the context or the statement holds for any simple drawing of K_n. Two (plane) subdrawings H and H' of a simple drawing D are said to be *compatible* if the union of H and H' is still plane.

Let $\mathcal{F}(\mathcal{T}_D)$ be the (abstract) graph that has a vertex for each plane spanning tree in $\mathcal{T}_D$ and two vertices are adjacent if and only if the corresponding trees are compatible. We call $\mathcal{F}(\mathcal{T}_D)$ the *compatibility graph* of $\mathcal{T}_D$. In this paper, we study properties of $\mathcal{F}(\mathcal{T}_D)$, focusing primarily on connectivity aspects:

Question 1. *Let n be an integer. Is the compatibility graph $\mathcal{F}(\mathcal{T}_D)$ connected for any simple drawing D of the complete graph K_n?*

Note that the notion of compatibility is closely related to the notion of edge flips: An *edge flip* in a plane spanning tree is the operation of removing an edge and replacing it with a new edge such that the resulting graph is again a plane spanning tree. In our setting, we further require this pair of edges to be non-crossing. In fact, one can simulate transformations via compatible trees in terms of crossing free edge flips: for two compatible trees T_1, T_2, successively add edges from T_2 to T_1, while removing an edge that is not in T_2 from the resulting cycle.

We observe that the compatibility graph of simple drawings that are not of the complete graph might not be connected even if the graph is dense. For example, Fig. 1 shows a simple drawing of the complete bipartite graph containing a plane tree that crosses all edges of the graph not belonging to the tree. Hence, this tree is an isolated vertex in the corresponding compatibility graph.

Related Work. The problem of transforming elements within a class of objects (e.g. plane spanning trees or matchings) into each other via a certain operation

Fig. 2. *Left to right:* cylindrical, monotone, strongly c-monotone drawing.

(e.g. edge flips or compatibility) has been studied extensively in a huge variety of contexts. Considering edge flips, some of the earliest results have been obtained on triangulations: Wagner [16] showed connectivity of the corresponding flip-graph in the combinatorial setting and Lawson [13] in the straight-line setting. For more details we refer the reader to the survey of Bose and Hurtado [9].

Considering the notion of compatibility, most of the work has been done in the straight-line setting, e.g., in the context of perfect matchings with [5,8] or without [1,2] vertex coloring, or for edge-disjoint compatibility [3,12]. Aichholzer et al. [4] showed, in the straight-line setting, that the compatibility graph of plane spanning trees is connected with diameter $O(\log k)$, where k denotes the number of convex layers of the point set. Buchin et al. [10] provided a corresponding worst case lower bound of $\Omega(\log n/\log\log n)$.

It is natural to extend this question to simple drawings, which however are inherently difficult to handle (even the existence of certain plane substructures is still unresolved in simple drawings; see e.g. [14]). On the positive side, García, Pilz and Tejel [11] proved that any maximal plane subgraph is 2-connected, which guarantees for any plane spanning tree the existence of a compatible plane spanning tree. In this paper, we aim to shed some light on this wide open topic of compatibility graphs of trees in simple drawings.

Contribution. We approach Question 1 from two directions, proving a positive answer for special classes of drawings (namely, cylindrical, monotone, and strongly c-monotone drawings) and for special classes of spanning trees (namely stars, double stars, and twin stars). We postpone the precise definitions of these classes of drawings and graphs to the later sections, however, Fig. 2 gives an illustration of these notions.

Theorem 1. *Let D be a cylindrical, monotone, or strongly c-monotone drawing of the complete graph K_n. Then, the compatibility graph $\mathcal{F}(\mathcal{T}_D)$ is connected.*

Theorem 2. *Let D be a simple drawing of the complete graph K_n and let $\mathcal{T}_D^*$ be the set of all plane spanning stars, double stars, and twin stars on D. Then, the compatibility graph $\mathcal{F}(\mathcal{T}_D^*)$ is connected.*

Section 2 is devoted to the proof of Theorem 1, while Sect. 3 is dedicated to the proof of Theorem 2. See Appendix A in the full version of this paper [6] for the details of the missing proofs.

2   Special Simple Drawings of K_n

In this section we prove connectedness of the compatibility graph for certain classes of drawings. Clearly, for any drawing of K_n that admits a plane spanning tree which is not crossed by any edge of D, the compatibility graph is connected with diameter at most 2. This is, for example, the case for 2-page book drawings, where the vertices are placed along a line and each edge lies entirely in one of the two open halfplanes defined by this line.

2.1   Cylindrical Drawings

Following the definition of Schaefer [15], in a *cylindrical drawing* of a graph the vertices are placed along two concentric circles, the *inner* and *outer* circle, and no edge is allowed to cross these circles.

Lemma 1. *Let D be a cylindrical drawing of K_n. Then $\mathcal{F}(\mathcal{T}_D)$ is connected with diameter at most 4.*

2.2   Monotone Drawings

A simple drawing in which no two vertices have the same x-coordinate and every edge is drawn as an x-monotone curve is called *monotone drawing*. Let $v_1, v_2, \ldots, v_n$ denote the sequence of vertices in increasing x-order. W.l.o.g. assume that these vertices are on the x-axis. Then, the plane spanning path $S = v_1, v_2, \ldots, v_n$ is called *spine* path. An edge that intersects the spine path is called *twiggly* edge.

We define a relation on the twiggly edges of D as follows: for two twiggly edges e, f we have $e \succ f$ if they are non-intersecting and admit a vertical line intersecting the relative interiors of both edges that intersects e at a larger y-coordinate than f. All other pairs of twiggly edges are incomparable. For a set E of pairwise non-intersecting twiggly edges, an edge $e \in E$ is *maximal* if there is no other edge $f \in E$ s.t. $f \succ e$. Note that this relation is acyclic, i.e., there are no twiggly edges $e_1, \ldots, e_k$ such that $e_1 \succ e_2 \succ \ldots \succ e_k \succ e_1$. And hence, any non-empty set of twiggly edges admits a maximal element.

Lemma 2. *For any monotone drawing D of K_n, the compatibility graph $\mathcal{F}(\mathcal{T}_D)$ is connected with diameter $O(n)$.*

Proof (Sketch). We show that any plane spanning tree T in D can be transformed to the spine path S. If T does not contain any twiggly edge, clearly it is compatible to S. Otherwise, we proceed as follows. Corresponding to a maximal twiggly edge e of T, we find a path P' connecting the vertices of e (see Fig. 3).

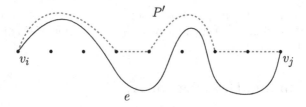

Fig. 3. The (maximal) twiggly edge $e = v_i v_j$ divides the vertices between v_i and v_j into two groups – above and below. The path P' is formed by joining the consecutive vertices lying above e including the vertices of e.

We can show that P' is compatible to T and lies strictly above e. Thus, we can add P' to T, which creates at least one cycle in T. Removing appropriate edges including e, we get a compatible tree with at least one twiggly edge less and repeating this process, we will eventually reach the spine path $\mathcal{S}$.   □

2.3   Strongly C-Monotone Drawings

A curve is called *c-monotone* (w.r.t. a point x) if every ray emanating from x intersects the curve at most once. A simple drawing is *c-monotone*, if all vertices are drawn along a circle and every edge is a c-monotone curve w.r.t. the center of the circle. A c-monotone drawing is *strongly* c-monotone if for any pair of edges e, e' there is a ray (rooted at the circle center) that neither intersects e nor e'.

In a (strongly) c-monotone drawing, we label the vertices $v_1, v_2, \ldots, v_n$ in cyclic order and denote the center of the circle by c. In the following, we often consider edges and their intersections with rays rooted at c; unless stated otherwise, any ray is rooted at c and edges are intersected in their relative interiors.

An edge e connecting two consecutive vertices v_i, v_{i+1} is called *cycle edge* and if e is drawn along the "shorter" side of the circle it is called *spine edge* (that is, no ray formed by the center and any vertex intersects e). All spine edges form the *spine* and any path consisting entirely of spine edges is called *spine path*.

Lemma 3. *Any strongly c-monotone drawing D of K_n either has all cycle edges as spine edges or is isomorphic to a monotone drawing.*

Again, we define *twiggly* edges to be those that intersect a spine edge. A crucial difference to the monotone setting is that an analogue to the relation '$\succ$' (adjusted with respect to the intersection with rays emanating from c) may now be cyclic and hence, we cannot guarantee the existence of a *maximal* twiggly edge anymore. We therefore need a different approach.

For a twiggly edge $e = uw$, let $x_1, \ldots, x_k$ be its crossings with the spine (note that these are not vertices of K_n) and assume the labeling to be in such a way that $u, x_1, \ldots, x_k, w$ appear in clockwise order. For $i \in \{1, \ldots, k\}$ denote the vertex (of K_n) in clockwise order before x_i by x_i^- and the one after by x_i^+. Furthermore, set $u = x_0^-$ and $w = x_{k+1}^+$. Then, for $i \in \{0, \ldots, k\}$, we call the

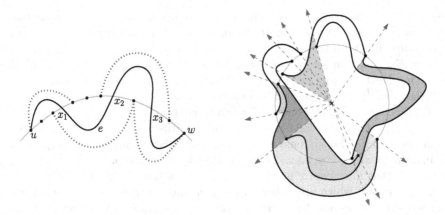

Fig. 4. *Left:* The red dotted edges are bumpy edges of the twiggly edge e. *Right:* A set of twiggly edges and some corridors; the dotted green is an inner corridor. (Color figure online)

edges $x_i^- x_{i+1}^+$ *bumpy* edges (see Fig. 4 (left)). Note that bumpy edges do not intersect the spine and for any twiggly edge there are at least two bumpy edges.

Clearly, we can identify any ray r with an angle θ, the angle it forms with the vertical ray (upwards). Two edges e, f are called *neighbours on an interval* $[\theta_1, \theta_2]$, if for any ray $r \in [\theta_1, \theta_2]$ the intersections of e and f with r appear consecutively on r. A *corridor* is a maximally connected region bounded by two neighbouring edges (along a maximal interval). Again, we identify corridors by an interval $[\theta_1, \theta_2]$ and usually we speak of corridors defined by the edges of a plane spanning tree. The *twiggly depth* (with respect to a plane spanning tree T) of a ray r is the number of twiggly edges (of T) that r intersects.

We extend our definition of neighbours (along an interval) also to the very inside and very outside by inserting a dummy edge at the circle center and one at infinity. More precisely, an edge e is the neighbor of the circle center c along an interval $[\theta_1, \theta_2]$ if for any ray $r \in [\theta_1, \theta_2]$ the intersection of r and e is closest to c (and furthest in the case of being a neighbor of infinity). We call the corresponding corridors *inner/outer* corridors. Note that the set of all corridors partitions the plane. See Fig. 4 (right) for an illustration.

We further remark that for any plane spanning tree T, any corridor $C = [\theta_1, \theta_2]$ (of edges of T) begins and ends at a vertex, i.e., the rays at θ_1 and θ_2 hit a vertex.

Lemma 4. *For any plane spanning tree T of a strongly c-monotone drawing D and any corridor C of T with start and end vertex s and t, there is a path P in D from s to t staying entirely in C, that does not intersect T. Furthermore, if C is an inner or outer corridor, P does not use any twiggly edge.*

Lemma 5. *For any strongly c-monotone drawing of K_n, the compatibility graph $\mathcal{F}(\mathcal{T}_D)$ is connected with diameter $O(n)$.*

Proof. Let D be a strongly c-monotone drawing of K_n and let T be a plane spanning tree. We show that T can be compatibly transformed to a spine path (by iteratively decreasing its twiggly depth). By Lemma 3, we may assume that all n spine edges are present in D. Again, if there is no twiggly edge in T, then T is compatible with the spine.

Let E_{twig} be the set of twiggly edges of T and construct the set $\mathcal{C}$ of all corridors. Next, for any corridor $C \in \mathcal{C}$ with start and end vertex s and t, we add the path P_C as guaranteed by Lemma 4 to T.

Clearly, we do not disconnect T when removing E_{twig} now. Indeed, let $e = uw \in E_{twig}$, then the collection of corridor paths below (and also above) e connects u and w. So we remove E_{twig} and potentially some further edges until T forms a spanning tree again (which by Lemma 4 is also plane). Furthermore, any ray r that intersects x previous twiggly edges (i.e., E_{twig}) intersects $x + 1$ corridors, two of which are either an inner or outer corridor. By Lemma 4 and the properties of c-monotone curves, r intersects at most $x - 1$ (new) twiggly edges. Hence, the twiggly depth of any ray decreased by at least one and we recursively continue this process until all rays have twiggly depth 0, in which case T is compatible to a spine path. As we have twiggly depth at most $n - 1$ in the beginning, $\mathcal{F}(\mathcal{T}_D)$ has diameter $O(n)$. □

Theorem 1 now follows from Lemma 1, 2, and 5.

3 Special Plane Spanning Trees

In this section, we are not restricting our drawing anymore, i.e., D will be a simple drawing of K_n throughout this section. Instead we focus on special classes of spanning trees and show that the subgraph $\mathcal{F}(\mathcal{T}_D^*)$ of $\mathcal{F}(\mathcal{T}_D)$ induced by the set of vertices corresponding to stars, double stars, and twin stars is connected.

A plane spanning tree with a fixed path P of length k such that all other vertices are incident to either the start or end vertex of P is called a k-star. A 0-star (i.e., P consists of a single vertex) is called *star*. A 1-star is called *double star* and a 2-star is called *twin star*.

The following relation, introduced in [7], will be very useful: Given a simple drawing of K_n with vertex set V and two vertices $g \neq r \in V$, for any two vertices $v_i, v_j \in V \setminus \{g, r\}$, we define $v_i \rightarrow_{gr} v_j$ if and only if the edge $v_i r$ crosses $v_j g$. In [7] it is shown that this relation is asymmetric and acylic.

We start by showing that stars can always be transformed into each other via a sequence of crossing free edge flips.

Lemma 6. *Any two stars in D have distance $O(n)$ in $\mathcal{F}(\mathcal{T}_D^*)$.*

Proof. Given a star T in g (i.e., g is incident to all other vertices of T), we can transform it into a star H in r via a sequence of crossing free edge flips, such that in every step, the graph is a double star with fixed path r,g, in the following way. We label the vertices in $V \setminus \{g, r\}$ such that $v_i \rightarrow_{gr} v_j$ implies $i < j$ (see Fig. 5). We iteratively replace an edge gv_i by rv_i starting from $i = n - 2$ and continuing in decreasing order. Clearly, all intermediate trees are double stars (with fixed path r,g) and hence, it remains to argue that the flips are compatible,

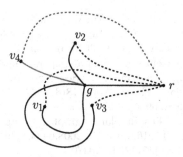

Fig. 5. Proof of Theorem 6: The solid edges represent a star in g, while the dotted edges form a star in r. The vertices are labeled conforming to the relation $\to_{gr}$. In order to transform the star in g to the star in r, the first step is adding the dotted blue edge $v_4 r$ and deleting the red edge $v_4 g$. (Color figure online)

i.e., for $i = n-2, \ldots, 1$ the edge gv_i does not cross any edge of the current T. By construction, in any step i, T contains edges of the form (a) rv_j for $j > i$ and (b) gv_k for $k < i$. The edge gv_i cannot cross edges in (a) by the definition of the relation $\to_{gr}$ and also not those in (b) due to the properties of simple drawings. As we need at most $n-2$ steps for the transformation, any two stars have distance $O(n)$ in $\mathcal{F}(T_D^*)$.   □

Theorem 2 then follows from Theorem 6 in combination with the following two lemmata.

Lemma 7. *Any double star in D has distance $O(n)$ to any star in $\mathcal{F}(T_D^*)$.*

Lemma 8. *Any twin star in D has distance $O(n)$ to any star in $\mathcal{F}(T_D^*)$.*

References

1. Aichholzer, O., et al.: Compatible geometric matchings. Comput. Geom. Theory Appl. **42**(6–7), 617–626 (2009). https://doi.org/10.1016/j.comgeo.2008.12.005
2. Aichholzer, O., García, A., Hurtado, F., Tejel, J.: Compatible matchings in geometric graphs. In: Proceedings of the XIV Encuentros de Geometría Computacional (EGC2011), pp. 145–148 (2011)
3. Aichholzer, O., Asinowski, A., Miltzow, T.: Disjoint compatibility graph of non-crossing matchings of points in convex position. Electron. J. Comb. **22**, 1 (2015)
4. Aichholzer, O., Aurenhammer, F., Huemer, C., Krasser, H.: Transforming spanning trees and pseudo-triangulations. Inf. Process. Lett. **97**(1), 19–22 (2006). https://doi.org/10.1016/j.ipl.2005.09.003. https://www.sciencedirect.com/science/article/pii/S0020019005002486
5. Aichholzer, O., Barba, L., Hackl, T., Pilz, A., Vogtenhuber, B.: Linear transformation distance for bichromatic matchings. Comput. Geom. Theory Appl. **68**, 77–88 (2018). https://doi.org/10.1016/j.comgeo.2017.05.003
6. Aichholzer, O., et al.: Compatible spanning trees in simple drawings of K_n. arXiv preprint (2022). http://arxiv.org/abs/2208.11875

7. Aichholzer, O., Parada, I., Scheucher, M., Vogtenhuber, B., Weinberger, A.: Shooting stars in simple drawings of $K_{m,n}$. In: Proceedings of 35th European Workshop on Computational Geometry EuroCG 2019, pp. 59:1–59:6. Utrecht, The Netherlands (2019). http://www.eurocg2019.uu.nl/papers/59.pdf

8. Aloupis, G., Barba, L., Langerman, S., Souvaine, D.: Bichromatic compatible matchings. Comput. Geom. Theory Appl. **48**(8), 622–633 (2015). https://doi.org/10.1016/j.comgeo.2014.08.009

9. Bose, P., Hurtado, F.: Flips in planar graphs. Comput. Geom. **42**(1), 60–80 (2009). https://doi.org/10.1016/j.comgeo.2008.04.001. https://www.sciencedirect.com/science/article/pii/S0925772108000370

10. Buchin, K., Razen, A., Uno, T., Wagner, U.: Transforming spanning trees: a lower bound. Comput. Geom. **42**(8), 724–730 (2009). https://doi.org/10.1016/j.comgeo.2008.03.005

11. García, A., Pilz, A., Tejel, J.: On plane subgraphs of complete topological drawings. ARS Math. Contemporanea **20**, 69–87 (2021). https://doi.org/10.26493/1855-3974.2226.e93

12. Ishaque, M., Souvaine, D.L., Tóth, C.D.: Disjoint compatible geometric matchings. Discrete Comput. Geom. **49**(1), 89–131 (2012). https://doi.org/10.1007/s00454-012-9466-9

13. Lawson, C.L.: Transforming triangulations. Discrete Math. **3**(4), 365–372 (1972)

14. Rafla, N.H.: The good drawings D_n of the complete graph K_n, Ph. D. thesis, McGill University, Montreal (1988). https://escholarship.mcgill.ca/concern/filesets/cv43nx65m?locale=en

15. Schaefer, M.: The graph crossing number and its variants: a survey. Electron. J. Comb. **1000** (2013). https://www.combinatorics.org/DS21

16. Wagner, K.: Bemerkungen zum Vierfarbenproblem. Jahresber. Dtsch. Mathematiker-Ver. **46**, 26–32 (1936). http://eudml.org/doc/146109

Mutual Witness Gabriel Drawings of Complete Bipartite Graphs

William J. Lenhart[1] and Giuseppe Liotta[2]([⊠])

[1] Williams College, Williamstown, USA
wlenhart@williams.edu
[2] Università degli Studi di Perugia, Perugia, Italy
giuseppe.liotta@unipg.it

Abstract. Let Γ be a straight-line drawing of a graph and let u and v be two vertices of Γ. The Gabriel disk of u, v is the disk having u and v as antipodal points. A pair $\langle \Gamma_0, \Gamma_1 \rangle$ of vertex-disjoint straight-line drawings form a mutual witness Gabriel drawing when, for $i = 0, 1$, any two vertices u and v of Γ_i are adjacent if and only if their Gabriel disk does not contain any vertex of Γ_{1-i}. We characterize the pairs $\langle G_0, G_1 \rangle$ of complete bipartite graphs that admit a mutual witness Gabriel drawing. The characterization leads to a linear time testing algorithm. We also show that when at least one of the graphs in the pair $\langle G_0, G_1 \rangle$ is complete k-partite with $k > 2$ and all partition sets in the two graphs have size greater than one, the pair does not admit a mutual witness Gabriel drawing.

Keywords: Proximity drawings · Gabriel drawings · Witness proximity drawings · Simultaneous drawing of two graphs

1 Introduction

Proximity drawings, including Delaunay triangulations, rectangle of influence drawings, minimum spanning trees, and unit disk graphs, are among the most studied geometric graphs. They are commonly used as descriptors of the "shape" of a point set and are used in a variety of applications, including machine learning, pattern recognition, and computer graphics (see, e.g., [14]). They have also been used to measure the faithfulness of large graph visualizations (see, e.g., [11]).

Proximity drawings are geometric graphs in which two vertices are adjacent if and only if they are deemed close by some measure. A common approach to define the closeness of two vertices u and v uses a *region of influence* of u and v, which is a convex region whose shape depends only on the relative position of

Work partially supported by: (i) MUR, grant 20174LF3T8 AHeAD: efficient Algorithms for HArnessing networked Data", (ii) Dipartimento di Ingegneria, Universita degli Studi di Perugia, grant RICBA21LG: Algoritmi, modelli e sistemi per la rappresentazione visuale di reti.

P. Angelini and R. von Hanxleden (Eds.): GD 2022, LNCS 13764, pp. 25–39, 2023.
https://doi.org/10.1007/978-3-031-22203-0_3

u with respect to v. Then we say that u and v are adjacent if and only if their region of influence does not contain some obstacle, often another vertex of the drawing. For example, a *Gabriel drawing* Γ is a proximity drawing where the region of influence u and v is the disk having u and v as antipodal points, called the Gabriel region of u and v; u and v are adjacent in Γ if and only if their Gabriel region does not contain any other vertex. See also [17] for a survey on different types of proximity regions and drawings.

An interesting generalization of proximity drawings is given in a sequence of papers by Aronov, Dulieu, and Hurtado who introduce and study *witness proximity drawings* and *mutual witness proximity drawings* [2–5]. In a witness proximity drawing the obstacles are points, called witnesses, that are suitably placed in the plane to impede the existence of edges between non-adjacent vertices; these points may or may not include some of the vertices of the drawing itself. A mutual witness proximity drawing is a pair of witness proximity drawings that are computed simultaneously and such that the vertices of one drawing are the witnesses of the other drawing. For example, Fig. 1 depicts a mutual witness Gabriel drawing (MWG-drawing for short) of two trees. In the figure, the Gabriel disk of v_0, v_1 of Γ_0 includes vertex v_2 but no vertices of Γ_1 and hence v_0, v_1 are adjacent in Γ_0; conversely, v_1 and v_2 are not adjacent in Γ_0 because their Gabriel disk contains vertex u_1 of Γ_1.

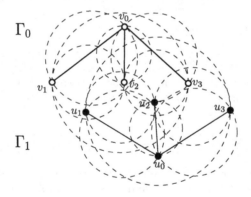

Fig. 1. A mutual witness Gabriel drawing of two trees (Gabriel disks are dotted).

In this paper we characterize those pairs of complete bipartite graphs that admit an MWG-drawing. While every complete bipartite graph has a witness Gabriel drawing [3], not all pairs of complete bipartite graphs admit an MWG-drawing. To characterize the drawable pairs we also investigate some properties of MWG-drawings that go beyond complete bipartiteness. More precisely:

– We show that if $\langle \Gamma_0, \Gamma_1 \rangle$ is an MWG-drawing such that both Γ_0 and Γ_1 have diameter two, then the set of vertices of Γ_0 is linearly separable from the set of vertices of Γ_1. This extends a result of [4], where linear separability is proved when the diameter is one, i.e. when the two graphs are complete.

- We show, perhaps surprisingly, that if $\langle G_0, G_1 \rangle$ is a pair of complete bipartite graphs that admits an MWG-drawing, then both must be planar.
- The above result let us characterize those pairs $\langle G_0, G_1 \rangle$ of complete bipartite graphs that admit an MWG-drawing and leads to a linear time testing algorithm. When the test returns that $\langle G_0, G_1 \rangle$ is drawable, an MWG-drawing can be constructed in linear time in the real RAM model.
- We show that relaxing the bipartiteness assumption does not significantly enlarge the class of representable graph pairs: We consider those pairs of complete multi-partite graphs each having all partition sets of size at least two and prove that if at least one of the graphs in the pair has more than two partition sets, then the pair does not admit an MWG-drawing.

We remark that our contribution not only fits into the rich literature devoted to proximity drawings, but it also relates to two other well studied topics in graph drawing, namely simultaneous embeddings (see, e.g., [8, 21] for references) and obstacle representations (see, e.g., [1, 6, 9, 10, 12, 15, 18–20]). As in simultaneous embeddings, the coordinates of the vertices of Γ_i in a mutual witness proximity drawing are defined by taking into account the (geometric and topological) properties of Γ_{1-i}; as in obstacle graph representations, the adjacency of the vertices Γ_i depends on whether their geometric interaction is obstructed by some external obstacles, namely the vertices of Γ_{1-i};. Finally, mutual witness proximity drawings are of interest in pattern recognition, where they have been used in the design of trained classifiers to convey information about the interclass structure of two sets of features (see, e.g. [13]).

2   Preliminaries

We assume familiarity with basic definitions and results of graph drawing [7]. We assume that all drawings occur in the Euclidean plane with standard x and y axes, and so concepts such as above/below a (non-vertical) line are unambiguous. Given two distinct points p and q in the plane, we denote by $\overline{pq}$ the straight-line segment whose endpoints are p and q. Also, let a, b, c be three distinct points in the Euclidean plane, we denote by $\Delta(abc)$ the triangle whose vertices are a, b, c. Given two non-axis-parallel lines ℓ_1 and ℓ_2 intersecting at a point b, those lines divide the plane into four *wedges: the top, bottom, left, and right wedges of b with respect to ℓ_1 and ℓ_2*. The top and bottom wedges lie entirely above and below the horizontal line through b, respectively; the left and right wedges lie entirely to the left and right of the vertical line through b. When the two lines are determined by providing a point (other than b) on each line, say a and c, we denote the wedges by $W_T[b, a, c]$, $W_B[b, a, c]$, $W_L[b, a, c]$, and $W_R[b, a, c]$ when we want to include the boundary of each wedge as part of that wedge and by $W_T(b, a, c)$, $W_B(b, a, c)$, $W_L(b, a, c)$, and $W_R(b, a, c)$ when we do not.

Note that exactly one of the four wedges will have both a and c on its boundary, we denote that wedge as $W[b, a, c]$ (or $W(b, a, c)$). See Fig. 2(a).

Let Γ be a straight line drawing of a graph G and let u and v be two vertices of Γ (and of G). Vertices u and v may either be adjacent in G and thus $\overline{uv}$ is an

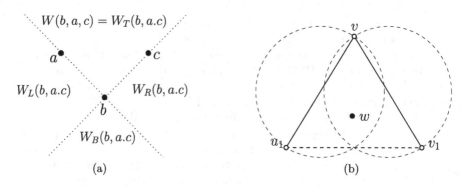

Fig. 2. (a) $W(b,a,c) = W_T(b,a,c)$, $W_B(b,a,c)$, $W_L(b,a,c)$, and $W_R(b,a,c)$ (b) If $w \in \Delta(vu_1v_1)$, at most one of $\overline{vu_1}$ and $\overline{vv_1}$ is an edge of a WG-drawing.

edge of Γ or u and v are not adjacent vertices, in which case $\overline{uv}$ is a *non-edge* of Γ. For example, $\overline{v_0v_1}$ in Fig. 1 is an edge while $\overline{v_1v_2}$ is a non-edge of Γ_1. Also, the *Gabriel disk* of p and q, denoted as $D[p,q]$ is the disk having p and q as antipodal points; $D[p,q]$ is a closed set.

Let V and P be two sets of distinct points in the plane. A *witness Gabriel drawing* (*WG-drawing*) with vertex set V and witness set P is a geometric graph Γ whose vertices are the points of V and such that any two vertices u and v form an edge if and only if $D[u,v] \cap P = \emptyset$. A graph G is *witness Gabriel drawable* (*WG-drawable*) if there exist two point sets V and P such that the witness Gabriel drawing with vertex set V and witness set P represents G (i.e., there is a bijection between the vertex set of G and the point set V and between the edge set of G and the edge set of Γ that is incidence-preserving). The following property can be proved with elementary geometric arguments (see also Fig. 2(b)).

Property 1. Let Γ be a WG-drawing with witness set P and let $\overline{vu_1}$ and $\overline{vv_1}$ be two edges of Γ incident on the same vertex v. Then $\Delta(vu_1v_1) \cap P = \emptyset$.

For a pair $\langle G_0, G_1 \rangle$ of WG-drawable graphs, a *mutual witness Gabriel drawing* (*MWG-drawing*) is a pair $\langle \Gamma_0, \Gamma_1 \rangle$ of straight-line drawings such that Γ_i is a WG-drawing of G_i with witness set the vertices of Γ_{1-i} ($i = 0, 1$). If $\langle G_0, G_1 \rangle$ admits an MWG-drawing we say that $\langle G_0, G_1 \rangle$ is *mutually witness Gabriel drawable* (*MWG-drawable*).

Some proofs have been omitted. They can be found in Lenhart and Liotta [16].

3   Linear Separability of Diameter-2 MWG-drawings

In this section we extend a result by Aronov et al. [4] about the linear separability of the MWG-drawings of complete graphs to graphs of diameter two.

Lemma 1. *Let $\langle \Gamma_0, \Gamma_1 \rangle$ be an MWG-drawing such that Γ_i has diameter at most 2 ($i = 0, 1$). Then no segment of Γ_0 intersects any segment of Γ_1.*

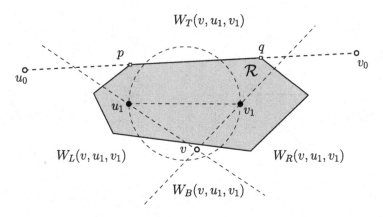

Fig. 3. Illustration for the proof of Theorem 1: a non-edge $\overline{u_1 v_1}$ of Γ_1 in a region $\mathcal{R}$ bounded by portions of edges and non-edges of Γ_0.

Proof. Note first that by Property 1, a vertex u of Γ_i cannot lie on a non-edge $\overline{u_1, v_1}$ of Γ_{1-i} since $\{u_1, v_1\}$ have at least one common neighbor v. Also no vertex of Γ_i can lie on an edge of Γ_{1-i}. Let $\overline{u_0 v_0}$ be an edge of Γ_0 and let $\overline{u_1 v_1}$ be an edge of Γ_1. Assume that they cross and consider the quadrilateral Q whose vertices are the end-points of the two crossing edges. Since some internal angle of Q must be at least $\frac{\pi}{2}$, either $D[u_0, v_0]$ contains one of $\{u_1, v_1\}$ or $D[u_1, v_1]$ contains one of $\{u_0, v_0\}$ contradicting the fact that both $\overline{u_0 v_0}$ is an edge of Γ_0 and $\overline{u_1 v_1}$ is an edge of Γ_1.

Let $\overline{u_0 v_0}$ be an edge of Γ_0 and let $\overline{u_1 v_1}$ be a non-edge of Γ_1. Since Γ_1 has diameter at most two, there is a vertex v in Γ_1 such that both $\overline{v u_1}$ and $\overline{v v_1}$ are edges of Γ_1. Since $\overline{u_1 v_1}$ crosses $\overline{u_0 v_0}$, but neither $\overline{v u_1}$ nor $\overline{v v_1}$ crosses $\overline{u_0 v_0}$, we have that one of $\{u_0, v_0\}$ is a point of $\Delta(u_1, v_1, v)$. However, Γ_1 is a WG-drawing whose witness set is the set of vertices of Γ_0 and, by Property 1, no vertex of Γ_0 can be a point of $\Delta(u_1, v_1, v)$. An analogous argument applies when $\overline{u_0 v_0}$ is a non-edge of Γ_0 while $\overline{u_1 v_1}$ is an edge of Γ_1.

It remains to consider the case that $\overline{u_0 v_0}$ is a non-edge of Γ_0 and $\overline{u_1 v_1}$ is a non-edge of Γ_1. Let v be a vertex such that both $\overline{v u_1}$ and $\overline{v v_1}$ are edges of Γ_1. By the previous case, neither of these two edges can cross $\overline{u_0 v_0}$. It follows that one of $\{u_0, v_0\}$ is a point of $\Delta(u_1, v_1, v)$ which, by Property 1, is impossible.

We are now ready to prove the main result of this section. We denote by $CH(\Gamma)$ the convex hull of the vertex set of a drawing Γ.

Theorem 1. *Let $\langle \Gamma_0, \Gamma_1 \rangle$ be an MWG-drawing such that each Γ_i has diameter 2. Then Γ_0 and Γ_1 are linearly separable.*

Proof. By Lemma 1, no vertex, edge or non-edge of Γ_i intersects any vertex, edge or non-edge of Γ_{1-i}. Hence, either $CH(\Gamma_0)$ and $CH(\Gamma_1)$ are linearly separable and we are done, or one of the convex hulls – say $CH(\Gamma_1)$ – is contained in a convex region $\mathcal{R}$ bounded by (portions of) edges and/or non-edges of Γ_0. We

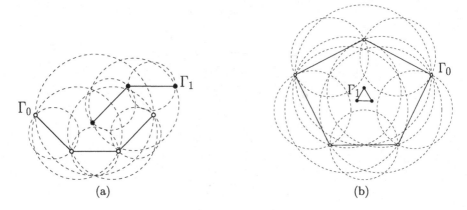

Fig. 4. Non-linearly separable MWG-drawings of: (a) a diameter two graph and a diameter three graph; (b) a diameter two graph and a diameter one graph.

prove that region $\mathcal{R}$ cannot exist, which implies the statement. Suppose for a contradiction that Γ_1 is contained in $\mathcal{R}$ and let $\overline{u_1 v_1}$ be a non-edge of Γ_1 with $x(u_1) \leq x(v_1)$. See Fig. 3 for a schematic illustration. Since u_1 and v_1 are not adjacent, there is some vertex v of Γ_0 such that $v \in D[u_1, v_1]$. Without loss of generality, assume that $\overline{u_1 v_1}$ is horizontal and that v is below the line through u_1 and v_1. Since u_1 and v_1 are points of $W_T[v, u_1, v_1]$ and Γ_1 is contained in $\mathcal{R}$, there is some segment $\overline{pq}$ of the boundary of $\mathcal{R}$ such that $\overline{pq}$ intersects $W_T(v, u_1, v_1)$ above the line through u_1 and v_1. Let u_0 and v_0 be the two vertices of Γ_0 such that $\overline{pq}$ is a subset of $\overline{u_0 v_0}$. For concreteness, we assume that the x-coordinates of u_0, p, q, and v_0 are such that $x(u_0) \leq x(p) \leq x(q) \leq x(v_0)$.

Claim. $u_0 \in W_L(v, u_1, v_1)$, $v_0 \in W_R(v, u_1, v_1)$ and $\overline{u_0 v_0}$ is a non-edge of Γ_0.

Proof of the claim: Suppose for a contradiction that a vertex of $\{u_0, v_0\}$ – say v_0 – were a point of $W_T[v, u_1, v_1]$. Since $\overline{pq}$ intersects $W_T(v, u_1, v_1)$ above the horizontal line through u_1 and v_1, we have that v_0 must also be above this horizontal line or else $\overline{u_0 v_0}$ and $\overline{u_1 v_1}$ would cross, contradicting Lemma 1. However, if v_0 is above the line through u_1 and v_1 we have that $\overline{u_1 v_1}$ and $\overline{v v_0}$ cross which again contradicts Lemma 1. Therefore, $v_0 \notin W_T[v, u_1, v_1]$ and, by the same argument, $u_0 \notin W_T[v, u_1, v_1]$. Note that this argument also precludes either point of $\{u_0, v_0\}$ from being in $W_B[v, u_1, v_1]$, since, because $\overline{pq}$ intersects $W_T[v, u_1, v_1]$, we would then have that the other point of $\{u_0, v_0\}$ lies in $W_T[v, u_1, v_1]$. Finally, observe that if u_0 and v_0 were both points of either $W_L(v, u_1, v_1)$ or $W_R(v, u_1, v_1)$, segment $\overline{pq}$ would not intersect $W_T(v, u_1, v_1)$. It follows that $u_0 \in W_L(v, u_1, v_1)$ and $v_0 \in W_R(v, u_1, v_1)$. Note that $\triangle(u_0 v v_0)$ contains both u_1 and v_1, so $\angle u_0 v v_0 > \angle u_1 v v_1 \geq \frac{\pi}{2}$, which implies that $\angle u_0 v_1 v_0 > \frac{\pi}{2}$ and so $\overline{u_0 v_0}$ is a non-edge of Γ_0. This concludes the proof of the claim.

By the claim above and by the assumption that Γ_0 has diameter two, there is some vertex z such that both $\overline{z u_0}$ and $\overline{z v_0}$ are edges of Γ_0. Vertex z may or may not coincide with v. If z coincides with v or if $z \in W_B[v, u_1, v_1]$, we have that $\triangle(zu_0v_0)$ contains both u_1 and v_1 and two of its sides are edges of Γ_0, which contradicts Property 1. If $z \in W_T[v, u_1, v_1]$ and z is above the line through u_1 and v_1, we have that $\overline{vz}$ and $\overline{u_1v_1}$ cross, which contradicts Lemma 1. If $z \in W_T[v, u_1, v_1]$ and z is below the line through u_1 and v_1, either $\overline{u_1v_1}$ crosses one of $\{\overline{zu_0}, \overline{zv_0}\}$ contradicting Lemma 1, or $\triangle(zu_0v_0)$ contains both u_1 and v_1 contradicting Property 1. If $z \in W_L(v, u_1, v_1)$, we consider three cases. If edge $\overline{zv_0}$ crosses $\overline{u_1v_1}$, we would violate Lemma 1. If edge $\overline{zv_0}$ is above $\overline{u_1v_1}$, then $\angle zvv_0 > \frac{\pi}{2}$ and since both u_1 and v_1 are in the interior of $\triangle(zvv_0)$, we have that u_1 and v_1 are in $D[z, v_0]$, contradicting the fact that $\overline{zv_0}$ is an edge of Γ_0. If edge $\overline{zv_0}$ is below $\overline{u_1v_1}$, $\triangle(u_0zv_0)$ contains both u_1 and v_1, which violates Property 1. By a symmetric argument, we have that z cannot be a point of $W_R(v, u_1, v_1)$ either. Since point z does not exist, it follows that $\mathcal{R}$ does not exist.

Theorem 1 shows that MWG-drawings with diameter two capture useful information about the interaction of two point sets. As pointed out by both Ichino and Slansky [13] and by Aronov et al. [4], the linear separability of mutual witness proximity drawings gives useful information about the interclass structure of two set of points. It is also worth noting that if at least one of the graphs in the pair has diameter different from two, a non-linearly separable drawing may exist. For example Fig. 4(a) and Fig. 4(b) show MWG-drawings of graph pairs in which the diameter two property is violated for one of the two graphs.

4   MWG-drawable Complete Bipartite Graphs

In this section we exploit Theorem 1 to characterize those pairs of complete bipartite graphs that admit an MWG-drawing. In Sect. 4.1 we prove that any two complete bipartite graphs that form an MWG-drawable pair are planar. The complete characterization is then given in Sect. 4.2. In what follows we shall assume without loss of generality that the line separating a drawing Γ from its set of witnesses is horizontal and it coincides with the line $y = 0$, with the witnesses in the negative half-plane. The proof of the following property is trivial and therefore omitted, but Fig. 5(a) and its caption illustrate it.

Property 2. Let Γ be a WG-drawing with witness set P, let $\overline{uv}$ be a non-edge of Γ with witness $p \in P$, and let z be a vertex of Γ such that both $\overline{zu}$ and $\overline{zv}$ are edges of Γ. Then $z \in W(p, u, v)$.

4.1   Planarity

Let Γ be a WG-drawing; an *alternating 4-cycle* in Γ consists of two vertex-disjoint edges $\overline{u_0u_1}$ and $\overline{v_0v_1}$ of Γ such that $\overline{u_0v_0}$ and $\overline{u_1v_1}$ are both non-edges in the drawing. For example, Fig. 5(b) shows a WG-drawing Γ whose witness set consists of points p_0 and p_1. In the figure, $\overline{u_0u_1}$ and $\overline{v_0v_1}$ are edges of Γ while $\overline{u_0v_0}$ and $\overline{u_1v_1}$ are non-edges of Γ: these two pairs of edges and non-edges (bolder in the figure) form an alternating 4-cycle in Γ.

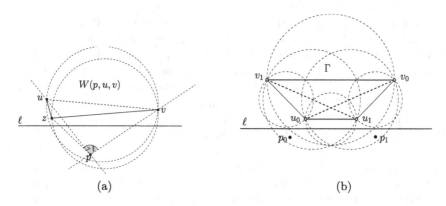

(a)   (b)

Fig. 5. (a) If $z \notin W(p, u, v)$, then $p \in D[z, v]$ and $\overline{zv}$ is not an edge of Γ. (b) A WG-drawing Γ with an alternating 4-cycle highlighted in bold.

Lemma 2. *Let Γ be a WG-drawing of a complete bipartite graph such that Γ is linearly separable from its witness set P and let C be an alternating 4-cycle defined on Γ. The two edges of Γ in C do not cross while the two non-edges of Γ in C do cross.*

Proof. Let ℓ be the line separating Γ from its witness set. Let u_0, u_1, v_1, v_0 be the four vertices of C such that $\overline{u_0 u_1}$ and $\overline{v_0 v_1}$ are two edges of Γ while $\overline{u_0 v_0}$ and $\overline{u_1 v_1}$ are two non-edges of Γ. Since the drawing is a complete bipartite graph, $\overline{u_0 v_1}$ and $\overline{v_0 u_1}$ are edges of Γ. We prove that $\overline{u_0 v_0}$ and $\overline{u_1 v_1}$ must cross in Γ, which implies that $\overline{u_0 u_1}$ and $\overline{v_0 v_1}$ do not cross.

Let $p_0 \in P$ such that $p_0 \in D[u_0, v_0]$. By Property 2, both u_1 and v_1 lie in the wedge $W(p_0, u_0, v_0) = W_T(p_0, u_0, v_0)$. Observe that p_0 cannot also be a witness for the pair u_1 and v_1 as otherwise, by Property 2, we should have that also u_0 and v_0 lie in the top wedge $W_T(p_0, u_1, v_1)$, which is impossible. So, let $p_1 \in P$ be distinct from p_0 and such that $p_1 \in D[u_1, v_1]$. If p_0 were a point in $W_T[p_1, u_1, v_1]$, p_0 would also be a point in $\triangle(v_1 p_1 u_1)$ and we would have $p_0 \in D[u_1, v_1]$, which we just argued is impossible (see, e.g. Figure 6). By analogous reasoning we have that p_1 cannot be a point of $W_T[p_0, u_0, v_0]$. Also, $p_1 \notin W_B[p_0, u_0, v_0]$ or else p_0 would be in $W_T[p_1, u_1, v_1]$ since $W_T(p_1, u_1, v_1)$ contains both u_0 and v_0. It follows that either $p_1 \in W_L(p_0, u_0, v_0)$ or $p_1 \in W_R(p_0, u_0, v_0)$. In either case, $W_T(p_1, u_1, v_1)$ can contain both u_0 and v_0 only if $\overline{u_0 v_0}$ and $\overline{u_1 v_1}$ cross. $\blacksquare$

The following corollaries are a consequence of Lemma 2 and of Theorem 1 .

Corollary 1. *Let G_0 and G_1 be two vertex disjoint complete bipartite graphs. If the pair $\langle G_0, G_1 \rangle$ is MWG-drawable, then both G_0 and G_1 are planar graphs.*

Corollary 2. *Let Γ be a WG-drawing of a complete bipartite graph such that Γ is linearly separable from its witness set. Any 4-cycle formed by edges of Γ is a convex polygon.*

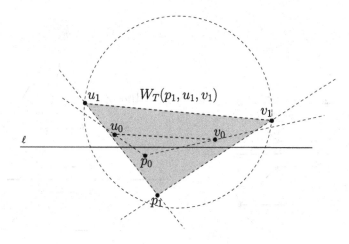

Fig. 6. If $p_0 \in W_T[p_1, u_1, v_1]$, then $p_0 \in D[u_1, v_1]$.

4.2   Characterization

We start with two technical lemmas.

Lemma 3. *Let $\langle G_0, G_1 \rangle$ be an MWG-drawable pair admitting a linearly separable MWG-drawing. Then the pair $\langle G_0 \cup \{v_0\}, G_1 \rangle$ also admits a linearly separable MWG-drawing, where v_0 is a vertex not in G_0 and is adjacent to all vertices in G_0—that is, a* universal vertex *of $G_0 \cup \{v_0\}$.*

Figure 7(b) shows the addition of the universal vertex v_0 to the MWG-drawing of Fig. 7(a). In fact, a universal vertex can be added to either drawing as long as it is positioned sufficiently far from the separating line.

Let u_0, u_1 be two points with $x(u_0) < x(u_1)$. The open *vertical strip* of u_0, u_1, denoted as $S(u_0, u_1)$, is the set of points (x, y) such that $x(u_0) < x < x(u_1)$. Assume now that u_0 and u_1 are vertices of a WG-drawing Γ such that Γ is linearly separable from its witness set by a line ℓ. Segment $\overline{u_0 u_1}$ divides $S(u_0, u_1)$ into two (open) *half-strips*: $S_N(u_0, u_1)$ is the (near) half-strip on the same side of $\overline{u_0 u_1}$ as ℓ and $S_F(u_0, u_1)$ is the other (far) half-strip. $S[u_0, u_1]$, $S_N[u_0, u_1]$, and $S_F[u_0, u_1]$ consist of $S(u_0, u_1)$, $S_N(u_0, u_1)$, and $S_F(u_0, u_1)$ along with their respective boundaries.

Lemma 4. *Let $\langle G_0, G_1 \rangle$ be an MWG-drawable pair admitting a linearly separable MWG-drawing. Then at least one of the pairs $\langle G_0 \cup \{v_0\}, G_1 \rangle$, $\langle G_0, G_1 \cup \{v_1\} \rangle$ also admits a linearly separable MWG-drawing, where, for $i = 0, 1$, v_i is a vertex not in G_i and has no edges to any vertex in G_i—that is, v_i is an isolated vertex of G_i.*

Figure 8(b) shows the addition of the isolated vertex v_0 to the MWG-drawing of Fig. 8(a). In fact, an isolated vertex can be added to the left (right) of whichever of the two drawings has the leftmost (rightmost) vertex, as long as it

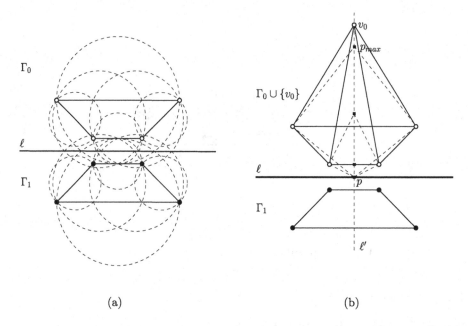

Fig. 7. (a) A linearly separable MWG-drawing $\langle \Gamma_0, \Gamma_1 \rangle$. (b)Adding a universal vertex v_0 to Γ_0 by placing it far enough from p on ℓ'.

is positioned sufficiently far enough to the left (right) of that vertex. Lemmas 3 and 4 are used in the following lemma, where we use colors to distinguish vertices in distinct partition sets.

Lemma 5. $\langle K_{1,n_0}, K_{1,n_1} \rangle$ *has a MWG-drawing if* $|n_0 - n_1| \leq 2$.

Proof. Observe that two independent sets whose sizes differ by at most 1, one consisting of red vertices and one consisting of blue vertices, admit a linearly separable MWG-drawing where the red vertices are above a horizonal separating line ℓ while the blue vertices are below ℓ: start with one red vertex with coordinates $(0, 1)$ and a blue vertex with coordinates $(-1, -1)$ and iteratively add red and blue vertices by applying the isolated vertex-addition procedure in the proof of Lemma 4. Let $G_i = K_{1,n_i}$, for $i = 0, 1$.

Denote by v_0 the non-leaf vertex of G_0 and by u_0 the non-leaf vertex of G_1. Assume first that $|n_0 - n_1| \leq 1$. By the previous observation, $\langle G_0 \backslash \{v_0\}, G_1 \backslash \{u_0\} \rangle$ admits a linearly separable MWG-drawing. Therefore, by Lemma 3 applied to $\langle G_0 \backslash \{v_0\}, G_1 \backslash \{u_0\} \rangle$ we have that $\langle G_0, G_1 \backslash \{u_0\} \rangle$ admits a a linearly separable MWG-drawing. By Lemma 3 applied to $\langle G_0, G_1 \backslash \{u_0\} \rangle$ we have that if $|n_0 - n_1| \leq 1$, the pair $\langle G_0, G_1 \rangle$ is MWG-drawable.

Consider now the case $|n_0 - n_1| = 2$ and assume that $n_0 > n_1$ (the proof when $n_1 > n_0$ is analogous). Let v_1 be a leaf of G_0. With the same reasoning as in the previous case, $\langle G_0 \backslash \{v_0, v_1\}, G_1 \backslash \{u_0\} \rangle$ admits a linearly separable MWG-drawing that we denote as $\langle \Gamma_0 \backslash \{v_0, v_1\}, \Gamma_1 \backslash \{u_0\} \rangle$. By the technique in the proof

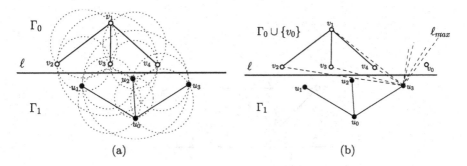

Fig. 8. (a) A linearly separable MWG-drawing $\langle \Gamma_0, \Gamma_1 \rangle$. (b)Adding an isolated vertex v_0 to Γ_0.

of Lemma 3, we add the universal vertex u_0 to $\Gamma_1 \setminus \{u_0\}$ in such a way that u_0 is the rightmost vertex of the linearly separable MWG-drawing $\langle \Gamma_0 \setminus \{v_0, v_1\}, \Gamma_1 \rangle$. We now exploit the construction of Lemma 4 to add the isolated vertex v_1 to $\langle \Gamma_0 \setminus \{v_0, v_1\}, \Gamma_1 \rangle$ and obtain a linearly separable MWG-drawing $\langle \Gamma_0 \setminus \{v_0\}, \Gamma_1 \rangle$. Finally, we use Lemma 3 to construct an MWG-drawing of $\langle G_0, G_1 \rangle$ also when $|n_0 - n_1| = 2$.

Lemma 6. *Let Γ be a WG-drawing of a graph such that Γ is linearly separable from its witness set P. If $\overline{uv}$ is an edge of Γ and $z \in S_F[u, v]$ is a vertex of Γ, then both $\overline{uz}$ and $\overline{vz}$ are edges of Γ.*

Proof. Consider, w.l.o.g., the segment $\overline{uz}$. If it is not an edge of Γ, then it must have a witness in $S_N(u, z)$. But any such point will also be in $D[u, v]$, contradicting the fact that $\overline{uv}$ is an edge of Γ.

Lemma 7. *Let Γ be a WG-drawing of a graph such that Γ is linearly separable from its witness set P. Let u_0, u_1, v_0, v_1 be such that u_0, v_0, u_1, v_1 induce a C_4 in Γ. We have that: (i) v_0 and v_1 are in opposite half-planes with respect to the line through u_0, u_1, and (ii) one of $\{v_0, v_1\}$ is a point of $S_N(u_0, u_1)$ and the other is not in $S[u_0, u_1]$.*

By means of Lemma 7 we can restrict the set of complete bipartite graph pairs that are MWG-drawable.

Lemma 8. *Let Γ be a WG-drawing of a complete bipartite graph such that Γ is linearly separable from its witness set. Then Γ does not have $K_{2,3}$ as a subgraph.*

We now characterize the MWG-drawable pairs of complete bipartite graphs. We recall that Aronov et al. prove that every complete bipartite graph admits a WG-drawing (Theorem 5 of [3]). The following theorem can be regarded as an analog of the result by Aronov et al. in the context of MWG-drawings.

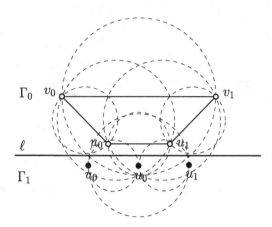

Fig. 9. An MWG-drawing of $K_{2,2}$ and of an independent set of size three.

Theorem 2. *Let* $\langle G_0, G_1 \rangle$ *be a pair of complete bipartite graphs such that* G_i *has* n_i *vertices. The pair* $\langle G_0, G_1 \rangle$ *admits an MWG-drawing if and only if, for* $i = 0, 1$, G_i *is either* K_{1,n_i-1} *or* $K_{2,2}$ *and* $|n_0 - n_1| \leq 2$.

Proof. By Theorem 1, any MWG-drawing $\langle \Gamma_0, \Gamma_1 \rangle$ of G_0 and G_1 is linearly separable, so any witnes for a non-edge $\overline{uv}$ in Γ_i must lie in $S(u, v)$. By Corollary 1 both G_0 and G_1 must be planar. By Lemma 8 each of the two graphs is either $K_{2,2}$ or a star (i.e. K_{1,n_i-1}, $i = 0, 1$). Together, these imply that the difference in the cardinalities of the vertex sets in the two graphs is at most two.

If $G_0 = K_{1,n_0-1}$ and $G_1 = K_{1,n_1-1}$, the theorem follows by Lemma 5. If $G_0 = K_{2,2}$ and $G_1 = K_{2,2}$ the pair $\langle G_0, G_1 \rangle$ has an MWG-drawing as shown, for example, in Fig. 7(a). By removing one of the bottom-most vertices of Γ_1 in Fig. 7(a) we obtain an MWG-drawing of $\langle K_{2,2}, K_{1,2} \rangle$ and by removing both the bottom-most vertices of Γ_1 in Fig. 7(a) we obtain an MWG-drawing of $\langle K_{2,2}, K_{1,1} \rangle$. To complete the proof we have to show that $\langle K_{2,2}, K_{1,3} \rangle$, $\langle K_{2,2}, K_{1,4} \rangle$, and $\langle K_{2,2}, K_{1,5} \rangle$ are also MWG-drawable pairs. To this end refer to Fig. 9 that shows an MWG-drawing $\langle \Gamma_0, \Gamma_1 \rangle$ where Γ_0 is $K_{2,2}$ while Γ_1 is an independent set consisting of three vertices. By applying Lemma 3 we can add a universal vertex to Γ_1, thus obtaining an MWG-drawing of $\langle K_{2,2}, K_{1,3} \rangle$. In order to construct MWG-drawings of $\langle K_{2,2}, K_{1,4} \rangle$ and of $\langle K_{2,2}, K_{1,5} \rangle$, notice that in Fig. 9 v_0 is the leftmost vertex and v_1 is the rightmost vertex of $\langle \Gamma_0, \Gamma_1 \rangle$. By Lemma 4 we can add either one isolated vertex or two isolated vertices to Γ_1. In the former case we obtain an MWG-drawing of $K_{2,2}$ and of an independent set of size four which can be extended to an MWG-drawing of $\langle K_{2,2}, K_{1,4} \rangle$ by means of Lemma 3. In the latter case, we again use Lemma 3 to add a universal vertex to the drawing of the independent set of size five and obtain an MWG-drawing of $\langle K_{2,2}, K_{1,5} \rangle$.

The following theorem is a consequence of Theorem 2 and of the constructive arguments of Lemma 5

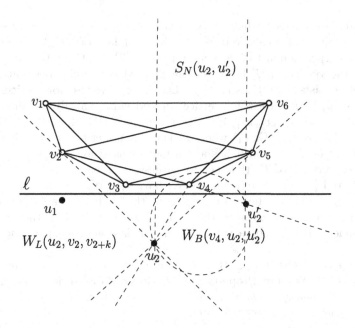

Fig. 10. $v_4 \in S_N(u_2, u_2')$ implies $W_L(u_2, v_2, v_{2+k}) \cap W_B(v_4, u_2, u_2') = \emptyset$.

Theorem 3. *Let $\langle G_0, G_1 \rangle$ be a pair of complete bipartite graphs such that G_0 has n_0 vertices and G_1 has n_1 vertices. There exists an $O(n_0+n_1)$-time algorithm that tests whether $\langle G_0, G_1 \rangle$ admits an MWG-drawing. In the affirmative case, there exists an $O(n_0 + n_1)$-time algorithms to compute an MWG-drawing of $\langle G_0, G_1 \rangle$ in the real RAM model of computation.*

5   MWG-drawable Complete k-partite Graphs

Aronov et al. also showed that there exists a complete multipartite graph, namely $K_{3,3,3,3}$, which does not admit a WG-drawing (Theorem 15 of [3]). We extend this result in the context of MWG-drawings by proving the following result.

Theorem 4. *Let $\langle G_0, G_1 \rangle$ be a pair of complete multi-partite graphs such that for each of the graphs every partition set has size at least two. The pair is mutually Gabriel drawable if and only if it is $\langle K_{2,2}, K_{2,2} \rangle$.*

Proof (Sketch). Let Γ be a WG-drawing of a complete k-partite graph, with $k \geq 2$ such that Γ is linearly separable from its witness set P by a separating line ℓ. Note that any induced subgraph G' of G admits a WG-drawing with witness set P, which can be derived from Γ by removing the vertices not in G'. By this observation, Theorem 1, and Lemma 8 we conclude that if $\langle G_0, G_1 \rangle$ is a pair of complete MWG-drawable multi-partite graphs, then neither G_0 nor G_1 can have $K_{2,3}$ as a subgraph. Therefore we can assume that all partition sets in each of the two graphs have size exactly two. Refer to Fig. 10.

The proof proceeds by first showing that $CH(\Gamma)$ is a *convex terrain* with respect to ℓ; that is for each vertex v on the boundary of $CH(\Gamma)$, the segment from v to ℓ perpendicular to ℓ does not intersect $CH(\Gamma)$. Using this property we can order the vertices of Γ by increasing x-coordinate and show that the i-th partite set consists of vertices v_i, v_{i+k}. Let p_i be a witness for vertices v_i and v_{k+i} and let p_j be a witness for vertices v_j and v_{k+j}. Thirdly, we show that if $i < j$ then $p_i \in W_L(p_j, v_j, v_{j+k})$; if $i > j$ then $p_i \in W_R(p_j, v_j, v_{j+k})$. Consider now an MWG-drawing $\langle \Gamma_0, \Gamma_1 \rangle$ where Γ_0 is in the upper half-plane with respect to the separating line ℓ and it has at least three partition sets. Let u_1, u_2 and u_3 be three vertices of Γ_1 that act as witnesses for partition sets $\{v_1, v_{1+k}\}$, $\{v_2, v_{2+k}\}$, and $\{v_3, v_{3+k}\}$, respectively. We have $u_1 \in W_L(u_2, v_2, v_{2+k})$ and $u_3 \in W_R(u_2, v_2, v_{2+k})$. Let u_2' be the vertex of Γ_1 that is in the same partition set as u_2 and assume that u_2' is to the right of u_2 (the proof in the other case being symmetric). Let v be a vertex of Γ_0 that is a witness of u_2 and u_2' (i.e. $v \in D[u_2, u_2']$).

By Property 2 either $v \in \{v_2, v_{2+k}\}$ or $v \in W_T(u_2, v_2, v_{2+k})$, hence $v \in W_T[u_2, v_2, v_{2+k}]$. Again by Property 2, all vertices of Γ_1 must lie in $W_B[v, u_2, u_2']$. Because $v \in W_T[u_2, v_2, v_{2+k}]$, $W_B[v, u_2, u_2']$ is disjoint from $W_L(u_2, v_2, v_{2+k})$. But $u_1 \in W_L(u_2, v_2, v_{2+k})$, a contradiction.

6 Open Problems

The results of this paper naturally suggest many interesting open problems. For example: (i) Can one give a complete characterization of those pairs of complete multipartite graphs that admit an MWG-drawing extending Theorem 4 by taking into account graphs some of whose partition sets have size one? It is not hard to see that the ideas of Lemmas 3 and 5 can be used to construct MWG-drawings of graph pairs of the form $\langle K_{1,\cdots,1,n_0}, K_{1,\cdots,1,n_1} \rangle$ as long as the number of partition sets of size one in the two graphs differ by at most two. However, this may not be a complete characterization. (ii) Which other pairs of diameter-2 graphs admit an MWG-drawing? (iii) Which pairs of (not necessarily complete) bipartite graphs admit an MWG-drawing? (iv) Finally, it would be interesting to study mutual witness drawings for other proximity regions.

References

1. Alpert, H., Koch, C., Laison, J.D.: Obstacle numbers of graphs. Discrete Comput. Geom. **44**(1), 223–244 (2009). https://doi.org/10.1007/s00454-009-9233-8
2. Aronov, B., Dulieu, M., Hurtado, F.: Witness (delaunay) graphs. Comput. Geom. **44**(6–7), 329–344 (2011). https://doi.org/10.1016/j.comgeo.2011.01.001
3. Aronov, B., Dulieu, M., Hurtado, F.: Witness Gabriel graphs. Comput. Geom. **46**(7), 894–908 (2013). https://doi.org/10.1016/j.comgeo.2011.06.004
4. Aronov, B., Dulieu, M., Hurtado, F.: Mutual witness proximity graphs. Inf. Process. Lett. **114**(10), 519–523 (2014). https://doi.org/10.1016/j.ipl.2014.04.001
5. Aronov, B., Dulieu, M., Hurtado, F.: Witness rectangle graphs. Graph. Comb. **30**(4), 827–846 (2013). https://doi.org/10.1007/s00373-013-1316-x

6. Balko, M., Cibulka, J., Valtr, P.: Drawing graphs using a small number of obstacles. Discrete Comput. Geom. **59**(1), 143–164 (2017). https://doi.org/10.1007/s00454-017-9919-2
7. Battista, G.D., Eades, P., Tamassia, R., Tollis, I.G.: Graph drawing: algorithms for the visualization of graphs. Prentice-Hall (1999)
8. Bläsius, T., Kobourov, S.G., Rutter, I.: Simultaneous embedding of planar graphs. In: Tamassia, R. (ed.) Handbook on Graph Drawing and Visualization, pp. 349–381. Chapman and Hall/CRC (2013)
9. Chaplick, S., Lipp, F., Park, J., Wolff, A.: Obstructing visibilities with one obstacle. In: Hu, Y., Nöllenburg, M. (eds.) GD 2016. LNCS, vol. 9801, pp. 295–308. Springer, Cham (2016). https://doi.org/10.1007/978-3-319-50106-2_23
10. Dujmovic, V., Morin, P.: On obstacle numbers. Electron. J. Comb. **22**(3), 3.1 (2015). http://www.combinatorics.org/ojs/index.php/eljc/article/view/v22i3p1
11. Eades, P., Hong, S., Nguyen, A., Klein, K.: Shape-based quality metrics for large graph visualization. J. Graph Algorithms Appl. **21**(1), 29–53 (2017). https://doi.org/10.7155/jgaa.00405
12. Firman, O., Kindermann, P., Klawitter, J., Klemz, B., Klesen, F., Wolff, A.: Outside-obstacle representations with all vertices on the outer face (2022). https://arxiv.org/abs/2202.13015. https://doi.org/10.48550/ARXIV.2202.13015
13. Ichino, M., Sklansky, J.: The relative neighborhood graph for mixed feature variables. Pattern Recognit. **18**(2), 161–167 (1985). https://doi.org/10.1016/0031-3203(85)90040-8
14. Jacob, J.O., Goodman, E., Toth, C.: Handbook of discrete and computational geometry, third edition. Wiley Series in Probability and Mathematical Statistics. Chapman and Hall/CRC (2017)
15. Johnson, M.P., Sariöz, D.: Representing a planar straight-line graph using few obstacles. In: Proceedings of the 26th Canadian Conference on Computational Geometry, CCCG 2014. Carleton University, Ottawa, Canada (2014)
16. Lenhart, W., Liotta, G.: Mutual witness Gabriel drawings of complete bipartite graphs. CoRR, 2209.01004 (2022). http://arxiv.org/abs/2209.01004
17. Liotta, G.: Proximity drawings. In: Tamassia, R. (ed.) Handbook on Graph Drawing and Visualization, pp. 115–154. Chapman and Hall/CRC (2013)
18. Mukkamala, P., Pach, J., Pálvölgyi, D.: Lower bounds on the obstacle number of graphs. Electron. J. Comb. **19**(2), 32 (2012)
19. Mukkamala, P., Pach, J., Sariöz, D.: Graphs with large obstacle numbers. In: Thilikos, D.M. (ed.) Graph Theoretic Concepts in Computer Science - 36th International Workshop, WG 2010, vol. 6410 of Lecture Notes in Computer Science, pp. 292–303 (2010). https://doi.org/10.1007/978-3-642-16926-7_27
20. Pach, J., Sariöz, D.: On the structure of graphs with low obstacle number. Graphs Comb. **27**(3), 465–473 (2011). https://doi.org/10.1007/s00373-011-1027-0
21. Rutter, I.: Simultaneous embedding. In: Hong, S.-H., Tokuyama, T. (eds.) Beyond Planar Graphs, pp. 237–265. Springer, Singapore (2020). https://doi.org/10.1007/978-981-15-6533-5_13

Empty Triangles in Generalized Twisted Drawings of K_n

Alfredo García[1] , Javier Tejel[1] , Birgit Vogtenhuber[2] ,
and Alexandra Weinberger[2(✉)]

[1] Departamento de Métodos Estadísticos and IUMA, Universidad de Zaragoza,
Zaragoza, Spain
{olaverri,jtejel}@unizar.es

[2] Institute of Software Technology, Graz University of Technology, Graz, Austria
{bvogt,weinberger}@ist.tugraz.at

Abstract. Simple drawings are drawings of graphs in the plane or on
the sphere such that vertices are distinct points, edges are Jordan arcs
connecting their endpoints, and edges intersect at most once (either in
a proper crossing or in a shared endpoint). Simple drawings are generalized twisted if there is a point O such that every ray emanating from O
crosses every edge of the drawing at most once and there is a ray emanating from O which crosses every edge exactly once. We show that all
generalized twisted drawings of K_n contain exactly $2n - 4$ empty triangles, by this making a substantial step towards proving the conjecture
that this is the case for every simple drawing of K_n.

Keywords: Simple drawings · Simple topological graphs · Empty
triangles

1 Introduction

Simple drawings are drawings of graphs in the plane or on the sphere such that
vertices are distinct points, edges are Jordan arcs connecting their endpoints, and
edges intersect at most once either in a proper crossing or in a shared endpoint.
The edges and vertices of a drawing partition the plane into regions, which are
called the *cells* of the drawing. A *triangle* in a simple drawing D is a subdrawing
of D which is a drawing of K_3. By the definition of simple drawings, any triangle
is crossing free and thus splits the plane (or the sphere) in two connected regions.

This project has received funding from the European Union's Horizon
2020 research and innovation programme under the Marie Skłodowska-
Curie grant agreement No 734922.

A.G. and J.T. partially supported by the Gobierno de Aragón project E41-17R. J.T.
partially supported by project PID2019-104129GB-I00 / AEI / 10.13039/501100011033
of the Spanish Ministry of Science and Innovation. B.V. partially supported by the Austrian Science Fund as FWF project I 3340-N35 within the collaborative DACH project
Arrangements and Drawings A.W. partially supported by the FWF project W1230.

P. Angelini and R. von Hanxleden (Eds.): GD 2022, LNCS 13764, pp. 40–48, 2023.
https://doi.org/10.1007/978-3-031-22203-0_4

We call those regions the *sides* of the triangle. If one side of a triangle does not contain any vertices of D, that side is called an *empty side* of the triangle, and the triangle is called *empty triangle*. Note that empty (sides of) triangles might be intersected by edges. We observe that simple drawings of K_3 consist of exactly one triangle, which has two empty sides. Triangles in simple drawings of graphs with $n \geq 4$ vertices have at most one empty side.

In this work, we study the number of empty triangles in simple drawings of K_n. Note that simple drawings are a topological generalization of straight-line drawings. The number $h_3(n)$ of empty triangles that every straight-line drawing of K_n contains has been subject of intensive research. It is easy to see that $h_3(n) = \Omega(n^2)$. The currently best known bounds are $n^2 - \frac{32}{7}n + \frac{22}{7} \leq h_3(n) \leq 1.6196n^2 + o(n^2)$ [2,6].

For simple drawings of complete graphs, the situation changes drastically. Harborth [9] showed in 1989 that there are simple drawings of K_n that contain only $2n - 4$ empty triangles; see Fig. 1b. This especially implies that most edges in these drawings are not incident to any empty triangles. On the other hand, Harborth observed that every vertex in these drawings is incident to at least two empty triangles, a property he conjectured to be true in general. This conjecture has been proven in 2013 [8,13]. The currently best lower bound on the number of empty triangles in simple drawings of K_n is n [4]. Further, it is conjectured that Harborth's upper bound should actually be the true lower bound.

Conjecture 1 ([4]). *For any $n \geq 4$, every simple drawing of K_n contains at least $2n - 4$ empty triangles.*

The drawings that Harborth used for his upper bound are now well known as *twisted drawings* [12] and have received considerable attention [1,5,7,10–12,14]. A generalization of twisted drawings was introduced in [3] as a special type of c-monotone drawings. A simple drawing D in the plane is *c-monotone* if there is a point O such that any ray emanating from O intersects any edge of D at most once. A c-monotone drawing D is generalized twisted if there exists a ray r emanating from O that intersects every edge of D.

As twisted drawings and the upper bound obtained by them are crucial in the study of empty triangles, it is natural to ask about the number of triangles in their generalization. The initial goal of this work was to prove Conjecture 1 for generalized twisted drawings. As the number of such drawings is exponential (in the number of vertices), one might expect that they contain different numbers of empty triangles. However, we show that surprisingly, the conjectured bound is tight for all of them.

Theorem 1. *For any $n \geq 4$, every generalized twisted drawing of K_n contains exactly $2n - 4$ empty triangles.*

Outline. In Sect. 2, we introduce some properties of generalized twisted drawings and empty triangles in general simple drawings. Then, in Sect. 3, we show several results about empty triangles in generalized twisted drawings, which we finally put together to obtain a proof of Theorem 1.

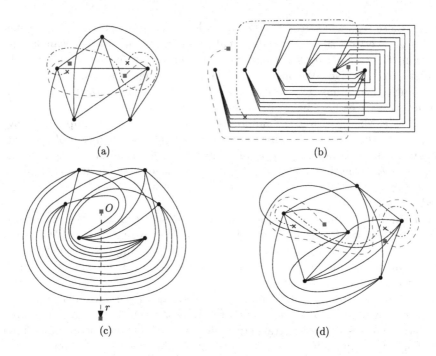

Fig. 1. All (up to weak isomorphism) generalized twisted drawings of K_5 and K_6. O and Z have to lie in cells marked with red squares or in cells with blue crosses. Curves OZ are drawn dashed or dash-dotted (and as ray in c). (Color figure online)

2   Preliminaries

Two simple drawings are weakly isomorphic if the same pairs of edges cross. It is well-known that the weak isomorphism class of a drawing of K_n completely determines which triangles are empty. To prioritize readability, several of our figures show drawings that are weakly isomorphic to generalized twisted (sub-)drawings rather than a generalized twisted drawing.

For a generalized twisted drawing D of K_n, we put a point Z into the unbounded cell of D, on the ray r that crosses everything. Similarly, for every drawing that is weakly isomorphic to a generalized twisted drawing, there exists a simple curve OZ corresponding to the part of the ray r from O to Z; see [3]. Note that, given a simple drawing D of K_n, there might be several cell pairs where O and Z could be placed such that D is weakly isomorphic to a generalized twisted drawing with the corresponding cells for O and Z. For instance, Fig. 1 shows all [3] generalized twisted drawings of K_5 and K_6 up to weak isomorphism, together with all possible cell pairs for O and Z and some curve OZ for each pair. With this addition of O and Z, we will use the following properties of generalized twisted drawings, which have been shown in [3].

Lemma 1 ([3]). *Let D be a simple drawing in the plane that is weakly isomorphic to a generalized twisted drawing of K_n. Then the following holds:*

1. *For each triangle of D, the cell containing O and the cell containing Z lie on different sides. In particular, this implies that D does not contain three interior-disjoint triangles.*
2. *The cells containing O and Z each have at least one vertex on their boundary.*
3. *Every subdrawing of D induced by four vertices contains a crossing. If p is such a crossing, then O and Z lie in two different cells that are incident to p and opposite to each other (see Fig. 2a for an illustration).*

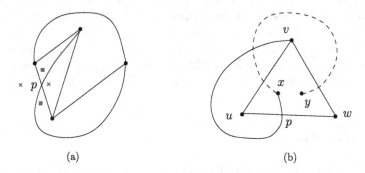

(a)   (b)

Fig. 2. (a) O and Z have to lie in cells marked with red squares or in cells with blue crosses. (b) The edge (x, y) cannot cross (v, u) and (v, w) simultaneously. (Color figure online)

We will also use the following technical lemma for simple drawings.

Lemma 2. *Let D be a simple drawing of K_n. Let Δ be a triangle of D with vertices u, v, w. Let x, y be two vertices on the same side of Δ. If the edge (x, v) crosses (u, w), then the edge (x, y) can cross at most one of (v, u) and (v, w).*

Proof. Assume that (x, y) crosses both (v, u) and (v, w). Since x and y are on the same side of Δ, the edge (x, y) must cross the boundary of Δ an even number of times. Thus, if (x, y) crosses (v, u) and (v, w), it cannot cross (u, w). Let p be the crossing point between (x, v) and (u, w). See Fig. 2b for an illustration. Suppose that (x, y) crosses first (v, u) and then (v, w). After crossing (v, u), the edge (x, y) and the vertex y are in different regions defined by the closed curve C consisting of (v, u), the part of the edge (u, w) from u to p and the part of the edge (x, v) from p to v. Then, after crossing (v, u), the edge (x, y) must cross C to reach y, and this is not possible without violating the simplicity of the drawing. Therefore, (x, y) cannot cross (v, u) and (v, w) simultaneously. An analogous analysis can be done if (x, y) crosses first (v, w) and then (v, u). $\quad\blacksquare$

In addition to the properties of generalized twisted drawings, we will use the concept of star triangles as introduced in [4]. A triangle Δ with vertices x, y, z is a *star triangle* at x if yz is not crossed by any edges incident to x. We will use the following properties of star triangles in simple drawings of K_n.

Lemma 3. *Let D be a simple drawing of K_n in the plane and x be a vertex of D. Then the following holds:*

1. *There are at least two empty star triangles at x.*
2. *A star triangle xyz at a vertex x is an empty triangle if and only if the vertices y and z are consecutive in the rotation around x.*
3. *For any two different empty star triangles at x, $xy'z'$ and xyz, the empty sides of $xy'z'$ and xyz are disjoint.*

Proof. Properties 1 and 2 have been shown in [4]. To prove Property 3, consider the boundary edges of the triangles xyz and $xy'z'$. Of these edges, the only pair that could cross is (y, z) and (y', z'). However, y' and z' lie on the same side of the triangle xyz, so (y', z') has to cross the boundary of xyz an even number of times, which is not possible if exactly (y, z) and (y', z') cross. Thus, no edges on the boundary of the star triangles cross, and therefore their empty sides are disjoint.

3   Proof of Theorem 1

In the following, we derive several lemmata about empty triangles in generalized twisted drawings. These lemmata put together will give the proof of Theorem 1.

Lemma 4. *Let D be a generalized twisted drawing of K_n in the plane with $n \geq 4$ and x be a vertex of D. Then x is incident to exactly two empty star triangles, one has O on the empty side and the other has Z on the empty side. Further, these star triangles have disjoint empty sides.*

Proof. By Lemma 3 (1 and 3), for every vertex x there are at least two empty star triangles at x and the empty sides of these triangles are disjoint. By Lemma 1(1), any triangle of a generalized twisted drawing has O on one side and Z on the other side, and D cannot contain three interior-disjoint triangles. Thus, for any vertex x in a generalized twisted drawing it holds that: (i) one empty star triangle at x has O on the empty side, (ii) another empty star triangle at x has Z on the empty side, and (iii) there cannot be a third empty star triangle at x.

Lemma 5. *Let D be a generalized twisted drawing of K_n with $n \geq 4$. Let C_O be the cell of D containing O and let v be a vertex on the boundary of C_O. Let Δ be an empty triangle in D that has O on the empty side. Then the following holds:*

1. *The vertex v is a vertex of Δ, that is, $\Delta = xyv$ for some x, y.*
2. *The triangle $\Delta = xyv$ is an empty star triangle at x or y or both.*
3. *If $\Delta = xyv$ is a star triangle at both x and y, then all edges emanating from v cross (x, y). Hence, Δ is a star triangle for at most two of its vertices.*

Proof. Since Δ has O on its empty side and since C_O is a cell of D, Δ also has C_O on the empty side. Therefore, since v belongs to C_O and Δ is empty, necessarily v must be one of the vertices of Δ and Property 1 is fulfilled.

To prove Property 2, assume to the contrary that $\Delta = xyv$ is not a star triangle at x or y. Then at least one edge (x, x') must cross (v, y) at a point q' and at least an edge (y, y') must cross (v, x) at a point q (see Fig. 3a for an illustration). Note that since Δ is empty and any edge incident to x or y can cross Δ at most once, (x, x') and (y, y') must emanate from x and y, respectively, at the empty side of Δ, and cross at a point p on that side of Δ. Without loss of generality, we may assume that O is very close to v. Consider the subdrawing D' induced by x, x', y and y'. Observe that since (x, y') cannot cross (x, v) or (y, y'), the cell of D' defined by x, p and y' cannot contain O, regardless of the shape of (x, y'). Thus, by Lemma 1(3) applied to D', O is in the region defined by x', p and y', and Z is in the region defined by x, p and y, contradicting that O and Z lie on different sides of Δ.

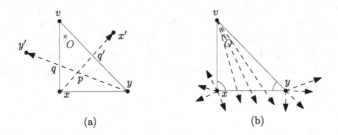

Fig. 3. (a) Illustrating the proof of Lemma 5(2). (b) Any empty triangle of D that has O on the empty side cannot be a star triangle at three vertices.

To prove Property 3, take a vertex w that is not a vertex of Δ. By Lemma 1(3), the subdrawing induced by v, x, y and w has a crossing. As Δ is a star triangle at x and y, any edge incident to x or y emanates from x or y on the non-empty side of Δ, so neither (x, w) nor (y, w) can cross Δ. Then (v, w) must cross (x, y), emanating from v on the empty side of Δ. Therefore, Property 3 follows.

Note that by Lemma 1(2), the cell containing O always has a vertex on its boundary. Hence, by Lemma 5, any empty triangle with O on the empty side is a star triangle at one or two vertices. The following lemma proves that there are exactly two such triangles that are star triangles at two vertices.

Lemma 6. *Let D be a generalized twisted drawing of K_n with $n \geq 4$. Then D contains exactly two empty triangles with O on the empty side that are star triangles at two vertices.*

Proof. Let C_O be the cell containing O and v be a vertex on the boundary of C_O, which exists by Lemma 1(2). By Lemma 4, there is an empty star triangle

$\Delta = vuw$ at v that has C_O on the empty side. By Lemma 5, Δ is a star triangle at exactly one of u or w, say w. Thus, Δ is an empty star triangle at two vertices with O on the empty side.

On the other hand, by Lemma 5(3), all edges emanating from u cross (v, w). Among all edges crossing (v, w), we choose the edge (a, b) that crosses (v, w) closest to v. Let p be the crossing point between (a, b) and (v, w), and consider the triangle vab. We will show that the triangle vab is a star triangle at a and b and that the side F of it in which O lies is empty.

If $a \neq u$, then (a, b) crosses (v, u) at a point q; see Fig. 4. Note that F is partitioned into three triangular shapes vaq, vqp, and vpb, where O lies in vqp. Assume, for a contradiction, that a vertex x lies in vbp. Since (a, x) can cross neither (a, b) nor (v, p), it has to cross (v, b). As no simple drawing of the K_4 can contain more than one crossing, the edges xv and xb have to stay completely in F. Since (v, p) is crossing-free, the edges (x, v) and (x, b) must be in vbp, one side of the triangle vxb is contained in vpb. As O is not on the side of vxb contained in F, then Z has to be on that side. This implies that both O and Z are in F, a contradiction. Therefore, vbp is empty. Using a similar argument, one can prove that vqa is also empty, so F is empty. Besides, any edge incident to a or b must emanate outside F because (v, p) is crossing free. As a consequence, F is the empty side of the triangle vab, which is a star triangle at a and b.

The reasoning for $a = u$ is similar (with two triangular shapes vup and vpb).

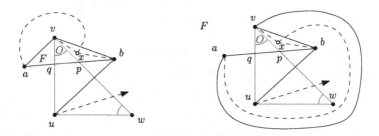

Fig. 4. Illustrating the proof of Lemma 6.

What remains to show is that there is no third empty triangle with O on the empty side that is a star triangle at two vertices. Assume for a contradiction that such a triangle $x'y'z'$ exists. By Lemma 5(1), one of x', y' and z' must be v, say $z' = v$. As v, a, and b are incident to at most one empty star triangle with O on the empty side, by Lemma 4, x' and y' are different from a and b, and $vx'y'$ is a star triangle at x' and y'. Consider the triangle vab. Since x' and y' are on the same side of vab and (v, x') crosses (a, b) by Lemma 5(3), the edge (x', y') cannot cross (v, a) and (v, b) by Lemma 2. But all edges emanating from v must cross (x', y') by Lemma 5(3), a contradiction. Hence $x'y'z'$ cannot exist.

We note that the lemmata above and their proofs hold for every choice of C_O (if there are many) and any vertex v on the boundary C_O. However, whether a triangle is empty and at how many vertices it is a star triangle does not change between weakly isomorphic drawings. As a consequence, the empty star triangles obtained in the previous lemmata and proofs must be the same, regardless of the choice of C_O and the vertex v on the boundary of C_O. We also note that for empty triangles having Z on the empty side, the reasoning in Lemmas 5 and 6 works to prove that these triangles are star triangles at one or two vertices, and that exactly two of them are star triangles at two vertices. By Lemma 6 and these observations, we get the following lemma.

Lemma 7. *Let D be a generalized twisted drawing of K_n with $n \geq 4$. Then D contains exactly four empty triangles that are star triangles at two vertices.*

Now, we can prove our main theorem.

Proof (of Theorem 1). When summing up the number of empty star triangles over all vertices, we obtain $2n$ empty star triangles by Lemma 4 (n triangles with O on the empty side and n with Z on the empty side). By Lemma 5, all empty triangles have been counted this way, but the triangles that are empty star triangles at two vertices have been counted twice. By Lemma 7, there are exactly four triangles that are empty star triangles at two vertices. Thus, there are exactly four triangles that have been counted exactly twice and the precise number of empty triangles in D is $2n - 4$.

References

1. Ábrego, B., Fernández-Merchant, S., Figueroa, A.P., Montellano-Ballesteros, J.J., Rivera-Campo, E.: Crossings in twisted graphs. In: Collection of Abstracts for the 22nd Japan Conference on Discrete and Computational Geometry, Graphs, and Games (JCDCG3), pp. 21–22 (2019). https://www.csun.edu/~sf70713/publications/2019_JCDCG3_Twisted_Crossings.pdf
2. Aichholzer, O., Fabila-Monroy, R., Hackl, T., Huemer, C., Pilz, A., Birgit, V.: Lower bounds for the number of small convex k-holes. Comput. Geom. Theor. Appl. **47**(5), 605–613 (2014). https://doi.org/10.1016/j.comgeo.2013.12.002
3. Aichholzer, O., García, A., Tejel, J., Vogtenhuber, B., Weinberger, A.: Twisted ways to find plane structures in simple drawings of complete graphs. In: Proceedings of the 38th International Symposium on Computational Geometry (SoCG 2022), pp. 1–18 (2022). https://doi.org/10.4230/LIPIcs.SoCG.2022.5
4. Aichholzer, O., Hackl, T., Pilz, A., Ramos, P., Sacristán, V., Vogtenhuber, B.: Empty triangles in good drawings of the complete graph. Graphs Comb. **31**(2), 335–345 (2015). https://doi.org/10.1007/s00373-015-1550-5
5. Arroyo, A., Richter, R.B., Salazar, G., Sullivan, M.: The unavoidable rotation systems. arXiv:1910.12834 (2019)
6. Bárány, I., Valtr, P.: Planar point sets with a small number of empty convex polygons. studia scientiarum mathematicarum hungarica. Q. Hung. Acad. Sci. **41**, 243–266 (2004). https://doi.org/10.1556/SScMath.41.2004.2.4
7. Figueroa, A.P., Fresán-Figueroa, J.: The biplanar tree graph. Bol. Soc. Matemática Mex. **26**(3), 795–806 (2020). https://doi.org/10.1007/s40590-020-00287-y

8. Fulek R., Ruiz-Vargas, A.J.: Topological graphs: empty triangles and disjoint matchings. In: Proceedings of the 29th Annual Symposium on Computational Geometry (SoCG2013), pp. 259–266. New York, ACM (2013). https://doi.org/10.1145/2462356.2462394

9. Harborth, H.: Empty triangles in drawings of the complete graph. Discret. Math. **191**, 109–111 (1998)

10. Kynčl, J., Valtr, P.: On edges crossing few other edges in simple topological complete graphs. In: Healy, P., Nikolov, N.S. (eds.) GD 2005. LNCS, vol. 3843, pp. 274–284. Springer, Heidelberg (2006). https://doi.org/10.1007/11618058_25

11. Omaña-Pulido, E., Rivera-Campo, E.: Notes on the twisted graph. In: Márquez, A., Ramos, P., Urrutia, J. (eds.) EGC 2011. LNCS, vol. 7579, pp. 119–125. Springer, Heidelberg (2012). https://doi.org/10.1007/978-3-642-34191-5_11

12. Pach, J., Solymosi, J., Tóth, G.: Unavoidable configurations in complete topological graphs. Discrete Comput. Geom. **30**(2), 311–320 (2003). https://doi.org/10.1007/s00454-003-0012-9

13. Ruiz-Vargas, A.J.: Empty triangles in complete topological graphs. Discrete Comput. Geom. **53**(4), 703–712 (2015). https://doi.org/10.1007/s00454-015-9671-4

14. Suk, A., Zeng, J.: Unavoidable patterns in complete simple topological graphs. arXiv:2204.04293 (2022)

Shooting Stars in Simple Drawings of $K_{m,n}$

Oswin Aichholzer[1], Alfredo García[2], Irene Parada[3,4(✉)],
Birgit Vogtenhuber[1], and Alexandra Weinberger[1]

[1] Institute of Software Technology, Graz University of Technology, Graz, Austria
{oaich,bvogt,weinberger}@ist.tugraz.at

[2] Departamento de Métodos Estadísticos and IUMA, Universidad de Zaragoza,
Zaragoza, Spain
olaverri@unizar.es

[3] Departament de Matemàtiques, Universitat Politècnica de Catalunya,
Barcelona, Spain
irene.maria.de.parada@upc.edu

[4] Department of Information and Computing Sciences, Utrecht University, Utrecht,
The Netherlands

Abstract. Simple drawings are drawings of graphs in which two edges have at most one common point (either a common endpoint, or a proper crossing). It has been an open question whether every simple drawing of a complete bipartite graph $K_{m,n}$ contains a plane spanning tree as a subdrawing. We answer this question to the positive by showing that for every simple drawing of $K_{m,n}$ and for every vertex v in that drawing, the drawing contains a *shooting star rooted at v*, that is, a plane spanning tree containing all edges incident to v.

Keywords: Simple drawing · Simple topological graph · Complete bipartite graph · Plane spanning tree · Shooting star

1 Introduction

A *simple drawing* is a drawing of a graph on the sphere S^2 or, equivalently, in the Euclidean plane where (1) the vertices are distinct points in the plane, (2) the edges are non-self-intersecting continuous curves connecting their incident

O.A., I.P., and A.W. were partially supported by the Austrian Science Fund (FWF) grant W1230. A.G. was supported by MINECO project MTM2015-63791-R and Gobierno de Aragón under Grant E41-17 (FEDER). I.P. and B.V. were partially supported by the Austrian Science Fund within the collaborative DACH project *Arrangements and Drawings* as FWF project I 3340-N35. I.P. was supported by the Margarita Salas Fellowship funded by the Ministry of Universities of Spain and the European Union (NextGenerationEU).

This work initiated at the 6th Austrian-Japanese-Mexican-Spanish Workshop on Discrete Geometry which took place in June 2019 near Strobl, Austria. We thank all the participants for the great atmosphere and fruitful discussions.

P. Angelini and R. von Hanxleden (Eds.): GD 2022, LNCS 13764, pp. 49–57, 2023.
https://doi.org/10.1007/978-3-031-22203-0_5

points, (3) no edge passes through vertices other than its incident vertices, (4) and every pair of edges intersects at most once, either in a common endpoint, or in the relative interior of both edges, forming a proper crossing. Simple drawings are also called *good drawings* [5,7] or *(simple) topological graphs* [11,12]. In *star-simple drawings*, the last requirement is softened so that edges without common endpoints are allowed to cross several times. Note that in any simple or star-simple drawing, there are no tangencies between edges and incident edges do not cross. If a drawing does not contain any crossing at all, it is called *plane*.

The search for plane subdrawings of a given drawing has been a widely considered topic for simple drawings of the complete graph K_n which still holds tantalizing open problems. For example, Rafla [14] conjectured that every simple drawing of K_n contains a plane Hamiltonian cycle, a statement which is by now known to be true for $n \leq 9$ [1] and several classes of simple drawings (e.g., 2-page book drawings, monotone drawings, cylindrical drawings), but still remains open in general. A related question concerns the least number of pairwise disjoint edges in any simple drawing of K_n. The currently best lower bound is $\Omega(n^{1/2})$ [3], which is improving over several previous bounds [8–10,12,13,15,16], while the trivial upper bound of $n/2$ would be implied by a positive answer to Rafla's conjecture. A structural result of Fulek and Ruiz-Vargas [10] implies that every simple drawing of K_n contains a plane sub-drawing with at least $2n - 3$ edges.

We will focus on plane trees. Pach et al. [12] proved that every simple drawing of K_n contains a plane drawing of any fixed tree with at most $c \log^{1/6} n$ vertices. For paths specifically, every simple drawing of K_n contains a plane path of length $\Omega(\frac{\log n}{\log \log n})$ [3,17]. Further, it is trivial that simple drawings and star-simple drawings of K_n contain a plane spanning tree, because every vertex is incident to all other vertices and adjacent edges do not cross. Thus, the vertices together with all edges incident to one vertex form a plane spanning tree. We call this subdrawing the *star* of that vertex.

In this work, we consider the search for plane spanning trees in drawings of complete bipartite graphs. Finding plane spanning trees there is more involved than for K_n. In fact, not every star-simple drawing of a complete bipartite graph contains a plane spanning tree; see Fig. 1.

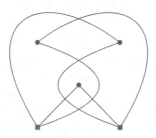

Fig. 1. Star-simple drawing of $K_{2,3}$ that does not contain a plane spanning tree.

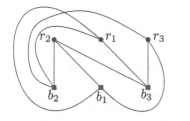 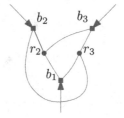

Fig. 2. Left: Simple drawing of $K_{3,3}$. Right: its stereographic projection from r_1.

It is not hard to see that straight-line drawings of complete bipartite graphs always contain plane spanning trees. Consider the star of an arbitrary vertex v. The prolongation of these edges creates a set of rays originating at v that partitions the plane into wedges, which we divide into two parts using the angle bisectors. We connect the vertices in each part of a wedge to the point on the ray that bounds it. These connections together with the star of v form a plane spanning tree of a special type called *shooting star*. A *shooting star rooted at v* is a plane spanning tree with root v that has height 2 and contains the star of vertex v. Aichholzer et al. showed in [4] that simple drawings of $K_{2,n}$ and $K_{3,n}$, as well as so-called *outer drawings* of $K_{m,n}$, always contain shooting stars. *Outer drawings of $K_{m,n}$* [6] are simple drawings in which all vertices of one bipartition class lie on the outer boundary.

Results. We show in Sect. 2 that every simple drawing of $K_{m,n}$ contains shooting stars rooted at an arbitrary vertex of $K_{m,n}$. The tightness of the conditions is shown in Sect. 3 and in Sect. 4 we discuss algorithmic aspects.

2 Existence of Shooting Stars

In this section, we prove our main result, the existence of shooting stars:

Theorem 1. *Let D be a simple drawing of $K_{m,n}$ and let r be an arbitrary vertex of $K_{m,n}$. Then D contains a shooting star rooted at r.*

Proof. We can assume that D is drawn on a point set $P = R \cup B$, $R = \{r_1, \ldots, r_m\}$, $B = \{b_1, \ldots, b_n\}$, in which the points in the two bipartition classes R and B are colored red and blue, respectively. Without loss of generality let $r = r_1$.

To simplify the figures, we consider the drawing D on the sphere and apply a stereographic projection from r onto a plane. In that way, the edges in the star of r are represented as (not necessarily straight-line) infinite rays; see Fig. 2. We will depict them in blue. In the following, we consider all edges oriented from their red to their blue endpoint. To specify how two edges cross each other, we introduce some notation. Consider two crossing edges $e_1 = r_i b_k$ and $e_2 = r_j b_l$ and let x be their crossing point. Consider the arcs xr_i and xb_k on e_1 and xr_j and xb_l on e_2. We say that e_2 crosses e_1 in *clockwise direction* if the clockwise cyclic

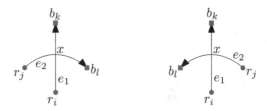

Fig. 3. Left: e_2 crosses e_1 in clockwise direction. Right: e_2 crosses e_1 in counterclockwise direction.

order of these arcs around the crossing x is xr_i, xr_j, xb_k, and xb_l. Otherwise, we say that e_2 crosses e_1 in *counterclockwise direction*; see Fig. 3.

We prove Theorem 1 by induction on n. For $n = 1$ and any $m \geq 1$, the whole drawing D is a shooting star rooted at any vertex, and in particular at r.

Assume that the existence of shooting stars rooted at any vertex has been proven for any simple drawing of $K_{m,n'}$ with $n' < n$. By the induction hypothesis, the subdrawing of D obtained by deleting the blue vertex b_1 and its incident edges contains at least one shooting star rooted at r. Of all such shooting stars, let S be one whose edges have the minimum number of crossings with rb_1, and let M be the set of edges of S that are not incident to r. We will show that $S \cup \{rb_1\}$ is plane and hence forms the desired shooting star. Note that it suffices to show that $M \cup \{rb_1\}$ is plane, since rb_1 cannot cross any edges of $\{\bigcup_{j=2}^{n} rb_j\}$ in any simple drawing.

Assume for a contradiction that rb_1 crosses at least one edge in M. When traversing rb_1 from b_1 to r, let x be the first crossing point of rb_1 with an edge $r_k b_t$ in M. W.l.o.g., when orienting rb_1 from r to b_1 and $r_k b_t$ from r_k to b_t, $r_k b_t$ crosses rb_1 in counterclockwise direction (otherwise we can mirror the drawing).

Suppose first that the arc $r_k x$ (on $r_k b_t$ and oriented from r_k to x) is crossed in counterclockwise direction by an edge incident to b_1 (and oriented from the red endpoint to b_1). Let $e = r_l b_1$ be such an edge whose crossing with $r_k x$ at a point y is the closest to x. Otherwise, let e be the edge $r_k b_1$ and y be the point r_k. In the remaining figures, we represent in blue the edges of the star of r, in red the edges in M, and in black the edge e.

We distinguish two cases depending on whether e crosses an edge of the star of r. The idea in both cases is to define a region Γ and, inside it, redefine the connections between red and blue points to reach a contradiction.

Case 1: e does not cross any edge of the star of r. Let Γ be the closed region of the plane bounded by the arcs yb_1 (on e), $b_1 x$ (on rb_1), and xy (on $r_k b_t$); see Fig. 4. Observe that all the blue points b_j lie outside the region Γ and that for all the red points r_i inside region Γ, the edge $r_i b_1$ must be in Γ. Let M_Γ denote the set of edges $r_i b_1$ with $r_i \in \Gamma$ and note that $r_k b_1 \in M_\Gamma$. Consider the set M' of red edges obtained from M by replacing, for each red point $r_i \in \Gamma$, the (unique) edge incident to r_i in M by the edge $r_i b_1$ in M_Γ, and keeping the other edges in M unchanged. In particular, the edge $r_k b_t$ has been replaced by the edge $r_k b_1$.

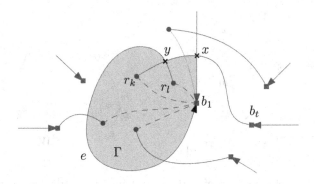

Fig. 4. Illustration of Case 1.

The edges in M_Γ neither cross each other nor cross any of the blue edges rb_j. Moreover, we now show that the non-replaced edges in M must lie completely outside Γ. These edges can neither cross $r_k b_t$ (by definition of M) nor the arc $b_1 x$ (on rb_1). Thus, if they are incident to b_1, they cannot cross the boundary of Γ; If they are not incident to b_1, both their endpoints lie outside Γ and they can only cross the boundary of Γ at most once (namely, on the arc $b_1 y$). Therefore, M' satisfies that $M' \cup \{\bigcup_{j=2}^n rb_j\}$ is plane and has fewer crossings with rb_1 than M, since at least the crossing x has been eliminated and no new crossings have been added. This contradicts the definition of M as the one with the minimum number of crossings with rb_1.

Case 2: e crosses the star of r. When traversing e from r_k or r_l (depending on the definition of e) to b_1, let $I = \{\alpha, \beta, \ldots, \rho\}$ be the indices of the edges of the star of r in the order as they are crossed by e and let $y_\alpha, \ldots, y_\rho$ be the corresponding crossing points on e. Note that, when orienting e from r_k or r_l to b_1, the edges $rb_\xi, \xi \in I$, oriented from r to b_ξ, cross e in counterclockwise direction, since they can neither cross $r_k b_t$ (by definition of M) nor rb_1.

The three arcs ry_α (on rb_α), $y_\alpha b_1$ (on e), and $b_1 r$ divide the plane into two (closed) regions, Π_{left}, containing vertex r_k, and Π_{right}, containing vertex b_t. For each $\xi \in I$, let M_ξ be the set of red edges of M incident to some red point in Π_{right} and to b_ξ. Note that all the edges in M_ξ (if any) must cross the edge e. When traversing e from r_k or r_l to b_1, we denote by x_ξ, z_ξ the first and the last crossing points of e with the edges of $M_\xi \cup rb_\xi$, respectively; see Fig. 5 for an illustration. We remark that both x_ξ and z_ξ might coincide with y_ξ and, in particular, if $M_\xi = \emptyset$ then $x_\xi = y_\xi = z_\xi$.

We now define some regions in the drawing D. Suppose first that there are edges in M (oriented from the red to the blue point) that cross rb_1 (oriented from r to b_1) in clockwise direction. Let $r_s b_\eta$ be the edge in M whose clockwise crossing with rb_1 at a point x' is the closest one to x (recall that the arc $b_1 x$ on rb_1 is not crossed by edges in M). Then, if $\eta \notin I$, we denote by W_η the region bounded by the arcs rx' (on rb_1), $x'b_\eta$ (on $r_s b_\eta$), and rb_η and not containing b_1; see Fig. 5 (left). If $\eta \in I$, we define W_η as the region bounded by the arcs rx' (on rb_1),

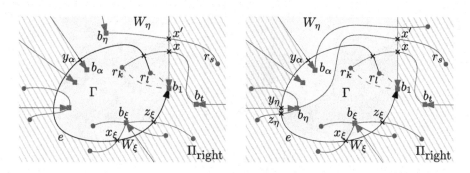

Fig. 5. Illustration of Case 2. Region Π_{right} is striped in gray, region Γ is shaded in blue, and regions in $\bigcup_{\xi \in I} W_\xi \cup W_\eta$ are shaded in yellow. Left: b_η does not cross e ($\eta \notin I$). Right: b_η crosses e ($\eta \in I$). (Color figure online)

$x'b_\eta$ (on $r_s b_\eta$), $b_\eta z_\eta$, $z_\eta y_\eta$ (on e), and $y_\eta r$ (on rb_η) and not containing b_1; see Fig. 5 (right). If no edges in M cross rb_1 in clockwise direction, then η is undefined and we set $W_\eta = \emptyset$ for convenience. Moreover, for each $\xi \in I \setminus \{\eta\}$, we define W_ξ as the region bounded by the arcs $x_\xi b_\xi$, $b_\xi z_\xi$, and $z_\xi x_\xi$ (and not containing b_1); see again Fig. 5.

We can finally define the region Γ for Case 2, which is the region obtained from Π_{left} by removing the interior of all the regions W_ξ, $\xi \in I$ plus region W_η if $\eta \notin I$ (otherwise it is already contained in $\bigcup_{\xi \in I} W_\xi$). Now consider the set of red and blue vertices contained in Γ. Let J denote the set of indices such that for all $j \in J$, the blue point b_j lies in Γ (note that $1 \in J$). Since b_t is not in Γ, we can apply the induction hypothesis to the subdrawing of D induced by the vertices in Γ plus r. Hence there exists a set of edges M_Γ connecting each red point in Γ with a blue point b_j, $j \in J$ such that $M_\Gamma \cup \{\bigcup_{j \in J} rb_j\}$ is plane. Moreover, all the edges in M_Γ lie entirely in Γ: An edge in M_Γ cannot cross any of the edges rb_j, with $j \in J$. Thus, it cannot leave Π_{left}, as otherwise it would cross e twice. Further, if it entered one of the regions in $\bigcup_{\xi \in I} W_\xi \cup W_\eta$, it would have to leave it crossing e, and then it could not re-enter Γ.

Consider the set M' of red edges obtained from M by replacing, for each red point $r_i \in \Gamma$, the edge $r_i b_\xi$ in M by the edge $r_i b_j$, $j \in J$, in M_Γ, and keeping the other edges in M unchanged. In particular, the edge $r_k b_t$ has been replaced by some edge $r_k b_j$, $j \in J$. The edges in M_Γ neither cross each other nor cross any of the blue edges rb_j, $j \in J$ nor any of the other ones, lying completely outside Γ. Moreover, the non-replaced edges in M cannot enter Γ since the only boundary part of Γ that they can cross are arcs on e. Therefore, M' satisfies that $M' \cup \{\bigcup_{j=2}^n rb_j\}$ is plane and has fewer crossings with rb_1 than M. This contradicts the definition of M as the one with the minimum number of crossings with rb_1. □

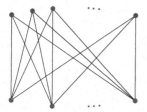 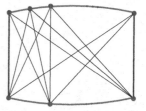

Fig. 6. Left: A simple drawing of $K_{m,n}$ where no plane subdrawing has more edges than a shooting star. Right: convex $(n + m)$-gon on the convex hull (green). (Color figure online)

3   Some Observations on Tightness

There exist simple drawings of $K_{m,n}$ in which every plane subdrawing has at most as many edges as a shooting star. For example, consider a straight-line drawing of $K_{m,n}$ where all vertices are in convex position such that all red points are next to each other in the convex hull; see Fig. 6 (left). The convex hull is an $(m + n)$-gon which shares only two edges with the drawing of $K_{m,n}$; see Fig. 6 (right). All other edges of the drawing of $K_{m,n}$ are diagonals of the polygon. As there can be at most $(m + n) - 3$ pairwise non-crossing diagonals in a convex $(m + n)$-gon, any plane subdrawing of this drawing of $K_{m,n}$ contains at most $m + n - 1$ edges.

Furthermore, both requirements from Theorem 1—the drawing being simple and containing a complete bipartite graph—are in fact necessary: As mentioned in the introduction, not all star-simple drawings of $K_{m,n}$ contain a plane spanning tree. Further, if in the example in Fig. 6 (left), we delete one of the two edges of $K_{m,n}$ on the boundary of the convex hull, then any plane subdrawing has at most $m + n - 2$ edges and hence it cannot contain any plane spanning tree.

4   Computing Shooting Stars

The proof of Theorem 1 contains an algorithm with which we can find shooting stars in given simple drawings. We start with constructing the shooting star for a subdrawing that is a $K_{m,1}$ and then inductively add more vertices. Every time we are adding a new vertex, the shooting star of the step before is a set fulfilling all requirements of $M_1 \cup \{\bigcup_{j=2}^{n} rb_j\}$ in the proof. By replacing edges as described in the proof, we obtain a new set with the same properties and fewer crossings. We continue replacing edges until we obtain a set of edges (M in the proof) that form a shooting star for the extended vertex set. We remark that the runtime of this algorithm might be exponential, as finding the edges of M_Γ might require solving the problem for the subgraph induced by Γ. However, we believe that there exists a polynomial-time algorithm for this task.

Open Problem 1. *Given a simple drawing of $K_{m,n}$, is there a polynomial-time algorithm to find a plane spanning tree contained in the drawing?*

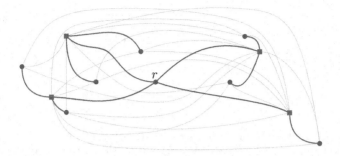

Fig. 7. An example of a shooting star rooted at r in a monotone drawing of $K_{9,5}$.

For some relevant classes of simple drawings of $K_{m,n}$ we can efficiently compute shooting stars. This is the case of outer drawings. In [4] it was shown that these drawings contain shooting stars and this existential proof leads directly to a polynomial-time algorithm to find shooting stars in outer drawings. In the full version of this paper [2] we show that *monotone drawings* of $K_{m,n}$, which are simple drawings in which all edges are x-monotone curves, admit an efficient algorithm for computing a shooting star. Figure 7 shows an illustration. The idea is as follows. Let the sides of the bipartition be R and B and let v be the leftmost vertex (without loss of generality assume $r \in R$). We first consider the star of r, which we denote by T. For each vertex $w \in R$ not in T we shoot two vertical rays, one up and one down. If only one of those vertical rays intersects T we connect w with the endpoint in B of the first intercepted edge. If both vertical rays intersect T we consider the endpoints in B of the first edge intercepted by the upwards and the downwards ray. We connect w with the horizontally closest one of the two. If neither of the rays intersects T we connect w with the horizontally closest vertex in B. In the full version [2] we prove that this indeed constructs a shooting star and we show how to efficiently compute it.

References

1. Ábrego, B.M., et al.: All good drawings of small complete graphs. In: Proceedings of the 31st European Workshop on Computational Geometry (EuroCG2015), pp. 57–60 (2015)
2. Aichholzer, O., García, A., Parada, I., Vogtenhuber, B., Weinberger, A.: Shooting stars in simple drawings of $K_{m,n}$. arXiv:2209.01190v1 (2022)
3. Aichholzer, O., García, A., Tejel, J., Vogtenhuber, B., Weinberger, A.: Twisted ways to find plane structures in simple drawings of complete graphs. In: Proceedings of the 38th International Symposium on Computational Geometry (SoCG2022), pp. 1–18 (2022). https://doi.org/10.4230/LIPIcs.SoCG.2022.5
4. Aichholzer, O., Parada, I., Scheucher, M., Vogtenhuber, B., Weinberger, A.: Shooting stars in simple drawings of $K_{m,n}$. In: Proceedings of the 34th European Workshop on Computational Geometry (EuroCG'19), pp. 1–6 (2019). http://www.eurocg2019.uu.nl/papers/59.pdf

5. Arroyo, A., McQuillan, D., Richter, R.B., Salazar, G.: Levi's Lemma, pseudolinear drawings of K_n, and empty triangles. J. Graph Theor. **87**(4), 443–459 (2018). https://doi.org/10.1002/jgt.22167

6. Cardinal, J., Felsner, S.: Topological drawings of complete bipartite graphs. J. Comput. Geom. **9**(1), 213–246 (2018). https://doi.org/10.20382/jocg.v9i1a7. https://doi.org/10.20382/jocg.v9i1a7

7. Erdős, P., Guy, R.K.: Crossing number problems. Am. Math. Mon. **80**(1), 52–58 (1973). https://doi.org/10.2307/2319261

8. Fox, J., Sudakov, B.: Density theorems for bipartite graphs and related Ramsey-type results. Combinatorica **29**(2), 153–196 (2009). https://doi.org/10.1007/s00493-009-2475-5

9. Fulek, R.: Estimating the number of disjoint edges in simple topological graphs via cylindrical drawings. SIAM J. Discret. Math. **28**(1), 116–121 (2014). https://doi.org/10.1137/130925554

10. Fulek, R., Ruiz-Vargas, A.J.: Topological graphs: empty triangles and disjoint matchings. In: Proceedings of the 29th Annual Symposium on Computational Geometry (SoCG2013), pp. 259–266. ACM, New York (2013). https://doi.org/10.1145/2462356.2462394

11. Kynčl, J.: Enumeration of simple complete topological graphs. Eur. J. Comb. **30**, 1676–1685 (2009). https://doi.org/10.1016/j.ejc.2009.03.005

12. Pach, J., Solymosi, J., Tóth, G.: Unavoidable configurations in complete topological graphs. Discrete Comput. Geom. **30**(2), 311–320 (2003). https://doi.org/10.1007/s00454-003-0012-9

13. Pach, J., Tóth, G.: Disjoint edges in topological graphs. In: Akiyama, J., Baskoro, E.T., Kano, M. (eds.) IJCCGGT 2003. LNCS, vol. 3330, pp. 133–140. Springer, Heidelberg (2005). https://doi.org/10.1007/978-3-540-30540-8_15

14. Rafla, N.H.: The good drawings D_n of the complete graph K_n, Ph. D. thesis, McGill University, Montreal (1988). http://digitool.library.mcgill.ca/thesisfile75756.pdf

15. Ruiz-Vargas, A.J.: Many disjoint edges in topological graphs. Comput. Geom. **62**, 1–13 (2017). https://doi.org/10.1016/j.comgeo.2016.11.003

16. Suk, A.: Disjoint edges in complete topological graphs. Discrete Comput. Geom. **49**(2), 280–286 (2012). https://doi.org/10.1007/s00454-012-9481-x

17. Suk, A., Zeng, J.: Unavoidable patterns in complete simple topological graphs. arXiv:2204.04293 (2022)

Stress-based Visualizations of Graphs

FORBID: Fast Overlap Removal by Stochastic GradIent Descent for Graph Drawing

Loann Giovannangeli$^{(\boxtimes)}$, Frederic Lalanne, Romain Giot, and Romain Bourqui

LaBRI, UMR CNRS 5800, University Bordeaux, 33405 Talence, France
{loann.giovannangeli,frederic.lalanne,romain.giot,
romain.bourqui}@u-bordeaux.fr

Abstract. While many graph drawing algorithms consider nodes as points, graph visualization tools often represent them as shapes. These shapes support the display of information such as labels or encode various data with size or color. However, they can create overlaps between nodes which hinder the exploration process by hiding parts of the information. It is therefore of utmost importance to remove these overlaps to improve graph visualization readability. If not handled by the layout process, Overlap Removal (OR) algorithms have been proposed as layout post-processing. As graph layouts usually convey information about their topology, it is important that OR algorithms preserve them as much as possible. We propose a novel algorithm that models OR as a joint stress and scaling optimization problem, and leverages efficient stochastic gradient descent. This approach is compared with state-of-the-art algorithms, and several quality metrics demonstrate its efficiency to quickly remove overlaps while retaining the initial layout structures.

Keywords: Layout adjustment · Overlap removal · Stress optimization · Stochastic gradient descent

1 Introduction

Most dimension reduction algorithms consider data as points (*e.g.*, Multi Dimensional Scaling [23,24], Graph Layout [10,16,26]). However, most visualization tools represent them by shapes with an area, whether to encode additional data within the screen representation of the nodes, or because it is simply more visually pleasing for end-users. Rendering such data that was laid out as points with shapes creates *overlaps* that can severely hinder the representation readability by hiding information. If not handled directly inside the dimension reduction algorithm (*e.g.*, [11,17]), it is then the responsibility of a post-processing *Overlap Removal* (OR) algorithm to remove these overlaps.

In this paper, we consider the OR problem in graph layouts context, meaning that our laid out data points are nodes positioned by a graph drawing algorithm. These nodes are represented as rectangles with a position and size in two dimensions. Other shapes can be considered as well, as long as it is possible to check

P. Angelini and R. von Hanxleden (Eds.): GD 2022, LNCS 13764, pp. 61–76, 2023.
https://doi.org/10.1007/978-3-031-22203-0_6

for overlaps and measure a distance between them. We consider that a pair of nodes overlaps if the intersection of their shapes representations is not null.

An OR algorithm takes a set of nodes positions and sizes as input and move these nodes to remove *all* overlaps while optimizing two main criteria: *compactness* and *initial layout preservation*. An algorithm that uniformly upscales the initial layout until there is no overlap perfectly works (*e.g.*, uniform Scaling [4]), but its result is not satisfactory as it produces very sparse layouts where the nodes visible areas are significantly reduced. Hence, a good OR algorithm should be able to optimize *compactness* by preserving the scale of the initial layout, using its empty spaces to move nodes apart. Initial layout preservation is also of utmost importance in a graph drawing context since the initial graph layout is often computed to emphasize the graph structure. It is then imperative to preserve the mental map a user has of the initial layout. In addition, graph representations also include visualization of their edges. In that regard, optimizing compactness only is not suited to the graph context since it ends up hiding the graph edges.

The main contribution of this paper is FORBID[1]: Fast Overlap Removal By stochastic gradIent Descent, a novel OR algorithm dedicated to graph drawings that produces overlap-free layouts balancing compactness and initial layout preservation. To the best of our knowledge, it is the first method to explicitly optimize the conjunction of these two aspects. It models the problem as a *stress* function: each pair of nodes in the overlap-free layout should be put at an *ideal distance* such that (i) there is no longer overlaps in the layouts and (ii) the distances between all the nodes are preserved. This stress is then optimized with an efficient state-of-the-art stochastic gradient descent algorithm [26]. Leveraging Chen *et al.* [3,4] evaluation protocol, FORBID is compared with major state-of-the-art OR algorithms on a set of quality metrics specifically selected for this purpose. It demonstrates great capabilities to preserve initial layouts while retaining a decent level of compactness.

The remainder of this paper is organized as follows. Section 2 presents related works principally centered around the description of OR algorithms and their evaluation as proposed in [3,4]. Section 3 describes FORBID algorithm, while Sect. 4 reports its evaluation. Finally, Sect. 5 discusses visual examples of overlap-free layouts from several OR algorithms as well as FORBID convergence.

Notations: Let $G = (V, E)$ be a graph with $V = \{v_1, v_2, ..., v_N\}$ its set of $N = |V|$ nodes and $E \subseteq V \times V$ its set of edges. A graph layout is defined as a tensor $X \in \mathbb{R}^{N \times 2}$ where X_i is the node v_i projection in 2D. A node v_i is defined by a rectangle of width and height (w_i, h_i) centered in X_i. Two nodes overlap each other if the intersection between their rectangles is not null. For convenience, we define the set of overlapping pairs of nodes in a graph by $O \subseteq V \times V$. A corresponding overlap-free layout is defined as $X' \in \mathbb{R}^{N \times 2}$. The euclidean distance between two nodes v_i and v_j is noted $||X_i - X_j||$.

---

[1] FORBID implementation: https://github.com/LoannGio/FORBID.

2   Related Works

This section presents major prior works in the Overlap Removal (OR) field. As described in Chen *et al.* survey [3,4], several efficient OR algorithms exist, many of which (*e.g.*, [12,14,19]) rely on *scan line* [6] to detect overlap presency in $\mathcal{O}(N \log N)$ and find all overlaps in $\mathcal{O}(|O|N(\log N + |O|))$ which *can* be faster than the pairwise search in $\mathcal{O}(N^2)$. These OR algorithms focus on different contexts (*e.g.*, graph or generic 2D representation) and optimize different criteria (*e.g.*, compactness, layout preservation). PFS [19], PFS' [12], FTA [14], RWordle-L [22] and uniform scaling [3,4] rely on the *scan line* algorithm to remove overlaps. PFS, PFS' and FTA are made of two passes handling horizontal and vertical movements separately, while RWordle-L moves nodes on both axes at the same time. In the end, they all have a quadratic complexity according to how nodes movements are computed. PRISM [8] models OR as a stress optimization problem in a *proximity graph* (*i.e.*, Delaunay triangulation of the initial layout) of a layout and runs in $\mathcal{O}(t(mkN + N \log N))$ where m, k are optimization hyper-parameters and t depends on the number of overlaps. GTREE [20] leverages PRISM proximity graph to remove overlaps, but constructs a minimum spanning tree upon it to reduce the number of forces to compute. They both propose a good level of initial layout preservation. As FORBID idea is close to that of PRISM in some way, it will be further discussed in Sect. 3. VPSC [6,7] models OR as a set of constraints to relax but tends to highly deform the initial layout. Its complexity is $\mathcal{O}(CN \log C)$ where C is the number of constraints in $\mathcal{O}(N)$ to relax; leading to a final complexity in $\mathcal{O}(N^2 \log N)$. Finally, Diamond [18] is another constraint programming-based OR algorithm in $\mathcal{O}(N^2)$ that optimizes orthogonal order preservation. Its originality is to propose to temporarily rotate nodes by 45°, representing them as *diamonds* to facilitate the constraints relaxations.

3   FORBID Algorithm

This section presents FORBID Overlap Removal (OR) algorithm. It is based on finding an optimal (*i.e.*, smallest) upscaling ratio while minimizing a stress function that models an overlap-free layout that preserves the initial one. The optimal scaling ratio is found with binary search, while the stress function is optimized with the S_GD^2 algorithm [26] that simulates stochastic gradient descent. An overview of the algorithm components is presented in Fig. 1 and its complexity is in $\mathcal{O}(s(N^2 + N \log N))$ where s is defined later in Sect. 3.3.

3.1   Stress Modelization for Overlap Removal

Preliminaries. Traditional graph layout algorithms (*e.g.*, [2,16,21,26]) often optimize a *stress* function that has been shown to lead to meaningful layouts and is defined as:

$$\sigma(X) = \sum_{i,j \in V} W_{ij}(\|X_i - X_j\| - \delta_{ij})^2 \tag{1}$$

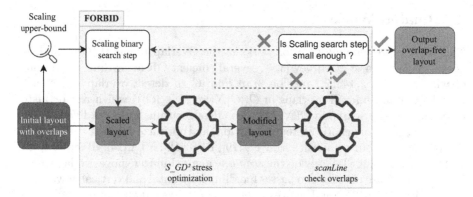

Fig. 1. Simplified schema of FORBID algorithm representing how the stress and upscaling criteria are organized together. Gears represent the algorithms FORBID relies on (*i.e.*, S_GD^2 [26] and *scan line* [6]), and colored boxes represent layouts while white boxes represent the search for the optimal scaling. (Color figure online)

where δ_{ij} is an *ideal distance* that the projected layout should preserve, and W_{ij} is a weight factor usually set to δ_{ij}^{-2}. In the graph layout context, δ_{ij} is set to the graph theoretical distances so that the projected representation of the graph enables end-users to apprehend the graph structure.

As proposed by Gansner and Hu [8], stress is also a good criterion for OR problems. In fact, optimizing stress comes down to fit a distribution of distances in a low dimensional space to that of a higher space, considered ideal but of too high dimensions to be represented. The adaptation to an overlap removal context simply changes the notion of *ideal distances*. Rather than fitting the projected positions to match graph theoretical distances, the ideal distances are two-folded here. For a pair of nodes $p_{ij} = (v_i, v_j)$, with $p_{ij} \notin O$, the ideal distance δ_{ij} is set to their distance $\|X_i - X_j\|$ in the original layout. On the other hand, if $p_{ij} \in O$, δ_{ij} is set to a distance such that v_i and v_j do not overlap anymore; that distance depending on the nodes *shapes* and some design choices. This two-folded definition enables the optimization of both the preservation of the original layout and the overlapped nodes movement at the same time.

PRISM [8] is an example of OR algorithm that optimizes stress. It constructs a *proximity graph* (Delaunay triangulation of the initial layout) and optimizes stress alongside its edges. Considering nodes are represented as rectangles, they define the ideal distance as $\delta_{ij} = s_{ij}\|X_i - X_j\|$ where s_{ij} is an expansion factor of the edge (v_i, v_j) computed so that both nodes would be side by side. PRISM main limitations comes from the use of the *proximity graph* that does not capture all the overlaps and only enables the preservation of distances between close nodes in the initial layout, ignoring longer distances preservation.

FORBID Stress Modelization. For every pair of overlapped nodes, δ_{ij} is set to the distance between v_i and v_j centers if they were tangent in their corner. It allows one of them to be placed on the circle centered on the other and which radius is the minimum so that the nodes do not overlap anymore regardless of their relative position. This distance ensures that two nodes do not overlap anymore and favors convergence by adding some margin space between them, which is necessary for two reasons. First, since stress is optimized by stochastic gradient descent in FORBID (see Sect. 3.2), the ideal distances will most likely get approximated, but never perfectly matched. Second, the PRISM definition of ideal distance means there is only one correct placement for every pair of overlapping nodes that satisfies the stress. However, as opposed to PRISM, FORBID optimizes the distances between all N^2 pairs of nodes, making the number of constraints to relax much higher. This margin enables FORBID to converge faster toward a solution. Formally, the stress is defined as Eq. 1 with δ_{ij} and W_{ij} set to:

$$\delta_{ij} = \begin{cases} \sqrt{\left(\frac{w_i+w_j}{2}\right)^2 + \left(\frac{h_i+h_j}{2}\right)^2}, \text{if } (v_i, v_j) \in O \\ \\ \|X_i - X_j\|, \text{if } (v_i, v_j) \notin O \end{cases} \qquad (2)$$

$$W_{ij} = \begin{cases} \delta_{ij}^{k*\alpha}, \text{if } (v_i, v_j) \in O \\ \\ \delta_{ij}^{\alpha}, \text{if } (v_i, v_j) \notin O \end{cases} \qquad (3)$$

where α is generally set to -2 and $k \in \mathbb{R}$ is an overlap-related factor to tailor the algorithm behavior to a desired initial layout preservation. The smaller the weight is given to overlapping pairs of nodes, the more the initial layout will be preserved at the cost of slower convergence or higher scaling.

3.2 FORBID Stress Optimization

FORBID optimizes stress by stochastic gradient descent by leveraging S_GD^2 [26] algorithm. S_GD^2 models stress as a set of constraints that are relaxed by individually moving pairs of nodes. The process is based on constrained graph layout [1,5], and is optimized by considering the constraints individually to efficiently model clothes movements [15]. By individually moving pairs of nodes to optimize the stress, the algorithm can create new overlaps. And since overlapping and non-overlapping pairs of nodes have different notions of ideal distances, δ_{ij} and W_{ij} are updated at each optimization iteration (everytime all pairs of nodes have been moved once) so that, at any time, the algorithm optimizes distances according to the current state of the layout.

S_GD^2 optimization convergence is based on an annealing step size schedule that mimicks a stochastic gradient descent. It computes $\mathcal{O}(N^2)$ movements, but in practice it converges in very few iterations, making it competitive with other algorithms (see Sect. 4.3). To make it even faster, we also stop the gradient descent if, at any iteration, the sum of all nodes movements in the current iteration is null. This explains why every execution of S_GD^2 is not necessarily of the same length (see Sect. 5).

3.3 Scaling to Ensure Convergence

Here, we define the *bounding box* as the minimum rectangle in which the initial layout fits. Removing all overlaps in a layout is sometimes not feasible without deforming its bounding box, *e.g.*, when the sum of the nodes areas exceeds the layout bounding box area. Two strategies can overcome this: (i) to allow the optimization algorithm to distort the bounding shape, or (ii) to scale up the drawing to create empty spaces that can be used to move nodes. These two choices boil down to considering that either "more space is needed" or "nodes must be smaller" to provide an overlap-free layout. The first strategy has the benefit of limiting the bounding box upscaling, but can result in strongly distorted layouts that makes it difficult to recognize the original graph structure. To guarantee that FORBID finds a solution, it uses the second method and searches for the optimal upscale ratio so that there is enough space to remove overlaps without deforming its aspect. The optimal upscale ratio is found by binary search between 1 and the minimum scaling ratio s_{max} for an overlap-free layout that has not moved any node (*i.e.*, scaling ratio of Scaling [4]). FORBID moves the nodes until there is no overlap (*scan line* [6]) *and* until the scaling ratio is optimal up to a given precision s_{step} (*e.g.*, 0.1).

We name *pass* a call to the optimization algorithm S_GD^2 (see Sect. 3.2) and *iteration* a step within S_GD^2 (*i.e.*, moving all pairs of nodes once). The maximum number of passes over S_GD^2 is defined by the binary search maximum depth $s \leq \log\left(\frac{s_{max}-1}{s_{step}}\right)$. Since the pass in S_GD^2 costs $\mathcal{O}(N^2)$ and that we test if there remains any overlap right after with the *scan line* [6] algorithm that executes in $\mathcal{O}(N \log N)$; the final complexity of FORBID is $\mathcal{O}(s(N^2 + N \log N))$.

As every pass in the optimization algorithm resets the annealing step size schedule (see Sect. 3.2), the nodes can be moved a lot at the beginning of every pass; meaning that FORBID is somewhat allowed to modify the initial layout more than expected. Hence, we also experiment a variant of FORBID, called FORBID', in which the model starts from the scaled initial layout at every pass of the optimization algorithm. We expect this variant to be able to preserve the initial layout even better, probably at the cost of convergence speed.

Algorithm 1 presents FORBID algorithm pseudo-code. It has been simplified by removing the first pass that occurs with the initial scaling of the layout if the sum of nodes areas is lower than the layout bounding box. In practice, it only enters the while loop (line 20) to search for the optimal scale *if* scaling is necessary. The only modification required to implement FORBID' is to change line 21 into $X' \leftarrow \text{scaleLayout}(X, curScale)$ so that the next pass in the optimization algorithm starts with the scaled initial layout.

4   FORBID Evaluation

For the sake of reproducibility, we used the same evaluation protocol as in the Chen *et al.* survey [3,4]. That includes quality metrics, datasets and algorithms. This section describes this protocol and presents the results of FORBID comparison with the selected algorithms on these datasets and metrics. Finally, FORBID execution time is also compared with some algorithms.

Algorithm 1. FORBID pseudo-code

1: **Methods**
2:   | getScalingRatio(*Layout, Sizes*): returns the minimum scaling ratio [3,4] so that there is no overlap anymore in *Layout*
3:   | containsOverlap(*Layout, Sizes*): [6] returns *true* if there is overlap in *Layout*, *false* otherwise
4:   | scaleLayout(*Layout, scaleFactor*): returns *Layout* scaled by *scaleFactor*
5:   | SGD2_StressOPT(*Layout'*, *Layout, Sizes*): One pass of S_GD^2 [26] to optimize overlap removal modeled as stress
6: **end Methods**
7:
8: **Variables**
9:   | X: Initial layout, 2D position of every node
10:   | S: Nodes sizes in 2D, $S_i = (w_i, h_i)$
11:   | *scaleStep*: Scaling step size to stop the search of optimal scale
12: **end Variables**
13:
14: **procedure** FORBID(X, S, *scaleStep*, *SGD2_HP*)
15:   | $lowScale \leftarrow 1$
16:   | $upScale \leftarrow$ getScalingRatio(X, S)
17:   | $curScale \leftarrow (lowScale + upScale)/2$
18:   | $X' \leftarrow X$
19:   | $thereIsOverlap \leftarrow$ containsOverlap(X', S)
20:   | **while** *thereIsOverlap* **or** ($upScale - lowScale > scaleStep$) **do**
21:     | $X' \leftarrow$ scaleLayout(X', *curScale*)
22:     | $X' \leftarrow$ SGD2_StressOPT(X', X, S)
23:     | $thereIsOverlap \leftarrow$ containsOverlap(X', S)
24:     | **if** *thereIsOverlap* **then**
25:       | $lowScale \leftarrow curScale$
26:     | **else**
27:       | $upScale \leftarrow curScale$
28:     | **end if**
29:     | $curScale \leftarrow (lowScale + upScale)/2$
30:   | **end while**
31:   | **return** X'
32: **end procedure**

4.1 Evaluation Protocol

Quality Metrics. The quality metrics used to compare FORBID with other algorithms from the literature were selected by Chen *et al.* [3,4]. All are oriented as *lower is better*, the optimal value being 0 unless specified otherwise.

oo_nni: stands for the *Orthogonal Ordering: Normalized Number of Inversions* and counts the number of times the nodes orthogonal order have been violated.

sp_ch_a [22]: is for *Spread Minimization: Convex Hull Area*. It measures by how much the convex hull area of the overlap-free layout is different from the one of the initial layout: $sp\_ch\_a = \frac{\text{convex\_hull\_area}(X')}{\text{convex\_hull\_area}(X)}$, the optimal value being 1. This metric mainly measures the layout scaling.

gs_bb_iar: means *Global Shape preservation: Bounding Box Improved Aspect Ratio* and is a variant of the aspect ratio between the bounding box of the initial and overlap-free layouts in which the minimal and target value is 1.

nm_dm_imse: stands for *Node Movement minimization: Distance Moved Improved Mean Squared Error*. It quantifies how much the nodes moved from their position in the initial layout to theirs in the overlap-free layout. To lessen the effect of positions value domains, the layouts are aligned as follows:

$$\text{nm\_dm\_imse} = \frac{1}{N} \sum_{v_i \in V} ||X_i' - \text{scale}(\text{shift}(X_i))||^2 \tag{4}$$

where *shift* and *scale* are moving and scaling the initial layout bounding box to match the center and dimensions of the overlap-free layout. It is important to mention that both in [3,4] and therefore in this paper, the bounding boxes computed to scale the layouts do not take the nodes sizes into account.

el_rsd: is for *Edge Length preservation: Relative Standard Deviation*. It measures by how much the lengths of edges in the Delaunay Triangulation graph of an initial layout are preserved, *i.e.*, how well short-distances are preserved.

Datasets. Still following Chen *et al.* [3,4] evaluation protocol, we use the Generated and Graphviz datasets available online[2].

Generated is a set of 840 synthetic graphs specifically generated for the benchmark in [3,4]. It is made of 120 graphs of each size $10, 20, 50, 100, 200, 500, 1000$, laid out with the FM^3 algorithm [10]. These layouts have $2770 \pm 7567(std)$ initial overlaps in average, ranging between 0 and 31843.

Graphviz is a set of 14 real-world graphs from the Graphviz suite. They have between 36 and 1463 nodes and are laid out with $SFDP$ algorithm [13] and have for between 4 and 11582 initial overlaps ($2118 \pm 4078(std)$ in average).

Baseline Algorithms. As already state (see Sect. 2), there are two main criteria to optimize in overlap removal algorithms: compactness and initial layout preservation. By design, FORBID belongs to the second category and is then only compared with its corresponding algorithms (*i.e.*, PFS [19], PFS' [12], PRISM [8], GTREE [20] and Diamond [18]). Other algorithms create embeddings so compact that it is not even possible to visualize the graph edges and structures anymore; meaning they are not suited to overlap removal for graph visualization. *Scaling* is also excluded since it does not look for balance and rawly upscales the layout; which is not a satisfactory solution on its own.

---

[2] Generated and Graphviz graphs: https://github.com/agorajs/agora-dataset, last consulted on May 2022.

4.2   Comparison with Baseline Algorithms on Quality Metrics

This section reports and discusses the performances of FORBID and the selected algorithms from the literature on the **Generated** and **Graphviz** datasets.

On the **Generated** dataset (see Fig. 2), every algorithm succeeds in minimizing oo_nni, with Diamond and FORBID having slightly higher scores than others on worst cases (Q3). On sp_ch_a and gs_bb_iar, FORBID and FORBID' have the best scores by a fair margin, especially on worst cases (Q3). This demonstrates a good capability to limit the upscaling of the drawing in complicated layouts. Both FORBID and FORBID' also minimize nodes movements nm_dm_imse more than other algorithms, especially FORBID' that barely moves the nodes even on complex cases ($nm\_dm\_imse = 50.23$ on Q3). Finally, FORBID and its variant both have high scores on el_rsd. As defined in Sect. 4.1, this metric measures the edge length preservation along the edges of the Delaunay Triangulation (DT) of a graph. Hence, it does not measures that the lengths of the actual graph edges are preserved, but rather quantifies the preservation of the distances between the closest nodes. On the other hand, FORBID focuses on the preservation of *all* the nodes pairwise distances. These two strategies do not have the same notion of *preservation* of the initial layout and by ignoring long distances, PRISM tends to break the overall layout aspect. In addition, since overlapped nodes are likely to be adjacent in the DT graph and since our ideal distance between them is

		FORBID	FORBID'	PFS	PFS'	PRISM	GTREE	Diamond
	Q1	0.00	0.00	0.00	0.00	0.00	0.00	0.04
oo_nni	Median	0.00	0.00	0.00	0.00	0.01	0.02	0.06
	Q3	0.06	0.03	0.00	0.00	0.02	0.03	0.09
	Q1	1.00	1.00	1.04	1.00	1.01	1.12	1.16
sp_ch_a	Median	1.03	1.03	1.69	1.22	1.12	1.49	1.96
	Q3	2.86	3.21	17.63	5.04	4.22	5.94	6.94
	Q1	1.00	1.00	1.01	1.00	1.00	1.01	1.04
gs_bb_iar	Median	1.01	1.00	1.19	1.04	1.04	1.04	1.07
	Q3	1.04	1.01	1.65	1.14	1.23	1.08	1.12
	Q1	3.90	3.90	6.50	1.65	9.82	28.22	535.65
nm_dm_imse	Median	34.69	25.03	694.83	116.06	131.57	292.56	1971.15
	Q3	419.88	50.23	7899.2	1782.80	688.1	2221.70	16571.10
	Q1	0.11	0.11	0.03	0.02	0.04	0.05	0.23
el_rsd	Median	0.37	0.32	0.19	0.11	0.16	0.16	0.66
	Q3	0.70	0.46	0.25	0.18	0.21	0.21	0.99

Fig. 2. Quality metrics quartile values for each algorithm on the **Generated** dataset (see Sect. 4.1). Cells color are selected based on the median value of the algorithms on each metric to enhance comparisons readability. The greener/lighter a cell color is, the better its quality metric score. (Color figure online)

	FORBID	FORBID'	PFS	PFS'	PRISM	GTREE	Diamond
oo_nni	0.05	0.02	0.00	0.00	0.02	0.02	0.05
sp_ch_a	3.71	5.39	210.28	6.92	2.18	4.02	30.63
gs_bb_iar	1.05	1.02	1.97	1.22	1.33	1.17	1.10
nm_dm_imse	43795.43	2697.03	9768157.94	63594.74	42919.66	37331.83	1348426.00
el_rsd	0.55	0.42	0.34	0.22	0.28	0.26	0.23

Fig. 3. Quality metrics mean values on the **Graphviz** dataset (see Sect. 4.1). The greener/lighter a cell color is, the better its quality metric score. (Color figure online)

not the shortest possible (see Sect. 3.1), it was expected that FORBID would distord these distances more than other algorithms (*e.g.*, PRISM).

The performances on the **Graphviz** dataset are reported in Fig. 3. On that dataset, the same trend can be observed with slight differences. Here again, every algorithm successfully minimizes oo_nni and both FORBID and FORBID' are still ahead by a fair margin on gs_bb_iar. However, they are no longer the bests on sp_ch_a and FORBID' is the only one to keep the lead on nm_dm_imse. On those two metrics, PRISM is slightly better than FORBID, while FORBID' has a deteriorated sp_ch_a for a much better node movement nm_dm_imse in comparison to other algorithms and to its own performances on the **Generated** dataset. In fact, since FORBID' focuses on preserving the initial layout, it tends to upscale the layouts more (worsening sp_ch_a) to minimize the nodes movements (improving nm_dm_imse). Finally, both have again the highest el_rsd scores, though the difference with other algorithms is less pronounced.

4.3 Execution Time Comparison with Some Baseline Algorithms

This section compares FORBID execution time with that of the best techniques according to the performances observed in Sect. 4.2: PFS', GTREE and PRISM.

Execution times on the **Graphviz** dataset are reported in Fig. 4 and enable to categorize three groups of difficulty. FORBID, FORBID' and PFS' are instantaneous on *easy* graphs, taking less than $20ms$ to solve the OR problem, being about ten to twenty times faster than GTREE and PRISM. On *medium* complexity graphs, PFS' achieves again the best performances. FORBID is slower, but remains faster than GTREE and PRISM by a fair margin, while FORBID' is slower than GTREE. On these graphs, PRISM is significantly slower than the four others. Finally, on the *hard*est graphs, PFS' is still almost instantaneous.

	$|V|$	$|E|$	$|O|$	Graph	FORBID	FORBID'	PFS'	GTREE	PRISM
Easy	36	107	4	*dpd*	0	0	0	29	34
	41	49	20	*unix*	0	0	0	46	54
	43	64	9	*rowe*	0	0	0	41	46
	47	55	33	*size*	1	1	0	49	91
	50	99	13	*ngk104*	0	0	0	40	46
	76	104	19	*nan*	1	1	0	58	115
	79	181	33	*b124*	4	5	0	77	229
	135	366	53	*b143*	13	19	0	126	275
Medium	213	269	1105	*mode*	166	420	1	270	1040
	302	611	268	*xx*	305	682	2	540	1450
	302	611	282	*b102*	268	824	2	540	1750
Hard	1054	1083	11582	*root*	3097	12989	22	7256	37734
	1235	1616	10540	*badvoro*	4577	18013	34	9694	27514
	1463	5806	5691	*b100*	11827	22880	48	13074	63877

Fig. 4. Execution time (in ms) of FORBID, FORBID', GTREE and PRISM on the **Graphviz** dataset (see Sect. 4.1). Rows are sorted by "graph complexity" measured with their number of nodes $|V|$, edges $|E|$ and overlaps $|O|$. The greener/lighter a cell color is, the better its quality metric score. (Color figure online)

FORBID is faster than the remaining methods, while FORBID' loses to GTREE. On *hard* graphs, PRISM becomes dramatically slower than other algorithms.

In the end, we can conclude that FORBID scales well with the OR problem complexity, while it is more difficult for FORBID'. For this last, the constraint to start from the scaled initial layout at each pass in the optimization algorithm significantly slows its convergence (see Sect. 5) and is responsible of the higher upscaling observed in Sect. 4.2 that enables a better layout preservation. In comparison with PRISM, which also optimizes OR modeled as stress, both FORBID variants are much faster, while PFS' faster handles large graphs.

5   Discussion

This section discusses FORBID behavior on some specific graph examples by comparing it to other algorithms. It also briefly discusses the convergence of FORBID and FORBID' variants, highlighting their different behaviors.

Visual Evaluation. This section focuses on the study of FORBID, FORBID', PFS', GTREE and PRISM on three graphs of the **Graphviz** dataset: *mode*, *badvoro* and *root*. Both initial and overlap-free layouts are presented in Fig. 5, while the methods quality metrics are reported in Table 1.

On *mode*, both FORBID and PRISM damaged the initial layout to produce more compact embeddings. On the other hand, the others preserved the initial layout structure, FORBID' being the most pleasing one. The quality metrics corroborate this observation, with FORBID and PRISM having lower sp_ch_a (*i.e.*, upscaling) scores while FORBID', PFS' and GTREE have smaller nm_dm_imse (*i.e.*, nodes movements) scores; FORBID' having the smallest one.

On both *badvoro* and *root*, the same trend can be observed between the Overlap Removal methods. PRISM produces more compact embeddings that makes it difficult to recognize the initial layout and even to visualize edges. On the other hand, GTREE produces less compact layouts, but distorts the initial graph structures. Finally, FORBID, FORBID' and PFS' are satisfactory, but PFS' upscales the initial layout more than necessary. Again, these observations are corroborated by the quality metrics presented in Table 1: PRISM is consistently better on sp_ch_a, GTREE is not the best on any metric, and the remaining three have less nodes movements but higher upscaling. The high nm_dm_imse scores of PFS' can be imputed to the fact that this metric is sensitive to the overlap-free layout scale, even though the graph layout structures seem visually preserved.

Overall, FORBID and FORBID' produce balanced layout that optimize both the initial layout preservation and the embedding compactness.

Convergence Analysis. This section studies how re-initializing the layout at every *pass* (*i.e.*, call to the optimization algorithm) in FORBID' affects its convergence speed in comparison to FORBID. Observations of both variants on

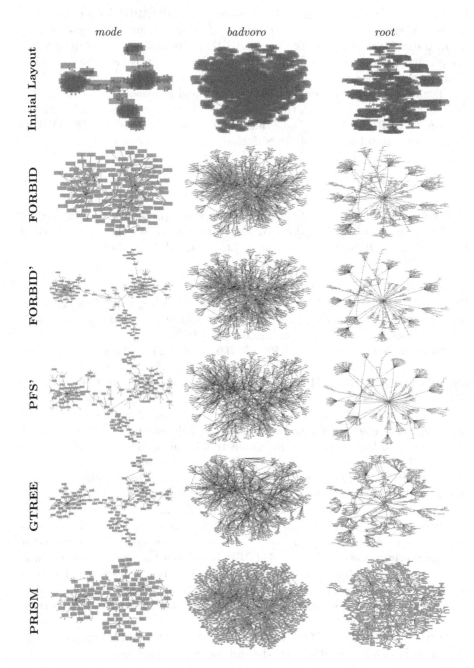

Fig. 5. Graph visualization of examples from the **Graphviz** dataset with FORBID, FORBID', PFS' GTREE and PRISM. Nodes are colored in transparent red if they overlap each other, and opaque blue otherwise. (Color figure online)

Table 1. Quality metrics of the drawings presented in Fig. 5. For each graph and quality metric, the best score is highlighted in bold.

		FORBID	FORBID'	PFS'	GTREE	PRISM
mode	*oo_nni*	0.20	**0.03**	0.54	0.05	0.06
	sp_ch_a	**2.98**	10.35	9.08	7.77	3.72
	gs_bb_iar	1.02	**1.01**	1.20	1.23	1.52
	nm_dm_imse	20417	**1657**	8655	10238	17442
	el_rsd	0.95	0.70	**0.39**	0.54	0.62
badvoro	*oo_nni*	0.01	**0.00**	0.48	0.02	0.02
	sp_ch_a	11.42	12.40	17.92	13.45	**6.87**
	gs_bb_iar	**1.00**	**1.00**	1.25	1.17	1.85
	nm_dm_imse	979	**451**	57267	67296	47708
	el_rsd	0.32	0.23	**0.20**	0.40	0.39
root	*oo_nni*	**0.01**	**0.01**	0.53	0.05	0.04
	sp_ch_a	19.17	29.99	48.34	14.29	**6.27**
	gs_bb_iar	**1.01**	**1.01**	1.34	1.30	2.04
	nm_dm_imse	24502	**12151**	662185	356632	450287
	el_rsd	0.90	0.72	**0.43**	0.67	0.72

the *mode*, *badvoro* and *root* **Graphviz** graphs are presented in Fig. 6. Each plot reports the number of passes, the length of each pass and the stress, number of overlaps and scaling ratio evolution against the total number of iterations.

For both FORBID and FORBID', the number of passes never exceeded 10. In these executions, the maximum number of iterations in the optimization algorithm was set to 30. Many passes stop before reaching this limit thanks to the stop condition on null nodes movements (see Sect. 3.2). As the stress and number of overlaps follow the same trends on every plot, it confirms that optimizing our stress effectively removes overlaps. The difference between FORBID and FORBID' is also distinctly observable. In FORBID, most overlaps are removed in the first few passes while the last ones are dedicated to the search for the optimal upscaling ratio while preserving the overlap-free layout. On the other hand, FORBID' has to restart from the scaled initial nodes every time a new pass begins, the problem being made simpler or harder through a different upscaling ratio. This explains why FORBID' is consistently slower than FORBID (see Sect. 4.3) but better preserves the initial layout (see Sect. 4.2).

Finally, we would like to discuss the choice of the binary search upper-bound s_{max}. In this paper, it was set to the minimum scaling ratio for an overlap-free layout that has not moved any node (see Sect. 3.3). Doing so guarantees that FORBID finds a overlap-free layout and solve the Overlap Removal task. Such an s_{max} value can be high when two nodes are almost perfectly overlapped, but in practice FORBID output layouts scaling remained far from this upper-bound during the benchmark (*e.g.*, green lines in Fig. 6). Nevertheless, it would be

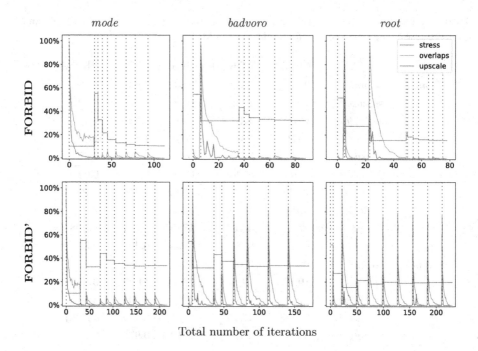

Fig. 6. Convergence plots of FORBID and FORBID'. It reports the evolution of stress, number of overlaps and scaling ratio against the total number of iterations. Vertical dashed lines represent beginning of new *passes*. Stress and number of overlaps are normalized by their respective maximum value, while upscaling ratio is normalized by its binary search maximum bound (see Sect. 3.3).

possible to set s_{max} to a lower value to enforce a smaller scaling in the produced layout, at the cost of initial layout preservation and the guarantee that it is overlap-free. For instance, s_{max} could be set to twice the sum of nodes areas, leading toward a layout which bounding box space is at least half occupied, but there would be no guarantee that this restrained space is enough to provide an overlap-free layout that preserves the initial one. The notion of *overlap* itself could also be approximated with a *tolerance* to speed up the convergence (*i.e.*, consider as *not* overlapped the pairs of nodes that do overlap by less than a *tolerance* margin). In the end, these choices mainly depend on the initial graph layout and the desired aspect of overlap-free layout. Although it can be slightly suboptimal, the s_{max} defined and used in this paper (see Sect. 3.3) is better adapted to the general case.

6   Conclusion

This paper has presented FORBID, an Overlap Removal (OR) algorithm that leverages upscaling and stress optimization by simulated stochastic gradient descent to minimize deformations in the initial layout. FORBID idea is based on combining upscaling and the preservation of all nodes pairwise distances to produce an overlap-free layout, focusing on the preservation of the initial graph layout structures while limiting the surface used. It has been compared to several state-of-the-art algorithms, and is among the best techniques to preserve the initial layout, which is critical in graph drawing to retain the readability of the graph layout structures. FORBID complexity is in $\mathcal{O}(s(N^2 + N \log N))$ and is among the fastest methods on the benchmark graphs (with up to 1463 nodes and 11 582 Overlaps). Future works leads include improvements of the algorithm complexity to better handle large graphs. The first idea to achieve that is to sub-sample the nodes pairs to process to remove overlaps. A multi-scale approach could enable to optimize the initial graph layout structures preservation while sampling the distances to preserve (*i.e.*, preserve distances between-clusters and within-cluster; ignore between nodes of different clusters). Finally, with the recent advances in Deep Learning for graph drawing [9,25], we plan to learn a Deep Learning model solve OR problems. By design, these models can scale to large graphs as they are capable of solving the task they have learned in almost constant time.

Acknowledgments. We would like to thank the Nouvelle-Aquitaine Region for supporting this work through its foundings OPE 2020–0408 and OPE 2020–0513.

References

1. Bostock, M., Ogievetsky, V., Heer, J.: D^3 data-driven documents. IEEE Trans. Visual Comput. Graphics **17**(12), 2301–2309 (2011)
2. Brandes, U., Pich, C.: An experimental study on distance-based graph drawing. In: Tollis, I.G., Patrignani, M. (eds.) GD 2008. LNCS, vol. 5417, pp. 218–229. Springer, Heidelberg (2009). https://doi.org/10.1007/978-3-642-00219-9_21
3. Chen, F., Piccinini, L., Poncelet, P., Sallaberry, A.: Node overlap removal algorithms: a comparative study. In: Archambault, D., Tóth, C.D. (eds.) GD 2019. LNCS, vol. 11904, pp. 179–192. Springer, Cham (2019). https://doi.org/10.1007/978-3-030-35802-0_14
4. Chen, F., Piccinini, L., Poncelet, P., Arnaud, S: Node overlap removal algorithms: an extended comparative study. J. Graph Algorithms Appl. **24**(4), 683–706 (2020)
5. Tim, D.: Scalable, versatile and simple constrained graph layout. Comput. Graphics Forum. **28**, 991–998 (2009). Wiley Online Library (2009)
6. Dwyer, T., Marriott, K., Stuckey, P.J.: Fast node overlap removal. In: Healy, P., Nikolov, N.S. (eds.) GD 2005. LNCS, vol. 3843, pp. 153–164. Springer, Heidelberg (2006). https://doi.org/10.1007/11618058_15
7. Dwyer, T., Marriott, K., Stuckey, P.J.: Fast node overlap removal—correction. In: Kaufmann, M., Wagner, D. (eds.) GD 2006. LNCS, vol. 4372, pp. 446–447. Springer, Heidelberg (2007). https://doi.org/10.1007/978-3-540-70904-6_44
8. Gansner, E., Hu, Y.: Efficient, proximity-preserving node overlap removal. J. Graph Algorithms Appl. **14**(1), 53–74 (2010)

9. Giovannangeli, L., Lalanne, F., Auber, D., Giot, R., Bourqui, R.: Deep neural network for drawing networks, $(DNN)^2$. In: Purchase, H.C., Rutter, I. (eds.) GD 2021. LNCS, vol. 12868, pp. 375–390. Springer, Cham (2021). https://doi.org/10.1007/978-3-030-92931-2_27

10. Hachul, S., Jünger, M.: Drawing large graphs with a potential-field-based multilevel algorithm. In: Pach, J. (ed.) GD 2004. LNCS, vol. 3383, pp. 285–295. Springer, Heidelberg (2005). https://doi.org/10.1007/978-3-540-31843-9_29

11. David, H., Yehuda, K.: Drawing graphs with non-uniform vertices. In: Proceedings of the Working Conference on Advanced Visual Interfaces, pp. 157–166 (2002)

12. Hayashi, K., Inoue, M., Masuzawa, T., Fujiwara, H.: A layout adjustment problem for disjoint rectangles preserving orthogonal order. In: Whitesides, S.H. (ed.) GD 1998. LNCS, vol. 1547, pp. 183–197. Springer, Heidelberg (1998). https://doi.org/10.1007/3-540-37623-2_14

13. Hu, Y.: Efficient, high-quality force-directed graph drawing. Math. J. **10**(1), 37–71 (2005)

14. Xiaodi, H., Wei, L., ASM, S., Junbin, G.: A new algorithm for removing node overlapping in graph visualization. Inf. Sci. **177**(14), 2821–2844 (2007)

15. Thomas, J.: Advanced character physics. In: Game Developers Conference. vol. 3, pp. 383–401. IO Interactive, Copenhagen Denmark (2001)

16. Tomihisa, K., Satoru, K.: An algorithm for drawing general undirected graphs. Inf. Process. Lett. **31**(1), 7–15 (1989)

17. Kamps, T., Kleinz, J., Read, J.: Constraint-based spring-model algorithm for graph layout. In: Brandenburg, F.J. (ed.) GD 1995. LNCS, vol. 1027, pp. 349–360. Springer, Heidelberg (1996). https://doi.org/10.1007/BFb0021818

18. Wouter, M.: Efficient optimal overlap removal: algorithms and experiments. Comput. Graphics Forum. **38**, 713–723 (2019). Wiley Online Library (2019)

19. Misue, K., Eades, P., Lai, W., Sugiyama, K.: Layout adjustment and the mental map. J. Vis. Lang. Comput. **6**(2), 183–210 (1995)

20. Nachmanson, L., Nocaj, A., Bereg, S., Zhang, L., Holroyd, A.: Node overlap removal by growing a tree. In: Hu, Y., Nöllenburg, M. (eds.) GD 2016. LNCS, vol. 9801, pp. 33–43. Springer, Cham (2016). https://doi.org/10.1007/978-3-319-50106-2_3

21. Ortmann, M., Klimenta, M., Brandes, U.: A sparse stress model. In: Hu, Y., Nöllenburg, M. (eds.) GD 2016. LNCS, vol. 9801, pp. 18–32. Springer, Cham (2016). https://doi.org/10.1007/978-3-319-50106-2_2

22. Hendrik, S., Marc, S., Andreas, S., Daniel, K., Oliver, D.: Rolled-out wordles: a heuristic method for overlap removal of 2D data representatives. Comput. Graphics Forum. **31**, 1135–1144 (2012). Wiley Online Library (2012)

23. Tipping, M.E., Bishop, C.M.: Probabilistic principal component analysis. J. Roy. Stat. Soc. Ser. B (Statistical Methodology) **61**(3), 611–622 (1999)

24. van der Maaten, L., Hinton, G.: Visualizing data using t-SNE. J. Mach. Learn. Res. 9(11), 2579–2605 (2008)

25. Wang, X., Yen, K., Hu, Y., Shen, H.-W.: DeepGD: a deep learning framework for graph drawing using GNN. IEEE Comput. Graphics Appl. **41**(5), 32–44 (2021)

26. Zheng, J.X., Pawar, S., Goodman, D.F.M.: Graph drawing by stochastic gradient descent. IEEE Trans. Vis. Comput. Graphics **25**(9), 2738–2748 (2018)

Spherical Graph Drawing
by Multi-dimensional Scaling

Jacob Miller[✉], Vahan Huroyan, and Stephen Kobourov

Department of Computer Science, University of Arizona, Tucson, USA
jacobmiller1@arizona.edu, vahanhuroyan@math.arizona.edu,
kobourov@cs.arizona.edu

Abstract. We describe an efficient and scalable spherical graph embedding method. The method uses a generalization of the Euclidean stress function for Multi-Dimensional Scaling adapted to spherical space, where geodesic pairwise distances are employed instead of Euclidean distances. The resulting spherical stress function is optimized by means of stochastic gradient descent. Quantitative and qualitative evaluations demonstrate the scalability and effectiveness of the proposed method. We also show that some graph families can be embedded with lower distortion on the sphere, than in Euclidean and hyperbolic spaces.

1   Introduction

Node-link diagrams are typically created by embedding the vertices and edges of a given graph in the Euclidean plane and different embedding spaces are rarely considered. Multi-Dimensional Scaling (MDS), realized via stress minimization or stress majorization, is one of the standard approaches to embedding graphs in Euclidean space. The idea behind MDS is to place the vertices of the graph in Euclidean space so that the distances between them are as close as possible to the

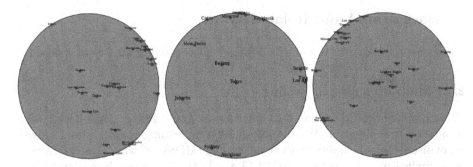

Fig. 1. Applying spherical MDS to embed 30 cities from around the Earth (given pairwise distances between the cities). The spherical MDS recovers the underlying geometry.

P. Angelini and R. von Hanxleden (Eds.): GD 2022, LNCS 13764, pp. 77–92, 2023.
https://doi.org/10.1007/978-3-031-22203-0_7

given graph theoretic distances. Due to the nature of Euclidean geometry, this cannot always be done without some distortion (e.g., while K_3 naturally lives in 2D, K_4 does not, as there are no four equidistant points in the Euclidean plane). Moreover, some graphs "live" naturally in manifolds other than the Euclidean plane. For example 3-dimensional polytopes, or triangulations of 3-dimensional objects can be better represented on the sphere, while trees and special lattices are well-suited to hyperbolic spaces (Fig. 1).

When visualizing graphs in Euclidean space, common techniques include adapting off-the-shelf dimensionality reduction algorithms to the graph setting. Such algorithms include the Multi-Dimensional Scaling (MDS) [38], Principal Component Analysis (PCA) [11], t-distributed Stochastic Neighbor Embedding (t-SNE) [25], and Uniform Manifold Approximation Projection (UMAP) [26]. The popularity of graph visualisation, and the fact that some of the underlying datasets are easier to embed in non-Euclidean spaces, led to some visualization techniques for spherical geometry [9,33] and hyperbolic geometry [19,27,36]. Most of the existing non-Euclidean graph visualization approaches, however, either lack in accuracy or do not scale to larger graphs.

With this in mind, we propose and analyze a stochastic gradient descent algorithm for spherical MDS. Specifically, we present a scalable technique to compute graph layout directly on the sphere, adapting previous work for general datasets [9] and applying stochastic gradient descent [34,41]. We provide an evaluation of the technique by comparing its speed and faithfulness to the exact gradient descent approach. We also investigate differences in graph layouts between the consistent geometries (Euclidean, spherical, hyperbolic) by first showing that *dilation or resizing* has a large effect on layouts in spherical and hyperbolic geometry, and second by showing some structures can be better represented in one geometry than the other two. All sourcecode, datasets and experiments, as well a web based visualization tool are available on GitHub: https://github.com/Mickey253/spherical-mds.

Note that the proposed method is not restricted to graphs, but is applicable to any dataset specifying a set of objects and pairwise distances between them.

2   Background and Related Work

We review related work in non-Euclidean geometry and graph layout methods.

2.1   Multi-dimensional Scaling

Using graph-theoretic distances to determine a graph layout dates back to the Kamada-Kawai algorithm [16]. A more general embedding approach from a given set of distances is the multi-dimensional scaling (MDS) [38] which has extensively been applied to graph layout; see [12,13,41]. Both the Kamda-Kawai and (metric) MDS algorithms aim to minimize the *stress* function, which is the sum of residual squares between the given and the embedded distances of each pair of datapoints. Formally, given a graph $G = (V, E)$ with the graph theoretic distances between its n vertices $(d_{ij})_{i,j=1}^{n,n}$, where the vertices are labeled $1, 2, \ldots, n$

MDS aims to embed the graph in $\mathbb{R}^d$ by minimizing the following *stress* function to find the locations for its vertices:

$$\sigma(X) = \sum_{i<j} w_{ij}(\|X_i - X_j\| - d_{ij})^2. \tag{1}$$

The resulting solution of $X_1, X_2, \ldots, X_n \in \mathbb{R}^d$ represents the coordinates of the embedded graph vertices.

Various forms of MDS have been analyzed. Metric MDS was first studied by Shepard [38] (see equation in (3)), and the related non-metric MDS by Kruskal [21]. Classical MDS is similar but uses an objective function called *strain*.

The classical MDS has a closed form solution while the metric MDS and non-metric MDS rely on solving an optimization problems to minimize the corresponding stress functions. Many approaches have been proposed to solve (metric) MDS including stress majorization [13] and (stochastic) gradient descent [2].

When used for the purposes of visualization, the embedding space for MDS is almost always 2 dimensional Euclidean, as that is the space of a flat sheet of paper, or the flat screen of a computer monitor. The natural measure of distance is then the Euclidean norm.

In this work we will focus on metric MDS, defined in (1) but instead of embedding the graph in Euclidean space, we embed it directly on the sphere. The MDS approach has already been applied to embed graphs on spherical [9] and hyperbolic [27] spaces. Our contribution is to solve the proposed optimization problem faster and be able to handle larger graphs, address the dilation/resizing problem, as well as analyze the approach on wider range of graphs and provide a working and easy to use implementation.

2.2   Non-Euclidean Geometry

Non-Euclidean geometries are a special case of Riemannian geometries, which are spaces that are locally "smooth": one can define an inner product on the tangent space at each point. Spherical and hyperbolic non-Euclidean geometries are similar to Euclidean geometry, except for one axiom.

Euclid's *Elements* specify five axioms/postulates upon which all true statements about geometry should be proved. The fifth axiom is significantly more involved than the first four and mathematicians attempted for centuries to prove it using only the first four. In 1892 Lobachevsky and Bolyai independently discovered and published their formulation of hyperbolic geometry by inverting an equivalent statement to Euclid's fifth axiom, Playfair's axiom: *In a plane, given a line and a point not on it, at most one line parallel to the given line can be drawn through the point.* Replace "*at most one line*" with "*at least two distinct lines*" to get hyperbolic geometry. Replace "*at most one line*" with "*there does not exist a line*" to arrive at spherical geometry.

Spherical geometry has benefits in the context of data visualization. In Euclidean (or hyperbolic) layouts, one is forced to choose a "center" of the embedding, intentionally or not, whereas on the sphere there is no notion of

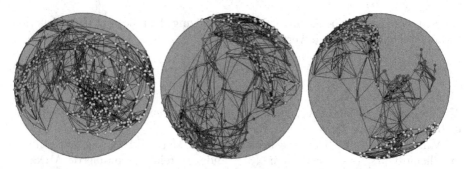

Fig. 2. The Maps of Science dataset [5] laid out using our SMDS algorithm, from three different perspectives. Each color represents a different field of science (nodes are subfields), and their relationships exhibit a ring-like structure. Any field can be placed in the center of the view.

a center (Fig. 5). A perceptual side effect of centered embeddings is that vertices near the center seem more important, while vertices away from the center seem more peripheral. This problem does not occur in spherical space, where simple rotation can place any vertex in the center of the view (a feature that is very useful when visualizing social networks, or networks of research fields); see Fig. 2. Additionally, many spherical projections into Euclidean space, such as the stereographic projection, provide a desirable focus+context effect.

Some focus+context type algorithms for visualizing large hierarchies by using hyperbolic geometry are discussed in [23,24].

Distances in non-Euclidean geometries generalize the concept of a straight line to that of a geodesic, defined as an arc of shortest length (not necessarily unique) that contains both endpoints. The distance between two endpoints is then the length of that curve.

A point on a sphere of radius R is uniquely represented by a pair of angles, (ϕ, λ), where $0 \leq \phi \leq \pi$ is known as the *latitude* and $0 \leq \lambda \leq 2\pi$ is the *longitude*. Given two points $(\phi_1, \lambda_1), (\phi_2, \lambda_2)$ on the sphere with radius R, the geodesic distance is then derived by the spherical law of cosines:

$$\delta((\phi_1, \lambda_1), (\phi_2, \lambda_2)) = R * \arccos(\sin \phi_1 \sin \phi_2 + \cos \phi_1 \cos \phi_2 \cos(\lambda_1 - \lambda_2)) \quad (2)$$

where $\delta(X_i, X_j)$ denotes the geodesic distance between points X_i and X_j, assuming X is an $n \times 2$ matrix whose rows correspond to spherical coordinates.

It is known that the surface of a sphere cannot be perfectly preserved in any 2-dimensional Euclidean drawing, due to its curvature. One can preserve various combintations of angles, areas, geodesics, or distances but not all of these simultaneously. The orthographic projection, or the "view from space" projects the sphere onto a tangent plane with point of perspective from outside the sphere. While half of the sphere is obscured and shapes and area are distorted near the boundary, geodesics through the origin are preserved and it gives the impression of a 3-dimensional globe. The stereographic projection is similar but

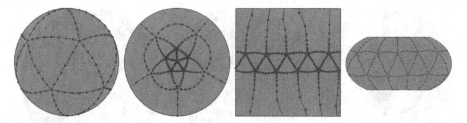

Fig. 3. A subdivided isocahedron graph embedded on the sphere, displayed with the orthographic, stereographic, Mercator, and equal Earth projections.

instead with a point of projection looking through the sphere, and preserves angles. The Mercator projection is a common cylindrical map projection with heavy area distortion near the poles. The equal Earth projection preserves area and gives the impression of a spherical shape. Examples are shown in Fig. 3. We primarily use the orthographic and equal Earth projections in this paper.

2.3   Graph Layout in Non-Euclidean Geometry

Non-Euclidean graph visualization has been studied by Munzner, with an emphasis on trees and hierarchies [28–31], and the following link treevis.net, provides several examples of hyperbolic and sphere based tree visualizations [37]. Spherical layouts have been investigated in an immersive setting such as virtual reality [22,40]. Self-organizing maps have been developed for both spherical and hyperbolic geometries [32]. Several other examples of spherical graph visualization include the Map of Science [5], the "Places and Spaces" [4], and "Worldprocessor" [15] exhibitions. Some limitations of the existing algorithms for hyperbolic graph visualization are discussed in [10].

Force-directed algorithms model the nodes and edges as a physical system, and provide a layout by minimizing the total energy. These algorithms are popular in part due to their conceptual simplicity and quality layouts [17]. A general technique for generalizing force-directed algorithms to non-Euclidean spaces is described in [18]. However, it only works for small graphs as for larger ones it is too computationally expensive and is unlikely to escape local minima.

There are several different approaches to embedding a graph on the sphere. A simple idea is to generate a 2D Euclidean layout and project it onto the sphere through a linear map [8,33], however, this embedding will not make full use of spherical geometry. Another approach is to embed the graph in 3D Euclidean space and modify it to force it on the surface of a sphere [7,33], but this is quite mathematically involved and complicates the optimization. A more natural method directly computes a 2D spherical embedding (in latitude and longitude) such that the geodesic distances on the sphere and graph-theoretic distances between pairs of vertices are closely matched [9]. We focus on this approach and make it scalable by adopting stochastic gradient descent for the optimization phase and by solving the dilation/resizing problem specified below.

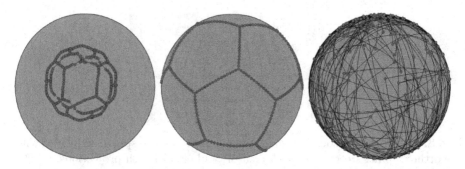

Fig. 4. A dodecahedron subdivision graph. Left: a small dilation factor forces the layout on a small patch of the sphere. Middle: a correct dilation factor using our heuristic discussed in Sect. 5.1, allows the graph to make use of the spherical geometry. Right: a large dilation factor makes the distances unachievable.

In the graph drawing literature, the normalized stress of a layout is a standard quality measure [12,20,42]. This is perfectly acceptable in Euclidean space where a layout is not meaningfully changed when the layout is resized. For non-Euclidean graph layouts there is a possible issue of *dilation or resizing*. Formally, a dilation is a function on a metric space M, $f : M \rightarrow M$ that satisfies $d(f(x), f(y)) = rd(x, y)$ for $x, y \in M$, $r > 0 \in \mathbb{R}$ and $d(x, y)$ being the distance between x and y. In non-Euclidean spaces, such as the sphere, the size of a layout can have drastic effects; see Fig. 4. At small dilation, a graph embedded on the sphere takes only a small patch and the sphere patch behaves like a piece of the Euclidean plane. At large dilation, a graph embedded on the sphere wraps over itself. At some optimal dilation the embedded graph fits on the sphere with low distortion. Choosing the size of the sphere is important to accurately represent the data. We are unaware of any work regarding this problem in spherical embedding, and propose a heuristic and optimization scheme to solve it in Sect. 5.1.

As stress is difficult to interpret between geometries, we use a more fair comparison metric called *distortion* [27,36] defined later in Sect. 5.

3   Algorithm

Our spherical multi-dimensional scaling algorithm (SMDS) resembles that of other stress based graph layout algorithms. That is, we first compute a graph-theoretic distance matrix via an all-pairs-shortest-paths algorithm and then use this distance matrix as an input to minimize the generalized stress function and compute vertex coordinates on the sphere. This differs from standard Euclidean MDS in that Euclidean distances between points are replaced by geodesic distances between the points on sphere. The corresponding stress function defined on the sphere is

$$\sigma_S(X) = \sum_{i<j} w_{ij}(\delta(X_i, X_j) - d_{ij})^2 \tag{3}$$

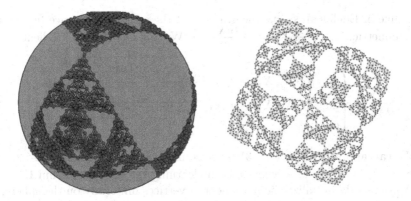

Fig. 5. The Sierpinski3d [20] graph on the sphere (left) and in the plane (right). While the Euclidean drawing on the right is aesthetically pleasing, it looks deceptively like a 2D structure and implies a center. The sphere more accurately captures the structure.

where $\delta(X_i, X_j)$ denotes the geodesic distance between vertices i and j, d_{ij} is the graph-theoretic distance between vertex i and j, and w_{ij} is a weight, typically set to d_{ij}^{-2}. However, one can also give preferred weights based on the importance of the points and based on the application. Another typical choice is binary weights, where w_{ij} is either 0 or 1. Unless otherwise specified, δ corresponds to the geodesic distances on the unit sphere and δ_R the geodesic distances on a sphere with radius R.

We minimize Eq. (3) by stochastic gradient descent (SGD), which we experimentally show converges in fewer iterations while achieving lower distortion than exact gradient descent for sufficiently large graphs. SGD is a minimization approach in the gradient descent family of algorithms. Fully computing the exact gradient can be too expensive and SGD instead repeatedly performs a constant time approximation of the gradient, by considering only a single term of the sum (or subset of terms for mini-batch stochastic), which allows it to make more updates. As a result, SGD tends to converge in fewer iterations while more consistently finding the global minimum [35].

To perform SGD on the stress function, we approximate the gradient by looking at only a single pair of vertices. Note that this corresponds to one summand of the full stress function. If we rewrite Eq. (3) as $\sigma_S(X) = \sum_{i<j} f_{i,j}(X)$ then we can more simply write the full gradient in terms of f. Apply the chain rule to see we will need to derive the partial gradient of the geodesic distance function:

$$\frac{\partial \delta(X_i, X_j)}{\partial \phi_i} = \frac{-\cos\phi_i \sin\phi_j - \sin\phi_i \cos\phi_j \cos(\lambda_i - \lambda_j)}{\sqrt{1 - \cos^2(\delta(X_i, X_j))}}$$

$$\frac{\partial \delta(X_i, X_j)}{\partial \lambda_i} = \frac{\cos\phi_i \cos\phi_j \sin(\lambda_i - \lambda_j)}{\sqrt{1 - \cos^2(\delta(X_i, X_j))}}$$

Unlike in Euclidean space, the gradient of the spherical distance function is not symmetric, i.e., $\frac{d\delta(X_i,X_j)}{d(\phi_i,\lambda_i)} \neq -\frac{d\delta(X_i,X_j)}{d(\phi_j,\lambda_j)}$. Writing out the full gradient:

$$
\frac{\partial f_{i,j}}{\partial X_k} = \begin{cases} 2w_{i,j}(\delta(X_i,X_j) - d_{ij})\left(\frac{\partial \delta(X_i,X_j)}{\partial \phi_i}, \frac{\partial \delta(X_i,X_j)}{\partial \lambda_i}\right) & k = i \\ 2w_{i,j}(\delta(X_i,X_j) - d_{ij})\left(\frac{\partial \delta(X_i,X_j)}{\partial \phi_j}, \frac{\partial \delta(X_i,X_j)}{\partial \lambda_j}\right) & k = j \\ 0 & \text{otherwise} \end{cases} \tag{4}
$$

We can apply SGD to Eq. (3) by selecting pairs i, j in random order, and updating X by $X - \eta \frac{\partial f_{i,j}}{\partial X_k}$ where η is the learning rate; see Algorithm 1.

We randomly initialize the placement of vertices uniformly on the sphere, as other work has shown that SGD is consistent across initialization strategies [1, 27,41].

Algorithm 1. Stochastic gradient descent algorithm for spherical MDS

> **procedure** STOCHASTIC GRADIENT DESCENT(d)
> $X \leftarrow$ random initialization
>   **while** $\Delta(\mathrm{stress}(X)) > \epsilon$ or max iterations is reached **do**
>   **for** (i,j) in random order **do**
> $X_i \leftarrow X_i - \eta \frac{df_{i,j}}{dX_i}$, according to (4)
> $X_j \leftarrow X_j - \eta \frac{df_{j,i}}{dX_j}$, according to (4)
>   **end for**
>   **end while**
>   **return** X
> **end procedure**

4   Evaluation

We first investigate the various parameters that effect SGD's optimization, then compare our results to exact gradient descent.

4.1   Hyper-parameters

There are several hyper-parameter choices to be made when using SGD. The learning rate η (also known as step size, annealing rate) has a large effect on the resulting embedding. If the learning rate is too large, the optimization will "overstep" and either fluctuate around a minimum, or diverge. If the learning rate is too small, the optimization may require many iterations to converge and is more likely to converge to a local minimum. A better strategy is to employ a learning rate schedule, where at early iterations the learning rate is large but decreases over time to allow for convergence. This is known to converge to a stationary point (could be a saddle) under certain conditions: $\sum g(t) = \infty$ and $\sum g(t)^2 < \infty$ [3].

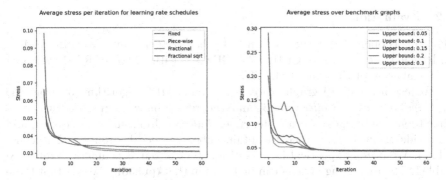

Fig. 6. (Left) Effect of the learning rate schedule on the optimization. The piece-wise schedule adapted from [41] arrives at a minimum faster on average. (Right) Effect of the upper bound on the learning rate on the optimization. An upper bound of 0.1 behaves predictably. Values for both are averaged over all graphs in our benchmark.

We investigate a limited subset of possible learning rate schedules, a fixed learning rate at 0.05, a piece-wise schedule similar to that of Zheng et al. [41], a fractional decay of $\Theta(t^{-1})$, and a slower fractional decay of $\Theta(t^{-.5})$. The piece-wise schedule begins with an exponential decay function, with large initial values and switches to $\Theta(t^{-1})$ once below a threshold. There are a few changes we needed to make to the piece-wise schedule. Firstly, while Zheng et al. [41] upper bound their learning rate by 1, this upper bound is too large for SMDS. The upper bound for the Euclidean algorithm was derived from the geometric structure, and a value of 1 reduces the stress between a single pair of vertices to 0. The latitude and longitude of the sphere are angles and so do not have this property. We instead need a relatively small upper bound, noting that large movements of a pair vertices on the sphere that need to be moved apart can actually bring them closer together (by wrapping around the sphere). We investigated values in the range 0 to 1, and settled on 0.1 as it achieves low stress quickly while not being so small as to fall into local minima; see Fig. 6.

Randomization is a to select pairs i, j in the stochastic optimization function. While SGD was originally done using sampling with replacement, random reshuffling has been shown to converge in fewer total updates [14]. To use random reshuffling in stress minimization, we enumerate all pairs $i < j$ of vertices in a ordered list and shuffle this list after every iteration. This ensures that the order in which we visit pairs is random, but that each pair is visited before we sample the same pair again.

A stopping condition is how the algorithm determines to terminate, either by converging or by reaching a maximum number of iterations. We measure convergence by tracking the change in the value of the optimization function between iterations. In the figures and evaluation results below, we set the convergence threshold to $1e^{-7}$ or a balance between speed and quality.

4.2   Evaluation

Our code is open source and written in Python. Experiments are performed on an Intel® Core™ i7-3770 CPU @ 3.40GHz × 8 with 32 GB of RAM with a 64 bit installation of Ubuntu 20.04.3 LTS.

We use a set of 40 graphs to evaluate our SMDS algorithm: 34 from the *SuiteSparse Matrix Collection* [6] and the remaining 6 from skeletons of 3-dimensional polytopes. We use the cube, dodecahedron, and isocahedron, and subdivide them 4 times each to obtain cube_4, dodecahedron_4 and isocahedron_4. We present spherical layouts of a subset of our benchmark graphs; see Table 1. The remaining layouts can be found in the Appendix. We see that there are several graphs particularly well suited to spherical layout: the 3D polytopes and their subdivisions have much lower distortion on the sphere than in the plane. 3-dimensional meshes and triangulations of surfaces such as dwt_307 and delaunay_n10 also have lower error on the sphere.

The SGD optimization scheme performs better than exact GD on both time to convergence and stress as the size of the data becomes large as we expect; see Fig. 7.

5   Geometry Comparison

Here we discuss some possible drawbacks of graph embedding in spherical space and compare graph embeddings between Euclidean, spherical and hyperbolic spaces. Stress works well for producing layouts, but directly comparing stress scores between geometries are difficult to interpret. Layouts are often uniformly scaled so that stress is minimum before reporting (see [12, 20]) which works fine in Euclidean space, but becomes a problem in spherical and hyperbolic spaces. In order to more fairly compare embedding error across geometries, we use the *distortion* [27, 36] metric, defined as

$$\text{distortion}(X) = \frac{1}{\binom{|V|}{2}} \sum_{i<j} \frac{|\delta(X_i, X_j) - d_{i,j}|}{d_{i,j}}. \tag{5}$$

5.1   Dilation of Distances

It is known that Euclidean MDS is invariant to dilation, that is if one multiplies the given distances by a positive real number, the corresponding MDS solution is the original MDS solution multiplied by the same scalar factor (up to rotation). However, this is not true for spherical and hyperbolic spaces. Moreover, spherical space is bounded, unlike Euclidean space. For example, on the 2D unit sphere the maximum distance that can be achieved between two points is π (assuming that between any two points we always take the shortest geodesic distance). Any graph with diameter (longest shortest path) longer than π cannot possibly be embedded on the unit sphere with zero error. A reasonable solution is to dilate

Table 1. Layouts

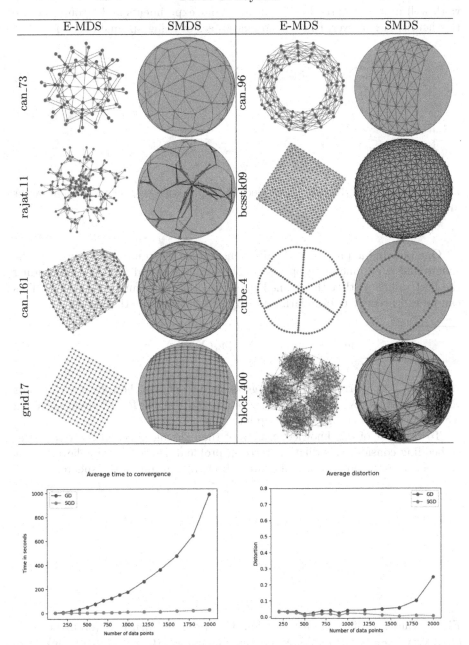

Fig. 7. How the SGD optimization scheme fairs compares to the exact GD in terms of time (left) and error (right). The larger the size of the graph, the more benefit is seen from the use of SGD.

the input distances so that all the given distances are less than or equal to π. That is, to multiply the distance matrix, d, by $\frac{\max d}{\pi}$. This heuristic appears to work well in practice; see Fig. 9. For all of our experiments and layouts, we use this heuristic. However, this has no guarantees of being optimal (Fig. 8).

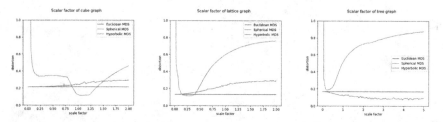

Fig. 8. Behavior of distortion on selected graphs with respect to dilation factor in each geometry.

One possible approach to the dilation problem is to make the radius of the sphere also a parameter to optimize. The problem would then become finding the best radius so that the defined stress function is as small as possible. This can be captured by reformulating Eq. (3) to also optimize the radius:

$$\arg\min_{R,X_1,...X_n \in S_R^2} \sum_{i,j=1}^{N} \left(\delta_R(X_i, X_j) - d_{ij}\right)^2 . \tag{6}$$

Here $\delta_R(X_i, X_j)$ corresponds to the geodesic distance on the sphere with radius R between points X_i and X_j. We derive the gradient for R and update it along with the vertex positions at each update step.

To the best of our knowledge none of the existing algorithms for spherical embedding consider this dilation/resizing problem. However, we believe that it is a crucial parameter while embedding/drawing a graph on the sphere.

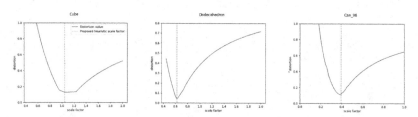

Fig. 9. Effect of dilation on distortion. Our proposed heuristic (orange line) is often very close to the minimum (of the blue curve). (Color figure online)

	E-MDS	SMDS	HMDS
cube	0.21494	0.11287	0.28334
isocahedron	0.23256	0.03905	0.2612
dodecahedron	0.21762	0.04262	0.48258
dwt_72	0.12862	0.14406	0.20609
can_73	0.21831	0.20114	0.35937
lesmis	0.22548	0.25536	0.1959
can_96	0.19802	0.11498	0.41292
rajat11	0.18752	0.20923	0.25801
can_144	0.11089	0.07133	0.28803
can_161	0.18532	0.16809	0.26348
dwt_162	0.09281	0.08898	0.32336
cube_4	0.20793	0.05715	0.54074
jazz	0.24723	0.25819	0.23984
dwt_221	0.08002	0.08593	0.50755
visbrazil	0.17988	0.20071	0.21925
grid1_dual	0.17266	0.18365	0.47681
grid17	0.10002	0.09049	0.28645
dwt_307	0.21835	0.11481	0.82275
dwt_361	0.07689	0.08097	0.51507
netscience	0.17617	0.19067	0.38224
block_400	0.27969	0.27782	0.25253

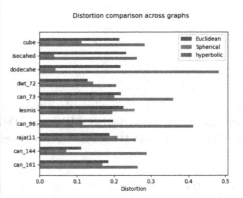

Fig. 10. The left subfigure shows a subset of results from the direct comparison for distortion in Euclidean, spherical and hyperbolic space. The right subfigure plots the first 10 rows. We note that 3D polytopes and meshes (the can graphs) are particularly well suited to the sphere, the LesMis graph is a complex network which is best embedded into hyperbolic space, and Euclidean space is better for the remaining ones.

5.2   Choosing Between Geometries

One reason to consider embedding graphs on different manifolds (Euclidean, hyperbolic, spherical) is to be able to preserve and visualize important properties of the given graph. Some graphs achieve lower distortion on the sphere, others in hyperbolic space. In this section we investigate how spherical graph layouts differ from other consistent geometries. We choose a selection of graphs from the sparse matrix collection, and lay them each out using the Euclidean, spherical, and hyperbolic variants of MDS and measure the distortion. We repeat the layout 5 times each, and report the average distortion for each graph in each geometry. We make use of [41] for the Euclidean MDS implementation and [27] for the hyperbolic MDS (HMDS) implementation.

The hypothesis we test here is that some graphs have a dramatically lower distortion in a particular geometry. For instance, rectangular lattices can be embedded with constant error in Euclidean space [39], regular 3D polytopes can be thought of as tesselations of the sphere, and trees have been described as "discrete hyperbolic spaces" [19]. The results are summarized in Fig. 10 with additional data can be found in the arXiv version: https://arxiv.org/abs/2209. 00191. We observe that spherical geometry is in fact able to embed polytopes and 3D meshes with lower distortion. Further, hyperbolic geometry is able to embed networks with "small-world" properties such as lesmis and block_400 with lower distortion. In graphs with 2D structure, Euclidean space is the clear winner.

In Fig. 11 we go beyond graphs to verify the different nature of the three geometries. We sample points randomly from each space, and use these points

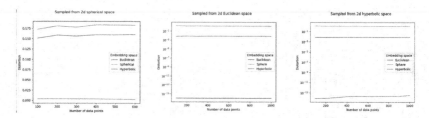

Fig. 11. Results from sampling data uniformly at random from each consistent geometry: as expected SMDS, MDS and HMDS perform dramatically better on data that comes from the geometry it embeds in.

to define the distance matrices. We expect the corresponding geometry's MDS to embed the data with much lower distortion and this is indeed the effect we see.

6   Conclusions and Future Work

We described an efficient method for embedding graphs in spherical space. The method generalizes beyond graphs to embedding high-dimensional data. We studied (quantitatively and qualitatively) the difference between spherical embeddings of graphs and embeddings in Euclidean and hyperbolic spaces. We discussed the issue of dilation and proposed an approach that seems to work well in practice. Furthermore, we compared how structures are preserved in different geometries. The algorithm is implemented and fully functional and we provide the source code, experimental data and results, and a web based visualization tool on GitHub: https://github.com/Mickey253/spherical-mds.

While our proposed algorithm is much faster than exact gradient descent (5 s for a 1000-vertex graph), it still requires an all-pairs-shortest-paths computation as a preprocessing step, which cannot be done faster than quadratic time in the number of vertices. This is a bottleneck computation for any graph-distance based approach and coming up with a strategy (e.g., sampling a subset of distances) is a problem whose solution can impact many existing algorithms. Another direction for future work is to quickly determine the best embedding space for a given graph. That is given a graph, decide the best manifold to embed it in: Euclidean, spherical or hyperbolic. We considered stress and distortion measures here, but exploring other graph drawing aesthetics across different geometries seems to be a worthwhile direction to explore.

References

1. Börsig, K., Brandes, U., Pasztor, B.: Stochastic gradient descent works really well for stress minimization. In: GD 2020. LNCS, vol. 12590, pp. 18–25. Springer, Cham (2020). https://doi.org/10.1007/978-3-030-68766-3_2

2. Bottou, L.: Large-scale machine learning with stochastic gradient descent. In: Lechevallier, Y., Saporta, G. (eds.) Proceedings of COMPSTAT 2010. Physica-Verlag HD, pp. 177–186. Springer (2010). https://doi.org/10.1007/978-3-7908-2604-3_16

3. Bottou, L.: Stochastic gradient descent tricks. In: Montavon, G., Orr, G.B., Müller, K.-R. (eds.) Neural Networks: Tricks of the Trade. LNCS, vol. 7700, pp. 421–436. Springer, Heidelberg (2012). https://doi.org/10.1007/978-3-642-35289-8_25

4. Börner, K.: (2012). http://scimaps.org/home.html

5. Börner, K., et al.: Design and update of a classification system: the UCSD map of science. PLoS ONE **7**(7), 1–10 (2012)

6. Davis, T.A., Hu, Y.: The University of Florida sparse matrix collection. ACM Trans. Math. Softw. **38**(1), 1:1-1:25 (2011)

7. De Leeuw, J., Mair, P.: Multidimensional scaling using majorization: SMACOF in R. J. Stat. Software **31**, 1–30 (2009)

8. Du, F., Cao, N., Lin, Y., Xu, P., Tong, H.: iSphere: focus+context sphere visualization for interactive large graph exploration. In: Proceedings of the 2017 CHI Conference on Human Factors in Computing, pp. 2916–2927. ACM (2017)

9. Elad, A., Keller, Y., Kimmel, R.: Texture mapping via spherical multi-dimensional scaling. In: Kimmel, R., Sochen, N.A., Weickert, J. (eds.) Scale-Space 2005. LNCS, vol. 3459, pp. 443–455. Springer, Heidelberg (2005). https://doi.org/10.1007/11408031_38

10. Eppstein, D.: Limitations on realistic hyperbolic graph drawing. In: Purchase, H.C., Rutter, I. (eds.) GD 2021. LNCS, vol. 12868, pp. 343–357. Springer, Cham (2021). https://doi.org/10.1007/978-3-030-92931-2_25

11. Frey, D., Pimentel, R.: Principal component analysis and factor analysis. In: Principal Component Analysis. Springer Series in Statistics. Springer, New York (1978). https://doi.org/10.1007/978-1-4757-1904-8_7

12. Gansner, E.R., Hu, Y., North, S.C.: A maxent-stress model for graph layout. IEEE Trans. Vis. Comput. Graph. **19**(6), 927–940 (2013)

13. Gansner, E.R., Koren, Y., North, S.: Graph drawing by stress majorization. In: Pach, J. (ed.) GD 2004. LNCS, vol. 3383, pp. 239–250. Springer, Heidelberg (2005). https://doi.org/10.1007/978-3-540-31843-9_25

14. Gürbüzbalaban, M., Ozdaglar, A., Parrilo, P.A.: Why random reshuffling beats stochastic gradient descent. Math. Program. **186**(1), 49–84 (2021)

15. Günther, I.: (2007). https://world-processor.com

16. Kamada, T., Kawai, S.: An algorithm for drawing general undirected graphs. Inf. Process. Lett. **31**(1), 7–15 (1989)

17. Kobourov, S.: Force-directed drawing algorithms. In: Handbook on Graph Drawing and Visualization, pp. 383–408. Chapman and Hall/CRC (2013)

18. Kobourov, S., Wampler, K.: Non-Euclidean spring embedders. IEEE Trans. Vis. Comput. Graph. **11**(6), 757–767 (2005)

19. Krioukov, D., Papadopoulos, F., Kitsak, M., Vahdat, A., Boguná, M.: Hyperbolic geometry of complex networks. Phys. Rev. E **82**(3), 036106 (2010)

20. Kruiger, J.F., Rauber, P.E., Martins, R.M., Kerren, A., Kobourov, S., Telea, A.C.: Graph layouts by t-SNE. Comput. Graph. Forum **36**(3), 283–294 (2017)

21. Kruskal, J.B.: Multidimensional scaling by optimizing goodness of fit to a nonmetric hypothesis. Psychometrika **29**(1), 1–27 (1964)

22. Kwon, O., Muelder, C., Lee, K., Ma, K.: A study of layout, rendering, and interaction methods for immersive graph visualization. IEEE Trans. Vis. Comput. Graph. **22**(7), 1802–1815 (2016)

23. Lamping, J., Rao, R.: The hyperbolic browser: a focus + context technique for visualizing large hierarchies. J. Vis. Lang. Comput. **7**(1), 33–55 (1996)
24. Lamping, J., Rao, R., Pirolli, P.: A focus+context technique based on hyperbolic geometry for visualizing large hierarchies. In: Human Factors in Computing Systems, CHI 1995 Conference Proceedings, pp. 401–408. ACM/Addison-Wesley (1995)
25. Van der Maaten, L., Hinton, G.: Visualizing data using t-SNE. J. Mach. Learn. Res. **9**(11), 2579–2605 (2008)
26. McInnes, L., Healy, J.: UMAP: uniform manifold approximation and projection for dimension reduction. CoRR abs/1802.03426 (2018)
27. Miller, J., Kobourov, S., Huroyan, V.: Browser based hyperbolic visualization of graphs. In: IEEE Pacific Visualization Symposium PacificVis. IEEE Computer Society (2022)
28. Munzner, T.: H3: laying out large directed graphs in 3d hyperbolic space. In: IEEE Symposium on Information Visualization, pp. 2–10. IEEE Computer Society (1997)
29. Munzner, T.: Exploring large graphs in 3D hyperbolic space. IEEE Comput. Graphics Appl. **18**(4), 18–23 (1998)
30. Munzner, T.: Interactive visualization of large graphs and networks. Ph.D. thesis, Stanford University (2000)
31. Munzner, T., Burchard, P.: Visualizing the structure of the world wide web in 3d hyperbolic space. In: Proceedings of the 1995 Symposium on Virtual Reality Modeling Language, pp. 33–38. ACM (1995)
32. Ontrup, J., Ritter, H.J.: Hyperbolic self-organizing maps for semantic navigation. In: Advances in Neural Information Processing Systems, vol. 14, pp. 1417–1424. MIT Press (2001)
33. Perry, S., Yin, M.S., Gray, K., Kobourov, S.: Drawing graphs on the sphere. In: AVI 2020: International Conference on Advanced Visual Interfaces, pp. 17:1–17:9. ACM (2020)
34. Robbins, H., Monro, S.: A stochastic approximation method. The Annals of Mathematical Statistics, pp. 400–407 (1951)
35. Ruder, S.: An overview of gradient descent optimization algorithms. CoRR abs/1609.04747 (2016)
36. Sala, F., Sa, C.D., Gu, A., Ré, C.: Representation tradeoffs for hyperbolic embeddings. In: Proceedings of the 35th International Conference on Machine Learning. Proceedings of Machine Learning Research, vol. 80, pp. 4457–4466. PMLR (2018)
37. Schulz, H.J.: Treevis.net: a tree visualization reference. IEEE Comput. Graphics Appl. **31**(6), 11–15 (2011)
38. Shepard, R.N.: The analysis of proximities: multidimensional scaling with an unknown distance function. I. Psychometrika **27**(2), 125–140 (1962)
39. Verbeek, K., Suri, S.: Metric embedding, hyperbolic space, and social networks. Comput. Geom. **59**, 1–12 (2016)
40. Yang, Y., Jenny, B., Dwyer, T., Marriott, K., Chen, H., Cordeil, M.: Maps and globes in virtual reality. Comput. Graph. Forum **37**(3), 427–438 (2018)
41. Zheng, J.X., Pawar, S., Goodman, D.F.M.: Graph drawing by stochastic gradient descent. IEEE Trans. Vis. Comput. Graph. **25**(9), 2738–2748 (2019)
42. Zhu, M., Chen, W., Hu, Y., Hou, Y., Liu, L., Zhang, K.: DRGraph: An efficient graph layout algorithm for large-scale graphs by dimensionality reduction. IEEE Trans. Vis. Comput. Graph. **27**(2), 1666–1676 (2021)

Shape-Faithful Graph Drawings

Amyra Meidiana[(✉)], Seok-Hee Hong, and Peter Eades

University of Sydney, Camperdown, Australia
amei2916@uni.sydney.edu.au, {seokhee.hong,peter.eades}@sydney.edu.au

Abstract. *Shape-based metrics* measure how faithfully a drawing D represents the structure of a graph G, using the *proximity graph S* of D. While some limited graph classes admit proximity drawings (i.e., optimally shape-faithful drawings, where $S = G$), algorithms for shape-faithful drawings of general graphs have not been investigated.

In this paper, we present the first study for shape-faithful drawings of *general graphs*. First, we conduct extensive comparison experiments for popular graph layouts using the shape-based metrics, and examine the properties of highly shape-faithful drawings. Then, we present $ShFR$ and $ShSM$, algorithms for shape-faithful drawings based on force-directed and stress-based algorithms, by introducing new *proximity forces/stress*. Experiments show that $ShFR$ and $ShSM$ obtain significant improvement over FR (Fruchterman-Reingold) and SM (Stress Majorization), on average 12% and 35% respectively, on shape-based metrics.

1 Introduction

Recently, *shape-based metrics* [7] have been introduced for evaluating the quality of large graph drawing. It measures how faithfully the "shape" of a drawing D represents the ground truth structure of a graph G, by comparing the similarity between the *proximity graph S* of the vertex point set of D and the graph G.

For a point set P in the plane, proximity graphs are defined as: two points are connected by an edge if they are "close enough". Specifically, a *proximity region* is defined for each pair of points, and if the proximity region is empty, the points are connected by an edge in the proximity graph [24].

Some limited graph classes always admit a *proximity drawing D*, where the graph G is realized as a proximity graph S in D. For such proximity drawable graph classes, some characterizations are known, and algorithms to construct such proximity drawings are available [1,3]. Consequently, such proximity drawings are optimally shape-faithful (i.e., shape-based metric of 1), since $S = G$.

However, such optimally shape-faithful drawings are only applicable for very limited graph classes. Algorithms to optimize shape-based metrics for *general graphs* (i.e., not proximity drawable graphs) have not been studied yet.

In this paper, we present the first study for shape-faithful drawings of general graphs. Specifically, our main contributions can be summarized as follows:

This work is supported by ARC grant DP190103301.

1. We evaluate the shape-faithfulness of popular graph drawing algorithms for various proximity drawable graph classes, including *strong* proximity drawable graphs (i.e., the best possible shape-based metric is 1), *almost* proximity drawable graphs with some forbidden subgraphs, *weak* proximity drawable graphs, and mesh graphs.

   Experiments show that $tsNET$ [16] obtains the highest shape-faithfulness on most large graph instances, for strong and almost proximity drawable graphs, and stress-based layouts [11] achieve good results on mesh graphs.

2. We present $ShFR$ and $ShSM$, algorithms for shape-faithful drawings for general graphs, based on the force-directed and stress-based layouts, by introducing new *proximity forces/stress*.

   Experiments with strong proximity drawable graphs, scale-free graphs and benchmark graphs show that $ShFR$ and $ShSM$ obtain significant improvement (on average, 12% and 35%) on the shape-based metrics over FR (Fruchterman-Reingold) [9] and SM (Stress Majorization) [11].

2   Related Work

2.1   Shape-Based Metrics

Shape-based metrics measure how faithfully the "shape" of a drawing D represents the ground truth structure of a graph G, by comparing the similarity between the *proximity graph S* of the vertex point set of D and the graph G [7].

Specifically, the shape-based metrics use proximity graphs such as the Gabriel Graph (GG) and Relative Neighborhood Graph (RNG) (defined in Sect. 2.2). To compute the similarity between G and S, both with vertex set V, the shape-based metrics use the *Jaccard Similarity (JS)* [15] as follows: $JS(G, S) = \frac{1}{|V|} \sum_{v \in V} \frac{N_G(v) \cap N_S(v)}{N_G(v) \cup N_S(v)}$, where $N_G(v)$ (resp., $N_S(v)$) is the set of neighbors of vertex v in G (resp., S). We denote the shape-based metrics computed with this formula using RNG (resp., GG) as Q_{RNG} (resp., Q_{GG}), having values between 0 and 1 where 1 means perfectly shape-faithful.

2.2   Proximity Graphs

For a point set P in the plane, a proximity graph S of P is roughly defined as follows: two points are connected by an edge if and only if they are "close enough". Namely, the *proximity region* defined for the two points should be empty (i.e., contains no other points) [24,25]. For example, *Gabriel Graph (GG)* [10] (resp., *Relative Neighborhood Graph (RNG)* [26]) is a proximity graph where two points x and y are connected by an edge if and only if the closed disk (resp., open lens) having line segment xy as its diameter contains no other points.

For *strong proximity*, two conditions must be fulfilled: (a) two points are connected by an edge only if their proximity region is empty, and (b) two points are not connected by an edge only if their proximity region is not empty [2].

A relaxation of condition (b) gives rise to the definition of *weak proximity*, where the proximity graph may omit an edge between points x and y even if their proximity region is empty [2]. Namely, while points need to be "close enough" to be connected by an edge in the proximity graph S, points can be made to be not connected by an edge in S even if they are "close enough".

2.3 Proximity Graph Drawing

Characterizations of *strong proximity drawable graphs* (i.e., graphs that admit a proximity drawing D, where the graph G is realized as a proximity graph $S = G$ in D) are known for RNG and GG [3,17]:

- RNG-drawable graphs: trees with maximum degree 5, maximal outerplanar graphs, biconnected outerplanar graphs
- GG-drawable graphs: trees with maximum degree 4 and no degree 4 vertex with all "wide" subtrees, maximal and biconnected outerplanar graphs

Moreover, forbidden subgraphs have also been characterized: no GG- and RNG-drawable graphs may contain K_4 and $K_{2,3}$ as subgraphs [10].

Characterizations of *weak proximity drawable graphs* include wider classes:

- trees (regardless of maximum degree): weak GG- and RNG-drawable [2]
- 1-connected outerplanar graphs with no vertex of degree 1: weak GG-drawable [8].

Algorithms to construct proximity drawings of both strong and weak proximity drawable graphs are available [2,3,17], although implementations are unavailable and challenging due to requiring precise geometric computations. For details on proximity graph drawing, see a survey [19].

3 Graph Layout Comparison Experiments

3.1 Experiment Design and Data Sets

In this Section, we present extensive experiments using the shape-based metrics Q_{RNG} and Q_{GG} to compare popular graph drawing algorithms:

- Force-directed layouts: Fruchterman-Reingold (FR) [9], Organic (OR) [28].
- Multi-level force-directed layouts: FM^3 [12], $sfdp$ [14].
- Backbone layout (BB) [23], which untangles hairballs in a drawing.
- LinLog layout (LL) [22], a force-directed algorithm displaying clusters.
- Stress-based layouts to minimize the *stress*: Stress Majorization (SM) [11], Stochastic Gradient Descent (SGD) [29].
- $tsNET$ layout [16], based on the t-SNE dimension reduction [20].
- Walker's level drawing algorithm (W) for trees [27].
- Chrobak and Kant algorithm (CK) [5] for convex grid drawings of triconnected planar graphs in quadratic area.

For data sets, we generate graphs with various sizes: *small* graphs with 50–250 vertices, *medium* graphs with 250–500 vertices, and *large* graphs with 500–1000 vertices. Furthermore, we consider graph types based on proximity drawability characterization: *strong* proximity drawable graphs, *almost* proximity drawable graphs, and *weak* proximity drawable graphs. We also use *mesh* graphs, which do not fall into known proximity drawability characterizations. For each graph type and size, we generate ten graph instances.

Strong Proximity Drawable Graphs: We generate strong proximity drawable graphs based on known characterizations [3,17]:

- *Maximum outerplanar graphs*, generated using the connected planar graph generator of OGDF [4].
- *Biconnected outerplanar graphs*: We start G as a cycle of random length $\leq$ the target size n. Then, select an edge (u, v) in G that is only involved in one cycle. Select a cycle length $x < n$, create a path p of length $x - 2$, and add an edge between u and the first vertex of p, and between v and the last vertex of p. Repeat while the number of vertices in G is less than n.
- Proximity drawable *trees*, generated using the random tree generator of OGDF: For RNG-drawable trees, we set the maximum vertex degree as 5; for GG-drawable trees, we set the maximum vertex degree as 4, and then prune forbidden subtrees until the tree contains no more forbidden subtrees.

Almost Proximity Drawable Graphs with Forbidden Subgraphs: We start with a strong proximity drawable graph G, and then add a few edges and/or vertices to create a forbidden subgraph. The number of edges (resp., vertices) added are limited to at most 10 (resp., 5). Specifically, we perform two types of forbidden subgraph augmentation:

- L-AUG (*Local* Augmentation) graphs: We choose a vertex v of G and add new vertices and edges around v to create a forbidden subgraph F.
- F-AUG (*Global* Augmentation) graphs: We select a subset of vertices of G, all separated by a shortest path length above a predefined threshold, and add edges between the selected vertices to create a forbidden subgraph F.

Weak Proximity Drawable Graphs: We also use *weak* proximity drawable graphs based on the weak proximity drawability characterization [2]:

- *1-connected* outerplanar graph with a minimum degree of 2, which are weak GG-drawable [8]: We generate the graphs in a similar way to the biconnected outerplanar graphs, however alternately appending the new cycle to a random vertex rather than a random edge.

Mesh Graphs: We use simple *mesh* graphs containing no chordless cycles of length > 3, from the *jagmesh* set of the SuiteSparse Matrix collection [6]. These graphs are not part of known proximity drawability characterizations, but can be drawn as an RNG drawing, by drawing each 3-cycle as an equilateral triangle.

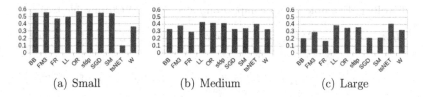

(a) Small   (b) Medium   (c) Large

Fig. 1. Average Q_{RNG} for trees. LL, OR, and $sfdp$ consistently perform well, with $tsNET$ performing much better on large trees. Even the highest-performing layouts are still far from optimal shape-faithfulness ($Q_{RNG} = 1$).

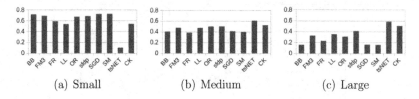

(a) Small   (b) Medium   (c) Large

Fig. 2. Average Q_{RNG} for maximum outerplanar graphs. $tsNET$ and CK are the top performing layouts on medium and large graphs. For highest-performing layouts, Q_{RNG} is slightly closer to optimal compared to RNG-drawable trees.

3.2   Results

Strong Proximity Drawable Graphs. On strong proximity drawable trees, all the drawing algorithms used fail to obtain shape-based metrics close to optimal.

Figure 1 shows the average Q_{RNG} for RNG-drawable trees. On small trees, the best performing layouts, OR, BB, multi-level layouts, and stress-based layouts, only obtain Q_{RNG} of 0.5–0.6 on average. $tsNET$ becomes the best performing layout on medium and large trees, with Q_{RNG} of about 0.4 on average. On large trees, the differences in Q_{RNG} between layouts are more pronounced, with $tsNET$ and LL performing the best, followed by $sfdp$ and OR.

For small proximity drawable outerplanar graphs (both GG- and RNG-drawable), the best performing layouts, stress-based layouts and BB, obtain Q_{RNG} of around 0.7 (see Fig. 2). This is notably closer to optimal compared to RNG-drawable trees, where all layouts obtain average Q_{RNG} of at most 0.6.

$tsNET$ and CK are the top performing layouts on medium and large outerplanar graphs, despite lower performance on small graphs: on medium and large maximum outerplanar graphs, $tsNET$ (resp., CK) obtains Q_{RNG} of 0.6 (resp., 0.5) on average. This is closer to optimal compared to large RNG-drawable trees, where $tsNET$, the best performing layout, only obtains average Q_{RNG} of 0.4.

For GG-drawable trees, the results on Q_{GG} are mostly similar to Q_{RNG}; similarly, for GG-drawable outerplanar graphs (same set of graphs as RNG-drawable outerplanar graphs), the results on Q_{GG} are similar to Q_{RNG}. For details, see Figures. 7 and 8 in Appendix B of the full version [21].

Table 1. Example layout comparison for a large *RNG*-drawable tree.

BB	*FM³*	*FR*	*LL*	*OR*
sfdp	*SGD*	*SM*	*tsNET*	*W*

Table 2. Example layout comparison for a large maximum outerplanar graph.

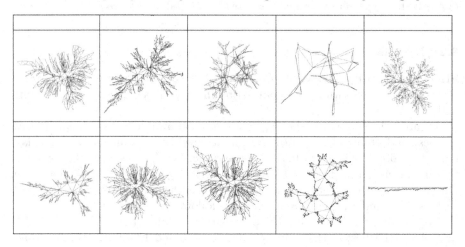

Table 1 shows a visual comparison of graph layouts on a large *RNG*-drawable tree. For the best performing layouts *tsNET* and *LL*, subtrees closer to the leaves are often "compacted" together, compared to the second best performing layouts such as OR and sfdp, where all branches are more "opened" up.

Table 2 shows a visual comparison of layouts on a maximal outerplanar graph. The best performing layout, *tsNET*, collapses the faces on the periphery, compared to the faces in the middle of the drawing. The "long" drawing of *CK* may have obtained a comparable effect, producing high shape-based metrics.

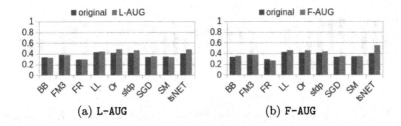

(a) L-AUG   (b) F-AUG

Fig. 3. Average Q_{RNG} for L-AUG and F-AUG graphs, compared to RNG-drawable trees. $tsNET$ obtains the highest shape-based metrics; surprisingly, Q_{RNG} is sometimes higher on L-AUG and F-AUG than on the strong proximity drawable graphs.

Table 3. Example layout comparison for a medium F-AUG graph.

BB	FM^3	FR	LL	OR
$sfdp$	SGD	SM	$tsNET$	

Almost Proximity Drawable Graphs. In general, the ranking of the graph drawing algorithms on the shape-based metrics do not change much between strong proximity drawable graphs and almost proximity drawable graphs.

Figure 3 shows comparisons on Q_{RNG} for the base RNG-drawable trees and the L-AUG and F-AUG graphs, where $tsNET$ still obtains the highest Q_{RNG}. LL also obtains the second highest Q_{RNG}, although with a smaller difference to the next best performing layouts OR and $sfdp$, compared to RNG-drawable trees.

Table 3 shows a visual comparison on a F-AUG graph, where the layouts with highest shape-based metrics, such as tsNET and LL, draw the "branches" in the periphery of the drawing in a more compact way, than other layout. This observation is consistent with the pattern also seen in the visual comparison for strong proximity drawable trees and outerplanar graphs.

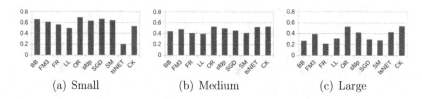

(a) Small   (b) Medium   (c) Large

Fig. 4. Average Q_{GG} for 1-connected outerplanar graphs. OR and CK performs the best on large 1-connected outerplanar graphs.

Table 4. Example layout comparison for a large 1-connected outerplanar graph.

BB	FM^3	FR	LL	OR

$sfdp$	SGD	SM	$tsNET$	CK

Weak Proximity Drawable Graphs. For weak GG-drawable 1-connected outerplanar graphs, OR surprisingly obtains the highest Q_{GG} on large 1-connected outerplanar graphs, followed by CK and tsNET; see Fig. 4.

Table 4 shows a visual comparison, where OR draws a number of chordless cycles with their vertices in a regular polygon configuration. In fact, this is the correct way to draw such cycles as GG, resulting in high Q_{GG}.

Mesh Graphs. On mesh graphs, the best performing layouts, stress-based layouts, obtain on average much higher shape-based metrics than on other strong proximity drawable graphs, see Fig. 5. In particular, SGD and SM obtain near-perfect shape-based metrics ($Q_{RNG} = 0.99$ on average), and OR and BB also obtain very high shape-based metrics ($Q_{RNG} = 0.98$ on average). On the other hand, $tsNET$ obtains comparatively lower shape-based metrics.

Table 5 shows a visual comparison on a mesh graph; most layouts manage to untangle the mesh. Furthermore, SGD and SM manage to untangle without twists or "distortions", where triangles in the periphery are more "squashed" compared to the triangles in the middle, as seen in $sfdp$ or $tsNET$ layouts.

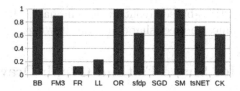

Fig. 5. Average Q_{RNG} for mesh graphs. Stress-based layouts obtain the best shape-based metrics, at almost perfect.

Table 5. Example layout comparison for mesh.

BB	FM^3	FR	LL	OR
$sfdp$	SGD	SM	$tsNET$	CK

3.3 Discussion and Summary

Overall, $tsNET$ performs the best on large strong proximity drawable graphs, followed by LL. Looking at the visual comparison, these layouts often "collapse" subgraphs on the periphery. This may have lead to fewer non-adjacent vertices being close to each other, leading to better shape-based metrics. Moreover, this improvement compared to other layouts is more apparent in larger graphs, where the larger number of vertices means more non-adjacent vertices being close to each other in drawings where subgraphs on the periphery are not "collapsed".

Most layout algorithms are better at computing drawings closer to optimal shape-faithfulness for *dense* strong proximity drawable graphs: the best-performing layouts, $tsNET$ and LL, obtain much higher average shape-based metrics on outerplanar graphs compared to trees. Lower density means more pairs of vertices are not adjacent in G, i.e., more proximity regions need to be non-empty in D.

The mesh graphs are drawn as RNG by drawing each face as equilateral triangles, i.e., having uniform edge lengths, a readability metric which is often used as a goal for a number of layout algorithms. This may be why more layout algorithms, especially stress-based layouts which emphasize distance faithfulness, are able to produce almost-perfect shape-faithful drawings for the mesh graphs.

4   Algorithms for Shape-Faithful Graph Drawings

In this Section, we present algorithms for shape-faithful drawings. Based on the qualitative observations from the layout comparison experiments in Sect. 3, high shape-based metrics are obtained often by "collapsing" subgraphs on the drawing's periphery - this keeps non-adjacent vertices in G distant from each other, and adjacent vertices in the collapsed subgraphs within close proximity. Therefore, our main idea for shape-faithful graph drawings is to "drive away" non-adjacent vertices in G that are geometrically too close in the drawing D.

Specifically, we present two algorithms $ShFR$ and $ShSM$ based on two popular graph drawing algorithms, force-directed and stress minimization algorithms. $ShFR$ and $ShSM$ aim to improve shape-based metrics by introducing two new types of *proximity forces/stress*. For a pair of adjacent vertices v and u in G and another vertex of t currently located in the proximity region of v and u in D:

– *proximity repulsion force/stress*: push t out of the proximity region of u, v;
– *proximity attraction force/stress*: pull v and u closer together.

4.1   *ShFR*: Force-Directed Layout for Shape-Faithful Drawings

We present $ShFR$, a force-directed layout for shape-faithful drawing, incorporating *proximity forces* with Fruchterman-Reingold (FR) [9].

To explain the design rationale for $ShFR$, consider the following case: for a pair of adjacent vertices u and v in a graph $G = (V, E)$, the edge (u, v) does not exist in the proximity graph $S = (V, E')$ of a drawing D of G, due to a vertex t located inside the proximity region of u and v in D. For such a case, to add back the edge (u, v) in the proximity graph S to achieve $S = G$, we introduce two new *proximity forces*: (1) repulsion force to repel t out of the proximity region of u and v; (2) attraction force on u and v to shrink the proximity region.

We first add a *proximity repulsion force* to drive t out of the proximity region of u and v in D. From the midpoint m between u and v, we add a repulsion force acting on t, with a magnitude proportional to how far t needs to be away from m in order to be driven out of the proximity region of u and v. Specifically, the x-displacement of t induced by the repulsion force can be computed as: $\frac{x_t - x_m}{||X_t - X_m||^2} f l^2 \frac{||X_v - X_u||}{||X_t - X_m||}$, where x_t is the x-coordinate of t, $||X_t - X_m||$ is the Euclidean distance between t and m, l is the parameter for ideal spring length (i.e., target edge length), and f is the parameter for spring stiffness.

Next, we add a *proximity attraction force* for a pair of adjacent vertices u and v in G with non-empty proximity regions. Specifically, we add an attraction force acting between u and v: $(x_u - x_v)(||X_v - X_u||)l^{-1}$.

The new proximity forces can be added to any force-directed algorithms. For our specific implementation, we add the proximity forces in conjunction with FR, where the proximity force computations are added to each force computation iteration of FR. For the details, see Appendix C of the full version [21].

GG and RNG are subgraphs of the Delaunay Triangulation, which can be computed in $O(n \log n)$ time [24]. The original FR algorithm runs in $O(n^2)$ time. Therefore, the total runtime of $ShFR$ is $O(n^2)$.

4.2 $ShSM$: Stress-Based Layout for Shape-Faithful Drawings

We now present $ShSM$ for shape-faithful drawing, incorporating *proximity stress* with Stress Majorization (SM) [11]. Similar to the force-directed case, for each case where in drawing D a vertex t lies in the proximity region of two neighboring vertices v and u, i.e. $(u,v) \in E$ but $(u,v) \notin E'$, we add two new types of stress: (1) repulsion stress to push t out of the proximity region; (2) attraction stress to pull v and u closer together.

We first add the *proximity repulsion stress* by exerting stress on t from the midpoint m of u and v. Specifically, we compute the x-displacement of t due to the stress between t and m as $w_{uv}(x_m) + d_{uv}(x_m - x_t)\|X_v - X_u\|/\|X_t - X_m\|)$, where d_{uv} is the shortest path distance between u and v and w_{uv} is the weight computed for the vertex pair u and v, often computed as $(d_{uv})^k$ for a constant k. Since m is not an actual vertex of G, there is no graph theoretic distance or weight between m and t; we instead use d_{uv} and w_{uv}, and then scale them based on the ratio of the Euclidean distances between u, v, and between t, m.

We next add the *proximity attraction stress* which has a weight lower than the standard stress of SM, to attract u and v closer in order to to reduce the distance between u and v. The x-displacement of v due to this additional stress is computed as $w'_{uv}(x_u) + d_{uv}(x_v - x_u)/\|X_v - X_u\|)$, where $w'_{uv} = w^{k'}_{uv}$ for $k' < 1$.

The new proximity stress can be added to any stress-based algorithms. For our specific implementation, we add the proximity stress in conjunction with SM, where the proximity stress computations are added to each stress computation iteration of SM. For details, see Appendix D in the full version [21].

As with $ShFR$, GG and RNG can be computed in $O(n \log n)$ time and the original stress computation of SM takes $O(n^2)$ time. The total runtime of $ShSM$ is therefore $O(n^2)$.

5 $ShFR$ and $ShSM$ Experiments

5.1 Experiment Design and Data Sets

In this experiment, we evaluate the effectiveness of $ShFR$ and $ShSM$ over FR and SM respectively, using shape-based metrics Q_{RNG} and Q_{GG}.

For data sets, we use *strong proximity drawable* graphs, as well as *scale-free* graphs and *benchmark* graphs:

- *strong proximity drawable* graphs, from Sect. 3.
- *scale-free* graphs: We generate synthetic scale-free graphs with density 2, 3, and 5, using the NetworkX [13] scale-free generator.

– *benchmark* graphs, including real-world scale-free graphs [6,18,30] with up
to 6000 vertices and 15000 edges. For details, see Appendix A in the full
version [21].

To measure the improvement of the shape-based metrics, for exam-
ple, on Q_{RNG} by $ShFR$ over FR, we define the formula $I(Q_{RNG}) = \frac{Q_{RNG}(ShFR) - Q_{RNG}(FR)}{Q_{RNG}(FR)}$. We use the same formula for Q_{GG}, and for the improvement by $ShSM$ over SM.

5.2   Results

$ShFR$ obtains notable improvement over FR on Q_{RNG} and Q_{GG} for large strong
proximity drawable graphs, obtaining average improvement of 15%, 12%, and
12% on maximum outerplanar graphs, biconnected outerplanar graphs, and trees
respectively, see Fig. 6 (a). $ShFR$ also obtains significant improvement over FR
on Q_{GG} for scale-free graphs, at on average 18%. For real-world benchmark
graphs, the improvement on Q_{RNG} and Q_{GG} average at around 10%.

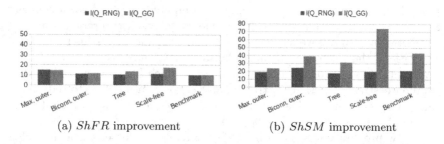

(a) $ShFR$ improvement   (b) $ShSM$ improvement

Fig. 6. Average shape-based metrics improvement (in percent) of $ShFR$ over FR and
$ShSM$ over SM on Q_{RNG} and Q_{GG}. $ShFR$ and $ShSM$ obtain significant improvement
over FR and SM respectively on all data sets.

$ShSM$ obtains significant improvement over SM for strong proximity drawable graphs, see Fig. 6 (b). For maximum outerplanar graphs, $ShSM$ obtains
significant improvement over SM (average 20% and 25%) on Q_{RNG} and Q_{GG}
respectively, which is much higher than the improvement by $ShFR$ over FR. For
biconnected outerplanar graphs, an even larger improvement of on average 40%
is achieved on Q_{GG}. For large trees, $ShSM$ also obtains significant improvement
over SM, on average 18% and 30% on Q_{RNG} and Q_{GG}, respectively.

$ShSM$ also obtains significant improvement over SM for scale-free graphs,
on average 20% improvement on Q_{RNG}. Notably, the largest improvement is
obtained by $ShSM$ on Q_{GG} for scale-free graphs, at over 70%, on scale-free
graphs. Note that $ShSM$ obtains on average 20% and 42% improvement over
SM for real-world benchmark graphs, on Q_{RNG} and Q_{GG} respectively.

Table 6 shows a visual comparison of FR and $ShFR$ on the benchmark
scale-free graph G_4. $ShFR$ untangles the "hairball" more clearly, compared

Table 6. Visual comparison of FR and $ShFR$, SM and $ShSM$ on benchmark graphs. $ShFR$ often untangles the hairballs better than FR, and $ShSM$ expands faces that are "collapsed" by SM.

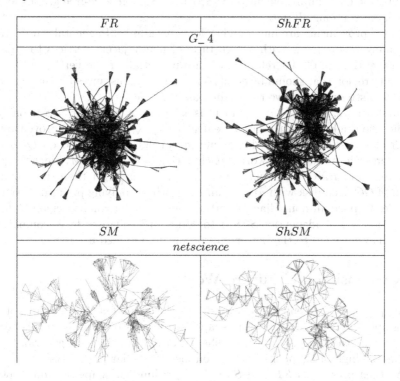

FR	$ShFR$
G_4	
SM	$ShSM$
netscience	

to FR. Table 6 also shows a visual comparison of SM and $ShSM$ on the benchmark scale-free graph *netscience*. $ShSM$ "expands" faces that are "squashed" in SM, showing the local neighborhood of some vertices more clearly. However, the expanded faces also leads to the drawing feeling more "crowded" compared to SM, thus increasing faithfulness but affecting readability. For more visual comparisons on other data sets, see Table 9 in Appendix E of the full version [21].

5.3   Discussion and Summary

Our extensive experiments demonstrate the effectiveness of $ShFR$ and $ShSM$ for shape-faithful drawings. $ShFR$ (resp., $ShSM$) obtains significant improvement over FR (resp., SM) of 11% and 13% (resp., 20% and 50%) on Q_{RNG} and Q_{GG} respectively, averaged over all data sets.

For strong proximity drawable graphs, $ShFR$ (resp., $ShSM$) obtains improvement over FR (resp., SM) of on average 13% and 13% (resp., 20% and 30%) on Q_{RNG} and Q_{GG} respectively. For real-world benchmark graphs, $ShFR$ (resp., $ShSM$) obtains improvement over FR (resp., SM) of on average 10% and 10% (resp., 20% and 43%) on Q_{RNG} and Q_{GG} respectively. For scale-free graphs, $ShFR$ (resp., $ShSM$) obtains improvement over FR (resp., SM) of on average 10% and 16% (resp., 17% and 70%) on Q_{RNG} and Q_{GG} respectively. Notably, the Q_{GG} improvement of $ShSM$ over SM on scale-free graphs at 70% is the largest among all data sets.

The improvements are much higher for *large* graphs. In general, large graphs have many vertex pairs, with a high ratio of non-adjacent vertices to adjacent pairs of vertices in G. Therefore, there are potentially more vertices located in proximity region that should be empty, creating more instances for the proximity forces and stress to improve the shape-based metrics.

Furthermore, the best improvement is achieved by $ShSM$ over SM on Q_{GG}, significantly higher than the improvement on Q_{RNG} and the improvements of $ShFR$ over FR. Specifically, larger improvements are obtained on Q_{GG} than Q_{RNG} on scale-free and real-world benchmark graphs by $ShSM$. Since the proximity region of RNG (i.e., lens at points u and v) is larger than the proximity region of GG (i.e., disk with uv as diameter), when applying proximity stress, it is harder to push all non-adjacent vertices out of the proximity region of RNG. In addition, the tendency for $ShSM$ to "open up" collapsed faces compared to $ShFR$ may have led to the better improvements obtained by $ShSM$.

6   Conclusion and Future Work

In this paper, we present the first study for the shape-faithful drawings of general graphs. We first evaluate the shape-faithfulness of existing graph layouts and examine the properties of good shape-faithful drawings. In general, tsNET obtains the highest shape-faithfulness on medium-to-large graphs.

We then present $ShFR$ and $ShSM$, algorithms for shape-faithful drawings of general graphs, based on force-directed and stress-based layouts, introducing new proximity forces/stress. Extensive experiments show that $ShFR$ and $ShSM$ achieve significant improvement over FR and SM, on average, 12% and 35% higher shape-based metrics respectively. Notably, $ShSM$ obtains a 70% average improvement on Q_{GG} over SM for scale-free graphs.

Future work includes shape-faithful layouts based on various other layouts.

References

1. Di Battista, G., Lenhart, W., Liotta, G.: Proximity drawability: a survey extended abstract. In: Tamassia, R., Tollis, I.G. (eds.) GD 1994. LNCS, vol. 894, pp. 328–339. Springer, Heidelberg (1995). https://doi.org/10.1007/3-540-58950-3_388
2. Di Battista, G., Liotta, G., Whitesides, S.: The strength of weak proximity (extended abstract). In: Brandenburg, F.J. (ed.) GD 1995. LNCS, vol. 1027, pp. 178–189. Springer, Heidelberg (1996). https://doi.org/10.1007/BFb0021802
3. Bose, P., Lenhart, W., Liotta, G.: Characterizing proximity trees. Algorithmica **16**(1), 83–110 (1996)
4. Chimani, M., Gutwenger, C., Jünger, M., Klau, G.W., Klein, K., Mutzel, P.: The open graph drawing framework (OGDF). In: Handbook of Graph Drawing and Visualization 2011, pp. 543–569 (2013)

5. Chrobak, M., Kant, G.: Convex grid drawings of 3-connected planar graphs. Int. J. Comput. Geom. Appl. **7**(03), 211–223 (1997)
6. Davis, T.A., Hu, Y.: The University of Florida sparse matrix collection. ACM Trans. Math. Softw. (TOMS) **38**(1), 1 (2011)
7. Eades, P., Hong, S.H., Nguyen, A., Klein, K.: Shape-based quality metrics for large graph visualization. J. Graph Algorithms Appl. **21**(1), 29–53 (2017)
8. Evans, W., Gansner, E.R., Kaufmann, M., Liotta, G., Meijer, H., Spillner, A.: Approximate proximity drawings. In: van Kreveld, M., Speckmann, B. (eds.) GD 2011. LNCS, vol. 7034, pp. 166–178. Springer, Heidelberg (2012). https://doi.org/10.1007/978-3-642-25878-7_17
9. Fruchterman, T.M.J., Reingold, E.M.: Graph drawing by force-directed placement. Softw. Pract. Experience **21**(11), 1129–1164 (1991)
10. Gabriel, K.R., Sokal, R.R.: A new statistical approach to geographic variation analysis. Syst. Zool. **18**(3), 259–278 (1969)
11. Gansner, E.R., Koren, Y., North, S.: Graph drawing by stress majorization. In: Pach, J. (ed.) GD 2004. LNCS, vol. 3383, pp. 239–250. Springer, Heidelberg (2005). https://doi.org/10.1007/978-3-540-31843-9_25
12. Hachul, S., Jünger, M.: Drawing large graphs with a potential-field-based multilevel algorithm. In: Pach, J. (ed.) GD 2004. LNCS, vol. 3383, pp. 285–295. Springer, Heidelberg (2005). https://doi.org/10.1007/978-3-540-31843-9_29
13. Hagberg, A., Swart, P., Chult, D.S.: Exploring network structure, dynamics, and function using networkx. Tech. rep., Los Alamos National Lab. (LANL), Los Alamos, NM (United States) (1 2008)
14. Hu, Y.: Efficient, high-quality force-directed graph drawing. Math. J. **10**(1), 37–71 (2005)
15. Jaccard, P.: The distribution of the flora in the alpine zone. 1. New Phytol. **11**(2), 37–50 (1912)
16. Kruiger, J.F., Rauber, P.E., Martins, R.M., Kerren, A., Kobourov, S., Telea, A.C.: Graph layouts by t-SNE. Comput. Graph. Forum **36**(3), 283–294 (2017). https://doi.org/10.1111/cgf.13187
17. Lenhart, W., Liotta, G.: Proximity drawings of outerplanar graphs (extended abstract). In: North, S. (ed.) GD 1996. LNCS, vol. 1190, pp. 286–302. Springer, Heidelberg (1997). https://doi.org/10.1007/3-540-62495-3_55
18. Leskovec, J., Krevl, A.: SNAP Datasets: stanford large network dataset collection. http://snap.stanford.edu/data (Jun 2014)
19. Liotta, G.: Proximity Drawings. chap. 4, pp. 115–154 (2013)
20. van der Maaten, L., Hinton, G.: Visualizing data using t-SNE. J. Mach. Learn. Res. **9**(86), 2579–2605 (2008)
21. Meidiana, A., Hong, S.H., Eades, P.: Shape-faithful graph drawings (2022). https://arxiv.org/abs/2208.14095
22. Noack, A.: An energy model for visual graph clustering. In: Liotta, G. (ed.) GD 2003. LNCS, vol. 2912, pp. 425–436. Springer, Heidelberg (2004). https://doi.org/10.1007/978-3-540-24595-7_40
23. Nocaj, A., Ortmann, M., Brandes, U.: Untangling the hairballs of multi-centered, small-world online social media networks. J. Graph Algorithms Appl. **19**(2), 595–618 (2015). https://doi.org/10.7155/jgaa.00370
24. Preparata, F.P., Shamos, M.I.: Computational geometry: an introduction. Springer Science and Business Media, New York (2012). https://doi.org/10.1007/978-1-4612-1098-6
25. Toth, C.D., O'Rourke, J., Goodman, J.E.: Handbook of Discrete and Computational Geometry. CRC Press, Boca Raton (2017)

26. Toussaint, G.T.: The relative neighbourhood graph of a finite planar set. Pattern Recogn. **12**(4), 261–268 (1980)
27. Walker, J.Q.: A node-positioning algorithm for general trees. Softw. Pract. Experience **20**(7), 685–705 (1990)
28. Wiese, R., Eiglsperger, M., Kaufmann, M.: yFiles - visualization and automatic layout of graphs. In: Jünger, M., Mutzel, P. (eds.) Graph Drawing Software. Mathematics and Visualization, pp. 173–191. Springer, Heidelberg (2004). https://doi.org/10.1007/978-3-642-18638-7_8
29. Zheng, J.X., Pawar, S., Goodman, D.F.: Graph drawing by stochastic gradient descent. IEEE Trans. Visual. Comput. Graphics **25**(9), 2738–2748 (2018)
30. Zitnik, M., Sosič, R., Maheshwari, S., Leskovec, J.: BioSNAP Datasets: stanford biomedical network dataset collection. http://snap.stanford.edu/biodata (2018)

Planar and Orthogonal Drawings

Planar Confluent Orthogonal Drawings of 4-Modal Digraphs

Sabine Cornelsen[ID] and Gregor Diatzko[✉][ID]

University of Konstanz, Konstanz, Germany
{sabine.cornelsen,gregor.diatzko}@uni-konstanz.de

Abstract. In a *planar confluent orthogonal drawing (PCOD)* of a directed graph (digraph) vertices are drawn as points in the plane and edges as orthogonal polylines starting with a vertical segment and ending with a horizontal segment. Edges may overlap in their first or last segment, but must not intersect otherwise. PCODs can be seen as a directed variant of Kandinsky drawings or as planar L-drawings of subdivisions of digraphs. The maximum number of subdivision vertices in an edge is then the *split complexity*. A PCOD is *upward* if each edge is drawn with monotonically increasing y-coordinates and *quasi-upward* if no edge starts with decreasing y-coordinates. We study the split complexity of PCODs and (quasi-)upward PCODs for various classes of graphs.

Keywords: Directed plane graphs · Kandinsky drawings · L-drawings · Curve complexity · Irreducible triangulations · (Quasi-)upward planar

1 Introduction

We consider *plane digraphs*, i.e., planar directed graphs with a fixed planar embedding and a fixed outer face. Directions of edges in node-link diagrams are usually indicated by arrow heads. Since this might cause clutter at vertices with high indegree, Angelini et al. [4] proposed L-drawings in which each edge is drawn with a 1-bend orthogonal polyline starting with a vertical segment at the tail. A plane digraph can only have an L-drawing without crossings if it is *4-modal*, where a plane digraph is *k-modal* if in the cyclic order around a vertex there are at most k pairs of consecutive edges that are neither both incoming nor both outgoing. However, not every 4-modal digraph admits a planar L-drawing. This motivates to extend the model to drawings with more than one bend per edge.

In a *planar confluent orthogonal drawing (PCOD)* of a digraph, vertices are represented as points in the plane with distinct x- and y-coordinates and each edge is represented as an orthogonal polyline starting with a vertical segment at the tail and ending with a horizontal segment at the head. Distinct edges may

S. Cornelsen—The work of Sabine Cornelsen was funded by the German Research Foundation DFG - Project-ID 50974019 - TRR 161 (B06).

P. Angelini and R. von Hanxleden (Eds.): GD 2022, LNCS 13764, pp. 111–126, 2023.
https://doi.org/10.1007/978-3-031-22203-0_9

overlap in a first or last segment, but must not intersect otherwise. For better readability bends have distinct coordinates and are drawn with rounded corners. A plane digraph has a PCOD if and only if it is 4-modal.

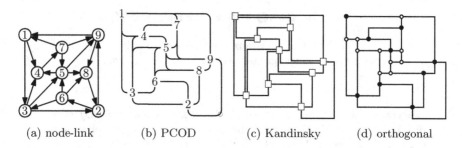

(a) node-link   (b) PCOD   (c) Kandinsky   (d) orthogonal

Fig. 1. Different representations of a 4-modal irreducible triangulation. (Color figure online)

A PCOD of a digraph G corresponds to a planar L-drawing of a subdivision of G. The number of subdivision vertices on an edge is its *split complexity*. See the red encircled vertex in Fig. 1b for the subdivision vertex. Since each edge starts with a vertical segment and ends with a horizontal segment, the number of bends on an edge is odd. An edge with split complexity k has $2k + 1$ bends. The *split complexity* of a PCOD is the maximum split complexity of any edge. The PCOD in Fig. 1b has split complexity one. A *planar L-drawing* [4,17] is a PCOD of split complexity zero. If the embedding is not fixed, then it is NP-complete to decide whether a digraph admits a planar L-drawing [17]. Every 2-modal digraph without 2-cycles has a planar L-drawing [3].

A PCOD of a digraph corresponds to a Kandinsky drawing [21] of the underlying undirected graph with the only difference that edges partially overlap instead of being drawn in parallel with a small gap. See Fig. 1c. While every simple planar graph has a Kandinsky drawing with one bend per edge [16], deciding whether a multigraph has a Kandinsky drawing with one bend per edge [16] or finding the minimum number of bends in a Kandinsky drawing of a plane graph [14] is NP-hard. For the bend-minimization problem in the Kandinsky model there are 2-approximization algorithms [5,20] and heuristics [7].

Among the results for orthogonal drawings of undirected graphs where edges must not overlap, we mention three: With one exception, every plane graph of maximum degree four admits an orthogonal drawing with at most two bends per edge [11]. In a bend-minimum drawing, however, there might have to be an edge with a linear number of bends [27]. An orthogonal drawing with the minimum number of bends can be computed by means of a min-cost flow approach [26] even if an upper bound on the number of bends per edge must be respected.

A PCOD is *upward* if each edge is drawn with monotonically increasing y-coordinates. A digraph is *upward-planar* if and only if it has an upward PCOD. A *plane st-graph*, i.e., a plane acyclic digraph with a single sink and a single source, both on the outer face, is always upward-planar; moreover, it has an upward-planar L-drawing if and only if it admits a so-called bitonic st-ordering [17]. Since it suffices to subdivide the edges of a plane st-graph at most once in order to obtain a digraph that admits a bitonic st-ordering [1,22], it follows that every plane st-graph admits an upward PCOD with split complexity one. Moreover, the minimum number of bends in an upward PCOD of a plane st-graph can be determined in linear time. In general, a digraph admits an upward-planar L-drawing, if and only if it is a subgraph of a plane st-graph admitting a bitonic st-ordering [2]. Not every 2-modal tree admits an upward-planar L-drawing [2].

In a *quasi-upward-planar drawing* [8] edges must be strictly monotonically increasing in y-direction in a small vicinity around the end vertices. A digraph has a 2-modal embedding if and only if it admits a quasi-upward-planar drawing. Every 2-modal graph without 2-cycles admits a quasi-upward planar drawing with at most two bends per edge and the curve complexity in such drawings can be minimized utilizing a min-cost flow approach [13]. We call a PCOD *quasi-upward* if no edge starts with decreasing y-coordinates.

Our Contribution. We show that PCODs of 4-modal trees have split complexity zero (Theorem 2), split complexity two is sufficient (Theorem 4) and sometimes necessary (Theorem 3) for PCODs of 4-modal digraphs with parallel edges or loops, while split complexity one suffices for 4-modal *irreducible triangulations* (Theorem 5), i.e., internally triangulated 4-connected graphs with an outer face of degree 4. Split complexity one also suffices for upward PCODs of upward-plane digraphs (Theorem 6) and for quasi-upward PCODs of 2-modal digraphs without 2-cycles (Theorem 7). Using an ILP, we conducted experiments that suggest that every simple 4-modal digraph without separating 2-cycles admits a PCOD with split complexity one (Sect. 8). Constant split complexity is not to be expected for bend-minimum PCODs (Theorem 1).

2   Preliminaries

Two consecutive incident edges of a vertex v are a *switch* if both edges are incoming or both outgoing edges of v. The drawing of a PCOD is determined by the coordinates of the vertices and the coordinates of every second bend of an edge. We call a bend *independent* if it is the second, fourth, etc. bend of an edge. Considering a PCOD as an L-drawing of a subdivision, the independent bends correspond to the subdivision vertices. The split complexity of an edge is the number of its independent bends. The total number of bends equals the number of edges plus twice the number of independent bends. The top, left, bottom, and right side of a vertex is its *North, West, South,* and *East port*, respectively.

An *st-ordering* of a biconnected (undirected) graph $G = (V, E)$ is a bijection $\pi : V \to \{1, \dots, |V|\}$ such that $\pi(s) = 1$, $\pi(t) = |V|$, and each vertex $v \in V \setminus \{s, t\}$ has neighbors u and w with $\pi(u) < \pi(v) < \pi(w)$. Let now $G = (V, E)$ be a plane st-graph. If (v, v_i), $i = 1, \dots, k$ are the outgoing edges of a vertex v from left to right then $S(v) = \langle v_1, \dots, v_k \rangle$ is the *successor list* of v. A *bitonic st-ordering* of

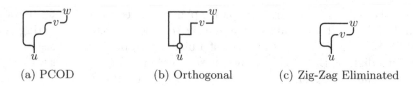

(a) PCOD   (b) Orthogonal   (c) Zig-Zag Eliminated

Fig. 2. Eliminating zig-zags.

G is a bijection $\pi : V \rightarrow \{1, \ldots, |V|\}$ such that $\pi(u) < \pi(v)$ for $(u, v) \in E$ and $S(v) = \langle v_1, \ldots, v_k \rangle$ is *bitonic* for each vertex v, i.e., there is a $1 \leq h \leq k$ such that $\pi(v_i) < \pi(v_{i+1})$, $i = 1, \ldots, h - 1$ and $\pi(v_i) > \pi(v_{i+1})$, $i = h, \ldots, k - 1$. The successor list $S(v) = \langle v_1, \ldots, v_k \rangle$ contains a *valley* with transitive edges $(v, v_{\ell-1})$ and (v, v_{r+1}) if there is a directed v_ℓ-$v_{\ell-1}$-path and a directed v_r-v_{r+1}-path for some $1 < \ell \leq r < k$. A plane st-graph admits a bitonic st-ordering if and only if it does not contain a valley [22].

3   Confluent Orthogonal Representation

Let Γ be a PCOD of a plane digraph G. We call a bend *covered* if it is contained in the drawing of another edge. We associate an orthogonal drawing of a plane graph G_Γ with Γ as follows [3]: Replace every covered bend in Γ by a dummy vertex. See Figs. 1d and 2b. A *zig-zag* is a pair of uncovered bends on an edge, one with a left turn, and one with a right turn. E.g., on the edge (u, v) in Fig. 2a there is a zig-zag, while on the edge (u, w) there is both a left and a right turn, but the left turn is covered, so there is no zig-zag. Since the number of bends in an orthogonal drawing can always be reduced by eliminating zig-zags, we will also do so in PCODs (see Fig. 2c) and, thus, the ordering of left- and right-turns at uncovered bends of an edge will not matter. Since planar (confluent) orthogonal drawings can be stretched independently in x- and y-directions, it is algorithmically often easier not to work with actual x- and y-coordinates, but rather with the shape of the faces in terms of bends on the edges and angles at the vertices. See also [21,26].

A *confluent orthogonal representation* R of a plane digraph $G = (V, E)$ is a set of circular lists $H(f)$, one for each face f of G. The elements of $H(f)$ are tuples $r = (e, v, a, s, b)$ associated with the edges e incident to f in counter-clockwise order. (a) v is the end vertex of e traversed immediately before e. (b) $a \in \{0, \frac{\pi}{2}, \pi, \frac{3\pi}{2}, 2\pi\}$ is the angle at v between e and its predecessor on f. It is a multiple of π if and only if it describes an angle at a switch. (c) s is the number of left turns (when traversing e starting from v) at bends in e. (d) $b \in \{L, N, R\}$ represents a covered bend on the segment of e incident to v, if any, with (L) a left bend, (R) a right bend, or (N) no such bend.

Let r_p be the predecessor of r in $H(f)$. If r is not clear from the context, we denote the entries by $e[r]$, $v[r]$, $a[r]$, $s[r]$, $b[r]$. Each edge is contained twice in a confluent orthogonal representation. Let $\bar{r}$ be the (other) entry containing $e[r]$. A confluent orthogonal representation is *feasible* if it fulfills the following.

(i) The rotation $\sum_{r\in H(f)}(2 - a[r]/\frac{\pi}{2} + s[r] - s[\overline{r}])$ of a face f is -4 if f is the outer face and 4 otherwise. (ii) The angular sum $\sum_{r;v[r]=v} a[r]$ around a vertex v is 2π. (iii) If $b[r] = L$ or $b[\overline{r}] = R$ then $s[r] \geq 1$ and if both $b[r] = L$ and $b[\overline{r}] = R$ then $s[r] \geq 2$. This ensures that covered bends are counted by s and that covered bends adjacent to the head or the tail of an edge must be distinct. (iv) The so called *bend-or-end property*, i.e., if $a[r] = 0$, then $b[r] = R$ or $b[\overline{r_p}] = L$. (v) The total number of bends $s[r] + s[\overline{r}]$ on $e[r]$ is odd.

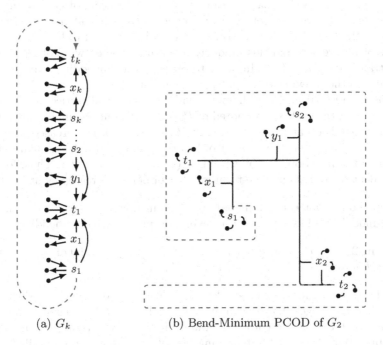

(a) G_k (b) Bend-Minimum PCOD of G_2

Fig. 3. Graphs with a linear number of bends in any bend-minimum PCOD. (Color figure online)

From a Representation to a PCOD. In order to construct a PCOD from a feasible confluent orthogonal representation R of a plane digraph G, we transform G into a graph G_R of maximum degree 4 and a feasible orthogonal representation R' without 0 or 2π angles. Using compaction for orthogonal representations [26] on G_R then yields a PCOD or a $\pi/2$-rotation of a PCOD in linear time. The idea for the construction of G_R is analogous to the construction of G_Γ from a PCOD Γ and is as follows: Consider a vertex $v \in V$ and let $e_1,\ldots,e_k$ be a maximum sequence of consecutive edges around v with 0 angles. Let r_i, $i = 1,\ldots,k$ be the entry with $e[r_i] = e_i$ and $v[r_i] = v$. Due to Property iv, there is a $1 \leq m \leq k$ such that $b[r_j] = L$, $j < m$ and $b[r_j] = R$, $m < j$. We subdivide the segment of e_m that is incident to v with $k - 1$ vertices $v_1,\ldots,v_{m-1}, v_k,\ldots,v_{m+1}$ in this order, starting from v. We attach $e_j, j \neq m$ to v_j instead of v. The representation R' is updated accordingly.

4   Some Initial Results

Theorem 1. *There is a family G_k, $k > 0$ of 4-modal digraphs with $14k - 3$ vertices and $16k - 4$ edges such that in any bend-minimum PCOD of G_k there is an edge with split complexity at least $k + 2$.*

Proof. Consider the digraphs G_k indicated in Fig. 3a. Let e be the red dashed edge. Let P_k be the path $s_1, x_1, t_1, y_1, \ldots, s_k, x_k, t_k$ of length $4k - 2$ in G_k that is drawn vertically in Fig. 3a. Consider a planar L-drawing of $G_k - e$ in which all edges of P_k (traversed from s_1 to t_k) bend to the left and the edge incident to s_1 is to the top of s_1. Such a drawing for G_2 is indicated in Fig. 3b. Since all vertices of P_k are 4-modal this uniquely determines the drawing of P_k and also of the transitive edges of P_k. In order to preserve the embedding, e can only be inserted into the drawing with split complexity at least $k + 2$.

Consider a PCOD of G_k with fewer bends on e. Since all vertices are 4-modal, the rotation of the cycle C composed of P_k and e can only be maintained, if the number of bends on at least one edge of P_k, say (s_i, x_i), is increased. But then we also must increment the number of bends on an edge (s_i, t_i) to maintain the rotation of the face bounded by the edges $(s_i, t_i), (s_i, x_i), (x_i, t_i)$. Thus, for each independent bend less on e the total number of bends increases by at least 2. □

Even though not every 2-modal tree has an upward-planar L-drawing [2], every 4-modal tree has a planar L-drawing, despite its fixed embedding.

Theorem 2. *Every 4-modal tree has a PCOD with split complexity zero. Moreover, such a drawing can be constructed in linear time.*

Proof. Let T be a 4-modal tree and let v be a leaf of T. We show by induction on the number m of edges that we can draw T as a PCOD Γ^α with split complexity zero such that v is in the corner α (lower left $(\ell\ell)$, lower right (ℓr), upper left $(u\ell)$, upper right (ur)) of the bounding box of Γ^α. We give the details for $\Gamma^{\ell\ell}$; the other cases are analogous. If $m = 1$, draw v at $(0,0)$ and its neighbor at $(1,1)$.

If $m > 1$, let the neighbor of v be v', and let the connected components of $T - v'$ be $v, T_1, \ldots, T_k$ in clockwise order around v'. See Fig. 4. Let T_0 be the subtree consisting of v only. Each tree $T_i + v'$, $i = 0, \ldots, k$ has at most $m - 1$ edges and the leaf v'; therefore, by the inductive hypothesis, we can construct PCODs Γ_i^α, $\alpha \in \{\ell\ell, \ell r, u\ell, ur\}$ of $T_i + v'$ with v' in the respective corner of the bounding box. W.l.o.g. let v be the tail of the edge connecting v and v', see Fig. 4a. Let $1 \leq a \leq b \leq c \leq d \leq k$ such that $T_{d+1}, \ldots, T_k, T_0, T_1, \ldots, T_a$ and $T_{b+1}, \ldots, T_c$ are connected to v' by an incoming edge and $T_{a+1}, \ldots, T_b$ and $T_{c+1}, \ldots, T_d$ by outgoing edges. Choose $\Gamma_0^{ur}, \Gamma_1^{\ell r}, \ldots, \Gamma_a^{\ell r}, \Gamma_{a+1}^{\ell\ell}, \ldots, \Gamma_c^{\ell\ell}, \Gamma_{c+1}^{ur}, \ldots, \Gamma_k^{ur}$ for $T_i + v'$, $i = 1, \ldots, k$. Finally, merge the drawings of the subtrees at v'.

In order to compute a confluent orthogonal representation, using dynamic programming, only $O(\deg(v'))$ steps are required for each vertex v'. Thus, the total time complexity is linear. □

5   Multi-Graphs

Theorem 3. *There are 4-modal multigraphs that need split complexity at least two in any PCOD.*

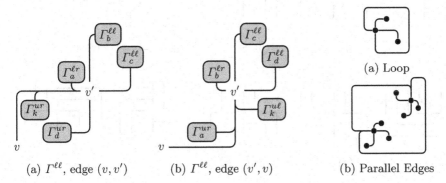

(a) $\Gamma^{\ell\ell}$, edge (v, v') (b) $\Gamma^{\ell\ell}$, edge (v', v)

Fig. 4. Constructing a tree from its subtrees.

(a) Loop

(b) Parallel Edges

Fig. 5. Multigraphs with split complexity two.

Proof. Consider the digraph containing a loop in Fig. 5a or the digraph containing two parallel edges in Fig. 5b. The incident 4-modal vertices on the one hand and the rotation of the outer face on the other hand, imply that the loop and one of the two parallel edges, respectively, must have split complexity two. In the case of the loop, the angle at the vertex is convex. Since the rotation of the outer face is -4, it follows that five concave bends on the loop are needed. In the case of two parallel edges, the angles at the vertices in the outer face are zero. Thus, the two edges together must have eight concave bends. Since each edge has an odd number of bends, there must be an edge with five bends. □

Theorem 4. *Every 4-modal multigraph has a PCOD with split complexity at most two. Moreover, such a drawing can be computed in linear time.*

Proof. The approach is inspired by [11]. Subdivide each loop. Let the resulting digraph be G. Then make the digraph biconnected maintaining its 4-modality [3]. Now compute in linear time [15] an *st*-ordering $v_1, \ldots, v_n$ of this biconnected graph G' (without taking into account the direction of the edges). Iteratively add the vertices with increasing y-coordinates in the order of the *st*-ordering, maintaining a column for each edge that has exactly one end vertex drawn.

Let v_k be a vertex. An edge e incident to v_k is incident to v_k *from below* if e has an end vertex that is before v_k in the *st*-ordering. Let $e_1, \ldots, e_j$ be the sequence of edges incident to v_k from below as they appear from left to right. Since v_k is 4-modal, $e_1, \ldots, e_j$ can be divided into at most five subsequences of edges consisting only of incoming $(-)$ or only of outgoing $(+)$ edges of v_k. Depending on the arrangement of these subsequences, we assign the bends around v_k. E.g., consider the Case $+-$ in Fig. 6b, i.e., among the edges incident to v_k from below,

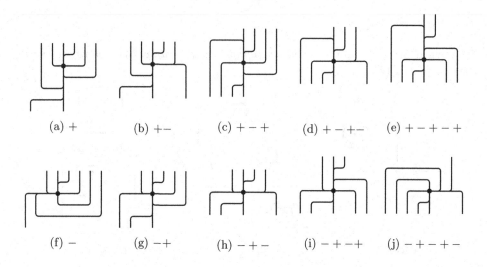

Fig. 6. Drawings around v_k in the proof of Theorem 4. A + represents outgoing edges from below. A − represents incoming edges from below.

there are first some outgoing edges, followed by some incoming edges. All outgoing edges from below are attached to the South port, while all incoming edges from below are attached to the East port. Mind that all outgoing edges except one need two bends near v_k. Consider now the edges incident to v_k to later vertices in the st-ordering. By 4-modality, there can be at most some incoming edges, followed by some outgoing, some incoming and again some outgoing edges in counter-clockwise order around v_k. We attach them to the East, North, West, and South port of v_k, respectively. The edges from below determine the position of v_k. In our case, v_k is drawn above one of the edges attached to its South port.

See Fig. 6 for the routing of the edges from below and the possible edges to later end vertices in the other cases. For v_1, we choose the assignment according to Fig. 6a or Fig. 6f. After all vertices are placed, we remove edges that are in G' but not in G. If x is a vertex that was inserted into a loop, we reroute the two incident edges near x such that the incoming edge of x has exactly one bend near x and the outgoing edge has no bend near x. Finally, we eliminate zig-zags.

By the st-ordering, the columns of the edges incident to v_k from below are consecutive among the edges with exactly one end vertex drawn [11]. This implies planarity if the columns for the new edges are inserted directly next to v_k. For each edge e, there are at most two bends near the tail of e and at most three bends near the head of e. Consider now a 2-cycle $(v, x), (x, v)$ replacing a loop at v. Since the subdivision vertex x is incident to exactly one incoming and one outgoing edge, it follows that near x there is no bend on (x, v) and one bend on (v, x). If (x, v) does not have three bends near v then in total there are at most six bends on the loop, namely the four bends near v plus one bend near x plus the bend on x. Since the number of bends on an edge must be odd, there are only five. Consider now the case that (x, v) has three bends near v (Figs. 6f and 6j). If in addition (v, x) has two bends near v, then there are seven bends on the loop. However, in this case, there is a zig-zag on (v, x) formed by the bend near x and the second bend near v. Thus, after eliminating zig-zags, the split complexity is at most two.   □

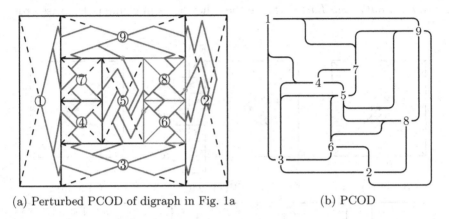

(a) Perturbed PCOD of digraph in Fig. 1a   (b) PCOD

Fig. 7. Perturbed PCOD and corresponding PCOD after zig-zag elimination. Red encircled bends are due to the change of the coordinate system and not real. (Color figure online)

6   Irreducible Triangulations

We prove that every 4-modal digraph whose underlying undirected graph is an irreducible triangulation has a PCOD with split complexity at most one.

Motivated by the approaches in [3,12], we use *rectangular duals*, a contact representation of an irreducible triangulation $G = (V, E)$ with the following properties. The vertices $v \in V$ are represented by internally disjoint axis-parallel rectangles $R(v)$. Two rectangles touch if and only if the respective vertices are adjacent in G. Moreover, no four rectangles representing a vertex meet at the same point and $\bigcup_{v \in V} R(v)$ is a rectangle. See the rectangles in Fig. 7a. A rectangular dual for an irreducible triangulation can be computed in linear time [9,10,23,24].

Given a rectangular dual, we perturb the coordinate system such that in each rectangle the axes correspond to the diagonals. The *perturbed x-axis* is the diagonal containing the bottommost-leftmost point of the rectangle, the other diagonal is the *perturbed y-axis*. A *perturbed orthogonal polyline* is a polyline such that in each rectangle the segments are parallel to one of the axes. A bend of a perturbed orthogonal polyline at the boundary of two rectangles is a *real bend* if among the two incident segments one is parallel to a perturbed x-axis and the other parallel to a perturbed y-axis. Bends inside a rectangle are always real. In a *perturbed PCOD* each vertex v is drawn at the center of $R(v)$. An edge (u, v)

is a perturbed orthogonal polyline in $R(u) \cup R(v)$ between u and v starting with a segment on the perturbed y-axis in $R(u)$ and ending with a segment on the perturbed x-axis in $R(v)$. The drawing of (u, v) must have at least one bend in the interior of both $R(u)$ and $R(v)$ and must cross the boundary of $R(u)$ and $R(v)$ exactly once. Distinct edges may overlap in a first or last segment, but must not intersect otherwise. No two bends have the same coordinates. See Fig. 7a. The *North port* of v is the port above and to the left of the center of R_v. The *West, South,* and *East ports* are the other ports in counter-clockwise order.

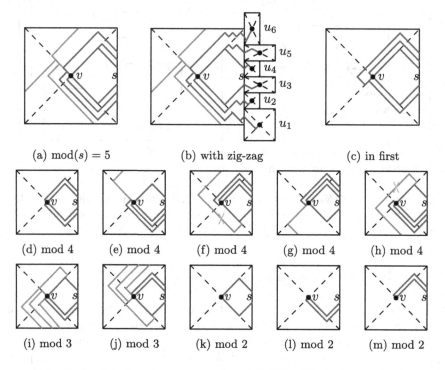

(a) $\mathrm{mod}(s) = 5$ (b) with zig-zag   (c) in first

(d) mod 4   (e) mod 4   (f) mod 4   (g) mod 4   (h) mod 4

(i) mod 3   (j) mod 3   (k) mod 2   (l) mod 2   (m) mod 2

Fig. 8. Routing the edges in a perturbed PCOD. (Color figure online)

In analogy to the arguments in [3], we obtain that a perturbed PCOD yields a confluent orthogonal representation where the number s of left turns counts only real bends. By the next theorem, we can derive a PCOD with split complexity one from a suitable perturbed PCOD after zig-zag elimination. See Fig. 7b.

Theorem 5. *Every 4-modal irreducible triangulation has a PCOD with split complexity at most one; and such a drawing can be computed in linear time.*

Proof. Let G be an irreducible triangulation. We construct a rectangular dual for G ignoring edge directions. Routing the edges inside any rectangle independently, we then construct a perturbed PCOD that yields a confluent orthogonal representation with split complexity at most one after zig-zag elimination.

Let v be a vertex of G. For a side s of $R(v)$ let u_i, $i = 1, \ldots, k$ be the adjacent vertices of v in counter-clockwise order such that s and $R(u_i)$ intersect in more than a point. Let e_i be the edge between v and u_i. Consider the division of $\langle e_1, \ldots, e_k \rangle$ into *mono-directed classes*, i.e., maximal subsequences such that any two edges in a subsequence are either both incoming or both outgoing edges. Let the modality $mod(s)$ of s be the number of these subsequences. Since G is 4-modal we have $mod(s) \leq 5$. Assume now that s is a side of $R(v)$ with maximal modality. Assume without loss of generality that s is the right side of $R(v)$.

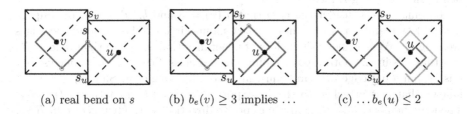

(a) real bend on s (b) $b_e(v) \geq 3$ implies ...   (c) ...$b_e(u) \leq 2$

Fig. 9. Eliminating zig-zags to reduce the number of bends per edge to three. (Color figure online)

$mod(s) = 5$. If e_1 is an outgoing edge of v, assign the mono-directed classes of edges crossing s from bottom to top in this order to (i) the North port bending three times to the left, (ii) to the West port bending twice to the left, (iii) to the South port bending once to the left, (iv) to the East port bending once to the right, and (v) to the North port bending twice to the right. Route the edges as indicated in blue in Fig. 8a to s. By adding zig-zags, it is always possible to route an edge e_i between v and u_i in such a way that the parts of e_i in $R(v)$ and $R(u_i)$ meet in s. See Fig. 8b. Edges crossing other sides of R_v are all outgoing edges of v and are assigned to the North port, bending once or twice in the direction of the side where they leave R_v. See the purple edges in Fig. 8a. If e_1 is an incoming edge, start analogously with the West port. See Fig. 8c.

$mod(s) \in \{1, \ldots, 4\}$. The assignment of edges to ports and the routing of the edges are contained in the drawing of the case $mod(s) = 5$. See the blue edges in the second and third row in Fig. 8. We make again sure that an edge to a side of $R(v)$ with modality one has at most two bends in the interior of $R(v)$. In order to do so, we have to take special care if $mod(s) = 4$ and the bottommost edge is an outgoing edge of v. Let s_t and s_b be the top and bottom side, respectively, of $R(v)$. If v is incident to an outgoing edge crossing s_t, we opt for the variant in Fig. 8e and otherwise (including the case that v is incident to an incoming edge crossing s_b) for the variant in Fig. 8g. No particular care has to be taken if $mod(s) = 2$ (Figs. 8l and 8m).

Let now e be an edge between two vertices u and v. We consider the number $b_e(u)$ and $b_e(v)$ of bends on e in $R(u)$ and $R(v)$, respectively, after eliminating zig-zags. We assume without loss of generality that $b_e(u) \leq b_e(v)$. Recall that then $1 \leq b_e(u) \leq b_e(v) \leq 3$. Let s_u and s_v be the sides of $R(u)$ and $R(v)$, respectively that contain the intersection s of $R(u)$ and $R(v)$.

We have to show that up to zig-zags there are in total at most three bends on e. This is clear if $b_e(u) = b_e(v) = 1$. Assume now that $b_e(v) \geq 2$. Since the number of real bends on e is odd it follows that the bend on s is real if and only if $b_e(u) + b_e(v)$ is even. In this case the bend on s bends in opposite direction as the next bend in $R(u)$ and $R(v)$ (otherwise e does not cross s). Since $b_e(v) \geq 2$, the real bend of e on s and the next bend of e in $R(v)$ form a zig-zag and can be eliminated. See Fig. 9a. Thus, if $b_e(u) + b_e(v) \leq 4$ then there are at most three bends on e after zig-zag elimination.

It remains to consider the case that $b_e(v) = 3$ and $b_e(u) \geq 2$. This implies $\mathrm{mod}(s_v) > 1$ and e is in the first or last mono-directed class among the edges crossing s_v. We assume without loss of generality that s_v is the right side of R_v and that e is in the bottommost mono-directed class. See Fig. 9b. It follows that e is an outgoing edge of v and thus, an incoming edge of u. Since $b_e(u) \geq 2$ it follows that e is attached to the East port of u. Assume first that $b_e(u) = 2$. Then the bends of e in $R(u)$ are in opposite direction as the bends of e in $R(v)$. Thus, there is at least one zig-zag consisting of a bend in $R(u)$ and a bend in $R(v)$. After eliminating this zig-zag there are only three bends left.

Assume now that $b_e(u) = 3$. This is only possible if $\mathrm{mod}(s_u) \geq 2$. Hence, $R(u)$ is the topmost or bottommost rectangle incident to the right of $R(v)$. Since e is in the bottommost class with respect to s_v, it must be the bottommost one. Thus, $R(v)$ is the topmost neighbor to the left of $R(u)$. Moreover, since $\mathrm{mod}(s_u) \geq 2$ there must be a port of u other than the East port that contains an edge e' (red edge in Fig. 9c) crossing s_u. But e' would have to bend at least four times in the interior of $R(u)$, which never happens according to our construction.   □

7   (Quasi-)Upward-Planar Drawings

Theorem 6. *Every upward-plane digraph admits an upward PCOD with split complexity at most one. Moreover, for plane st-graphs both the split complexity and the total number of bends can be minimized simultaneously in linear time.*

Proof. Let G be an upward-plane digraph. Then G can be augmented to a plane st-graph by adding edges [6]. Subdividing each edge once yields a plane st-graph with a bitonic st-ordering [1] and, thus, with an upward-planar L-drawing [17]. This corresponds to an upward PCOD of split complexity one for G.

If G is a plane st-graph it can be decided in linear time whether G has an upward-planar L-drawing [17], and thus, an upward PCOD of split complexity zero. Otherwise, the minimum number of edges that has to be subdivided in order to obtain a digraph that has a bitonic st-ordering can be computed in linear time [1]. Thus, a PCOD with the minimum number of bends among all upward PCODs of G with split complexity one can be computed in linear time. Observe that the total number of bends cannot be reduced by increasing the split complexity, since the subdivision of edges is only performed in order to break one of the transitive edges in a valley.   □

Theorem 7. *Every 2-modal digraph without 2-cycles admits a quasi-upward PCOD with split complexity at most one. Moreover, such a drawing can be computed in linear time.*

Proof. Let G be a 2-modal graph without 2-cycles. G has a planar L-drawing [3], say Γ. Process the vertices v of G top-down in Γ. If there are edges attached to the South port of v, we reroute them such that they are attached to the North port. Consider first the case that at least one among the East or the West port – say the East port – of v does not contain edges. Then we can reroute the edges as indicated in Fig. 10b. Edges attached to the South port of v in Γ are now attached to the North port of v and get two additional bends near their tail.

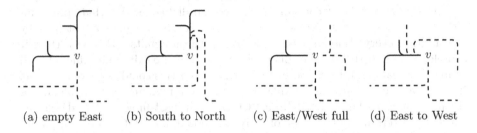

(a) empty East   (b) South to North   (c) East/West full   (d) East to West

Fig. 10. From a planar L-drawing of a 2-modal graph to a quasi-upward PCOD.

Assume now that both the East and the West port of v contain edges. Then, by 2-modality, no edge is attached to the North port of v. Those edges incident to the East port that bend upward are reattached to the West port without adding any additional bend. This is possible since these edges have already been rerouted near their other end vertex and two new bends have been inserted. So there are enough bends for the bend-or-end property. The rotation of the faces and the angular sum around the vertices are also maintained. The edges incident to the East port bending downward are rerouted from their original drawing to the West port with two new bends. See Fig. 10d. Now we can reroute the edges attached to the South port as in the first case.

An edge attached to the South port in Γ gets at most two new bends near its tail; an edge attached to the North port at most two new bends near its head. Thus, in the end each edge has at most three bends, i.e., split complexity 1. □

8   Experiments Using an ILP

Based on the definition of confluent orthogonal representations and the fact that each 4-modal multigraph has a PCOD with split complexity at most two, we developed an ILP to compute PCODs with minimum split complexity for 4-modal graphs. See [19] for details. Since each simple 4-modal graph without 2-cycles can be extended to a triangulated 4-modal graph [3], we first sampled several thousand upward-planar triangulations for various numbers $n \leq 500$ of vertices with two different methods: sampling (a) undirected triangulations uniformly at random [25] orienting the edges according to an st-ordering [15] and (b) with an OGDF method [18]. Then we flipped the direction of each edge with probability 0.5 maintaining 4-modality. Finally, we added as many 2-cycles as 4-modality allowed. The resulting digraphs contained $(\frac{3}{4} \pm \frac{1}{4})n$ separating triangles, roughly n 2-cycles, but no separating 2-cycles. All digraphs had split complexity one.

9 Conclusion and Future Work

We examined the split complexity of PCODs of various graph classes. In particular, we have shown that every 4-modal digraph admits a PCOD with split complexity two even if it contains loops and parallel edges and that split complexity two is sometimes necessary. For simple digraphs, we made a first step, by proving that every 4-modal irreducible triangulation admits a PCOD with split complexity one. It still remains open whether split complexity one suffices for all simple 4-modal digraphs. Experiments suggest that this could very well be true. It would also be interesting to know whether the minimum split complexity or the minimum number of bends in a PCOD or a (quasi-)upward PCOD can be efficiently determined in the case of a given 4-modal, 2-modal, or upward-planar embedding, respectively, as well as in the case when no embedding is given.

References

1. Angelini, P., Bekos, M.A., Förster, H., Gronemann, M.: Bitonic st-orderings for upward planar graphs: the variable embedding setting. In: Adler, I., Müller, H. (eds.) WG 2020. LNCS, vol. 12301, pp. 339–351. Springer (2020). https://doi.org/10.1007/978-3-030-60440-0_27
2. Angelini, P., Chaplick, S., Cornelsen, S., Da Lozzo, G.: On upward-planar L-drawings of graphs. In: Szeider, S., Ganian, R., Silva, A. (eds.) MFCS 2022. LIPIcs, vol. 241, pp. 1–15. Schloss Dagstuhl - Leibniz-Zentrum für Informatik (2022). https://doi.org/10.4230/LIPIcs.MFCS.2022.10
3. Angelini, P., Chaplick, S., Cornelsen, S., Da Lozzo, G.: Planar L-drawings of bimodal graphs. JGAA 26(3), 307–334 (2022). https://doi.org/10.7155/jgaa.00596
4. Angelini, P., et al.: Algorithms and bounds for L-drawings of directed graphs. Int. J. Found. Comput. Sci. 29(4), 461–480 (2018). https://doi.org/10.1142/S0129054118410010
5. Barth, W., Mutzel, P., Yıldız, C.: A new approximation algorithm for bend minimization in the Kandinsky model. In: Kaufmann, M., Wagner, D. (eds.) GD 2006. LNCS, vol. 4372, pp. 343–354. Springer (2007). https://doi.org/10.1007/978-3-540-70904-6_33
6. Battista, G.D., Tamassia, R.: Algorithms for plane representations of acyclic digraphs. Theor. Comput. Sci. 61, 175–198 (1988). https://doi.org/10.1016/0304-3975(88)90123-5
7. Bekos, M.A., Kaufmann, M., Krug, R., Siebenhaller, M.: The effect of almost-empty faces on planar Kandinsky drawings. In: Bampis, E. (ed.) SEA 15. LNCS, vol. 9125, pp. 352–364. Springer (2015). https://doi.org/10.1007/978-3-319-20086-6_27

8. Bertolazzi, P., Battista, G.D., Didimo, W.: Quasi-upward planarity. Algorithmica **32**(3), 474–506 (2002). https://doi.org/10.1007/s00453-001-0083-x
9. Bhasker, J., Sahni, S.: A linear algorithm to find a rectangular dual of a planar triangulated graph. Algorithmica **3**, 247–278 (1988). https://doi.org/10.1007/BF01762117
10. Biedl, T., Derka, M.: The (3,1)-ordering for 4-connected planar triangulations. JGAA **20**(2), 347–362 (2016). https://doi.org/10.7155/jgaa.00396
11. Biedl, T., Kant, G.: A better heuristic for orthogonal graph drawings. Comp. Geom. **9**(3), 159–180 (1998). https://doi.org/10.1016/S0925-7721(97)00026-6
12. Biedl, T., Mondal, D.: A note on plus-contacts, rectangular duals, and box-orthogonal drawings. Tech. Rep. arXiv:1708.09560v1. Cornell University Library (2017). https://doi.org/10.48550/arXiv.1708.09560
13. Binucci, C., Giacomo, E.D., Liotta, G., Tappini, A.: Quasi-upward planar drawings with minimum curve complexity. In: Purchase, H.C., Rutter, I. (eds.) GD 2021. LNCS, vol. 12868, pp. 195–209. Springer (2021). https://doi.org/10.1007/978-3-030-92931-2_14
14. Bläsius, T., Brückner, G., Rutter, I.: Complexity of higher-degree orthogonal graph embedding in the Kandinsky model. In: Schulz, A.S., Wagner, D. (eds.) ESA 2014. LNCS, vol. 8737, pp. 161–172. Springer (2014). https://doi.org/10.1007/978-3-662-44777-2_14
15. Brandes, U.: Eager *st*-ordering. In: Möhring, R., Raman, R. (eds.) ESA 2002. LNCS, vol. 2461, pp. 247–256. Springer (2002). https://doi.org/10.1007/3-540-45749-6_25
16. Brückner, G.: Higher-degree orthogonal graph drawing with flexibility constraints. Bachelor thesis, Department of Informatics, KIT (2013). https://i11www.iti.kit.edu/_media/teaching/theses/ba-brueckner-13.pdf
17. Chaplick, S., et al.: Planar L-drawings of directed graphs. In: Frati, F., Ma, K.L. (eds.) GD 2017. LNCS, vol. 10692, pp. 465–478. Springer (2018). https://doi.org/10.1007/978-3-319-73915-1_36
18. Chimani, M., Gutwenger, C., Jünger, M., Klau, G.W., Klein, K., Mutzel, P.: The open graph drawing framework (OGDF). In: Tamassia, R. (ed.) Handbook on Graph Drawing and Visualization, chap. 17, pp. 543–569. Chapman and Hall/CRC (2013). https://cs.brown.edu/people/rtamassi/gdhandbook/
19. Cornelsen, S., Diatzko, G.: Planar confluent orthogonal drawings of 4-modal digraphs. Tech. Rep. arXiv:2208.13446. Cornell University Library (2022). https://doi.org/10.48550/arXiv.2208.13446
20. Eiglsperger, M.: Automatic Layout of UML Class Diagrams: A Topology-Shape-Metrics Approach. Ph.D. thesis, Eberhard-Karls-Universität zu Tübingen (2003), https://publikationen.uni-tuebingen.de/xmlui/handle/10900/48535
21. Fößmeier, U., Kaufmann, M.: Drawing high degree graphs with low bend numbers. In: Brandenburg, F.J. (ed.) GD 1995. LNCS, vol. 1027, pp. 254–266. Springer, Heidelberg (1996). https://doi.org/10.1007/BFb0021809
22. Gronemann, M.: Bitonic *st*-orderings for upward planar graphs. In: Hu, Y., Nöllenburg, M. (eds.) GD 2016. LNCS, vol. 9801, pp. 222–235. Springer (2016). https://doi.org/10.1007/978-3-319-50106-2_18
23. He, X.: On finding the rectangular duals of planar triangular graphs. SIAM J. Comput. **22**(6), 1218–1226 (1993). https://doi.org/10.1137/0222072
24. Kant, G., He, X.: Two algorithms for finding rectangular duals of planar graphs. In: van Leeuwen, J. (ed.) WG 1993. LNCS, vol. 790, pp. 396–410. Springer (1994). https://doi.org/10.1007/3-540-57899-4_69

25. Poulalhon, D., Schaeffer, G.: Optimal coding and sampling of triangulations. Algorithmica **46**(3), 505–527 (2006). https://doi.org/10.1007/s00453-006-0114-8

26. Tamassia, R.: On embedding a graph in the grid with the minimum number of bends. SIAM J. Computing **16**(3), 421–444 (1987). https://doi.org/10.1137/0216030

27. Tamassia, R., Tollis, I.G., Vitter, J.S.: Lower bounds for planar orthogonal drawings of graphs. Inf. Process. Lett. **39**(1), 35–40 (1991). https://doi.org/10.1016/0020-0190(91)90059-Q

Unit-length Rectangular Drawings of Graphs

Carlos Alegría[iD], Giordano Da Lozzo[(✉)][iD], Giuseppe Di Battista[iD],
Fabrizio Frati[iD], Fabrizio Grosso[iD], and Maurizio Patrignani[iD]

Department of Engineering, Roma Tre University, Rome, Italy
{carlos.alegria,giordano.dalozzo,giuseppe.dibattista,fabrizio.frati,
fabrizio.grosso,maurizio.patrignani}@uniroma3.it

Abstract. A *rectangular drawing* of a planar graph G is a planar drawing of G in which vertices are mapped to grid points, edges are mapped to horizontal and vertical straight-line segments, and faces are drawn as rectangles. Sometimes this latter constraint is relaxed for the outer face. In this paper, we study rectangular drawings in which the edges have unit length. We show a complexity dichotomy for the problem of deciding the existence of a unit-length rectangular drawing, depending on whether the outer face must also be drawn as a rectangle or not. Specifically, we prove that the problem is NP-complete for biconnected graphs when the drawing of the outer face is not required to be a rectangle, even if the sought drawing must respect a given planar embedding, whereas it is polynomial-time solvable, both in the fixed and the variable embedding settings, if the outer face is required to be drawn as a rectangle.

Keywords: Rectangular drawings · Rectilinear drawings · Matchstick graphs · Grid graphs · SPQR-trees · Planarity

1 Introduction

Among the most celebrated aesthetic criteria in Graph Drawing we have: (i) planarity, (ii) orthogonality of the edges, (iii) unit length of the edges, and (iv) convexity of the faces. We focus on drawings in which all the above aesthetics are pursued at once. Namely, we study orthogonal drawings where the edges have length one and the faces are rectangular.

Throughout the paper, any considered graph drawing has the vertices mapped at *distinct* points of the plane. Orthogonal representations are a classic research topic in graph drawing. A rich body of literature is devoted to orthogonal drawings of planar [16,21,25,50] and plane [14,40,41,45,46] graphs with the minimum number of bends in total or per edge [10,32,33]. An orthogonal drawing with no bend is a *rectilinear* drawing. Several papers address rectilinear

This research was partially supported by MIUR Project "AHeAD" under PRIN 20174LF3T8, and by H2020-MSCA-RISE project 734922 – "CONNECT".

P. Angelini and R. von Hanxleden (Eds.): GD 2022, LNCS 13764, pp. 127–143, 2023.
https://doi.org/10.1007/978-3-031-22203-0_10

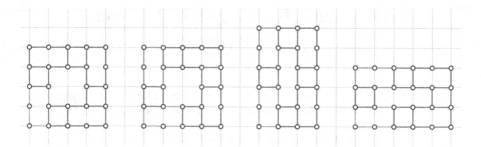

Fig. 1. Unit-length embedding-preserving rectangular drawings of a plane graph.

drawings of planar [13,24,26,29,37,38] and plane [20,24,43,49] graphs. When all the faces of a rectilinear drawing have a rectangular shape the drawing is *rectangular*. Maximum degree-3 plane graphs admitting rectangular drawings were first characterized in [47,48]. A linear-time algorithm to find a rectangular drawing of a maximum degree-3 plane graph, provided it exists, is described in [39] and extended to maximum degree-3 planar graphs in [42]. Surveys on rectangular drawings can be found in [23,35,36]. If only the internal faces are constrained to be rectangular, then the drawing is called *inner-rectangular*. In [34] it is shown that a plane graph G has an inner-rectangular drawing Γ if and only if a special bipartite graph constructed from G has a perfect matching. Also, Γ can be found in $O(n^{1.5}/\log n)$ time if G has n vertices and a "sketch" of the outer face is prescribed, i.e., all the convex and concave outer vertices are prescribed.

Computing straight-line drawings whose edges have constrained length is another core topic in graph drawing [1,2,4,7,12,22,44]. The graphs admitting planar straight-line drawings with all edges of the same length are also called *matchstick graphs*. Recognizing matchstick graphs is NP-hard for biconnected [22] and triconnected [12] graphs, and in fact, even strongly ∃ℝ-complete [1]; see also [44].

A *unit-length grid drawing* maps vertices to grid points and edges to horizontal or vertical segments of unit Euclidean length. A *grid graph* is a graph that admits a unit-length grid drawing[1]. Recognizing grid graphs is NP-complete for ternary trees of pathwidth 3 [9], for binary trees [27], and for trees of pathwidth 2 [28], but solvable in polynomial time on graphs of pathwidth 1 [28]. An exponential-time algorithm to compute, for a given weighted planar graph, a rectilinear drawing in which the Euclidean length of each edge is equal to the edge weight has been presented in [7].

Let G be a planar graph. The UNIT-LENGTH INNER-RECTANGULAR DRAWING RECOGNITION (for short, UIR) problem asks whether a unit-length inner-rectangular drawing of G exists. Similarly, the UNIT-LENGTH RECTANGULAR DRAWING RECOGNITION (for short, UR) problem asks whether a unit-length

---

[1] Note that in some literature the term "grid graph" denotes an "induced" graph, i.e., there is an edge between any two vertices at distance one. See, for example, [31].

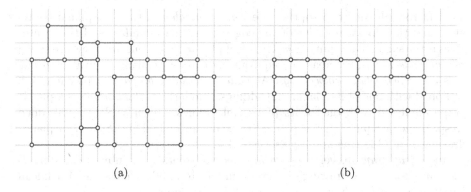

Fig. 2. (a) A planar rectilinear grid drawing of a graph. (b) A unit-length rectangular grid drawing of the same graph.

rectangular drawing of G exists. Let now H be a plane *or* planar embedded (i.e., no outer face specified) graph. The UNIT-LENGTH INNER-RECTANGULAR DRAWING RECOGNITION WITH FIXED EMBEDDING (for short, UIRFE) problem asks whether a unit-length inner-rectangular embedding-preserving drawing of H exists. Similarly, the UNIT-LENGTH RECTANGULAR DRAWING RECOGNITION WITH FIXED EMBEDDING (for short, URFE) problem asks whether a unit-length rectangular embedding-preserving drawing of H exists; see Fig. 1.

Our Contribution. In Sect. 3 we show NP-completeness for the UIRFE and UIR problems when the input graph is biconnected, which is surprising since a biconnected graph has degrees of freedom that are more restricted than those of a tree. In Sect. 4 we provide a linear-time algorithm for the UIRFE and URFE problems if the drawing of the outer face is given. In Sect. 5 we first show that the URFE problem is cubic-time solvable; the time bound becomes linear if all internal faces of the input graph have maximum degree 6. These results hold both when the outer face is prescribed and when it is not. Second, we show a necessary condition for an instance of the UR problem to be positive in terms of its SPQR-tree. Exploiting the above condition, we show that the UR problem is cubic-time solvable; the running time becomes linear when the SPQR-tree of the input graph satisfies special conditions. Finally, as a by-product of our research, we provide the first polynomial-time algorithm to test whether a planar graph G admits a rectangular drawing, for general instances of maximum degree 4.

Missing details for the proofs of the statements marked with a $(\star)$ are given in [3].

2   Preliminaries

For basic graph drawing terminology and definitions refer, e.g., to [15,35].

Drawings and Embeddings. Two planar drawings of a connected graph are *planar equivalent* if they induce the same counter-clockwise ordering of the edges incident to each vertex. Also, they are *plane equivalent* if they are planar equivalent and the clockwise order of the edges along the boundaries of their outer faces is the same. The equivalence classes of planar equivalent drawings are called *planar embeddings*, whereas the equivalence classes of plane equivalent drawings are called *plane embeddings*. A *planar embedded graph* is a planar graph equipped with one of its planar embeddings. Similarly, a *plane graph* is a planar graph equipped with one of its plane embeddings. Given a planar embedded (resp. plane) graph G and a planar (resp. plane) embedding $\mathcal{E}$ of G, a planar drawing Γ of G is *embedding-preserving* if $\Gamma \in \mathcal{E}$.

In a *grid drawing*, vertices are mapped to points with integer coordinates (i.e., *grid points*). A drawing of a graph in which all edges have unit Euclidean length is a *unit-length drawing* (see Fig. 2 for an example).

Observation 1. *A unit-length grid drawing is rectilinear and planar.*

Observation 2. *A unit-length rectangular (or inner-rectangular) drawing is planar and it is a grid drawing, up to a rigid transformation.*

The following simple property has been proved in [6, Lemma 1].

Property 1. Every cycle that admits a unit-length grid drawing has even length.

Since (inner) rectangular drawings exist only for maximum-degree-4 graphs, in the remainder, we assume that all considered graphs satisfy this requirement.

Connectivity. A *biconnected component* (or *block*) of a graph G is a maximal (in terms of vertices and edges) biconnected subgraph of G. A block is *trivial* if it consists of a single edge and *non-trivial* otherwise. A *split pair* of G is either a pair of adjacent vertices or a *separation pair*, i.e., a pair of vertices whose removal disconnects G. The *components* of G *with respect to* a split pair $\{u, v\}$ are defined as follows. If (u, v) is an edge of G, then it is a component of G with respect to $\{u, v\}$. Also, let $G_1, \ldots, G_k$ be the connected components of $G \backslash \{u, v\}$. The subgraphs of G induced by $V(G_i) \cup \{u, v\}$, minus the edge (u, v), are components of G with respect to $\{u, v\}$, for $i = 1, \ldots, k$. Due to space limitations, we refer the reader to [3] and to [17,18] for the definition of SPQR-tree.

3   NP-Completeness of the UIRFE and UIR Problems

In this section we show NP-completeness for both the UIRFE and UIR problems when the input graph is biconnected. We start with the following theorem.

Theorem 1. *The* UIRFE *problem is* NP-*complete, even for biconnected plane graphs whose internal faces have maximum size 6.*

Let ϕ be a Boolean formula in conjunctive normal form with at most three literals in each clause. We denote by G_ϕ the *incidence graph* of ϕ, i.e., the graph that has a vertex for each clause of ϕ, a vertex for each variable of ϕ, and an edge

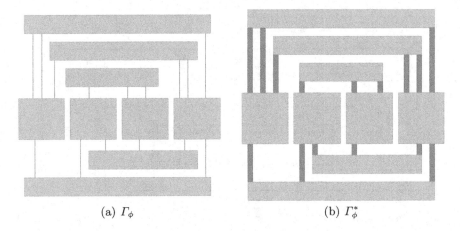

(a) Γ_ϕ (b) Γ_ϕ^*

Fig. 3. (a) The monotone rectilinear representations Γ_ϕ of G_ϕ. The rectangles representing variables and clauses are red, whereas the line segments and rectangles representing the edges of ϕ are blue. (b) The auxiliary representation Γ_ϕ^*. (Color figure online)

(c, v) for each clause c that contains the *positive literal* v or the *negated literal* $\bar{v}$. The formula ϕ is an instance of PLANAR MONOTONE 3-SAT if G_ϕ is planar and each clause of ϕ is either positive or negative. A *positive clause* contains only positive literals, while a *negative clause* contains only negated literals. Hereafter, w.l.o.g., we assume that all the clauses of ϕ contain *exactly* three literals.

A *monotone rectilinear representation* of G_ϕ is a drawing that satisfies the following properties (refer to Fig. 3a). **P1:** Variables and clauses are represented by axis-aligned rectangles with the same height. **P2:** The bottom sides of all rectangles representing variables lie on the same horizontal line. **P3:** The rectangles representing positive (resp. negative) clauses lie above (resp. below) the rectangles representing variables. **P4:** Edges connecting variables and clauses are represented by vertical segments. **P5:** The drawing is crossing-free.

The PLANAR MONOTONE 3-SAT problem is known to be NP-complete, even when the incidence graph G_ϕ of ϕ is provided along with a monotone rectilinear representation Γ_ϕ of G_ϕ [8]. We prove Theorem 1 by showing how to construct a plane graph H_ϕ that is biconnected, has internal faces of maximum size 6, and admits a unit-length inner-rectangular drawing *if and only if* ϕ is satisfiable. Our strategy is to modify Γ_ϕ to create a suitable auxiliary representation Γ_ϕ^* (see Fig. 3) and then to use the geometric information of Γ_ϕ^* as a blueprint to construct H_ϕ. Because of the lack of space, we describe in detail how to obtain Γ_ϕ^* from Γ_ϕ and how to construct H_ϕ in the full version [3]. We provide below a high-level description of the logic behind the reduction.

Overview of the Reduction. The reduction is based on three main types of gadgets. A variable $v \in \phi$ is modeled by means of a *variable gadget*, a clause $c \in \phi$

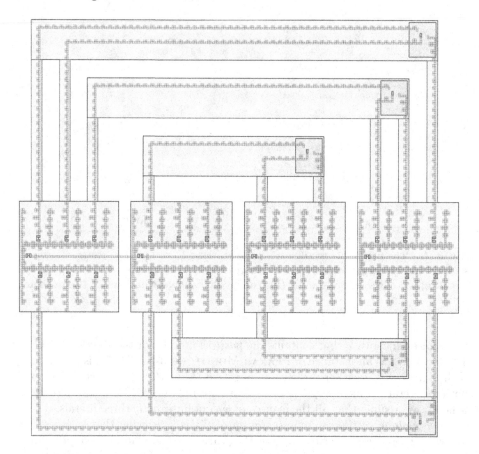

Fig. 4. The graph H_ϕ. Variable and clause gadgets are enclosed in light red boxes, while transmission gadgets are enclosed in light blue boxes. (Color figure online)

by means of an (α, β)-*clause gadget*, and an edge $(v, c) \in G_\phi$ by means of a λ-*transmission gadget*. We use the geometric properties of Γ_ϕ^* to determine the size and structure of each gadget, as well as how to combine the gadgets together to form H_ϕ. The width and height of the rectangles representing variables, clauses, and edges are used to construct variable gadgets and to compute the auxiliary parameters α, β and λ, which in turn are used to construct (α, β)-clause gadgets and λ-transmission gadgets. Finally, the incidences between the rectangles are used to decide how to join the gadgets to construct a single connected graph.

An example of a unit-length inner-rectangular drawing of H_ϕ is shown in Fig. 4; some faces of H_ϕ are omitted. All these missing faces are part of *domino components*, which admit a constant number of unit-length inner-rectangular drawings, see Fig. 5; some of these faces are shown filled in white or blue in Fig. 4.

The logic behind the construction is as follows. A variable gadget admits two unit-length inner-rectangular drawings (see Fig. 6), which differ from each other

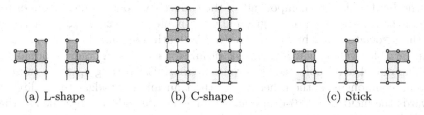

(a) L-shape   (b) C-shape   (c) Stick

Fig. 5. The unit-length grid drawings of the domino components. Domino component faces are filled blue (size 6) and white (size 4). (Color figure online)

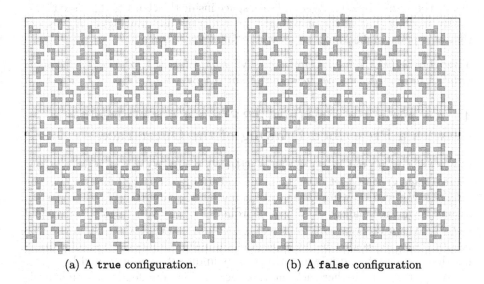

(a) A **true** configuration.   (b) A **false** configuration

Fig. 6. The variable gadget.

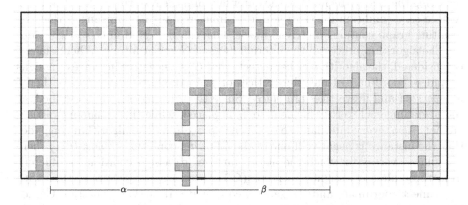

Fig. 7. In every unit-length inner-rectangular drawing of an (α, β)-clause gadget, at least one L-shape domino component crosses the red rectangle. (Color figure online)

on whether the domino components of the gadget stick out of the bottom or top side of the red enclosing rectangle, and correspond to a true/false assignment for the associated variable, respectively. The truth assignments are propagated from variable to clause gadgets via λ-transmission gadgets. A domino component sticking out of a variable gadget invades a transmission gadget, which causes a domino component at the other end of the transmission gadget to be directed towards the incident (α, β)-clause gadget. The clause gadget is designed so that it admits a unit-length inner-rectangular drawing if and only if at least one of the extremal domino components of its three incident transmission gadgets is not directed towards it; this allows a domino component of the clause gadget to invade the transmission gadget and save space inside the clause gadget; see Fig. 7.

By showing that all the unit-length inner-rectangular drawings of H_ϕ respect the same plane embedding, we prove the following theorem.

Theorem 2 (⋆). *The* UIR *problem is* NP-*complete, even for biconnected plane graphs whose internal faces have maximum size* 6.

4   An Algorithm for the UIRFE and URFE Problems with a Prescribed Drawing of the Outer Face

Consider a connected instance of the UIRFE problem, i.e., an n-vertex connected plane graph G; let $\mathcal{E}$ be the plane embedding prescribed for G. Let Γ_o be a unit-length grid drawing of the walk bounding the outer face f_o of $\mathcal{E}$. W.l.o.g, assume that the smallest x- and y- coordinates of the vertices of Γ_o are equal to 0. Next, we describe an $O(n)$-time algorithm, called RECTANGULAR-HOLES ALGORITHM, to decide whether G admits a unit-length inner-rectangular drawing that respects $\mathcal{E}$ and in which the walk bounding f_o is represented by Γ_o.

We first check whether each internal face of $\mathcal{E}$ is bounded by a simple cycle of even length, as otherwise the instance is negative by Property 1. This can be trivially done in $O(n)$ time. Consider the plane graph obtained from G by removing the bridges incident to the outer face and the resulting isolated vertices. A necessary condition for G to admit an inner-rectangular drawing is that the resulting graph contains no trivial block. This can be tested in $O(n)$ time [30].

The algorithm processes the internal faces of G one at a time. When a face f is considered, the algorithm either detects that G is a negative instance or assigns x- and y- coordinates to all the vertices of f. In the latter case, we say that f is *processed* and its vertices are *placed*. Since the drawing of f_o is prescribed, at the beginning each vertex incident to f_o is placed, while the remaining vertices are not. Also, every internal face of $\mathcal{E}$ is not processed. The algorithm concludes that the instance is negative if one of the following conditions holds: **(C1)** there is a placed vertex to which the algorithm tries to assign coordinates different from those already assigned to it, or **(C2)** there are two placed vertices with the same x-coordinate and the same y-coordinate. If neither Condition C1 nor C2 occurs, after processing all the internal faces the vertex placement provides a unit-length inner-rectangular drawing of the input instance.

To process faces, the algorithm maintains some auxiliary data structures:

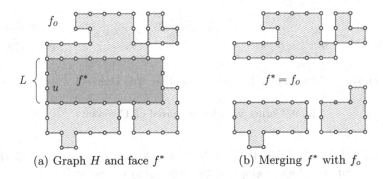

(a) Graph H and face f^* (b) Merging f^* with f_o

Fig. 8. A step of the RECTANGULAR-HOLES ALGORITHM.

- A graph H, called the CURRENT GRAPH, which is the subgraph of G composed of the vertices and of the edges incident to non-processed (internal) faces. Initially, we have $H = G$. In particular, we will maintain the invariant that each biconnected component of H is non-trivial. We will also maintain the outer face of the restriction $\mathcal{E}_H$ of $\mathcal{E}$ to H, which we will still denote by f_o.
- An **array** A, called the CURRENT OUTER-SORTER, that contains $M_x + 1$ buckets, each implemented as a double-linked list, where M_x is the largest x-coordinate of a vertex in Γ_o. The bucket $A[i]$ contains the placed vertices of H (i.e., those incident to the outer face of H) whose x-coordinate is equal to i. Moreover, A is equipped with the index $x_{\min}$ of the first non-empty bucket. To allow removals of vertices in $O(1)$ time, we enrich each placed vertex with x-coordinate i with a pointer to the corresponding list-item in the list $A[i]$.
- A **set of pointers** for the edges of H: Each edge (u, v) is equipped with two pointers ℓ_{uv} and ℓ_{vu}, that reference the faces of $\mathcal{E}$ lying to the left of (u, v), when traversing such an edge from u to v and from v to u, respectively.

At each iteration the algorithm performs the following steps; see Fig. 8.
Retrieve: It retrieves an internal face f^* with at least one vertex u with minimum x-coordinate (i.e., $x_{\min}$) among the placed vertices of H; such a vertex is incident to the outer face of H. **Draw:** It assigns coordinates to all the vertices incident to f^* in such a way that f^* is drawn as a rectangle R^*. Note that such a drawing is unique as the left side of R^* in any unit-length grid drawing of H with the given drawing of f_o coincides with the maximal path L containing u that is induced by all the placed vertices of f^* with x-coordinate equal to $x_{\min}$. **Merge:** It merges f^* with f_o by suitably changing the pointers of every edge incident to f^*, and by removing each edge (u, v) with pointers $\ell_{uv} = \ell_{vu} = f_o$, as well as any resulting isolated vertex. Further, it updates A consequently. Note that, after the merge step, the outer face f_o of the new current graph H is completely drawn. This invariant is maintained through each iteration of the algorithm. In [3], we describe each step in detail.

The proof of the next theorem exploits the RECTANGULAR-HOLES ALGORITHM.

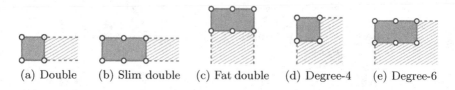

(a) Double   (b) Slim double   (c) Fat double   (d) Degree-4   (e) Degree-6

Fig. 9. Corner faces for the proof of Theorem 4.

Theorem 3 ($\star$). *The* UIRFE *and* URFE *problems are* $O(n)$-*time solvable for an n-vertex connected plane graph, if the drawing of the outer face is prescribed.*

Since any unit-length grid drawing of a cycle with 4 or 6 vertices is a rectangle, the previous theorem implies the following result, which contrasts with the NP-hardness of Theorem 1, where the drawing of the outer face is not prescribed.

Corollary 1. *The* UIRFE *problem is linear-time solvable if the drawing of the outer face is prescribed and all internal faces have maximum degree 6.*

5   Algorithms for the URFE and UR Problems

In this section we study the UR problem. Since rectangular drawings are convex, the input graphs for the UR problem must be biconnected [19].

Fixed Embedding. We start by considering instances with either a prescribed plane embedding (Theorem 4) or a prescribed planar embedding (Theorem 5).

Theorem 4 ($\star$). *The* URFE *problem is cubic-time solvable for a plane graph G and it is linear-time solvable if all internal faces of G have maximum degree 6.*

Proof (sketch). To solve the problem in cubic time, we examine the quadratically-many drawings of the outer face f_o, and invoke Theorem 3 for each of them.

Assume now that all internal faces have maximum degree 6. We efficiently determine $O(1)$ possible rectangular drawings of f_o and then invoke Theorem 3 for each of them. If G is a 4-cycle or a 6-cycle, then the instance is trivially positive. Refer to Fig. 9. A *double corner face* is a degree-4 face with three edges incident to f_o. A *slim double corner face* is a degree-6 face with five edges incident to f_o. A *fat double corner face* is a degree-6 face with four edges incident to f_o. Note that each of such faces must provide two consecutive 270° angles incident to f_o. Hence, if G has at least one of the above faces, the drawing of f_o is prescribed, and hence RECTANGULAR-HOLES ALGORITHM can be invoked.

Suppose now that none of the above cases holds. A *corner face* is a degree-4 (degree-6) face that has two (resp. three) edges incident to f_o. Each corner face provides a 270° angle incident to any realization of f_0 as a rectangle. Hence, there must be exactly four corner faces in order for G to be a positive instance. These faces can be computed in linear time, and determine $O(1)$ possible drawings of the outer face on which we invoke RECTANGULAR-HOLES ALGORITHM.   □

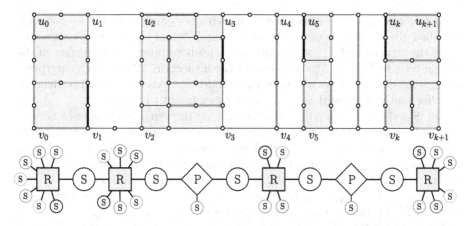

Fig. 10. A rectangular unit-length grid drawing of a planar graph and its pruned SPQR-tree T^*. S-, P-, and R-nodes are circles, rhombuses and squares, respectively. The subgraphs corresponding to S-nodes that are leaves of T^* are thick.

By showing that any planar embedding has a unique candidate outer face supporting a unit-length rectangular drawing, we get the following.

Theorem 5 ($\star$)**.** The URFE problem is cubic-time solvable for a planar embedded graph G, and it is linear-time solvable if all but at most one face of G have maximum degree 6.

Variable Embedding. Now, we turn our attention to instances with a variable embedding. We start by providing some relevant properties of the graphs that admit a rectangular (not necessarily unit-length or grid) drawing. Let G be one such graph. To avoid degenerate cases, in what follows, we assume that G is not a cycle (cfr. Property 1). Let Γ be a rectangular drawing of G and let Γ_o be the rectangle delimiting the outer face of Γ. Refer to Fig. 10. Consider the plane graph G_Γ corresponding to Γ. Since Γ is convex, then G_Γ is a subdivision of an *internally triconnected* plane graph [5, Theorem 1]. That is, every separation pair $\{u, v\}$ of G_Γ is such that u and v are incident to the outer face and each connected component of $G_\Gamma \backslash \{u, v\}$ contains a vertex incident to the outer face.

A *caterpillar* is a tree such that removing its leaves results in a path, called *spine*. The *pruned SPQR-tree* of a biconnected planar graph G, denoted by T^*, is the tree obtained from the SPQR-tree T of G, after removing the Q-nodes of T.

Lemma 1 ($\star$)**.** Let G be a graph that admits a rectangular drawing. Then the pruned SPQR-tree T^* of G is a caterpillar with the following properties: **(i)** All its leaves are S-nodes; **(ii)** its spine contains no two adjacent R-nodes; **(iii)** its spine contains no two adjacent nodes μ and ν, such that μ is a P-node and ν is an R-node; **(iv)** each P-node μ has exactly 3 neighbors; and **(v)** the skeleton of each S-node of the spine of T^* contains two chains of virtual edges corresponding to Q-nodes, separated by two virtual edges each corresponding to either a P- or an R-node.

Proof (sketch). Let Γ be a rectangular drawing of G and let Γ_o be the rectangle bounding the outer face of Γ. By inspecting Γ "from left to right", we argue about the structure of T^*, which ultimately leads to prove the statement of the lemma; refer to Fig. 10. At each point of the inspection, T^* will be a caterpillar whose spine does not have a P-node as an end-point. Also, a leaf will be denoted as *active* and will be used as an attachment endpoint to extend T^*.

Let $S = [\{u_1, v_1\}, \{u_2, v_2\}, \ldots, \{u_k, v_k\}]$ be the separation pairs of G such that both u_i and v_i lie on opposite sides of Γ_o, have degree 3, and share the same x-coordinate, for $i = 1, \ldots, k$, sorted in increasing order of their x-coordinate. In [3], we provide properties of rectangular drawings that show that these pairs are the only ones that correspond to poles of P- and R-nodes of T^*. We set $L = \{u_0, v_0\} \circ S \circ \{u_{k+1}, v_{k+1}\}$, where u_0, u_{k+1}, v_{k+1}, and v_0 are the vertices on the top-left, top-right, bottom-right, and bottom-left corner of Γ_o.

Consider any two consecutive pairs $\{u_i, v_i\}$ and $\{u_{i+1}, v_{i+1}\}$, for $i = 0, \ldots, k$. We can define a cycle C_i in G that contains u_i, u_{i+1}, v_{i+1}, and v_i, and that is drawn as a rectangle in Γ. Moreover, any two cycles C_i and C_{i+1} share a path P_{i+1} that is drawn as a straight-line segment in Γ. We denote by G_i the subgraph of G induced by the vertices in the interior and along the boundary of C_i.

We skip the discussion for the consecutive pairs $\{u_0, v_0\}$ and $\{u_1, v_1\}$. For $i = 1, \ldots, k$, consider the separation pair $\{u_i, v_i\}$. Let ξ be the active endpoint of the spine. In the following, we denote by $\mathrm{sk}(\mu)$ the skeleton of a node μ of T^*. Two cases are possible: ξ is either an S- or an R-node.

Suppose that $G_i = C_i$. If ξ is an S-node, then we introduce a P-node $\mu_{i,1}$ in T^* adjacent to ξ and to two new S-nodes $\mu_{i,2}$ and $\mu_{i,3}$. We have that $\mathrm{sk}(\mu_{i,1})$ is a bundle of three parallel edges (u_i, v_i), $\mathrm{sk}(\mu_{i,2})$ is a cycle containing one virtual edge for each edge of the path P_i plus a virtual edge (u_i, v_i), and $\mathrm{sk}(\mu_{i,3})$ is a cycle consisting of a virtual edge (u_i, v_i), followed by one virtual edge for each horizontal edge in the top side of C_i, followed by one virtual edge (u_{i+1}, v_{i+1}), followed by one virtual edge for each horizontal edge in the bottom side of C_i. We set S-node $\mu_{1,3}$ as the active node of T^*. If ξ is an R-node, then we introduce an S-node μ_i in T^* adjacent to ξ whose skeleton is a cycle consisting of a virtual edge (u_i, v_i), followed by one virtual edge for each horizontal edge in the top side of C_i, followed by a path P^* of virtual edges defined below, followed by one virtual edge for each horizontal edge in the bottom side of C_i. If $i < k$, then P^* consists of the single virtual edge (u_{i+1}, v_{i+1}); otherwise, if $i = k$, then P^* contains a virtual edge for each real edge incident to the right side of Γ_o. We set the S-node μ_i as the active endpoint of T^*, unless $i = k$.

Suppose now that $G_i \neq C_i$. In this case, G_i is the subdivision of a triconnected planar graph. We introduce an R-node μ_i in T^* adjacent to ξ and to the S-nodes corresponding to the components of G_i, with respect to its split pairs, that are simple paths. We add to $\mathrm{sk}(\mu_i)$ a virtual edge for each of such paths, as well as (u_i, v_i) and (u_{i+1}, v_{i+1}), unless $i = k$. We set the R-node μ_i as the active endpoint of T^*, unless $i = k$. $\square$

Consider a graph G that satisfies the conditions of Lemma 1. If the spine of the pruned SPQR-tree of G contains at least two nodes or at least one P-node,

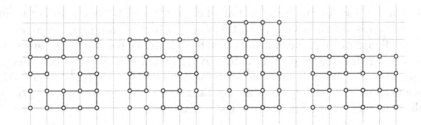

Fig. 11. Four plane embeddings of a graph G that support a rectangular drawing of G, obtained by selecting one of the plane embeddings $\mathcal{E}_1$ and $\mathcal{E}_4$ of the subgraph G_0 of G and one of the the plane embeddings $\mathcal{E}_2$ and $\mathcal{E}_3$ of the subgraph G_4 of G. Only the embeddings $\mathcal{E}_1$ and $\mathcal{E}_2$ support a unit-length rectangular drawing.

we say that G is *flat*; otherwise, G is the subdivision of a triconnected planar graph. Exploiting Lemma 1, we can prove the following; refer to Fig. 11.

Lemma 2 ($\star$). *Let G be an n-vertex graph. The following hold:*

- *All the unit-length rectangular drawings of G, if any, have the same plane embedding $\mathcal{E}$ (up to a reflection), which can be computed in $O(n)$ time.*
- *If G is flat, all the rectangular drawings of G, if any, have at most four possible plane embeddings (up to a reflection), which can be computed in $O(n)$ time.*

The next theorem shows that the UR problem is polynomial-time solvable. Surprisingly, the problem seems to be harder for non-flat instances.

Theorem 6 ($\star$). *Let G be a planar graph. The UR problem is cubic-time solvable for G. Also, if G is flat, then the UR problem is linear-time solvable.*

Proof (sketch). First, we test whether G satisfies the conditions of Lemma 1, which can clearly be done in $O(n)$ time by computing and visiting T^*, and reject the instance if this test fails. Then, by Lemma 2, we compute in $O(n)$ time the unique candidate plane embedding $\mathcal{E}$ of G that may support a unit-length rectangular drawing of G, if any, and reject the instance if such an embedding does not exist. Let f_o be the outer face of $\mathcal{E}$. If the spine of T^* consists of a single R-node, then $\mathcal{E}$ coincides with the unique planar embedding of G, and we test for the existence of such a drawing using Lemma 5 in $O(n^3)$ time. If G is flat, then we can show that there exists a unique candidate drawing Γ_o of f_o. Then, we use Theorem 3 to test in $O(n)$ time whether a unit-length rectangular drawing of G exists that respects $\mathcal{E}$ and such that f_o is drawn as Γ_o.   □

Theorem 7 ($\star$). *Let G be an n-vertex planar graph. The problem of testing for the existence of a rectangular drawing of G is solvable in $O(n^2 \log^3 n)$ time. Also, if G is flat, then this problem is solvable in $O(n \log^3 n)$ time.*

Proof (sketch). Assume that G satisfies the conditions of Lemma 1. If G is flat, then Lemma 2 guarantees the existence of only up to four plane embeddings of G that are candidates for a rectangular drawing of G that respects them. Otherwise, G is the subdivision of a triconnected planar graph, and there exists $O(n)$ candidate plane embeddings. For each of them, we test for the existence of a rectangular drawing respecting it by solving a max-flow problem on a linear-size planar network with multiple sources and sinks in $O(n \log^3 n)$ time [11]. Such a network can be defined following Tamassia's [15] classic approach to test for the existence of rectilinear drawings of plane graphs.   □

6   Conclusions and Open Problems

We studied the recognition of graphs admitting the beautiful drawings that require unit-length and orthogonality of the edges, planarity, and convexity of the faces. We show that, if the outer face is drawn as a rectangle, the problem is polynomial-time solvable, while it is NP-hard if the outer face is an arbitrary polygon (even if the input is biconnected), unless such a polygon is specified in advance. These results hold both in the fixed-embedding and in the variable-embedding settings. A byproduct of our results is a polynomial-time algorithm to recognize graphs admitting a rectangular (non-necessarily unit-length) drawing.

It is worth remarking that if the input is a subdivision of a triconnected planar graph, then our algorithms pay an extra time to handle the outer face. Specifically, for the rectangular unit-length setting, an extra quadratic time is used to guess a rectangular drawing of the unique candidate outer face, while, for the general rectangular setting, an extra linear time is used to determine the actual candidate outer face. Hence, it is appealing to study efficient algorithms for this specific case. Further, it is interesting to determine the complexity of the grid graph recognition problem for trees with a given embedding, even for the case of trees that are as simple as caterpillars. Observe that the NP-hardness results on trees in [9, 27] heavily rely on the variable embedding setting.

References

1. Abel, Z., Demaine, E.D., Demaine, M.L., Eisenstat, S., Lynch, J., Schardl, T.B.: Who needs crossings? Hardness of plane graph rigidity. In: Fekete, S.P., Lubiw, A. (eds.) 32nd International Symposium on Computational Geometry (SoCG 2016). LIPIcs, vol. 51, pp. 3:1–3:15. Schloss Dagstuhl - Leibniz-Zentrum für Informatik (2016). https://doi.org/10.4230/LIPIcs.SoCG.2016.3
2. Alegría, C., Borrazzo, M., Da Lozzo, G., Di Battista, G., Frati, F., Patrignani, M.: Planar straight-line realizations of 2-trees with prescribed edge lengths. In: Purchase, H.C., Rutter, I. (eds.) GD 2021. LNCS, vol. 12868, pp. 166–183. Springer, Cham (2021). https://doi.org/10.1007/978-3-030-92931-2_12
3. Alegría, C., Da Lozzo, G., Di Battista, G., Frati, F., Grosso, F., Patrignani, M.: Unit-length rectangular drawings of graphs. CoRR abs/2208.14142 (2022). https://arxiv.org/abs/2208.14142
4. Angelini, P., et al.: Anchored drawings of planar graphs. In: Duncan, C., Symvonis, A. (eds.) GD 2014. LNCS, vol. 8871, pp. 404–415. Springer, Heidelberg (2014). https://doi.org/10.1007/978-3-662-45803-7_34

5. Angelini, P., Da Lozzo, G., Frati, F., Lubiw, A., Patrignani, M., Roselli, V.: Optimal morphs of convex drawings. In: Arge, L., Pach, J. (eds.) 31st International Symposium on Computational Geometry, SoCG 2015, 22–25 June 2015, Eindhoven, The Netherlands. LIPIcs, vol. 34, pp. 126–140. Schloss Dagstuhl - Leibniz-Zentrum für Informatik (2015). https://doi.org/10.4230/LIPIcs.SOCG.2015.126
6. Asgharian Sardroud, A., Bagheri, A.: Embedding cycles and paths on solid grid graphs. J. Supercomputing 73(4), 1322–1336 (2016). https://doi.org/10.1007/s11227-016-1811-y
7. Beck, M., Storandt, S.: Puzzling grid embeddings. In: Blelloch, G.E., Finocchi, I. (eds.) Proceedings of the Symposium on Algorithm Engineering and Experiments, ALENEX 2020, Salt Lake City, UT, USA, 5–6 January 2020, pp. 94–105. SIAM (2020)
8. de Berg, M., Khosravi, A.: Optimal binary space partitions for segments in the plane. Int. J. Comput. Geom. Appl. 22(03), 187–205 (2012). https://doi.org/10.1142/S0218195912500045
9. Bhatt, S.N., Cosmadakis, S.S.: The complexity of minimizing wire lengths in VLSI layouts. Inf. Process. Lett. 25(4), 263–267 (1987). https://doi.org/10.1016/0020-0190(87)90173-6
10. Biedl, T.C., Kant, G.: A better heuristic for orthogonal graph drawings. Comput. Geom. 9(3), 159–180 (1998). https://doi.org/10.1016/S0925-7721(97)00026-6
11. Borradaile, G., Klein, P.N., Mozes, S., Nussbaum, Y., Wulff-Nilsen, C.: Multiple-source multiple-sink maximum flow in directed planar graphs in near-linear time. SIAM J. Comput. 46(4), 1280–1303 (2017). https://doi.org/10.1137/15M1042929
12. Cabello, S., Demaine, E.D., Rote, G.: Planar embeddings of graphs with specified edge lengths. J. Graph Algorithms Appl. 11(1), 259–276 (2007). https://doi.org/10.7155/jgaa.00145
13. Chang, Y., Yen, H.: On bend-minimized orthogonal drawings of planar 3-graphs. In: 33rd International Symposium on Computational Geometry, SoCG 2017, 4–7 July 2017, Brisbane, Australia, pp. 29:1–29:15 (2017). https://doi.org/10.4230/LIPIcs.SoCG.2017.29
14. Cornelsen, S., Karrenbauer, A.: Accelerated bend minimization. J. Graph Algorithms Appl. 16(3), 635–650 (2012). https://doi.org/10.7155/jgaa.00265
15. Di Battista, G., Eades, P., Tamassia, R., Tollis, I.G.: Graph Drawing: Algorithms for the Visualization of Graphs. Prentice-Hall (1999)
16. Di Battista, G., Liotta, G., Vargiu, F.: Spirality and optimal orthogonal drawings. SIAM J. Comput. 27(6), 1764–1811 (1998). https://doi.org/10.1137/S0097539794262847
17. Di Battista, G., Tamassia, R.: On-line maintenance of triconnected components with SPQR-trees. Algorithmica 15(4), 302–318 (1996). https://doi.org/10.1007/BF01961541
18. Di Battista, G., Tamassia, R.: On-line planarity testing. SIAM J. Comput. 25(5), 956–997 (1996)
19. Di Battista, G., Tamassia, R., Vismara, L.: Incremental convex planarity testing. Inf. Comput. 169(1), 94–126 (2001). https://doi.org/10.1006/inco.2001.3031
20. Didimo, W., Kaufmann, M., Liotta, G., Ortali, G.: Rectilinear planarity testing of plane series-parallel graphs in linear time. In: GD 2020. LNCS, vol. 12590, pp. 436–449. Springer, Cham (2020). https://doi.org/10.1007/978-3-030-68766-3_34
21. Didimo, W., Liotta, G., Ortali, G., Patrignani, M.: Optimal orthogonal drawings of planar 3-graphs in linear time. In: Chawla, S. (ed.) Proceedings of ACM-SIAM Symposium on Discrete Algorithms (SODA 2020), pp. 806–825. ACM-SIAM (2020). https://doi.org/10.1137/1.9781611975994.49

22. Eades, P., Wormald, N.C.: Fixed edge-length graph drawing is NP-hard. Disc. Appl. Math. **28**(2), 111–134 (1990). https://doi.org/10.1016/0166-218X(90)90110-X
23. Felsner, S.: Rectangle and square representations of planar graphs. In: Pach, J. (ed.) Thirty Essays on Geometric Graph Theory, pp. 213–248. Springer, New York (2013). https://doi.org/10.1007/978-1-4614-0110-0_12
24. Frati, F.: Planar rectilinear drawings of outerplanar graphs in linear time. In: Auber, D., Valtr, P. (eds.) GD 2020. LNCS, vol. 12590, pp. 423–435. Springer, Cham (2020). https://doi.org/10.1007/978-3-030-68766-3_33
25. Garg, A., Liotta, G.: Almost bend-optimal planar orthogonal drawings of biconnected degree-3 planar graphs in quadratic time. In: Kratochvíyl, J. (ed.) GD 1999. LNCS, vol. 1731, pp. 38–48. Springer, Heidelberg (1999). https://doi.org/10.1007/3-540-46648-7_4
26. Garg, A., Tamassia, R.: On the computational complexity of upward and rectilinear planarity testing. SIAM J. Comput. **31**(2), 601–625 (2001). https://doi.org/10.1137/S0097539794277123
27. Gregori, A.: Unit-length embedding of binary trees on a square grid. Inf. Process. Lett. **31**(4), 167–173 (1989). https://doi.org/10.1016/0020-0190(89)90118-X
28. Gupta, S., Sa'ar, G., Zehavi, M.: Grid recognition: classical and parameterized computational perspectives (2021). https://doi.org/10.48550/ARXIV.2106.16180
29. Hasan, M.M., Rahman, M.S.: No-bend orthogonal drawings and no-bend orthogonally convex drawings of planar graphs (extended abstract). In: Du, D.-Z., Duan, Z., Tian, C. (eds.) COCOON 2019. LNCS, vol. 11653, pp. 254–265. Springer, Cham (2019). https://doi.org/10.1007/978-3-030-26176-4_21
30. Hopcroft, J., Tarjan, R.: Algorithm 447: efficient algorithms for graph manipulation. Commun. ACM **16**(6), 372–378 (1973). https://doi.org/10.1145/362248.362272
31. Itai, A., Papadimitriou, C.H., Szwarcfiter, J.L.: Hamilton paths in grid graphs. SIAM J. Comput. **11**(4), 676–686 (1982). https://doi.org/10.1137/0211056
32. Kant, G.: Drawing planar graphs using the canonical ordering. Algorithmica **16**(1), 4–32 (1996). https://doi.org/10.1007/BF02086606
33. Liu, Y., Morgana, A., Simeone, B.: A linear algorithm for 2-bend embeddings of planar graphs in the two-dimensional grid. Discrete Appl. Math. **81**(1–3), 69–91 (1998). https://doi.org/10.1016/S0166-218X(97)00076-0
34. Miura, K., Haga, H., Nishizeki, T.: Inner rectangular drawings of plane graphs. Int. J. Comput. Geom. Appl. **16**(2–3), 249–270 (2006). https://doi.org/10.1142/S0218195906002026
35. Nishizeki, T., Rahman, M.S.: Planar Graph Drawing, Lecture Notes Series on Computing, vol. 12. World Scientific (2004). https://doi.org/10.1142/5648
36. Nishizeki, T., Rahman, M.S.: Rectangular drawing algorithms. In: Tamassia, R. (ed.) Handbook on Graph Drawing and Visualization, pp. 317–348. Chapman and Hall/CRC (2013)
37. Rahman, M.S., Egi, N., Nishizeki, T.: No-bend orthogonal drawings of series-parallel graphs. In: Healy, P., Nikolov, N.S. (eds.) GD 2005. LNCS, vol. 3843, pp. 409–420. Springer, Heidelberg (2006). https://doi.org/10.1007/11618058_37
38. Rahman, M.S., Egi, N., Nishizeki, T.: No-bend orthogonal drawings of subdivisions of planar triconnected cubic graphs. IEICE Trans. **88-D**(1), 23–30 (2005)
39. Rahman, M.S., Nakano, S., Nishizeki, T.: Rectangular grid drawings of plane graphs. Comput. Geom. **10**(3), 203–220 (1998). https://doi.org/10.1016/S0925-7721(98)00003-0

40. Rahman, M.S., Nakano, S., Nishizeki, T.: A linear algorithm for bend-optimal orthogonal drawings of triconnected cubic plane graphs. J. Graph Algorithms Appl. **3**(4), 31–62 (1999). http://www.cs.brown.edu/publications/jgaa/accepted/99/SaidurNakanoNishizeki99.3.4.pdf

41. Rahman, M.S., Nishizeki, T.: Bend-minimum orthogonal drawings of plane 3-graphs. In: Graph-Theoretic Concepts in Computer Science, 28th International Workshop, WG 2002, Cesky Krumlov, Czech Republic, 13–15 June 2002, Revised Papers, pp. 367–378 (2002). https://doi.org/10.1007/3-540-36379-3_32

42. Rahman, M.S., Nishizeki, T., Ghosh, S.: Rectangular drawings of planar graphs. J. Algorithms **50**(1), 62–78 (2004). https://doi.org/10.1016/S0196-6774(03)00126-3

43. Rahman, M.S., Nishizeki, T., Naznin, M.: Orthogonal drawings of plane graphs without bends. J. Graph Algorithms Appl. **7**(4), 335–362 (2003). http://jgaa.info/accepted/2003/Rahman+2003.7.4.pdf

44. Schaefer, M.: Realizability of graphs and linkages. In: Pach, J. (ed.) Thirty Essays on Geometric Graph Theory, pp. 461–482. Springer, New York (2013). https://doi.org/10.1007/978-1-4614-0110-0_24

45. Storer, J.A.: The node cost measure for embedding graphs on the planar grid (extended abstract). In: Miller, R.E., Ginsburg, S., Burkhard, W.A., Lipton, R.J. (eds.) Proceedings of the 12th Annual ACM Symposium on Theory of Computing, 28–30 April 1980, Los Angeles, California, USA, pp. 201–210. ACM (1980). https://doi.org/10.1145/800141.804667

46. Tamassia, R.: On embedding a graph in the grid with the minimum number of bends. SIAM J. Comput. **16**(3), 421–444 (1987). https://doi.org/10.1137/0216030

47. Thomassen, C.: Plane representations of graphs. In: Bondy, J., Murty, U. (eds.) Progress in Graph Theory, pp. 43–69. Academic Press, Toronto, Orlando (1987)

48. Ungar, P.: On Diagrams Representing Maps. J. London Math. Soc. **s1−28**(3), 336–342 (1953). https://doi.org/10.1112/jlms/s1-28.3.336

49. Vijayan, G., Wigderson, A.: Rectilinear graphs and their embeddings. SIAM J. Comput. **14**(2), 355–372 (1985)

50. Zhou, X., Nishizeki, T.: Orthogonal drawings of series-parallel graphs with minimum bends. SIAM J. Discrete Math. **22**(4), 1570–1604 (2008). https://doi.org/10.1137/060667621

Strictly-Convex Drawings of 3-Connected Planar Graphs

Michael A. Bekos[1] , Martin Gronemann[2] , Fabrizio Montecchiani[3](✉) ,
and Antonios Symvonis[4]

[1] Department of Mathematics, University of Ioannina, Ioannina, Greece
bekos@uoi.gr
[2] Algorithms and Complexity Group, TU Wien, Vienna, Austria
mgronemann@ac.tuwien.ac.at
[3] Department of Engineering, University of Perugia, Perugia, Italy
fabrizio.montecchiani@unipg.it
[4] School of Applied Mathematical & Physical Sciences, NTUA, Athens, Greece
symvonis@math.ntua.gr

Abstract. Strictly-convex straight-line drawings of 3-connected planar
graphs in small area form a classical research topic in Graph Drawing.
Currently, the best-known area bound for such drawings is $O(n^2) \times O(n^2)$,
as shown by Bárány and Rote by means of a sophisticated technique
based on perturbing (non-strictly) convex drawings. Unfortunately, the
hidden constants in such area bound are in the 10^4 order.

We present a new and easy-to-implement technique that yields
strictly-convex straight-line planar drawings of 3-connected planar
graphs on an integer grid of size $2(n-1) \times (5n^3 - 4n^2)$.

Keywords: Strictly-convex drawings · Area bounds · Planar graphs

1 Introduction

Drawing planar graphs is a fundamental topic in Graph Drawing with several
important contributions over the last few decades [8,14,17,21,24]. One of the
most influential is due to de Fraysseix, Pach and Pollack [8], who back in 1988
showed that every n-vertex planar graph admits a straight-line planar drawing
on a $(2n-4) \times (n-2)$ grid, which can be computed in $O(n)$ time [7]. Since then,
several improvements on the size of the underlying grid have been proposed in the
literature [12,15,17,18,21,25]. The best-known upper bound is $(n-2) \times (n-2)$
by Chrobak and Kant [6] and by Schnyder [21], who propose two conceptually
different approaches to derive this bound. The former is an incremental drawing
algorithm inspired by [8], while the latter is based on a face counting technique.

Straight-line drawings of planar graphs have also been extensively studied
by requiring convexity [22], that is, the boundary of every face must be a convex polygon. Such drawings are called *convex* and always exist for 3-connected

Research of FM partially supported by Dip. Ingegneria, Univ. of Perugia, grant
RICBA21LG "Algoritmi, modelli e sistemi per la rappresentazione visuale di reti".

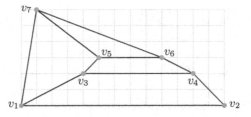

Fig. 1. A strictly-convex drawing of a 3-connected planar graph on 7 vertices.

planar graphs [23,24]. Again the aim is to keep the size of the underlying grid as small as possible; see [10] for a survey. Early results date back to Schnyder and Trotter [22], Chrobak and Kant [6], Di Battista et al. [11] and Felsner [12]. The latter guarantees the existence of a convex drawing of a 3-connected planar graph on a $(f-1) \times (f-1)$ integer grid, where $f = O(n)$ is the number of faces.

Note that in a convex drawing three vertices on the boundary of a face can be collinear. If this is not allowed, then the corresponding drawings are called *strictly-convex*. Since an n-vertex cycle cannot be drawn strictly-convex on a grid of size $o(n^3)$ [5], it follows that strictly-convex drawings are more demanding in terms of required area. As an adaptation of the standard incremental drawing algorithms or the face-counting methods is rather difficult, the only approach that has been exploited so far to obtain strictly-convex drawings is to perturb convex drawings. This idea was pioneered by Chrobak, Goodrich and Tamassia [5], who claimed (without giving details) that every 3-connected planar graph admits a strictly-convex drawing on an $O(n^3) \times O(n^3)$ grid. The area bound was improved to $O(n^{7/3}) \times O(n^{7/3})$ by Rote [19] and to $O(n^2) \times O(n^2)$ by Bárány and Rote [1], which is currently the best-known asymptotic upper bound. However, as the authors mention "the constants hidden in the O-notation are on the order of 100 for the width and on the order of 10,000 for the height. This is far too much for applications where one wants to draw graphs on a computer screen" [1].

Our Contribution. We continue the research on strictly-convex drawings of 3-connected planar graphs. Our contribution is a new technique that computes strictly-convex drawings of 3-connected n-vertex planar graphs on an integer grid of size $2(n-1) \times (5n^3 - 4n^2)$, as outlined in the following theorem. Although the asymptotic area bound is the same as the one in [1], the multiplicative constants are significantly smaller. Also, the proposed technique is elegant and can be readily implemented to run in linear time. On the other hand, the aspect-ratio of the produced drawings is quadratic rather than constant.

Theorem 1. *Every 3-connected planar graph with n vertices admits a strictly-convex planar straight-line drawing on an integer grid of size $2(n-1) \times (5n^3 - 4n^2)$. Also, the drawing can be computed in $O(n)$ time.*

Our technique starts with a convex drawing computed by Kant's algorithm [17]. We rely on properties of such a drawing to show that shifting vertices upwards

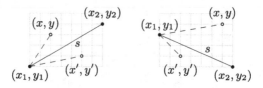

Fig. 2. In both drawings, (x, y) is above s, while (x', y') is below s.

by using a strictly-increasing and strictly-convex function preserves planarity; a property of independent interest. Also, the obtained planar drawing is convex and collinear vertices in a face, if any, are horizontally aligned. For such vertices, a second shifting yields an internally strictly-convex drawing. A suitable augmentation guarantees that the outer face is also strictly-convex.

Paper Structure. Section 2 contains basic definitions and tools. In Sect. 3, we introduce properties of Kant's algorithm that we leverage in our technique. Section 4 describes our algorithm. Section 5 concludes the paper with a brief discussion and open problems. For space reasons, some proofs are omitted (the corresponding statements are marked with $\star$) and can be found in [4].

2   Preliminaries

Basic Definitions. Let $f : \mathbb{R} \to \mathbb{R}$ be a function. If $f(a) < f(b)$ for every pair $a, b \in \mathbb{R}$ with $a < b$, then f is *strictly-increasing*. Function f is *strictly-convex* if for all t, $0 < t < 1$, and all $a, b \in \mathbb{R}$, it holds $f(ta + (1-t)b) < tf(a) + (1-t)f(b)$. Consider three points $(x_1, y_1), (x, y), (x_2, y_2)$, with $x_1 < x < x_2$ and let s be the line-segment connecting (x_1, y_1) and (x_2, y_2). We say that (x, y) is *above* (resp., *below*) s if the slope of s is smaller (resp., larger) than the slope of the line-segment connecting (x_1, y_1) and (x, y); see Fig. 2. The next lemma easily follows from the chordal slope lemma [20].

Lemma 1 ($\star$). *Let (x_1, y_1), (x, y) and (x_2, y_2) be three collinear points with $x_1 < x < x_2$ that are not horizontally aligned. If $f : \mathbb{R} \to \mathbb{R}$ is a strictly-convex function, then point $(x, f(y))$ is below the line-segment with endpoints $(x_1, f(y_1))$ and $(x_2, f(y_2))$.*

Drawings and Embeddings. We assume familiarity with basic graph drawing concepts [9]. In particular, a *plane graph* is a graph with a prescribed planar embedding. Unless otherwise specified, we consider drawings that are straight-line, planar and whose vertices are on an integer grid. A drawing is *convex* (*strictly-convex*) if the boundary of each face is a convex (strictly-convex) polygon. Similarly, a drawing is *internally convex* (*internally strictly-convex*) if the boundary of each inner face is a convex (strictly-convex) polygon. Given a drawing Γ of a graph G, denote by (x_u, y_u) the coordinates of vertex u in Γ. For two vertices u and v in Γ, we denote by Δ_{uv} the interior of the right triangle whose corners are u, v and the intersection of the vertical line though the vertex having the lowest y-coordinate with the horizontal line through the vertex having the highest y-coordinate (among u, v). For example, in Fig. 3a, the Δ_{uv} triangle of the endpoints of each edge (u, v) is striped.

Canonical Order. Let G be a 3-connected plane graph with n vertices. Let $\delta = (P_0, \ldots, P_m)$ be a partition of the vertices of G into paths, such that $P_0 = \{v_1, v_2\}$, $P_m = \{v_n\}$, and edges (v_1, v_2) and (v_1, v_n) exist and belong to the outer face. For $k = 0, \ldots, m$, let G_k be the subgraph induced by $\cup_{i=0}^{k} P_i$. Let C_k be the *contour* of G_k defined as follows: If $k = 0$, then C_0 is the edge (v_1, v_2), while if $k > 0$, then C_k is the path from v_1 to v_2 obtained by removing (v_1, v_2) from the cycle delimiting the outer face of G_k. Partition δ is a *canonical order* [17] of G if for each $k = 1, \ldots, m-1$ the following conditions hold: (i) G_k is biconnected and internally 3-connected, (ii) all neighbors of P_k in G_{k-1} are on C_{k-1}, (iii) either P_k is a *singleton* (i.e., $|P_k| = 1$), or P_k is a *chain* (i.e., $|P_k| > 1$) and the degree of each vertex of P_k is 2 in G_k, (iv) all vertices of P_k with $0 \leq k < m$ have at least one neighbor in P_j for some $j > k$. For example, a canonical order for the graph of Fig. 1 is $P_0 = \{v_1, v_2\}$, $P_1 = \{v_3, v_4\}$, $P_2 = \{v_5, v_6\}$ and $P_3 = \{v_7\}$. A canonical order of G can be computed in $O(n)$ time [17].

Kant's Algorithm. Kant [17] describes an incremental drawing algorithm that, in linear time, computes a convex straight-line planar drawing Γ of an n-vertex plane graph G on an integer grid of size $(2n - 4) \times (n - 2)$. The drawing Γ has the same planar embedding as the input graph G. The algorithm is based on a canonical order δ of G and works as follows: Initially, vertices v_1 and v_2 of P_0 are placed at points $(0,0)$ and $(1,0)$, respectively. For $k = 1, \ldots, m$, assume that a convex drawing Γ_{k-1} of G_{k-1} has been constructed in which the edges of contour C_{k-1} are drawn with slopes 0 and ± 1 (*contour condition*; see Fig. 3a). Let $(w_1, \ldots, w_p)$ be the vertices of C_{k-1} from left to right in Γ_{k-1}, where $w_1 = v_1$ and $w_p = v_2$. Each vertex v in G_{k-1} has been associated with a *shift-set* $S(v)$, such that Γ_{k-1} is *stretchable*, that is, for each $i = 1, \ldots, p$ the result of shifting $S(w_i), \ldots, S(w_p)$ by one (or more) units to the right is a convex drawing of G_{k-1}. Let $P_k = \{z_1, \ldots, z_p\}$ be the next path in δ. Let w_ℓ and w_r be the leftmost and rightmost neighbors of P_k on C_{k-1} in Γ_{k-1}, where $1 \leq \ell < r \leq p$. To introduce P_k and to avoid edge-overlaps, the algorithm first identifies two so-called critical vertices $w_{\ell'}$ and $w_{r'}$ with $\ell \leq \ell', r' \leq r$ and then shifts (i) by one unit to the right each vertex in $\bigcup_{i=\ell'}^{p} S(w_i)$ and then (ii) by one unit to the right each vertex in $\bigcup_{i=r'}^{p} S(w_i)$. Then, z_1 is placed at intersection of the line of slope $+1$ through w_ℓ with the line through of slope -1 point w_r; see Fig. 3b. If P_k is a chain, then for $i = 2, \ldots, p$, vertex z_i is placed one unit to the right of z_{i-1} by shifting each vertex in $\bigcup_{i=r'}^{p} S(w_i)$ one unit to the right. Finally, the shift-sets of the vertices of P_k are defined accordingly to ensure that Γ_k is stretchable.

3 Properties of Kant's Algorithm

We provide properties of drawings computed by Kant's algorithm that we leverage in the next section; some of these properties are indirectly mentioned also

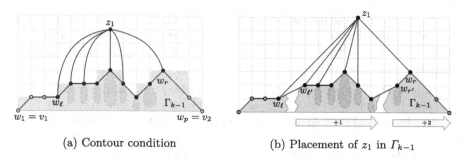

(a) Contour condition   (b) Placement of z_1 in Γ_{k-1}

Fig. 3. Introducing a singleton $P_k = \{z_1\}$ in Γ_{k-1} in the algorithm by Kant [17].

in [16]. To ease the presentation, we first introduce a 4-coloring for the edges of G similar to the one by Schnyder [13,21]. We color edge (v_1, v_2) of G_0 black. Given a 4-coloring for G_{k-1} with $k = 1, \ldots, m$, we extend it for G_k as follows (see Figs. 1 and 3a). We first color the edges of G_k that do not belong to G_{k-1} and are on contour C_k. Namely, the first such edge encountered in a clockwise walk of C_k from v_1 to v_2 is blue, the last one is green and all remaining ones (that is, those having both endpoints in P_k when P_k is a chain) are black. The remaining edges of G_k not in G_{k-1} are red (i.e., those that are incident to P_k and are not part of C_k; this case only arises if P_k is a singleton by Condition (iii) of the canonical order), which implies that C_k has no red edges.

Since a shift to introduce a path of δ in the incremental construction of Γ can only decrease the slope of a blue edge, increase the slope of a green edge, while the black and the red edges maintain their slope [17], we have that:

- the slope of each blue edge ranges in $(0, 1]$,
- the slope of each black edge is 0,
- the slope of each green edge ranges in $[-1, 0)$, and
- the slope of each red edge ranges in the complement of $[-1, 1]$.

Since each inner face in Γ is formed when a path of δ is introduced during the incremental construction, part of it belongs to the contour, while its remaining part is formed by the introduced path, which gives rise to the following property.

Property 1. *Let x be the leftmost vertex of an inner face g in Γ (in case of more than one such vertices select the bottommost one). A counterclockwise walk of g starting from x consists of the following boundary parts (see Fig. 4):*

  i. a (possibly empty) strictly descendant path of green edges,
 ii. at most one black edge,
iii. a (possibly empty) strictly ascendant path of blue edges,
 iv. a green or red edge,
  v. a (possibly empty) horizontal path of black edges, and
 vi. a blue or red edge.

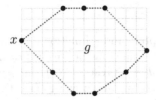

Fig. 4. Illustration for the shape of a face.

Boundary parts (iv)–(vi) (dotted in Fig. 4) are introduced when a path is added during the incremental construction of Γ (which implies that at least one of boundary parts (i)–(iii) is part of the contour and thus is non empty in g). So, boundary parts (iv)–(vi) cannot simultaneously contain black and red edges (by the edge-coloring and Condition (iii) of canonical order).

Property 2 ($\star$). *Each vertex w of a path P_i, with $0 < i \leq m$, has at least two incident edges (a, w) and (b, w), such that $y_a \leq y_w$, $y_b \leq y_w$, and $x_a < x_w < x_b$.*

Property 3 ($\star$). *Every face in Γ has at most one edge drawn vertical.*

Property 4 ($\star$). *Let u, v and w be three consecutive vertices encountered in this order in a counterclockwise walk along the boundary of an inner face of Γ. If they are collinear and the line through them has zero slope, then they are part of a chain. If they are collinear and the line through them has positive (negative) slope, then $y_u < y_v < y_w$ ($y_u > y_v > y_w$). If they are not collinear, then $v \notin \Delta_{uw}$.*

4   Algorithm Description

We now describe our approach to compute a strictly-convex drawing of a 3-connected plane graph, assuming that its outer face has at most 5 vertices (such a face always exists). We start with Sect. 4.1, in which we describe the properties of what we call lifting functions and liftable drawings. A key property is that applying a non-affine transformation to a liftable drawing by means of a lifting function preserves planarity. As this tool might be of independent interest, we state it as general as possible. In Sect. 4.2, we prove that drawings computed by Kant's algorithm are indeed liftable and that a transformation via a liftable function makes them internally strictly-convex except for possible horizontally-aligned paths. Up to this point, it was not needed to choose a particular lifting function; in Sect. 4.3 we unveil our choice. We also design a second transformation targeted to faces containing paths of horizontally-aligned vertices. The output of this step is an internally strictly-convex drawing. The last step of the algorithm is described in Sect. 4.4, namely a simple preprocessing in which the outer face of the input graph, which by our assumption has at most 5 vertices, is suitably augmented with dummy vertices, whose removal from the computed drawing guarantees that all faces (including the outer one) are strictly-convex.

4.1   Lifting Functions

Given a drawing Γ of a graph G and a function $f : \mathbb{R} \mapsto \mathbb{R}$, we refer to the drawing Γ_f obtained by applying the transformation $(x_u, y_u) \mapsto (x_u, f(y_u))$ to Γ as the *transformed drawing* of Γ with respect to f. In view of Theorem 2 below, we focus on lifting functions and liftable drawings (see Definitions 1 and 2).

Definition 1. *A function $f : \mathbb{R} \mapsto \mathbb{R}$ is* lifting *if and only if (i) f is strictly-convex and strictly-increasing; (ii) $f(r) \geq r, \forall r \in \mathbb{R}$; (iii) $r \in \mathbb{N} \Rightarrow f(r) \in \mathbb{N}$.*

The next property follows directly from Definition 1.

Observation 1. *Let Γ be a drawing of a graph G. Given a lifting function f, three vertices are horizontally (vertically) aligned in Γ if and only if they are horizontally (vertically) aligned in the transformed drawing Γ_f.*

Definition 2. *A planar straight-line grid drawing of a graph is called* liftable *if for every edge (u, v) there is no vertex of G in Δ_{uv}.*

Theorem 2. *Let Γ be a planar straight-line grid drawing of a plane graph G. If Γ is liftable, then its transformed drawing Γ_f with respect to a lifting function f is a planar straight-line grid drawing of G with the same planar embedding as Γ.*

Proof. Condition (iii) of Definition 1 trivially implies that Γ_f is a grid drawing. We next prove that Γ and Γ_f have the same planar embedding. (Note that, if Γ is not liftable, Γ_f is not necessarily planar.) Since, by Condition (ii) of Definition 1, Γ_f is obtained from Γ by shifting vertices upwards, the existence in Γ_f of an edge crossing or of a vertex having a circular order of its incident edges different than the one in Γ, implies that there exist a vertex w and an edge (u, v) in G, such that w is below (above) (u, v) in Γ and above (below) (u, v) in Γ_f.

We next argue that the situation described above is not possible. Consider a vertex w and an edge (u, v) of G and let s and s' be the line-segments representing (u, v) in Γ and Γ_f. Clearly, it suffices to consider the case in which $x_u \leq x_w \leq x_v$. Let p and p' be the vertical projection of w on s and s', respectively. Also, let y_p and $y_{p'}$ be the y-coordinates of p in Γ and of p' in Γ_f, respectively. Suppose $y_u \leq y_v$; the case in which $y_u > y_v$ is symmetric.

Firstly, consider the case in which w is above s, i.e., $y_w > y_p$. Since Γ is liftable, vertex w does not belong to Δ_{uv}. Hence, $y_u \leq y_p \leq y_v \leq y_w$. Since f is strictly-increasing, it follows $f(y_u) \leq f(y_p) \leq f(y_v) \leq f(y_w)$. However, $f(y_w) > y_{p'}$ implies that w is above s', as desired. Secondly, consider the case in which w is below s in Γ. Here, we distinguish three cases: $x_w = x_u$, $x_w = x_v$ and $x_u < x_w < x_v$. In the first case, we have $y_w < y_p = y_u$ and $y_{p'} = f(y_u)$. Since f is strictly-increasing, it holds $f(y_w) < f(y_u) = y_{p'}$, i.e., w is below s', as desired. The second case is analogous. For the third case, we know that $y_w < y_p$. Since p lies on s, by Lemma 1, $f(y_p) < y_{p'}$ holds. Also, since f is strictly-increasing, it follows $f(y_w) < f(y_p)$. Thus, $f(y_w) < y_{p'}$ holds, i.e., w is below s', as desired. □

4.2   Application to Kant's Drawings

We now show that applying a lifting function to a drawing computed by Kant's algorithm (see Sect. 3) yields a drawing with several important properties.

Lemma 2. *Let Γ be a drawing of a 3-connected plane graph G computed by Kant's algorithm. Drawing Γ is liftable.*

Proof. Consider an edge (u, v) of G and w.l.o.g. assume $y_u < y_v$ in Γ. We prove that there is no vertex w in Δ_{uv}. This is obvious when $x_u = x_v$, since $\Delta_{uv} = \emptyset$. Hence, either $x_u < x_v$ or $x_u > x_v$. Consider the former case; the latter can be treated symmetrically. Suppose for a contradiction that there exists at least one vertex (other than u and v) in Δ_{uv}. Let w be the rightmost vertex out of those in Δ_{uv}. Since w is in Δ_{uv} we know $y_u < y_w < y_v$. Since $y_u \neq y_v$, it follows that (u, v) is not the edge (v_1, v_2) of P_0, which, in turn, implies that vertex w belongs to a path P_i with $i > 0$, since $y_u < y_w$. Hence, by Property 2, w has at least two incident edges (a, w) and (b, w), such that $y_a \leq y_w$, $y_b \leq y_w$, and $x_a < x_w < x_b$. Since b is to the right of w, the way we selected w implies that b does not belong to Δ_{uv}; consequently, (w, b) crosses (u, v), contradicting the planarity of Γ.   □

By combining Theorem 2 and Lemma 2, we conclude the following.

Theorem 3 ($\star$). *Given a 3-connected plane graph G and a lifting function f, let Γ_f be the transformed drawing of a drawing Γ of G computed by Kant's algorithm. Then, Γ_f is internally-convex and planar with the same embedding as Γ. Also, if two consecutive edges of an inner face of Γ_f form an angle π inside this face, then these edges are horizontal.*

Proof sketch. Since, by Lemma 2, Γ is liftable, by Theorem 2 Γ_f has the same planar embedding as Γ. Consider a counterclockwise walk along the boundary of an inner face g in Γ and let u, v and w be three consecutive vertices along this walk. Let α and α' be the angle at v formed by the edges (u, v) and (v, w) inside g in Γ and in Γ_f, respectively. Since Γ is convex, $\alpha \leq \pi$. We claim that $\alpha' \leq \pi$ and that if $\alpha' = \pi$, then (u, v) and (v, w) are horizontally aligned in Γ_f. We prove the claim when u, v and w are collinear in Γ. By Property 3, vertices u, v and w are not vertically aligned in Γ. If they are horizontally aligned in Γ, then by Observation 1 they are horizontally aligned also in Γ_f, as desired. Suppose now the line ℓ through u, v and w in Γ is either of positive or of negative slope, which by Property 4 implies that either $y_u < y_v < y_w$ or $y_u > y_v > y_w$ holds, respectively. Then, by Lemma 1, it follows that v is below the line-segment connecting u and w in Γ_f, hence $\alpha' < \pi$.   □

4.3   Putting Everything Together

We are now ready to put all pieces together. Let G be a 3-connected plane graph with n vertices. Without loss of generality, we can assume that the outer face of G contains at most 5 vertices (which will be useful in the next subsection), since

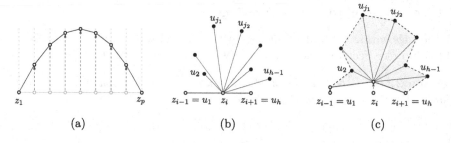

Fig. 5. (a) Illustration of the procedure of shifting the vertices of a chain upwards, and (b-c) cases that arise in the proof of Theorem 4.

such a face always exists. Let Γ be a convex drawing of G computed by Kant's algorithm and let $f : \mathbb{R} \to \mathbb{R}$ be the function $f(y) = 5(n-2)^2 y + y^2$. Clearly, f is a lifting function. Hence, by Theorem 3, in the transformed drawing Γ_f, each inner face g that is not strictly-convex contains at least three horizontally-aligned vertices. By property 4, these vertices are part of a chain in the canonical order δ. Hence, by Condition (iv) of canonical order, each of the vertices of this chain has at least one neighbor placed above it. By definition of f it follows that each of these neighbors is positioned at least $5(n-2)^2$ units above the chain in Γ_f. We exploit this property to turn Γ_f into an internally strictly-convex drawing by shifting all vertices of each chain upwards while keeping Γ_f planar.

To this end, let $P_k = \{z_1, \ldots, z_p\}$ with $p \geq 2$ be a chain in the canonical order δ used to construct Γ. For $i = 1, \ldots, p$, we shift vertex z_i of P_k by $(x_{z_i} - x_{z_1})(x_{z_p} - x_{z_i})$ units upwards; see Fig. 5a. It follows that if λ is the total width of P_k in Γ (and thus also in Γ_f), then each vertex in P_k is shifted by at most $\lambda^2/4$ units of length upwards, which is in turn at most $(n-2)^2$, since the total width of Γ is at most $2n - 4$, and therefore $\lambda \leq 2n - 4$. Also, note that only the *internal vertices* of P_k are shifted (if any), i.e., only the vertices z_i with $2 \leq i \leq p - 1$. Let $\widehat{\Gamma_f}$ be the drawing obtained from Γ_f by applying the aforementioned procedure to each chain, which we call the *curved drawing* of Γ_f. Clearly, $\widehat{\Gamma_f}$ is a grid drawing, we can prove that it is planar and internally strictly-convex.

Theorem 4 ($\star$). *Given a 3-connected n-vertex plane graph G and the lifting function $f : \mathbb{R} \mapsto \mathbb{R}$ with $f(y) = 5(n-2)^2 y + y^2$, let Γ_f be the transformed drawing of a drawing Γ of G computed by Kant's algorithm. Then, the curved drawing $\widehat{\Gamma_f}$ of Γ_f is an internally strictly-convex grid drawing of G with the same planar embedding as Γ.*

Proof sketch. Concerning the planarity of $\widehat{\Gamma_f}$, consider any internal vertex of a chain and its neighbors. At high-level, we have that the envelope through such neighbors is a polygon whose boundary is formed by a left and a right path that are y-monotone (see Fig. 5b). Since the vertex remains below its lowest successor in the canonical order (see Fig. 5c), no edge crossing can be introduced. Thus, $\widehat{\Gamma_f}$ is planar with the same embedding as Γ_f (and thus as Γ, by Theorem 3).

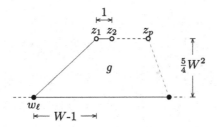

Fig. 6. Illustration for the proof of Theorem 4; W denotes the width of Γ_f.

We next prove that $\widehat{\Gamma_f}$ is internally strictly-convex. Let g be an inner face in $\widehat{\Gamma_f}$. Recall that only internal chain-vertices are shifted in the transition from Γ_f to $\widehat{\Gamma_f}$. Hence, if g does not contain internal chain-vertices, then it is strictly-convex by Theorem 3. Thus, we may assume that g contains at least one such vertex z. By Property 1, z is either in the topmost chain of black edges of g, or it is one of the (at most two) bottommost vertices of g. Consider the former, as the latter is similar. Let $P_k = \{z_1, \ldots, z_p\}$ be the chain containing z. We argue that any angle inside g incident to these vertices is smaller than π. By construction, this is the case for vertices $z_2, \ldots, z_{p-1}$. Hence, it remains to consider the angles at z_1 and z_p. Since the two cases are symmetric, consider the angle at z_1. Let w_ℓ be the neighbor of z_1 along C_{k-1}, i.e., w_ℓ is the vertex preceding z_1 in a clockwise walk of g starting from z_1. We will prove that the slope of (w_ℓ, z_1) is strictly greater than the one of (z_1, z_2), hence the angle at z_1 is less than π.

Refer to Fig. 6. By the way the vertices of P_k are shifted in the transition from Γ_f to $\widehat{\Gamma_f}$, it follows that the maximum of the slope of (z_1, z_2) is $2(n-2)-1$ (i.e., achieved when P_k is of maximum x-length in $\widehat{\Gamma_f}$ and the x-distance of z_1 and z_2 in $\widehat{\Gamma_f}$ is 1). We next argue for the slope of the edge (w_ℓ, z_1). Recall that vertex z_1 is not an interior vertex of P_k, which implies that it has not been shifted in the transition from Γ_f to $\widehat{\Gamma_f}$. The same, however, does not necessarily hold for w_ℓ. As a matter of fact, this vertex may be part of a chain, i.e., when g does not contain boundary part (i) but boundary part (ii) of Property 1. This implies that it may have been shifted upwards by at most $(n-2)^2$ units in the transition from Γ_f to $\widehat{\Gamma_f}$. The minimum of the slope of (w_ℓ, z_1) in Γ_f is achieved, when (w_ℓ, z_1) is of maximum x-length in Γ_f and of minimum y-length. Since the former is at most $2(n-2)-1$, while the latter is at least $5(n-2)^2$, it follows that the minimum of the slope of (w_ℓ, z_1) is potentially $\frac{5(n-2)^2}{2(n-2)-1}$ in Γ_f. Since in the transition from Γ_f to $\widehat{\Gamma_f}$ vertex w_ℓ may be shifted by at most $(n-2)^2$ units, it follows that the slope of (w_ℓ, z_1) may reduce further to $\frac{5(n-2)^2-(n-2)^2}{2(n-2)-1}$, which is its minimum value. Therefore, the slope of the edge (w_ℓ, z_1) is strictly greater than the one of (z_1, z_2), since the following trivially holds:

$$\frac{5(n-2)^2-(n-2)^2}{2(n-2)-1} > 2(n-2)-1 \iff 2(n-2) > 2(n-2)-1. \qquad \square$$

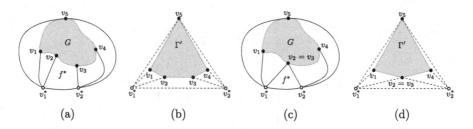

Fig. 7. Treating the outer face when its degree is five (a–b) and four (c–d).

4.4   Outer Face and Final Analysis

To complete the description of our algorithm, it remains to guarantee that the outer face of the computed drawings is strictly-convex. To this aim, we slightly augment the input graph G and suitably choose the canonical order to give as input to Kant's algorithm. Consider a planar embedding of G and let $v_1, v_2, \ldots, v_h$ be the vertices on the outer face (see Fig. 7a); recall that we have assumed $h \leq 5$. If $h = 3$, then the boundary of the outer face is a triangle and hence strictly-convex. So, assume $4 \leq h \leq 5$. To ease the presentation, we let $h = 5$ (see Figs. 7a and 7b), as the case $h = 4$ is simpler (see Figs. 7c and 7d). We proceed by adding two vertices $v_1^\star$ and $v_2^\star$ in the outer face of G and edges $(v_1^\star, v_2^\star)$, $(v_1^\star, v_5)$, $(v_1^\star, v_1)$, $(v_1^\star, v_2)$, $(v_2^\star, v_3)$, $(v_2^\star, v_4)$, and $(v_2^\star, v_5)$. The resulting graph $G^\star$ is still planar and 3-connected. In particular, its outer face is a 3-cycle formed by $v_1^\star, v_2^\star, v_5$. We compute a canonical order $\delta^\star$ of $G^\star$ with $P_0 = (v_1^\star, v_2^\star)$ and $P_m = \{v_5\}$. The key observation is that the second set of $\delta^\star$ is the chain $P_1 = \{v_2, v_3\}$, since it forms the inner face $f^\star$ of $G^\star$ with $(v_1^\star, v_2^\star)$ on its boundary.

Next, we apply the algorithm supporting Theorem 4 to $G^\star$ using the aforementioned canonical order $\delta^\star$ and obtain a drawing of it that is internally strictly-convex. We next prove that the removal of $v_1^\star$ and $v_2^\star$ from this drawing yields a drawing of G that is strictly-convex. By Theorem 4 and by our augmentation, it suffices to guarantee that the outer face of the obtained drawing is strictly-convex. Consider first the inner angle at v_5 of the polygon bounding the outer face; this angle is strictly less than π, because v_5 is the topmost vertex of the drawing (and no other vertex is horizontally aligned with it). A similar argument applies for the angles at v_1 and v_4; in particular, after the removal of $v_1^\star$ and $v_2^\star$, vertices v_1 and v_4 are the leftmost and the rightmost neighbors of v_5, respectively, and therefore they are the leftmost and the rightmost vertices in the drawing, respectively. Concerning v_2 and v_3, they are horizontally aligned and, after the removal of $v_1^\star$ and $v_2^\star$, they are the bottommost vertices of the drawing. Thus, their angles are also strictly less that π completing the proof of our claim. To conclude the proof of Theorem 1, it remains to discuss the area required by the drawing obtained as above and the time complexity to compute it.

Area Bound. The drawing Γ computed by Kant's algorithm for $G^\star$ fits on an integer grid of size $(2n^\star - 4) \times (n^\star - 2)$, where $n^\star = n + 2$ ($G^\star$ has two more vertices than G). The transformed drawing Γ_f of Γ by means of the lifting function $f : \mathbb{R} \mapsto \mathbb{R}$ with $f(y) = 5(n^\star - 2)^2 y + y^2$ has the same width as Γ, while the vertices $v_1^\star$ and $v_2^\star$ have y-coordinate 0 in Γ_f. On the other hand, vertex v_5 has y-coordinate $5(n^\star - 2)^2(n^\star - 2) + (n^\star - 2)^2$, which is also the height of Γ_f. Since no vertex of the outer face of Γ_f is further shifted upwards, the curved drawing $\widehat{\Gamma_f}$ of Γ_f has the same width and height as Γ_f. After removing $v_1^\star$ and $v_2^\star$, the width of the final drawing of G is at least two units less than the one of $\widehat{\Gamma_f}$, while its height is at least $5(n^\star - 2)^2$ units less. Since $n^\star = n + 2$, the final drawing lies on a grid of size $((2(n+2) - 4) - 2) \times (5((n+2) - 2)^3 - 4((n+2) - 2)^2) = 2(n-1) \times (5n^3 - 4n^2)$.

Time Complexity. Each step of our algorithm can be implemented in $O(n)$ time: (i) finding a planar embedding of G with a face of degree at most 5, (ii) computing $G^\star$, a canonical order of it, and applying Kant's algorithm to $G^\star$, (iii) computing the transformed drawing with respect to our lifting function f and updating the position of the internal chain-vertices. This completes the proof of Theorem 1.

5   Conclusions and Open Problems

We have provided a linear-time algorithm that computes a strictly-convex drawing of a 3-connected planar graph on an integer grid of size $2(n-1) \times (5n^3 - 4n^2)$. Compared to the previously best-known upper bound for such drawings [1], we largely improve the multiplicative constants by means of an arguably simpler algorithm, which therefore has the potential to be of practical use. Along the way, we proved tools that can be of independent interest (see in particular Theorem 2). Some problems that stem from our research are the following:

– Can we achieve a similar area bound together with a constant aspect ratio?
– Is $\Omega(n^4)$ a lower bound for the area requirement of strictly-convex drawings?
– Can we compute strictly-convex drawings in small area with good edge-vertex resolution [2,3]?

References

1. Bárány, I., Rote, G.: Strictly convex drawings of planar graphs. Doc. Math. **11**, 369–391 (2006)
2. Bekos, M.A., Gronemann, M., Montecchiani, F., Pálvölgyi, D., Symvonis, A., Theocharous, L.: Grid drawings of graphs with constant edge-vertex resolution. Comput. Geom. **98**, 101789 (2021)
3. Bekos, M.A., Gronemann, M., Montecchiani, F., Symvonis, A.: Convex grid drawings of planar graphs with constant edge-vertex resolution. In: IWOCA 2022. LNCS, vol. 13270, pp. 157–171. Springer (2022)
4. Bekos, M.A., Gronemann, M., Montecchiani, F., Symvonis, A.: Strictly-convex drawings of 3-connected planar graphs. CoRR abs/2208.13388 (2022)
5. Chrobak, M., Goodrich, M.T., Tamassia, R.: Convex drawings of graphs in two and three dimensions (preliminary version). In: Whitesides, S. (ed.) SoCG, pp. 319–328. ACM (1996)

6. Chrobak, M., Kant, G.: Convex grid drawings of 3-connected planar graphs. Int. J. Comput. Geom. Appl. **7**(3), 211–223 (1997)
7. Chrobak, M., Payne, T.H.: A linear-time algorithm for drawing a planar graph on a grid. Inf. Process. Lett. **54**(4), 241–246 (1995)
8. de Fraysseix, H., Pach, J., Pollack, R.: How to draw a planar graph on a grid. Combinatorica **10**(1), 41–51 (1990)
9. Di Battista, G., Eades, P., Tamassia, R., Tollis, I.G.: Graph Drawing: Algorithms for the Visualization of Graphs. Prentice-Hall (1999)
10. Di Battista, G., Frati, F.: Drawing trees, outerplanar graphs, series-parallel graphs, and planar graphs in a small area. In: Pach, J. (ed.) Thirty Essays on Geometric Graph Theory, pp. 121–165. Springer, New York (2013)
11. Di Battista, G., Tamassia, R., Vismara, L.: Output-sensitive reporting of disjoint paths. Algorithmica **23**(4), 302–340 (1999)
12. Felsner, S.: Convex drawings of planar graphs and the order dimension of 3-polytopes. Order **18**(1), 19–37 (2001)
13. Felsner, S.: Geometric Graphs and Arrangements. Advanced Lectures in Mathematics, Vieweg (2004)
14. Fáry, I.: On straight lines representation of planar graphs. Acta Sci. Math. (Szeged) **11**, 229–233 (1948)
15. He, X.: Grid embedding of 4-connected plane graphs. Discret. Comput. Geom. **17**(3), 339–358 (1997)
16. Kant, G.: Hexagonal grid drawings. In: Mayr, E.W. (ed.) WG 1992. LNCS, vol. 657, pp. 263–276. Springer, Heidelberg (1993). https://doi.org/10.1007/3-540-56402-0_53
17. Kant, G.: Drawing planar graphs using the canonical ordering. Algorithmica **16**(1), 4–32 (1996)
18. Miura, K., Nakano, S., Nishizeki, T.: Grid drawings of 4-connected plane graphs. Discret. Comput. Geom. **26**(1), 73–87 (2001)
19. Rote, G.: Strictly convex drawings of planar graphs. In: SODA, pp. 728–734. SIAM (2005)
20. Royden, H.L., Fitzpatrick, P.M.: Real analysis. NY, fourth. edn, Pearson modern classic, Pearson, New York (2010)
21. Schnyder, W.: Embedding planar graphs on the grid. In: SODA, pp. 138–148. SIAM (1990)
22. Schnyder, W., Trotter, W.T.: Convex embeddings of 3-connected plane graphs. Abstracts AMS **13**, 502 (1992)
23. Thomassen, C.: A refinement of Kuratowski's theorem. J. Comb. Theory, Ser. B **D**(3), 245–253 (1984)
24. Tutte, W.T.: How to draw a graph. Proc. London Math. Soc. **13**, 743–768 (1963)
25. Zhang, H., He, X.: Compact visibility representation and straight-line grid embedding of plane graphs. In: Dehne, F., Sack, J.-R., Smid, M. (eds.) WADS 2003. LNCS, vol. 2748, pp. 493–504. Springer, Heidelberg (2003). https://doi.org/10.1007/978-3-540-45078-8_43

Rectilinear Planarity of Partial 2-Trees

Walter Didimo[1][iD], Michael Kaufmann[2][iD], Giuseppe Liotta[1][iD], and Giacomo Ortali[1(✉)][iD]

[1] Università degli Studi di Perugia, Perugia, Italy
{walter.didimo,giuseppe.liotta,giacomo.ortali}@unipg.it
[2] University of Tübingen, Tübingen, Germany
mk@informatik.uni-tuebingen.de

Abstract. A graph is rectilinear planar if it admits a planar orthogonal drawing without bends. While testing rectilinear planarity is NP-hard in general, it is a long-standing open problem to establish a tight upper bound on its complexity for partial 2-trees, i.e., graphs whose biconnected components are series-parallel. We describe a new $O(n^2 \log^2 n)$-time algorithm to test rectilinear planarity of partial 2-trees, which improves over the current best bound of $O(n^3 \log n)$. Moreover, for series-parallel graphs where no two parallel-components share a pole, we are able to achieve optimal $O(n)$-time complexity. Our algorithms are based on an extensive study and a deeper understanding of the notion of orthogonal spirality, introduced in 1998 to describe how much an orthogonal drawing of a subgraph is rolled-up in an orthogonal drawing of the graph.

Keywords: Rectilinear planarity testing · Variable embedding · Series-parallel graphs · Partial 2-trees · Orthogonal drawings

1 Introduction

In an *orthogonal drawing* of a graph each vertex is a distinct point of the plane and each edge is a chain of horizontal and vertical segments. Rectilinear planarity testing asks whether a planar 4-graph (i.e., with vertex-degree at most four) admits a planar orthogonal drawing without edge bends. It is a classical subject of study in graph drawing, partly for its theoretical beauty and partly because it is at the heart of the algorithms that compute bend-minimum orthogonal drawings, which find applications in several domains (see, e.g., [3,6,12,19–21]). Rectilinear planarity testing is NP-hard [15], it belongs to the XP-class when parameterized by treewidth [5], and it is FPT when parameterized by the number of degree-4 vertices [10]. Polynomial-time solutions exist for restricted versions of the problem. Namely, if the algorithm must preserve a given planar embedding, rectilinear planarity testing can be solved in subquadratic time

Work partially supported by: (i) MIUR, grant 20174LF3T8 AHeAD: "efficient Algorithms for HArnessing networked Data", (ii) Dipartimento di Ingegneria, Universita degli Studi di Perugia, grant RICBA21LG: Algoritmi, modelli e sistemi per la rappresentazione visuale di reti.

P. Angelini and R. von Hanxleden (Eds.): GD 2022, LNCS 13764, pp. 157–172, 2023.
https://doi.org/10.1007/978-3-031-22203-0_12

for general graphs [2,14], and in linear time for planar 3-graphs [23] and for biconnected series-parallel graphs (SP-graphs for short) [7]. When the planar embedding is not fixed, linear-time solutions exist for (families of) planar 3-graphs [11,17,22,25] and for outerplanar graphs [13]. A polynomial-time solution for SP-graphs has been known for a long time [4], but establishing a tight complexity bound for rectilinear planarity testing of SP-graphs remains a long-standing open problem.

In this paper we provide significant advances on this problem. Our main contribution is twofold:

- We present an $O(n^2 \log^2 n)$-time algorithm to test rectilinear planarity of partial 2-trees, i.e., graphs whose biconnected components are SP-graphs. This result improves the current best known bound of $O(n^3 \log n)$ [5].
- We give an $O(n)$-time algorithm for those SP-graphs where no two parallel-components share a pole. We also show a logarithmic lower bound on the spirality that an orthogonal component of a graph in this family can take.

Our algorithms are based on an extensive study and a deeper understanding of the notion of orthogonal spirality, introduced in 1998 to describe how much an orthogonal drawing of a subgraph is rolled-up in an orthogonal drawing of the graph [4]. In the concluding remarks we also mention some of the pitfalls behind an $O(n)$-time algorithm for partial 2-trees. For reasons of space several details are omitted and can be found in [9].

2   Preliminaries

A planar orthogonal drawing can be computed in linear time from a so-called *planar orthogonal representation*, which describes the sequences of bends along the edges and the angles at the vertices [24]. Hence, rectilinear planarity testing is equivalent to asking whether a graph has a planar *rectilinear representation*, i.e., a planar orthogonal representation without bends. We study graphs that are not simple cycles, as otherwise rectilinear planarity testing is trivial. We just use the term "rectilinear representation" in place of "planar rectilinear representation".

A biconnected graph G is a *series-parallel graph* (or *SP-graph*) when none of its triconnected components is a triconnected graph. A *partial 2-tree* is a graph whose biconnected components are SP-graphs. If G is an SP-graph, the *SPQ-tree* T of G describes the decomposition of G into its triconnected components. It can be computed in linear time [3,16,18] and consists of three types of nodes: *S-*, *P-*, and *Q-nodes*. The degree-1 nodes of T are Q-nodes, each corresponding to a distinct edge of G. If ν is an S-node (resp. a P-node) it represents a series-component (resp. a parallel-component), denoted as skel(ν) and called the *skeleton* of ν. If ν is an S-node, skel(ν) is a simple cycle of length at least three; if ν is a P-node, skel(ν) is a bundle of at least three multiple edges. Any two S-nodes (resp. P-nodes) are never adjacent in T. A *real edge* (resp. *virtual edge*) in skel(ν) corresponds to a Q-node (resp. an S- or a P-node) adjacent to ν in T.

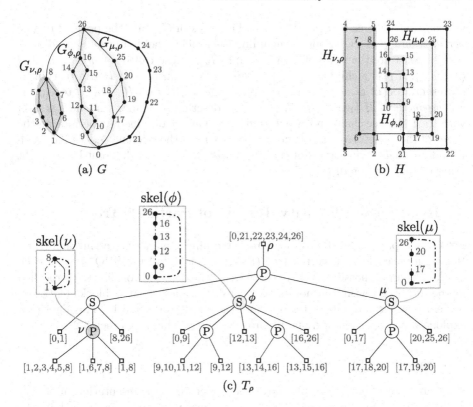

Fig. 1. (a) An SP-graph G. (b) A rectilinear representation H of G. (c) The SPQ*-tree T_ρ of G, where ρ corresponds to the thick chain; Q*-nodes are small squares. The components and the skeletons of the nodes ν, μ, ϕ are shown: virtual edges are dashed and the reference edge is thicker.

We use a variant of SPQ-tree called *SPQ*-tree* (refer to Fig. 1). In an SPQ*-tree, each degree-1 node of T is a *Q*-node*, and represents a maximal chain of edges of G starting and ending at vertices of degree larger than two and passing through a sequence of degree-2 vertices only. If ν is an S- or a P-node, an edge of skel(ν) corresponding to a Q*-node μ is virtual if μ is a chain of at least two edges, else it is a real edge.

For any given Q*-node ρ of T, denote by T_ρ the tree T rooted at ρ. The chain of edges represented by ρ is the *reference chain* of G w.r.t. T_ρ. If ν is an S- or a P-node distinct from the root child of T_ρ, then skel(ν) contains a virtual edge that has a counterpart in the skeleton of its parent; this edge is the *reference edge* of skel(ν). If ν is the root child, the reference edge of skel(ν) is the edge corresponding to ρ. For any S- or P-node ν of T_ρ, the end-vertices of the reference edge of skel(ν) are the *poles* of ν and of skel(ν). Note that skel(ν) does not change if we change ρ. For any S- or P-node ν of T_ρ, the *pertinent graph* $G_{\nu,\rho}$ of ν is the subgraph of G formed by the union of the chains represented by the leaves in the subtree of T_ρ rooted at ν. The *poles* of $G_{\nu,\rho}$ are the poles of ν. The *pertinent graph* of a Q*-node ν (including the root) is the chain represented by ν, and its *poles* are the poles of ν. Any graph $G_{\nu,\rho}$ is also called a *component* of G (w.r.t. ρ). If μ is a child of ν, $G_{\mu,\rho}$ a *child component* of ν. If H is a rectilinear representation of G, for any node ν of T_ρ, the restriction $H_{\nu,\rho}$ of H to $G_{\nu,\rho}$ is a *component* of H (w.r.t. ρ). We use T_ρ to describe all planar embeddings of G with the reference chain on the external face; they are obtained by permuting the edges of the skeletons of the P-nodes distinct from the reference edges. For each P-node ν, each permutation of the edges in $\mathrm{skel}(\nu)$ gives a different left-to-right order of the children of ν in T_ρ.

3 Rectilinear Planarity Testing of Partial 2-Trees

Let G be a partial 2-tree. We describe a rectilinear planarity testing algorithm that visits the block-cutvertex tree (BC-tree) of G and the SPQ*-tree of each block of G, for all possible roots of these trees (the definition of BC-tree is recalled in the [9]). We revisit the notion of "spirality values" for the blocks of G and present new ideas to efficiently compute these values (Sect. 3.1). The algorithm exploits a combination of dynamic programming techniques (Sect. 3.2).

3.1 Spirality of SP-graphs

Let G be a degree-4 SP-graph and H be a rectilinear representation of G. Let T_ρ be a rooted SPQ*-tree of G, $H_{\nu,\rho}$ be a component in H, and $\{u, v\}$ be the poles of ν, ordered according to some st-numbering of G, where s and t are the poles of ρ. For each pole $w \in \{u, v\}$, let $\mathrm{indeg}_\nu(w)$ and $\mathrm{outdeg}_\nu(w)$ be the degree of w inside and outside $H_{\nu,\rho}$, respectively. Define two (possibly coincident) *alias vertices* of w, denoted by w' and w'', as follows: (i) if $\mathrm{indeg}_\nu(w) = 1$, then $w' = w'' = w$; (ii) if $\mathrm{indeg}_\nu(w) = \mathrm{outdeg}_\nu(w) = 2$, then w' and w'' are dummy vertices, each splitting one of the two distinct edges incident to w outside $H_{\nu,\rho}$; (iii) if $\mathrm{indeg}_\nu(w) > 1$ and $\mathrm{outdeg}_\nu(w) = 1$, then $w' = w''$ is a dummy vertex that splits the edge incident to w outside $H_{\nu,\rho}$. Let A^w be the set of distinct alias vertices of a pole w. Let P^{uv} be any simple path from u to v inside $H_{\nu,\rho}$ and let u' and v' be the alias vertices of u and of v, respectively. The path $S^{u'v'}$ obtained concatenating (u', u), P^{uv}, and (v, v') is called a *spine* of $H_{\nu,\rho}$. Denote by $n(S^{u'v'})$ the number of right turns minus the number of left turns encountered along $S^{u'v'}$ while moving from u' to v'. The *spirality* $\sigma(H_{\nu,\rho})$ of $H_{\nu,\rho}$, introduced in [4], is either an integer or a semi-integer number, defined as follows (see Fig. 2): (i) If $A^u = \{u'\}$ and $A^v = \{v'\}$ then $\sigma(H_\nu) = n(S^{u'v'})$. (ii) If $A^u = \{u'\}$ and $A^v = \{v', v''\}$ then $\sigma(H_\nu) = \frac{n(S^{u'v'}) + n(S^{u'v''})}{2}$. (iii) If $A^u = \{u', u''\}$ and $A^v = \{v'\}$ then $\sigma(H_\nu) = \frac{n(S^{u'v'}) + n(S^{u''v'})}{2}$. (iv) If $A^u = \{u', u''\}$ and $A^v = \{v', v''\}$ assume, w.l.o.g., that (u, u') precedes (u, u'') counterclockwise around u and that (v, v') precedes (v, v'') clockwise around v; then $\sigma(H_\nu) = \frac{n(S^{u'v'}) + n(S^{u''v''})}{2}$.

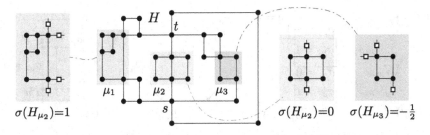

Fig. 2. A rectilinear representation H and three components w.r.t. the reference chain from s to t. Alias vertices (white squares) and spiralities are shown.

It has been proved that the spirality of $H_{\nu,\rho}$ does not depend on the choice of P^{uv} [4]. Also, a component $H_{\nu,\rho}$ of H can always be substituted by any other representation $H'_{\nu,\rho}$ of the pertinent graph $G_{\nu,\rho}$ with the same spirality, getting a new valid orthogonal representation with the same set of bends on the edges of H that are not in $H_{\nu,\rho}$ (see [4] and also Theorem 1 in [8]). For brevity, we shall denote by σ_ν the spirality of a rectilinear representation of $G_{\nu,\rho}$. Lemmas 1 to 3 relate, for any S- or P-node ν, the spirality values for a rectilinear representation of $G_{\nu,\rho}$ to those of the rectilinear representations of the child components of $G_{\nu,\rho}$ [4]. See Fig. 3.

Lemma 1. [4] *Let ν be an S-node of T_ρ with children $\mu_1,\ldots,\mu_h$ ($h \geq 2$). $G_{\nu,\rho}$ has a rectilinear representation with spirality σ_ν if and only each G_{μ_i} ($1 \leq i \leq h$) has a rectilinear representation with spirality σ_{μ_i}, such that $\sigma_\nu = \sum_{i=1}^{h} \sigma_{\mu_i}$.*

Lemma 2. [4] *Let ν be a P-node of T_ρ with three children μ_l, μ_c, and μ_r. $G_{\nu,\rho}$ has a rectilinear representation with spirality σ_ν, where $G_{\mu_l,\rho}$, $G_{\mu_c,\rho}$, $G_{\mu_r,\rho}$ are in this left-to-right order, if and only if there exist values σ_{μ_l}, σ_{μ_c}, σ_{μ_r} such that: (i) $G_{\mu_l,\rho}$, $G_{\mu_c,\rho}$, $G_{\mu_r,\rho}$ have rectilinear representations with spirality σ_{μ_l}, σ_{μ_c}, σ_{μ_r}, respectively; and (ii) $\sigma_\nu = \sigma_{\mu_l} - 2 = \sigma_{\mu_c} = \sigma_{\mu_r} + 2$.*

Let ν be a P-node of T_ρ with two children, H be a rectilinear representation of G with the reference chain on the external face. By orienting upward the edges of H according to the st-numbering, we can naturally talk about leftmost and rightmost incoming (outgoing) edges of every vertex. For a pole $w \in \{u, v\}$ of ν, the angle formed by the two leftmost (rightmost) edges incident to w (one incoming and one outgoing) is the *leftmost angle* (*rightmost angle*) at w in $H_{\nu,\rho}$. Let α_w^l and α_w^r be variables defined as: $\alpha_w^l = 0$ ($\alpha_w^r = 0$) if the leftmost (rightmost) angle at w in H is 180°, while $\alpha_w^l = 1$ ($\alpha_w^r = 1$) if this angle is 90°. Also let k_w^l and k_w^r be variables defined as: $k_w^d = 1$ if $\mathrm{indeg}_{\mu_d}(w) = \mathrm{outdeg}_\nu(w) = 1$, while $k_w^d = 1/2$ otherwise, for $d \in \{l, r\}$.

Lemma 3. [4] *Let ν be a P-node of T_ρ with two children μ_l and μ_r, and poles u and v. $G_{\nu,\rho}$ has a rectilinear representation with spirality σ_ν, where $G_{\mu_l,\rho}$ and $G_{\mu_r,\rho}$ are in this left-to-right order, if and only if there exist values σ_{μ_l}, σ_{μ_r}, α_u^l, α_u^r, α_v^l, α_v^r such that: (i) $G_{\mu_l,\rho}$ and $G_{\mu_r,\rho}$ have rectilinear representations with*

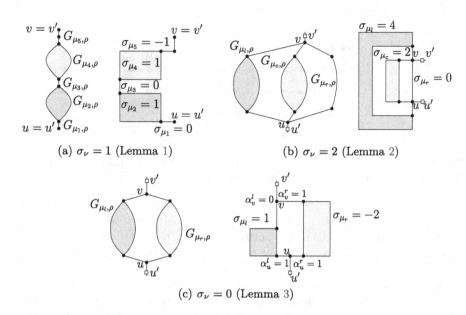

Fig. 3. Illustrations for Lemmas 1 to 3 (alias vertices are small squares).

spirality σ_{μ_l} and σ_{μ_r}, respectively; (ii) $\alpha_w^l, \alpha_w^r \in \{0, 1\}$, $1 \leq \alpha_w^l + \alpha_w^r \leq 2$ with $w \in \{u, v\}$; and (iii) $\sigma_\nu = \sigma_{\mu_l} - k_u^l \alpha_u^l - k_v^l \alpha_v^l = \sigma_{\mu_r} + k_u^r \alpha_u^r + k_v^r \alpha_v^r$.

Spirality sets. Let G be an SP-graph with n vertices, T_ρ be a rooted SPQ*-tree of G, and ν be a node of T_ρ. We say that $G_{\nu,\rho}$, or directly ν, admits spirality σ_ν in T_ρ if there exists a rectilinear representation $H_{\nu,\rho}$ with spirality σ_ν in some rectilinear representation H of G. The rectilinear spirality set $\Sigma_{\nu,\rho}$ of ν in T_ρ (and of $G_{\nu,\rho}$) is the set of spirality values for which $G_{\nu,\rho}$ admits a rectilinear representation. $\Sigma_{\nu,\rho}$ is representative of all "shapes" that $G_{\nu,\rho}$ can take in a rectilinear representation of G with the reference chain on the external face. If $G_{\nu,\rho}$ is not rectilinear planar, $\Sigma_{\nu,\rho}$ is empty. If N_ν is the number of vertices of $G_{\nu,\rho}$, we have $|\Sigma_{\nu,\rho}| = O(N_\nu) = O(n)$, as the spirality of any rectilinear representation of $G_{\nu,\rho}$ cannot exceed the length of the longest spine in this component. A key-ingredient for our testing algorithm is how to efficiently compute the spirality sets of each node in a bottom-up visit of the SPQ*-tree of G for each possible choice of its root. We give a core lemma regarding S-nodes.

Lemma 4. Let G be an SP-graph, T be the SPQ*-tree of G, and ν be an S-node of T with n_ν children. Let $\rho_1, \rho_2, \ldots, \rho_h$ be a sequence of Q*-nodes of T such that, for each child μ of ν in T_{ρ_i}, the set Σ_{μ,ρ_i} is given. The set Σ_{ν,ρ_i} can be computed in $O(n_\nu n \log^2 n)$ time for $i = 1$ and in $O(n \log n)$ time for $2 \leq i \leq h$.

Sketch of proof: Let A and B be two sets of numbers. The Cartesian sum $A + B$ is the set $\{a + b \mid a \in A, b \in B\}$. If A and B are sets of $O(n)$ numbers of $O(n)$ size, $A + B$ can be computed in $O(n \log n)$ time (see, e.g., [1]). Even for the cases

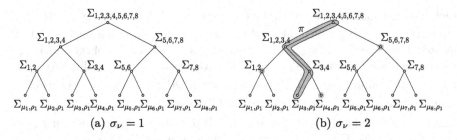

Fig. 4. (a) Schematic illustration of the equipped tree τ. (b) A path π from the leaf corresponding to μ_3 to the root; we have $\Delta_{\nu,\rho_1}^{(-3)} = \Sigma_{1,2} + \Sigma_{\mu_4,\rho_1} + \Sigma_{5,6,7,8}$.

of negative integers or semi-integers, simple modifications allow us to apply this technique. We extensively apply Cartesian sums between the spirality sets of the children of ν. Indeed, based on Lemma 1, the series composition of two components admits a set of spirality values that is the Cartesian sum of the sets of the combined components. Since the spirality sets resulting from this composition still contain $O(n)$ distinct values of size $O(n)$ and each spirality value is either an integer or a semi-integer, we compute each Cartesian sum between two of these sets in $O(n \log n)$ time. Let $\mu_1, \dots, \mu_s$ be the children of ν in T_{ρ_1}, with $s = n_\nu$.

Case $i = 1$. When we process ν in T_{ρ_1} we construct the following sets: The spirality set Σ_{ν,ρ_1} of ν and, for each $1 \le j \le s$, a set $\Delta_{\nu,\rho_1}^{(-j)}$ corresponding to the Cartesian sum of the spirality sets of all children of ν except μ_j. Set $\Delta_{\nu,\rho_1}^{(-j)}$ will be used to efficiently compute the spirality set of ν when μ_j becomes the parent of ν in some tree T_{ρ_i} $i > 1$. To compute Σ_{ν,ρ_1} and $\Delta_{\nu,\rho_1}^{(-j)}$ we apply this procedure:

Step 1. Let d be the minimum integer such that $s + d = 2^k$, for an integer $k > 0$. Note that $d < s$. If $d > 0$, temporarily add d dummy children $\mu_{s+1}, \dots, \mu_{s+d}$ to ν, with spirality sets $\Sigma_{\mu_{s+1},\rho_1} = \dots = \Sigma_{\mu_{s+d},\rho_1} = \{0\}$.

Step 2. Construct a complete rooted binary tree τ such that (see Fig. 4(a)): (i) the leaves of τ correspond to $\mu_1, \dots, \mu_{s+d}$, in this left-to-right order; (ii) the leaf of τ corresponding to μ_j is equipped with the set Σ_{μ_j,ρ_i} $(1 \le j \le s+d)$; (iii) each internal node ψ of τ is equipped with the Cartesian sum of the two sets stored at its children. Denote by $\Sigma_{j_1,j_2,\dots,j_g}$ the set stored at ψ, where $\mu_{j_1}, \mu_{j_2}, \dots, \mu_{j_g}$ are the leaves of the subtree rooted at ψ $(j_{a+1} = j_a + 1$, for $1 \le a \le g-1)$. Hence, the root is equipped with $\Sigma_{1,\dots,s+d} = \Sigma_{\mu_1,\rho_1} + \dots + \Sigma_{\mu_s,\rho_1} = \Sigma_{\nu,\rho_1}$.

Step 3. For each $j = 1, \dots, s$, compute $\Delta_{\nu,\rho_1}^{(-j)}$ as follows (see Fig. 4(b)). Let π be the path in τ from the root to the leaf corresponding to μ_j, and let ψ_ℓ be the node along π at level ℓ of τ $(\psi_0$ being the root). For each non-leaf node ψ_ℓ, let $\phi_{\ell+1}$ be the sibling of $\psi_{\ell+1}$ in τ. $\Delta_{\nu,\rho_1}^{(-j)}$ is computed as the Cartesian sum of all sets stored at the nodes $\phi_{\ell+1}$, for each non-leaf node ψ_ℓ.

About the time complexity, **Step 1** takes $O(s) = O(n_\nu)$ time. The data structure of **Step 2** is constructed in $O(n_\nu n \log n)$ time, namely τ has $O(n_\nu)$ nodes; for each internal node of τ we execute a Cartesian sum between two sets with $O(n)$ numbers of size $O(n)$, which takes $O(n \log n)$ time. To compute a single set $\Delta_{\nu,\rho_1}^{(-j)}$ in Step 3, we execute $O(\log n_\nu)$ Cartesian sums between pairs of sets (picking a set for each level of the tree except at the root level), still spending $O(n \log n)$ time for each sum. Hence, $\Delta_{\nu,\rho_1}^{(-j)}$ is computed in $O(\log n_\nu \cdot n \log n) = O(n \log^2 n)$ time, and all sets $\Delta_{\nu,\rho_1}^{(-j)}$ are computed in $O(n_\nu n \log^2 n)$ time.

Case $i > 1$. Let μ be the parent of ν in T_{ρ_1}. If μ is also the parent of ν in T_{ρ_i}, we do not recompute Σ_{ν,ρ_i}, as $\Sigma_{\nu,\rho_i} = \Sigma_{\nu,\rho_1}$. Else, μ is a child of ν in T_{ρ_i}, and one of the children of ν in T_{ν,ρ_1}, say μ_j, is the parent of ν in T_{ρ_i}. To compute Σ_{ν,ρ_i} we execute, in $O(n \log n)$ time, the Cartesian sum between $\Delta_{\nu,\rho_1}^{(-j)}$ and Σ_{μ,ρ_i}. □

3.2   Testing Algorithm

Lemma 5. *Let G be an n-vertex SP-graph. There exists an $O(n^2 \log^2 n)$-time algorithm that tests whether G is rectilinear planar and that computes a rectilinear representation of G in the positive case.*

Sketch of proof: We sketch here the testing phase, which elaborates on ideas in [4] and exploits Lemma 4. Let T be the SPQ*-tree of G and let $\{\rho_1, \ldots, \rho_h\}$ be a sequence of its Q*-nodes. Denote by ℓ_i the length of the chain corresponding to ρ_i; the spirality set of ρ_i consists of all integer values in the interval $[-(\ell_i - 1), (\ell_i - 1)]$. For each $i = 1, \ldots, h$, the testing algorithm performs a post-order visit of T_{ρ_i}. For every visited non-root node ν of T_{ρ_i} the algorithm computes the set Σ_{ν,ρ_i} by combining the spirality sets of the children of ν, based on Lemmas 1–3. If $\Sigma_{\nu,\rho_i} = \emptyset$, the algorithm stops the visit, discards T_{ρ_i}, and starts visiting $T_{\rho_{i+1}}$ (if $i < h$). If the algorithm achieves the root child ν and if $\Sigma_{\nu,\rho_i} \neq \emptyset$, it checks whether G is rectilinear planar by verifying if there exists a value $\sigma_\nu \in \Sigma_{\nu,\rho_i}$ and a value $\sigma_{\rho_i} \in \Sigma_{\rho_i,\rho_i} = [-(\ell_i - 1), (\ell_i - 1)]$ such that $\sigma_\nu - \sigma_{\rho_i} = 4$ (we show in [9] that this condition is necessary and sufficient). If so, the test is positive and the algorithm does not visit the remaining trees.

We now explain how to perform the testing algorithm in $O(n^2 \log^2 n)$ time. Tree T is computed in $O(n)$ time [3,16,18]. Let T_{ρ_i} be the currently visited tree, and let ν be a node of T_{ρ_i}. Denote by n_ν the number of children of ν. If the parent of ν in T_{ρ_i} coincides with the parent of T_{ρ_j} for some $j \in \{1, \ldots, i-1\}$, and if Σ_{ν,ρ_j} was previously computed, then the algorithm does not need to compute Σ_{ν,ρ_i}, because $\Sigma_{\nu,\rho_i} = \Sigma_{\nu,\rho_j}$. Hence, for each node ν, the number of computations of its rectilinear spirality sets that are performed over all possible trees T_{ρ_i} is at most $n_\nu + 1 = O(n_\nu)$ (one for each different way of choosing the parent of ν).

If ν is a Q*-node, Σ_{ν,ρ_i} is easily computed in $O(1)$ time. If ν is a P-node with three children, it is sufficient to check, for each of the six permutations of the children of ν and for each value in the rectilinear spirality set of one of the three children, whether the sets of the other two children contain the values that satisfy condition (ii) of Lemma 2. Hence, Σ_{ν,ρ_i} can be computed in $O(n)$ time. If ν is a P-node with two children, Σ_{ν,ρ_i} is computed in $O(n)$ with a similar approach, using Lemma 3. Since $\sum_\nu \deg(\nu) = O(n)$, the computation of the rectilinear spirality sets of all P-nodes takes $O(n^2)$ time, over all T_{ρ_i} $(i = 1, \ldots, h)$.

Suppose now that ν is an S-node. By Lemma 4, in the first rooted tree for which we are able to compute the rectilinear spirality set of ν (i.e., in the first rooted tree for which all children of ν admit a rectilinear representation), we spend $O(n_\nu n \log^2 n)$ time to compute such a set. For each of the other $O(n_\nu)$ trees in which the parent of ν changes, we spend $O(n \log n)$ time to recompute the rectilinear spirality set of ν. Thus, for each S-node ν we spend in total $O(n_\nu n \log^2 n)$ time over all visits of T_{ρ_i} $(i = 1, \ldots, h)$, and hence, since $\sum_\nu \deg(\nu) = O(n)$, we spend $O(n^2 \log^2 n)$ time over all S-nodes of the tree. $\quad\square$

Theorem 1. *Let G be an n-vertex partial 2-tree. There exists an $O(n^2 \log^2 n)$ time algorithm that tests whether G is rectilinear planar and that computes a rectilinear representation of G in the positive case.*

Sketch of proof: Let $\mathcal{T}$ be the BC-tree of G, and let $B_1, \ldots, B_q$ be the blocks of G $(q \geq 2)$. We denote by $\beta(B_i)$ the block-node of $\mathcal{T}$ corresponding to B_i $(1 \leq i \leq q)$ and by $\mathcal{T}_{B_i}$ the tree $\mathcal{T}$ rooted at $\beta(B_i)$. For a cutvertex c of G, we denote by $\chi(c)$ the node of $\mathcal{T}$ that corresponds to c. Each $\mathcal{T}_{B_i}$ describes a class of planar embeddings of G such that, for each non-root node $\beta(B_j)$ $(1 \leq j \leq q)$ with parent node $\chi(c)$, the cutvertex c lies in the external face of B_j. We say that G is *rectilinear planar with respect to* $\mathcal{T}_{B_i}$ if it is rectilinear planar for some planar embedding in the class described by $\mathcal{T}_{B_i}$. To check whether G is rectilinear planar with respect to $\mathcal{T}_{B_i}$, we have to perform a constrained rectilinear planarity testing for every block $B_1, \ldots, B_q$ so to guarantee that the rectilinear representations of the different blocks can be merged together at the shared cutvertices. The different types of constraints for a cutvertex c of B_j are defined based on the degree of c in G and in B_j, as well as on the number of blocks sharing c with B_j. The testing for B_j must consider the planar embeddings in which none of the cutvertices is forced to be on the external face (when $\beta(B_j)$ is the root of the BC-tree) and those in which a prescribed cutvertex (the one corresponding to the parent of $\beta(B_j)$) must be on the external face. By adapting the technique in Lemma 5, a constrained rectilinear planarity testing for B_j can still be executed in $O(n_{B_j}^2 \log^2 n_{B_j})$ over all these planar embeddings, where n_{B_j} is the number of vertices of B_j (see [9] for details). For a block B_j, the choice of which cutvertex-node is the parent (if any) of $\beta(B_j)$ and of which cutvertex-nodes are the children of $\beta(B_j)$ is a *configuration* of B_j. Note that, each rooted BC-tree defines a configuration for each block.

In a pre-processing phase, we perform the constrained rectilinear planarity testing for each block of the BC-tree and for each configuration of this block. Also, for each configuration of the cutvertex-nodes incident to $\beta(B_j)$, we store at $\beta(B_j)$ a Boolean *local label* that is either true if B_j is rectilinear planar for that configuration or false otherwise. Since each block B_j is processed in $O(n_{B_j}^2 \log^2 n_{B_j})$, the pre-processing phase is executed in $O(n^2 \log^2 n)$ time. After the pre-processing phase, we first visit $\mathcal{T}_{B_1}$ bottom-up. For each node of $\mathcal{T}_{B_1}$ (either a block-node or a cutvertex-node) we compute a Boolean *cumulative*

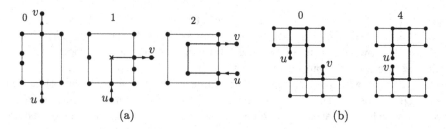

Fig. 5. Two components that are: (a) rectilinear planar for spiralities 0 and 2, but not 1 (which requires a bend, shown as a cross); (b) rectilinear planar only for spiralities 0 and 4. In bold, an arbitrary path from the pole u to the pole v.

label that is the conjunction of those of its children; the cumulative label of a leaf coincides with its local label. This is easily done in $O(n)$ time for all nodes of T_{B_1}.

After the visit of T_{B_1} one of the following three cases holds: (i) The cumulative label of the root is true. In this case the test is positive. (ii) There are two block-nodes γ_1 and γ_2 in T_{B_1} with cumulative label false and they are along two distinct paths from a leaf to the root. In this case the test is negative, as for any other T_{B_i} $(i = 2, \ldots, q)$, at least one of the subtrees rooted at γ_1 and γ_2 remains unchanged. (iii) Let $\beta(B_j)$ be the deepest node along the unique path that contains of block-nodes with cumulative label false. It is sufficient to test all rooted BC-trees whose root $\beta(B_i)$ is a leaf of the subtree rooted at $\beta(B_j)$. For each of these trees we repeat the procedure above and compute the cumulative label of a node γ of T_{B_i} only if the subtree of γ has changed with respect to any previous visits. For a node γ whose parent has changed, its cumulative label is computed in $O(1)$ time (γ has at most one child whose cumulative label is false).

Therefore: For each node γ of T of degree n_γ, the cumulative label of γ is computed in $O(n_\gamma)$ time for T_{B_1} and in $O(1)$ for T_{B_i} $(i > 1)$. Since each node γ changes its parent $O(n_\gamma)$ times, summing up over all γ, the testing phase that follows the pre-processing phase takes in total $O(n)$ time. □

4   Independent-Parallel SP-graphs

By a deeper investigation of spirality, we devise a linear-time algorithm for the *independent-parallel* SP-graphs, namely those in which no two P-components share a pole (the graph in Fig. 1(a) is an independent-parallel SP-graph). To improve the time complexity in Lemma 5, we ask whether the components of an independent-parallel SP-graph have spirality sets of constant size, as for the case of planar 3-graphs [11,25]. Unfortunately, this is not the case for SP-graphs with degree-4 vertices, even when they are independent-parallel. Namely, in Sect. 4.1 we describe an infinite family of independent-parallel SP-graphs whose rectilinear representations require that some components have spirality $\Omega(\log n)$.

Moreover, it is not obvious how to describe the spirality sets for independent-parallel SP-graphs with degree-4 vertices in $O(1)$ space. See for example the irregular behavior of the spirality sets of the components in Fig. 5(a) and Fig. 5(b).

Indeed, the absence of regularity is an obstacle to the design of a succinct description based on whether a component is rectilinear planar for consecutive spirality values. By carefully analyzing the spirality properties of independent-parallel SP-graphs, in Sects. 4.2 and 4.3 we show how to overcome these difficulties and design a linear-time rectilinear planarity testing algorithm for this graph family.

4.1 Spirality Lower Bound

Theorem 2. *For infinitely many integer values of n, there exists an n-vertex independent-parallel SP-graph for which every rectilinear representation has a component with spirality $\Omega(\log n)$.*

Sketch of proof: We describe an infinite family of graphs, schematically illustrated in Fig. 6, that have components whose spirality is not bounded by a constant in any rectilinear representation. For any even integer $N \geq 2$, we construct an independent-parallel SP-graph G with $n = O(3^N)$ vertices whose rectilinear representations require a component with spirality larger than N. Namely, let $L = \frac{N}{2} + 1$. For $k \in \{0, \dots, L\}$, let G_k be the SP-graph inductively defined as follows: (i) G_0 is a chain of $N + 4$ vertices; (ii) G_1 is a parallel composition of three copies of G_0, with coincident poles (Fig. 6(a)); (iii) for $k \geq 2$, G_k is a parallel composition of three series composition, each starting and ending with an edge, and having G_{k-1} in the middle (Fig. 6(b)). Graph G is obtained by composing in a cycle two chains p_1 and p_2 of length three, with two copies of G_L (Fig. 6(c)). The graph G_L for $N = 4$ is in Fig. 6(d). In any representation of G, at least one of the G_0 components has spirality larger than N; see Fig. 6(e). $\qquad\square$

4.2 Rectilinear Spirality Sets

Let G be an independent-parallel SP-graph, T be the SPQ*-tree of G, and ρ be a Q*-node of T. Each pole w of a P-node ν of T_ρ is such that $\mathrm{outdeg}_\nu(w) = 1$; if ν is an S-node, either $\mathrm{indeg}_\nu(w) = 1$ or $\mathrm{outdeg}_\nu(w) = 1$. In all cases, $\mathrm{outdeg}_\nu(w) = 1$ when $\mathrm{indeg}_\nu(w) > 1$. For any node ν of T_ρ, denote by $\Sigma^+_{\nu,\rho}$ (resp. $\Sigma^-_{\nu,\rho}$) the subset of non-negative (resp. non-positive) values of $\Sigma_{\nu,\rho}$. Clearly, $\Sigma_{\nu,\rho} = \Sigma^+_{\nu,\rho} \cup \Sigma^-_{\nu,\rho}$. Note that, $\sigma_\nu \in \Sigma^+_{\nu,\rho}$ if and only if $-\sigma_\nu \in \Sigma^-_{\nu,\rho}$ (just flip the embedding of $G_{\nu,\rho}$ around its poles), thus we can restrict the study of the properties of $\Sigma_{\nu,\rho}$ to $\Sigma^+_{\nu,\rho}$, which we call the *non-negative rectilinear spirality set* of ν in T_ρ (or of $G_{\nu,\rho}$).

The main result of this subsection is Theorem 3, which proves that if G is an independent-parallel SP-graph, there is a limited number of possible structures for the sets $\Sigma^+_{\nu,\rho}$ (see also Fig. 7). Let m and M be two non-negative integers such that $m < M$: (i) $[M]$ denotes the singleton $\{M\}$; (ii) $[m, M]^1$ denotes the set of all integers in the interval $[m, M]$, i.e., $\{m, m+1, \dots, M-1, M\}$; (iii) If m and M have the same parity, $[m, M]^2$ denotes the set $\{m, m+2, \dots, M-2, M\}$. The proof of Theorem 3 exploits several key technical results (see [9]).

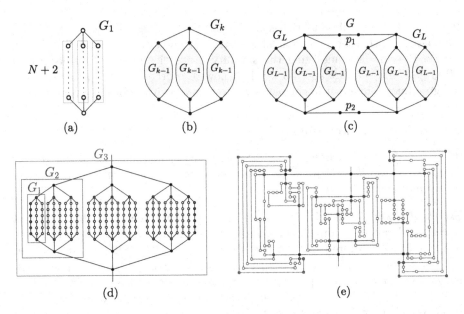

Fig. 6. (a)–(c) The graph family of Theorem 2, where $L = \frac{N}{2} + 1$. (d)–(e) Graph G_L for $N = 4$ and a rectilinear representation of G_L; the two G_0 components with blue vertices have spirality $N + 2 = 6$ (left) and $-(N + 2) = -6$ (right), respectively.

Theorem 3. *Let G be a rectilinear planar independent-parallel SP-graph and let $G_{\nu,\rho}$ be a component of G. The non-negative rectilinear spirality set $\Sigma^+_{\nu,\rho}$ of $G_{\nu,\rho}$ has one the following six structures:* $[0]$, $[1]$, $[1,2]^1$, $[0,M]^1$, $[0,M]^2$, $[1,M]^2$.

4.3   Rectilinear Planarity Testing

Let G be an independent-parallel SP-graph, T be its SPQ*-tree, and $\{\rho_1, \ldots, \rho_h\}$ be the Q*-nodes of T. To test whether G is rectilinear planar, we exploit a similar strategy as in Lemma 5. For each possible choice of the root $\rho \in \{\rho_1, \ldots, \rho_h\}$, the algorithm visits T_ρ bottom-up and computes, for each visited node ν, the non-negative spirality set $\Sigma^+_{\nu,\rho}$, based on the sets of the children of ν. Using Theorem 3, we prove that this computation can be done in $O(1)$ time for each type of node, including the S-nodes which require the most expensive operations for general SP-graphs. At the level of the root, when ν is the root child, we prove that it is sufficient to test whether there exist two values $\sigma_\nu \in \Sigma^+_{\nu,\rho}$ and $\sigma_\rho \in \Sigma^+_{\rho,\rho}$, such that $\sigma_\nu + \sigma_\rho = 4$ (see [9] for details). Thus we spend $O(1)$ time also at the root level. An $O(n)$-time testing algorithm over all choices of the root ρ is achieved through the same reusability principle described in Lemma 5, where again we can prove that updating the spirality set of each type of node can be done in $O(1)$ by Theorem 3.

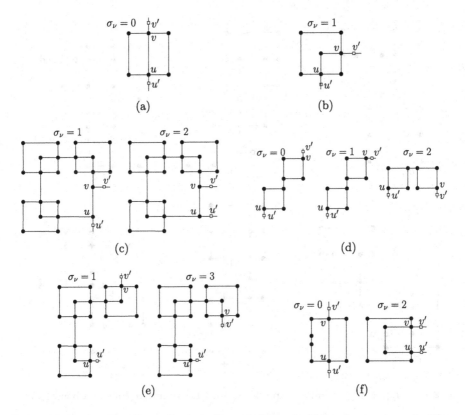

Fig. 7. Examples of non-negative spirality sets for each of the six structures in Theorem 3: (a) $[0]$; (b) $[1]$; (c) $[1,2]^1$; (d) $[0,2]^1$; (e) $[1,3]^2$; (f) $[0,2]^2$.

Theorem 4. *Let G be an n-vertex independent-parallel SP-graph. There exists an $O(n)$-time algorithm that tests whether G is rectilinear planar and that computes a rectilinear representation of G in the positive case.*

5   Final Remarks and Open Problems

We proved that rectilinear planarity can be tested in $O(n^2 \log^2 n)$ time for general partial 2-trees and in $O(n)$ time for independent-parallel SP-graphs. Establishing a tight bound on the complexity of rectilinear planarity testing algorithm for partial 2-trees remains an open problem. A pitfall to achieve $O(n)$-time complexity in the general case is that, in contrast with the independent-parallel SP-graphs, the spirality set of a component may not exhibit a regular behavior. Even extending Theorem 4 to non-biconnected partial 2-trees whose blocks are independent-parallel is not immediate, as the constraints for the angles at the cutvertices affect the regularity of the spirality sets.

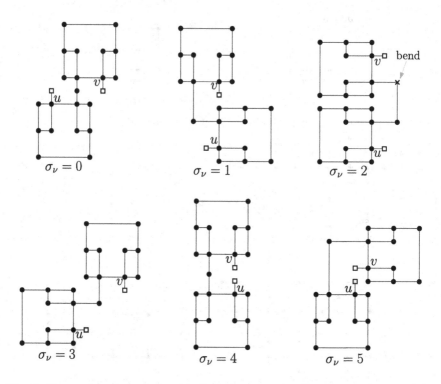

Fig. 8. Component that admits spiralities 0,1,3,4,5. Spirality 2 needs a bend (×).

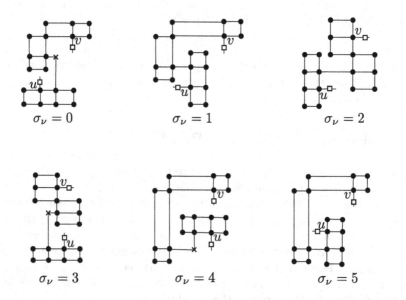

Fig. 9. Component that admits spiralities 1,2,5; other values require bends (×).

References

1. Cormen, T.H., Leiserson, C.E., Rivest, R.L., Stein, C.: Introduction to Algorithms. MIT Press, Cambridge (2009)
2. Cornelsen, S., Karrenbauer, A.: Accelerated bend minimization. J. Graph Algorithms Appl. **16**(3), 635–650 (2012). https://doi.org/10.7155/jgaa.00265
3. Di Battista, G., Eades, P., Tamassia, R., Tollis, I.G.: Graph Drawing: Algorithms for the Visualization of Graphs. Prentice-Hall (1999)
4. Di Battista, G., Liotta, G., Vargiu, F.: Spirality and optimal orthogonal drawings. SIAM J. Comput. **27**(6), 1764–1811 (1998). https://doi.org/10.1137/S0097539794262847
5. Di Giacomo, E., Liotta, G., Montecchiani, F.: Orthogonal planarity testing of bounded treewidth graphs. J. Comput. Syst. Sci. **125**, 129–148 (2022)
6. Didimo, W., Liotta, G.: Mining graph data. In: Cook, D.J., Holder, L.B. (eds.) Graph Visualization and Data Mining, pp. 35–64. Wiley (2007)
7. Didimo, W., Kaufmann, M., Liotta, G., Ortali, G.: Rectilinear planarity testing of plane series-parallel graphs in linear time. In: GD 2020. LNCS, vol. 12590, pp. 436–449. Springer, Cham (2020). https://doi.org/10.1007/978-3-030-68766-3_34
8. Didimo, W., Kaufmann, M., Liotta, G., Ortali, G.: Computing bend-minimum orthogonal drawings of plane series-parallel graphs in linear time. CoRR arXiv:2205.07500v1 (2022)
9. Didimo, W., Kaufmann, M., Liotta, G., Ortali, G.: Rectilinear planarity of partial 2-trees. CoRR arXiv:2208.12558v3 (2022)
10. Didimo, W., Liotta, G.: Computing orthogonal drawings in a variable embedding setting. In: Chwa, K.-Y., Ibarra, O.H. (eds.) ISAAC 1998. LNCS, vol. 1533, pp. 80–89. Springer, Heidelberg (1998). https://doi.org/10.1007/3-540-49381-6_10
11. Didimo, W., Liotta, G., Ortali, G., Patrignani, M.: Optimal orthogonal drawings of planar 3-graphs in linear time. In: Chawla, S. (ed.) Proceedings of the 2020 ACM-SIAM Symposium on Discrete Algorithms, SODA 2020, Salt Lake City, UT, USA, January 5–8, 2020. pp. 806–825. SIAM (2020). https://doi.org/10.1137/1.9781611975994.49
12. Duncan, C.A., Goodrich, M.T.: Planar orthogonal and polyline drawing algorithms. In: Tamassia, R. (ed.) Handbook on Graph Drawing and Visualization, pp. 223–246. Chapman and Hall/CRC (2013). https://www.crcpress.com/Handbook-of-Graph-Drawing-and-Visualization/Tamassia/9781584884125
13. Frati, F.: Planar rectilinear drawings of outerplanar graphs in linear time. Comput. Geom. **103**, 101854 (2022)
14. Garg, A., Tamassia, R.: A new minimum cost flow algorithm with applications to graph drawing. In: North, S. (ed.) GD 1996. LNCS, vol. 1190, pp. 201–216. Springer, Heidelberg (1997). https://doi.org/10.1007/3-540-62495-3_49
15. Garg, A., Tamassia, R.: On the computational complexity of upward and rectilinear planarity testing. SIAM J. Comput. **31**(2), 601–625 (2001). https://doi.org/10.1137/S0097539794277123
16. Gutwenger, C., Mutzel, P.: A linear time implementation of SPQR-trees. In: Graph Drawing, 8th International Symposium, GD 2000, Colonial Williamsburg, VA, USA, September 20-23, 2000, Proceedings. Lecture Notes in Computer Science, vol. 1984, pp. 77–90. Springer, Heidelberg (2000). https://doi.org/10.1007/3-540-44541-2

17. Hasan, M.M., Rahman, M.S.: No-bend orthogonal drawings and no-bend orthogonally convex drawings of planar graphs (Extended abstract). In: Du, D.-Z., Duan, Z., Tian, C. (eds.) COCOON 2019. LNCS, vol. 11653, pp. 254–265. Springer, Cham (2019). https://doi.org/10.1007/978-3-030-26176-4_21

18. Hopcroft, J.E., Tarjan, R.E.: Dividing a graph into triconnected components. SIAM J. Comput. **2**(3), 135–158 (1973). https://doi.org/10.1137/0202012

19. Jünger, M., Mutzel, P. (eds.): Graph Drawing Software. Springer (2004). https://doi.org/10.1007/978-3-642-18638-7

20. Kaufmann, M., Wagner, D. (eds.): Drawing Graphs. LNCS, vol. 2025. Springer, Heidelberg (2001). https://doi.org/10.1007/3-540-44969-8

21. Nishizeki, T., Rahman, M.S.: Planar Graph Drawing. Lecture Notes Series on Computing, vol. 12. World Scientific, Singapore (2004)

22. Rahman, M.S., Egi, N., Nishizeki, T.: No-bend orthogonal drawings of subdivisions of planar triconnected cubic graphs. IEICE Trans. Inf. Syst. 88-D(1), 23–30 (2005). http://search.ieice.org/bin/summary.php?id=e88-d_1_23&category=D&year=2005&lang=E&abst=

23. Rahman, M.S., Nishizeki, T., Naznin, M.: Orthogonal drawings of plane graphs without bends. J. Graph Algorithms Appl. **7**(4), 335–362 (2003). http://jgaa.info/accepted/2003/Rahman+2003.7.4.pdf

24. Tamassia, R.: On embedding a graph in the grid with the minimum number of bends. SIAM J. Comput. **16**(3), 421–444 (1987). https://doi.org/10.1137/0216030

25. Zhou, X., Nishizeki, T.: Orthogonal drawings of series-parallel graphs with minimum bends. SIAM J. Discret. Math. **22**(4), 1570–1604 (2008). https://doi.org/10.1137/060667621

Drawings and Properties of Directed Graphs

Testing Upward Planarity of Partial 2-Trees

Steven Chaplick[1], Emilio Di Giacomo[2], Fabrizio Frati[3](✉),
Robert Ganian[4], Chrysanthi N. Raftopoulou[5], and Kirill Simonov[4]

[1] Maastricht University, Maastricht, The Netherlands
s.chaplick@maastrichtuniversity.nl
[2] Università degli Studi di Perugia, Perugia, Italy
emilio.digiacomo@unipg.it
[3] Roma Tre University, Rome, Italy
frati@dia.uniroma3.it
[4] Technische Universität Wien, Wien, Austria
rganian@ac.tuwien.ac.at
[5] National Technical University of Athens, Athens, Greece
crisraft@mail.ntua.gr

Abstract. We present an $O(n^2)$-time algorithm to test whether an n-vertex directed partial 2-tree is upward planar. This result improves upon the previously best known algorithm, which runs in $O(n^4)$ time.

1 Introduction

A digraph is *upward planar* if it admits a drawing that is at the same time *planar*, i.e., it has no crossings, and *upward*, i.e., all edges are drawn as curves monotonically increasing in the vertical direction. Upward planarity is a natural variant of planarity for directed graphs and finds applications in those domains where one wants to visualize networks with a hierarchical structure.

Upward planarity is a classical research topic in Graph Drawing since the early 90s. Garg and Tamassia have shown that recognizing upward planar digraphs is NP-complete [13], however polynomial-time algorithms have been proposed for various cases, including digraphs with fixed embedding [1], single-source digraphs [2,3,16,17], outerplanar digraphs [18]. The case of directed partial 2-trees, which is of central interest to this paper and includes, among others, series-parallel digraphs, has been investigated by Didimo et al. [10] who presented an $\mathcal{O}(n^4)$-time testing algorithm. The parameterized complexity of the upward planarity testing problem has also been investigated [4,5,10,15].

In this paper, we present an $\mathcal{O}(n^2)$-time algorithm to test upward planarity of directed partial 2-trees, improving upon the $\mathcal{O}(n^4)$-time algorithm by Didimo et al. [10]. There are two main ingredients that allow us to achieve such result.

First, following the approach in [5], our algorithm traverses the SPQ-tree of the input digraph G while computing, for each component of G, the possible "shapes" of its upward planar embeddings. The algorithm in [5] only works for *expanded* digraphs, i.e., digraphs such that every vertex has at most one incoming

P. Angelini and R. von Hanxleden (Eds.): GD 2022, LNCS 13764, pp. 175–187, 2023.
https://doi.org/10.1007/978-3-031-22203-0_13

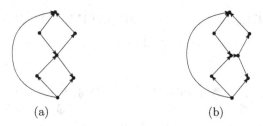

Fig. 1. Splitting a vertex in a non-expanded directed partial 2-tree (a) might result in an expanded digraph (b) which is not a directed partial 2-tree.

or outgoing edge. Although every digraph can be made expanded while preserving its upward planarity by "splitting" its vertices [2], this modification might not maintain that the digraph is a directed partial 2-tree; see Fig. 1. We present a novel algorithm that is applicable to non-expanded digraphs. We propose a new strategy to process P-nodes, which is simpler than the one of [5] and allows us to compute some additional information that is needed by the second ingredient. Further, we give a more efficient procedure than the one of [5] to process the S-nodes; this is vital for the overall running time of our algorithm.

Second, the traversal of the SPQ-tree T of G tests the upward planarity of G with the constraint that the edge corresponding to the root of T is incident to the outer face. Then $\mathcal{O}(n)$ traversals with different choices for the root of T can be used to test the upward planarity of G without that constraint. However, following a recently developed strategy [11,12], in the first traversal of T we compute some information additional to the possible shapes of the upward planar embeddings of the components of G. A clever use of this information allows us to handle P-nodes more efficiently in later traversals. Our testing algorithms can be enhanced to output an upward planar drawing, if one exists, although we do not describe the process explicitly.

Paper Organization. In Sect. 2 we give some preliminaries. In Sect. 3 we describe the algorithm for biconnected digraphs with a prescribed edge on the outer face, while in Sect. 4 we deal with general biconnected digraphs. Section 5 extends our result to simply connected digraphs. Future research directions are presented in Sect. 6. Lemmas and theorems whose proofs are omitted are marked with a ($\star$) and can be found in the full version of the paper [6].

2   Preliminaries

In a digraph, a *switch* is a source or a sink. The *underlying graph* of a digraph is the undirected graph obtained by ignoring the edge directions. When we mention connectivity of a digraph, we mean the connectivity of its underlying graph.

A *planar embedding* of a connected graph is an equivalence class of planar drawings, where two drawings are equivalent if: (i) the clockwise order of the edges incident to each vertex is the same; and (ii) the sequence of vertices and edges along the boundary of the outer face is the same.

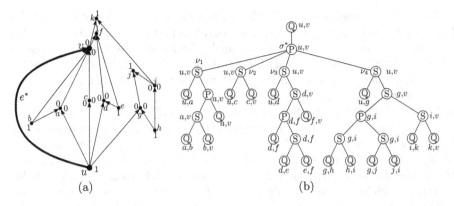

(a)   (b)

Fig. 2. (a) Labels at the angles of an upward planar embedding $\mathcal{E}$ of a biconnected directed partial 2-tree G; the missing labels are -1. (b) An SPQ-tree T of
G with respect to e^*. The restriction of $\mathcal{E}$ to G_{σ^*} is a uv-external upward pla-
nar embedding $\mathcal{E}_{\sigma^*}$ with shape description $\langle 0, 1, 1, 0, \text{out}, \text{out}, \text{in}, \text{out}\rangle$. The shape
sequence of $\mathcal{E}_{\sigma^*}$ is $[\langle 0, 0, 1, 1, \text{out}, \text{out}, \text{in}, \text{in}\rangle, \langle 0, 0, 1, 1, \text{out}, \text{out}, \text{in}, \text{in}\rangle, \langle -1, 1, 1, 1,$
$\text{out}, \text{out}, \text{out}, \text{out}\rangle, \langle -1, 1, 1, 1, \text{out}, \text{out}, \text{out}, \text{out}\rangle]$. The contracted shape sequence of
$\mathcal{E}_{\sigma^*}$ is $[\langle 0, 0, 1, 1, \text{out}, \text{out}, \text{in}, \text{in}\rangle, \langle -1, 1, 1, 1, \text{out}, \text{out}, \text{out}, \text{out}\rangle]$.

A drawing of a digraph is *upward* if every edge is represented by a Jordan arc
whose y-coordinates monotonically increase from the source to the sink of the
edge. A drawing of a digraph is *upward planar* if it is both upward and planar.
An upward planar drawing of a graph determines an assignment of labels to the
angles of the corresponding planar embedding, where an *angle* α at a vertex u
in a face f of a planar embedding represents an incidence of u on f. Specifically,
α is *flat* and gets label 0 if the edges delimiting it are one incoming and one
outgoing at u. Otherwise, α is a *switch* angle; in this case, α is *small* (and gets
label -1) or *large* (and gets label 1) depending on whether the (geometric) angle
at f representing α is smaller or larger than $180°$, respectively, see Fig. 2(a). An
upward planar embedding is an equivalence class of upward planar drawings of a
digraph G, where two drawings are equivalent if they determine the same planar
embedding $\mathcal{E}$ for G and the same label assignment for the angles of $\mathcal{E}$.

Theorem 1 ([1,10]). *Let G be a digraph with planar embedding $\mathcal{E}$, and λ be
a label assignment for the angles of $\mathcal{E}$. Then $\mathcal{E}$ and λ define an upward planar
embedding of G if and only if the following hold:*

(UP0) *If α is a switch angle then α is small or large, otherwise it is flat.*
(UP1) *If v is a switch vertex, the number of small, flat and large angles incident
to v is equal to $\deg(v) - 1$, 0, and 1, respectively.*
(UP2) *If v is a non-switch vertex, the number of small, flat and large angles
incident to v is equal to $\deg(v) - 2$, 2, and 0, respectively.*
(UP3) *If f is an internal face (the outer face) of $\mathcal{E}$, the number of small angles
in f is equal to the number of large angles in f plus 2 (resp. minus 2).*

The class of *partial 2-trees* can be defined equivalently as the graphs with
treewidth at most two, or as the graphs that exclude K_4 as a minor, or as the
subgraphs of the 2-trees. Notably, it includes the class of *series-parallel graphs*.

Let G be a biconnected partial 2-tree and let e^* be an edge of G. An *SPQ-tree* T of G *with respect to* e^* (see Fig. 2(b)) is a tree that describes a recursive decomposition of G into its "components". SPQ-trees are a specialization of *SPQR-trees* [8,14]. Each node μ of T represents a subgraph G_μ of G, called the *pertinent graph* of μ, and is associated with two special vertices of G_μ, called *poles* of μ. The nodes of T are of three types: a Q-node μ represents an edge whose end-vertices are the poles of μ, an S-node μ with children ν_1 and ν_2 represents a series composition in which the components G_{ν_1} and G_{ν_2} share a pole to form G_μ, and a P-node μ with children $\nu_1, \ldots, \nu_k$ represents a parallel composition in which the components $G_{\nu_1}, \ldots, G_{\nu_k}$ share both poles to form G_μ. The root of T is the Q-node representing the edge e^*. By our definition, every S-node has exactly two children that can also be S-nodes; because of this assumption, the SPQ-tree of a biconnected partial 2-tree is not unique. However, from an SPQ-tree T, we can obtain an SPQ-tree of G with respect to another reference edge e^{**} by selecting the Q-node representing e^{**} as the new root of T (see Fig. 3).

A *directed partial 2-tree* is a digraph whose underlying graph is a partial 2-tree. When talking about an SPQ-tree T of a biconnected *directed* partial 2-tree G, we always refer to an SPQ-tree of its underlying graph, although the edges of the pertinent graph of each node of T are oriented as in G. Let μ be a node of T with poles u and v. A *uv-external upward planar embedding* of G_μ is an upward planar embedding of G_μ in which u and v are incident to the outer face. In our algorithms, when testing the upward planarity of G, choosing an edge e^* of G as the root of T corresponds to requiring e^* to be incident to the outer face of the sought upward planar embedding $\mathcal{E}$ of G. For each node μ of T with poles u and v, the restriction of $\mathcal{E}$ to G_μ is a *uv-external upward planar embedding* $\mathcal{E}_\mu$ of G_μ. In [5], the possible "shapes" of the cycle bounding the outer face f_μ of $\mathcal{E}_\mu$ have been described by the concept of *shape description*. This is the tuple $\langle \tau_l, \tau_r, \lambda(u), \lambda(v), \rho_u^l, \rho_u^r, \rho_v^l, \rho_v^r \rangle$, defined as follows. Let the *left outer path* P_l (the *right outer path* P_r) of $\mathcal{E}_\mu$ be the path that is traversed when walking from u to v in clockwise (resp. counterclockwise) direction along the boundary of f_μ. The value τ_l, called *left-turn-number* of $\mathcal{E}_\mu$, is the sum of the labels of the angles at the vertices of P_l different from u and v in f_μ; the *right-turn-number* τ_r of $\mathcal{E}_\mu$ is defined similarly. The values $\lambda(u)$ and $\lambda(v)$ are the labels of the angles at u and v in f_μ, respectively. The value ρ_u^l is **in** (**out**) if the edge incident to u in P_l is incoming (outgoing) at u; the values ρ_u^r, ρ_v^l, and ρ_v^r are defined similarly. The values of a shape description depend on each other, as in the following.

Observation 1 ([5]). *The shape description* $\langle \tau_l, \tau_r, \lambda(u), \lambda(v), \rho_u^l, \rho_u^r, \rho_v^l, \rho_v^r \rangle$ *of* $\mathcal{E}_\mu$ *satisfies the following properties:*

(i) ρ_u^l *and* ρ_u^r *have the same value if* $\lambda(u) \in \{-1, 1\}$, *while they have different values if* $\lambda(u) = 0$;

(ii) ρ_v^l *and* ρ_v^r *have the same value if* $\lambda(v) \in \{-1, 1\}$, *while they have different values if* $\lambda(v) = 0$;

(iii) ρ_u^l *and* ρ_v^l *have the same value if* τ_l *is odd, while they have different values if* τ_l *is even;*

(iv) $\tau_l + \tau_r + \lambda(u) + \lambda(v) = 2$.

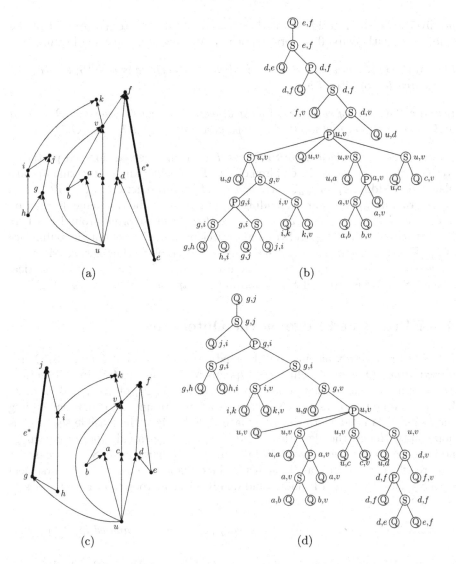

Fig. 3. Two different choices for the root of the SPQ-tree of Fig. 2. The reference edge is shown in bold.

A set $\mathcal{S}$ of shape descriptions is *n-universal* if, for every n-vertex biconnected directed partial 2-tree G, for every rooted SPQ-tree T of G, for every node μ of T with poles u and v, and for every uv-external upward planar embedding $\mathcal{E}_\mu$ of G_μ, the shape description of $\mathcal{E}_\mu$ belongs to $\mathcal{S}$. Thus, an n-universal set is a superset of the *feasible set* $\mathcal{F}_\mu$ of μ, that is, the set of shape descriptions s such that G_μ admits a uv-external upward planar embedding with shape description s. Our algorithm will determine $\mathcal{F}_\mu$ by inspecting each shape description s in an

n-universal set and deciding whether G_μ admits a uv-external upward planar embedding with shape description s or not. We have the following lemmas.

Lemma 1 ($\star$). *An n-universal set $\mathcal{S}$ of shape descriptions with $|\mathcal{S}| \in \mathcal{O}(n)$ can be constructed in $\mathcal{O}(n)$ time.*

Lemma 2 ($\star$). *Any subset $\mathcal{F}$ of an n-universal set can be stored in $\mathcal{O}(n)$ time and space and querying whether a shape description is in $\mathcal{F}$ takes $\mathcal{O}(1)$ time.*

Consider a P-node μ in an SPQ-tree T of a biconnected directed partial 2-tree G. Let $\nu_1, \ldots, \nu_k$ be the children of μ in T. Consider any uv-external upward planar embedding $\mathcal{E}_\mu$ of G_μ. For $i = 1, \ldots, k$, the restriction of $\mathcal{E}_\mu$ to G_{ν_i} is a uv-external upward planar embedding $\mathcal{E}_{\nu_i}$ of G_{ν_i}; let σ_i be the shape description of $\mathcal{E}_{\nu_i}$. Assume that $\mathcal{E}_{\nu_1}, \ldots, \mathcal{E}_{\nu_k}$ appear in this clockwise order around u, where the left outer path of $\mathcal{E}_{\nu_1}$ and the right outer path of $\mathcal{E}_{\nu_k}$ delimit the outer face of $\mathcal{E}_\mu$. We call $\sigma = [\sigma_1, \ldots, \sigma_k]$ the *shape sequence* of $\mathcal{E}_\mu$. Further, consider the sequence $S = [s_1, \ldots, s_x]$ obtained from σ by identifying consecutive identical shape descriptions. We call S the *contracted shape sequence* of $\mathcal{E}_\mu$; see Fig. 2.

3   A Prescribed Edge on the Outer Face

Let G be an n-vertex biconnected directed partial 2-tree and T be its SPQ-tree rooted at any Q-node ρ^*, which corresponds to an edge e^* of G. In this section, we show an algorithm that computes the feasible set $\mathcal{F}_\mu$ of every node μ of T. Let u and v be the poles of μ. Note that G admits an upward planar embedding such that e^* is incident to the outer face if and only if the feasible set of ρ^* is non-empty. Hence, the algorithm could be applied repeatedly (once for each Q-node as the root) to test the upward planarity of G; however, in Sect. 4 we devise a more efficient way to handle multiple choices for the root of T. We first deal with S-nodes, then with P-nodes, and finally with the root of T. For Q-nodes, it is easy to show the following lemma.

Lemma 3 ([5]). *For a non-root Q-node μ, $\mathcal{F}_\mu$ can be computed in $\mathcal{O}(1)$ time.*

S-nodes. We improve an algorithm from [5]. Let ν_1 and ν_2 be the children of μ in T, let $n_1^\mu = |V(G_{\nu_1})|$ and $n_2^\mu = |V(G_{\nu_2})|$, and let w be the vertex shared by G_{ν_1} and G_{ν_2}. Furthermore, let n_3^μ be the number of vertices in the subgraph H_μ of G induced by $V(G) \setminus V(G_\mu) \cup \{u, v\}$. Note that $n_3^\mu = |V(G)| - (n_1^\mu + n_2^\mu) + 3$. We distinguish two cases, depending on which of n_1^μ, n_2^μ, and n_3^μ is largest.

If $n_3^\mu \geq \max(n_1^\mu, n_2^\mu)$, we proceed as in [5, Lemma 6], by combining every shape description in $\mathcal{F}_{\nu_1}$ with every shape description in $\mathcal{F}_{\nu_2}$; for every such combination, the algorithm assigns the angles at w in the outer face with every possible label in $\{-1, 0, 1\}$. If the combination and assignment result in a shape description s of μ (the satisfaction of the properties of Theorem 1 are checked here), the algorithm adds s to $\mathcal{F}_\mu$. This allows us to compute the feasible set $\mathcal{F}_\mu$ of μ in time $\mathcal{O}(n + |\mathcal{F}_{\nu_1}| \cdot |\mathcal{F}_{\nu_2}|)$, which is in $\mathcal{O}(n + n_1^\mu \cdot n_2^\mu)$, as $|\mathcal{F}_{\nu_1}| \in \mathcal{O}(n_1^\mu)$ and $|\mathcal{F}_{\nu_2}| \in \mathcal{O}(n_2^\mu)$ by Lemma 1.

The most interesting case is when, say, $n_1^\mu \geq \max(n_2^\mu, n_3^\mu)$. Here, in order to keep the overall runtime in $\mathcal{O}(n^2)$, we cannot combine every shape description in $\mathcal{F}_{\nu_1}$ with every shape description in $\mathcal{F}_{\nu_2}$. Rather, we proceed as follows. Note that every shape description in $\mathcal{F}_\mu$ whose absolute value of the (left- or right-) turn-number exceeds $n_3^\mu + 4$ does not result in an upward planar embedding of G, by Property UP3 of Theorem 1 and since the absolute value of the turn-number of any path in any upward planar embedding of H_μ does not exceed n_3^μ. We hence construct an $(n_3^\mu + 4)$-universal set $\mathcal{S}$ in $\mathcal{O}(n_3^\mu)$ time by Lemma 1, and then test whether each shape description s in $\mathcal{S}$ belongs to the feasible set $\mathcal{F}_\mu$ of μ. In order to do that, we consider every shape description s_2 in $\mathcal{F}_{\nu_2}$ individually. There are $\mathcal{O}(1)$ shape descriptions in $\mathcal{F}_{\nu_1}$ which combined with s_2 might result in s, since the turn numbers add to each other when combining the shape descriptions in $\mathcal{F}_{\nu_1}$ and $\mathcal{F}_{\nu_2}$, with a constant offset. Hence, by Lemma 2, we check in $\mathcal{O}(1)$ time if there is a shape description s_1 in $\mathcal{F}_{\nu_1}$ which combined with s_2 leads to s. The running time of this procedure is hence $\mathcal{O}(n + n_2^\mu \cdot n_3^\mu)$, as $|\mathcal{F}_{\nu_2}| \in \mathcal{O}(n_2^\mu)$ and $|\mathcal{S}| \in \mathcal{O}(n_3^\mu)$ by Lemma 1. This yields the following.

Lemma 4. *Let μ be an S-node of T with children ν_1 and ν_2. Given the feasible sets $\mathcal{F}_{\nu_1}$ and $\mathcal{F}_{\nu_2}$ of ν_1 and ν_2, respectively, the feasible set $\mathcal{F}_\mu$ of μ can be computed in $\mathcal{O}(n + \min\{n_1^\mu \cdot n_2^\mu, \ n_2^\mu \cdot n_3^\mu, \ n_1^\mu \cdot n_3^\mu\})$ time.*

P-nodes. To compute the feasible set $\mathcal{F}_\mu$ of a P-node μ from the feasible sets $\mathcal{F}_{\nu_1}, \ldots, \mathcal{F}_{\nu_k}$ of its children, the algorithm constructs an n-universal set $\mathcal{S}$ in $\mathcal{O}(n)$ time by Lemma 1. Then it examines every shape description $s \in \mathcal{S}$ and decides whether it belongs to $\mathcal{F}_\mu$. Hence, we focus on a single shape description s and give an algorithm that decides in $\mathcal{O}(k)$ time whether s belongs to $\mathcal{F}_\mu$.

The basic structural tool we need for our algorithm is the following lemma. We call *generating set* $\mathcal{G}(s)$ of a shape description s the set of contracted shape sequences that the pertinent graph of any P-node with poles u and v can have in a uv-external upward planar embedding with shape description s.

Lemma 5 ($\star$). *For any shape description s, $\mathcal{G}(s)$ has size $\mathcal{O}(1)$ and can be constructed in $\mathcal{O}(1)$ time. Also, any sequence in $\mathcal{G}(s)$ has $\mathcal{O}(1)$ length.*

A contracted shape sequence $S \in \mathcal{G}(s)$ is *realizable* by μ if there exists a uv-external upward planar embedding of G_μ whose contracted shape sequence is a subsequence of S containing the first and last elements of S.

We now describe an algorithm that decides in $O(k)$ time whether s belongs to $\mathcal{F}_\mu$. Also, for each contracted shape sequence $S = [s_1, \ldots, s_x]$ in the generating set $\mathcal{G}(s)$ of s, the algorithm computes and stores the following information:

- Three labels $\mathrm{F}_1(\mu, S)$, $\mathrm{F}_2(\mu, S)$, and $\mathrm{F}_3(\mu, S)$ which reference three distinct children ν_i of μ such that $s_1 \in \mathcal{F}_{\nu_i}$.

- Three labels $L_1(\mu, S)$, $L_2(\mu, S)$, and $L_3(\mu, S)$ which reference three distinct children ν_i of μ such that $s_x \in \mathcal{F}_{\nu_i}$.
- Two labels $UF_1(\mu, S)$ and $UF_2(\mu, S)$ which reference two distinct children ν_i of μ such that $\mathcal{F}_{\nu_i}$ does not contain any shape description in S.

For each label type, if the number of children with the described properties is smaller than the number of labels, then labels with larger indices are NULL. We call *the set of relevant labels for μ and S* the set of labels described above.

The algorithm is as follows. First, by Lemma 5, we construct $\mathcal{G}(s)$ in $\mathcal{O}(1)$ time. Then we consider each sequence $S = [s_1, \ldots, s_x]$ in $\mathcal{G}(s)$. By Lemma 5, there are $\mathcal{O}(1)$ such sequences, each with length $\mathcal{O}(1)$. We decide whether S is realizable by μ and compute the set of relevant labels for μ and S as follows.

We initialize all the labels to NULL and process $\nu_1, \ldots, \nu_k$ one by one. For each ν_i, by Lemma 2 we test in $\mathcal{O}(1)$ time which of the shape descriptions $s_1, \ldots, s_x$ belong to $\mathcal{F}_{\nu_i}$ and update the labels accordingly. For example, if $s_1 \in \mathcal{F}_{\nu_i}$, then we update $F_j(\mu, S) = \nu_i$ for the smallest $j \in \{1, 2, 3\}$ with $F_j(\mu, S) = $ NULL.

After processing $\nu_1, \ldots, \nu_k$, we decide whether S is realizable by μ as follows. If $UF_1(\mu, S) \neq$ NULL, then S is not realizable by μ. Otherwise, each feasible set $\mathcal{F}_{\nu_i}$ contains a shape description among $s_1, \ldots, s_x$. Still, we have to check whether $\mathcal{F}_{\nu_i}$ contains s_1 and $\mathcal{F}_{\nu_j}$ contains s_x, for two distinct nodes ν_i and ν_j. If $F_1(\mu, S) = $ NULL or $L_1(\mu, S) = $ NULL, then S is not realizable by μ, as the feasible set of no child contains s_1 or s_x, respectively. Otherwise, if $F_1(\mu, S) \neq L_1(\mu, S)$, then S is realizable by μ, as $F_1(\mu, S)$ can be assigned with s_1 and $L_1(\mu, S)$ with s_x. Otherwise, if $F_2(\mu, S) \neq$ NULL or $L_2(\mu, S) \neq$ NULL, then S is realizable by μ, as $F_2(\mu, S)$ can be assigned with s_1 and $L_1(\mu, S)$ with s_x, or $F_1(\mu, S)$ can be assigned with s_1 and $L_2(\mu, S)$ with s_x, respectively. Otherwise, S is not realizable by μ, as s_1 and s_x are in the feasible set of a single child $F_1(\mu, S) = L_1(\mu, S)$ of μ.

Finally, we have that s belongs to $\mathcal{F}_\mu$ if and only if there exists a contracted shape sequence S in the generating set $\mathcal{G}(s)$ of s which is realizable by μ.

Lemma 6 ($\star$). *Let μ be an P-node of T with children $\nu_1, \ldots, \nu_k$. Given their feasible sets $\mathcal{F}_{\nu_1}, \ldots, \mathcal{F}_{\nu_k}$, the feasible set $\mathcal{F}_\mu$ of μ can be computed in $\mathcal{O}(nk)$ time. Further, for every shape description s in an n-universal set $\mathcal{S}$ and every contracted shape sequence S in the generating set $\mathcal{G}(s)$ of s, the set of relevant labels for μ and S can be computed and stored in overall $\mathcal{O}(nk)$ time and space.*

Root. As in [5], the root ρ^* of T is treated as a P-node with two children, whose pertinent graphs are e^* and the pertinent graph of the child σ^* of ρ^* in T.

Lemma 7 ([5]). *Given the feasible set $\mathcal{F}_{\sigma^*}$, the feasible set $\mathcal{F}_{\rho^*}$ of the root ρ^* of T can be computed in $\mathcal{O}(n)$ time.*

4   No Prescribed Edge on the Outer Face

In this section, we show an $\mathcal{O}(n^2)$-time algorithm to test the upward planarity of a biconnected directed partial 2-tree G. Let $e_1, \ldots, e_m$ be any order of the edges of G. For $i = 1, \ldots, m$, let ρ_i be the Q-node of the SPQ-tree T of G corresponding to e_i and T_i be the rooted tree obtained by selecting ρ_i as the root of T. For a node μ of T, distinct choices for the root of T define different pertinent graphs G_μ of μ. Thus, we change the previous notation and denote by $G_{\mu \to \tau}$ and $\mathcal{F}_{\mu \to \tau}$ the pertinent graph and the feasible set of a node μ when its parent is a node τ. We denote by $\mathcal{F}_{\rho_i}$ the feasible set of the root ρ_i of T_i.

Our algorithm performs traversals of $T_1, \ldots, T_m$. The traversal of T_1 is special; it is a bottom-up traversal using the results from Sect. 3 to compute the feasible set $\mathcal{F}_{\mu \to \tau}$ of every node μ with parent τ in T_1, as well as auxiliary information that is going to be used by later traversals. For $i = 2, \ldots, m$, we perform a top-down traversal of T_i that computes the feasible set $\mathcal{F}_{\mu \to \tau}$ of every node μ with parent τ in T_i. Due to the information computed by the traversal of T_1, this can be carried out in $\mathcal{O}(n)$ time for each P-node. Further, the traversal of T_i visits a subtree of T_i only if that has not been visited "in the same direction" during a traversal T_j with $j < i$. We start with two auxiliary lemmas.

Lemma 8 ($\star$). *Suppose that, for some $i \in \{1, \ldots, m\}$, a node μ with parent τ has a child ν_j in T_i such that $\mathcal{F}_{\nu_j \to \mu} = \emptyset$. Then $\mathcal{F}_{\mu \to \tau} = \emptyset$.*

Lemma 9 ($\star$). *Suppose that a node μ has two neighbors ν_j and ν_k such that $\mathcal{F}_{\nu_j \to \mu} = \mathcal{F}_{\nu_k \to \mu} = \emptyset$. Then G admits no upward planar embedding.*

Bottom-up Traversal of T_1. The first step of the algorithm consists of a bottom-up traversal of T_1. This step either rejects the instance (i.e., it concludes that G admits no upward planar embedding) or computes and stores, for each non-root node μ of T_1 with parent τ, the feasible set $\mathcal{F}_{\mu \to \tau}$ of μ, as well as the feasible set $\mathcal{F}_{\rho_1}$ of the root ρ_1. Further, if μ is an S- or P-node, it also computes the following information.

- A label $\mathrm{P}(\mu)$ referencing the parent τ of μ in T_1.
- A label $\mathrm{UC}(\mu)$ referencing a node ν such that $\mathcal{F}_{\nu \to \mu}$ has not been computed. Initially this is τ, and once $\mathcal{F}_{\tau \to \mu}$ is computed, this label changes to NULL.
- A label $\mathrm{B}(\mu)$ referencing any neighbor ν of μ such that $\mathcal{F}_{\nu \to \mu} = \emptyset$. This label remains NULL until such neighbor is found.

Finally, if μ is a P-node, for each shape description s in an n-universal set $\mathcal{S}$ and each contracted shape sequence $S = [s_1, \ldots, s_x]$ in the generating set $\mathcal{G}(s)$ of s, the algorithm computes and stores the set of relevant labels for μ and S.

The bottom-up traversal of T_1 computes the feasible set $\mathcal{F}_{\mu \to \tau}$ in $\mathcal{O}(1)$ time by Lemma 3, for any Q-node $\mu \neq \rho_1$ with parent τ. When an S- or P-node μ with parent τ is visited, the algorithm stores in $\mathrm{P}(\mu)$ and $\mathrm{UC}(\mu)$ a reference to τ. Then it considers $\mathrm{B}(\mu)$. Suppose that $\mathrm{B}(\mu) \neq$ NULL (the label $\mathrm{B}(\mu)$ might have been assigned a value different from NULL when visiting a child of μ). By Lemma 8 we have $\mathcal{F}_{\mu \to \tau} = \emptyset$, hence if $\mathrm{B}(\tau) \neq$ NULL, then by Lemma 9, the algorithm rejects the instance, otherwise it sets $\mathrm{B}(\tau) = \mu$ and concludes the visit of μ. Suppose next that $\mathrm{B}(\mu) =$ NULL. Then we have $\mathcal{F}_{\nu_j \to \mu} \neq \emptyset$, for every child ν_j of μ, thus $\mathcal{F}_{\mu \to \tau}$ is computed using Lemma 4 or 6, if μ is an S-node or a P-node, respectively. If

$\mathcal{F}_{\mu \to \tau} = \emptyset$, then the algorithm checks whether $\mathrm{B}(\tau) \neq \mathrm{NULL}$ (and then it rejects the instance) or not (and then it sets $\mathrm{B}(\tau) = \mu$). This concludes the visit of μ. Finally, when the algorithm reaches ρ_1, it checks whether $\mathrm{B}(\rho_1) = \mathrm{NULL}$ and if the test is positive, then it concludes that $\mathcal{F}_{\rho_1} = \emptyset$. Otherwise, it computes $\mathcal{F}_{\rho_1}$ by means of Lemma 7 and completes the traversal of T_1.

Top-Down Traversal of T_i. The top-down traversal of T_i computes $\mathcal{F}_{\mu \to \tau}$, for each non-root node μ with parent τ in T_i, as well as $\mathcal{F}_{\rho_i}$. For each S- or P-node μ, the labels $\mathrm{UC}(\mu)$ and $\mathrm{B}(\mu)$ might be updated during the traversal of T_i, while $\mathrm{P}(\mu)$ and the sets of relevant labels are never altered after the traversal of T_1. The traversal of T_i visits a node μ with parent τ only if $\mathcal{F}_{\mu \to \tau}$ has not been computed yet; this information is retrieved in $\mathcal{O}(1)$ time from the label $\mathrm{UC}(\tau)$.

When the traversal visits an S- or P-node μ with parent τ and children $\nu_1, \ldots, \nu_k$, it proceeds as follows. Note that $\mathrm{P}(\mu) \neq \tau$, as otherwise $\mathcal{F}_{\mu \to \tau}$ would have been already computed. Then we have $\mathrm{P}(\mu) = \nu_{j*}$, for some $j^* \in \{1, \ldots, k\}$.

If $\mathrm{UC}(\mu) = \nu_{j*}$, then before computing $\mathcal{F}_{\mu \to \tau}$, the algorithm descends in ν_{j*} in order to compute $\mathcal{F}_{\nu_{j*} \to \mu}$. Otherwise, $\mathcal{F}_{\nu_j \to \mu}$ has been computed for $j = 1, \ldots, k$.

If $\mathrm{B}(\mu) = \nu_j$, for some $j \in \{1, \ldots, k\}$, then by Lemma 8 we have $\mathcal{F}_{\mu \to \tau} = \emptyset$, hence if $\mathrm{B}(\tau) \neq \mathrm{NULL}$ and $\mathrm{B}(\tau) \neq \mu$, then the algorithm rejects the instance by Lemma 9, otherwise it sets $\mathrm{B}(\tau) = \mu$ and concludes the visit of μ. Conversely, if $\mathrm{B}(\mu) = \mathrm{NULL}$ or $\mathrm{B}(\mu) = \tau$, then $\mathcal{F}_{\nu_j \to \mu} \neq \emptyset$ for $j = 1, \ldots, k$. The algorithm then computes $\mathcal{F}_{\mu \to \tau}$, as described below. Afterwards, if $\mathrm{UC}(\tau) = \mu$, the algorithm sets $\mathrm{UC}(\tau) = \mathrm{NULL}$. Further, if $\mathcal{F}_{\mu \to \tau} = \emptyset$, the algorithm checks whether $\mathrm{B}(\tau) \neq \mathrm{NULL}$ (and then rejects the instance) or not (and then sets $\mathrm{B}(\tau) = \mu$).

The computation of $\mathcal{F}_{\mu \to \tau}$ distinguishes the case when μ is an S-node or a P-node. If μ is an S-node, then the computation of $\mathcal{F}_{\mu \to \tau}$ is done by means of Lemma 4. The running time of the procedure for the S-nodes sums up to $\mathcal{O}(n^2)$, over all S-nodes and all traversals of T. If μ is a P-node, then the computation of $\mathcal{F}_{\mu \to \tau}$ cannot be done by just applying the algorithm from Lemma 6, as that would take $\Theta(n^3)$ time for all P-nodes and all traversals of T. Instead, the information computed when traversing T_1 allows us to determine in $\mathcal{O}(1)$ time whether any shape description is in $\mathcal{F}_{\mu \to \tau}$. This results in an $\mathcal{O}(n)$ time for processing μ in T_i, which sums up to $\mathcal{O}(nk)$ time over all traversals of T, and thus in a $\mathcal{O}(n^2)$ total running time for the entire algorithm.

The algorithm determines $\mathcal{F}_{\mu \to \tau}$ by examining each shape description s in an n-universal set $\mathcal{S}$, which has $\mathcal{O}(n)$ elements and is constructed in $\mathcal{O}(n)$ time by Lemma 1, and deciding whether it is in $\mathcal{F}_{\mu \to \tau}$ or not. This is done as follows. We construct in $\mathcal{O}(1)$ time the generating set $\mathcal{G}(s)$ of s, by Lemma 5. Recall that $\mathcal{G}(s)$ contains $\mathcal{O}(1)$ contracted shape sequences, each with length $\mathcal{O}(1)$. For each sequence $S = [s_1, \ldots, s_x]$ in $\mathcal{G}(s)$, we test whether S is realizable by μ as follows.

- If $\mathrm{UF_2}(\mu, S) \neq \mathrm{NULL}$, or if $\mathrm{UF_1}(\mu, S) \neq \mathrm{NULL}$ and $\mathrm{UF_1}(\mu, S) \neq \tau$, then there exists a child ν_j of μ in T_i such that $\mathcal{F}_{\nu_j \to \mu}$ does not contain any shape description in S. Then we conclude that S is not realizable by μ.
- Otherwise, we test whether $\mathcal{F}_{\nu_{j*} \to \mu}$ contains any shape description among the ones in S. If not, S is not realizable by μ. Otherwise, for $j = 1, \ldots, k$,

$\mathcal{F}_{\nu_j \to \mu}$ contains a shape description in S. However, this does not imply that S is realizable by μ, as we need to ensure that $s_1 \in \mathcal{F}_{\nu_j \to \mu}$ and $s_x \in \mathcal{F}_{\nu_l \to \mu}$ for two distinct children ν_j and ν_l of μ in T_i. This can be tested as follows. We construct a bipartite graph $\mathcal{B}_{\mu \to \tau}(S)$ in which one family has two vertices labeled s_1 and s_x. The other one has a vertex for each child of μ in the set $\{F_1(\mu, S), F_2(\mu, S), F_3(\mu, S), L_1(\mu, S), L_2(\mu, S), L_3(\mu, S), \nu_{j*}\}$. The graph $\mathcal{B}_{\mu \to \tau}(S)$ contains an edge between the vertex representing a child ν_j of μ and a vertex representing s_1 or s_x if s_1 or s_x belongs to $\mathcal{F}_{\nu_j \to \mu}$, respectively. We now have that $s_1 \in \mathcal{F}_{\nu_j \to \mu}$ and $s_x \in \mathcal{F}_{\nu_l \to \mu}$ for two distinct children ν_j and ν_l of μ in T_i (and thus S is realizable by μ) if and only if $\mathcal{B}_{\mu \to \tau}(S)$ contains a size-2 matching, which can be tested in $\mathcal{O}(1)$ time.

Testing whether S is realizable by μ can be done in $\mathcal{O}(1)$ time, as it only requires to check $\mathcal{O}(1)$ labels, to find a size-2 matching in a $\mathcal{O}(1)$-size graph, and to check $\mathcal{O}(1)$ times whether a shape description belongs to a feasible set. The last operation requires $\mathcal{O}(1)$ time by Lemma 2. We conclude that s is in $\mathcal{F}_{\mu \to \tau}$ if and only if at least one contracted shape sequence S in $\mathcal{G}(s)$ is realizable by μ. This concludes the description of how the algorithm handles a P-node.

Finally, $\mathcal{F}_{\rho_i}$ is computed in $\mathcal{O}(n)$ time by Lemma 7. We get the following.

Lemma 10 ($\star$). *The described algorithm runs in $\mathcal{O}(n^2)$ time and either correctly concludes that G admits no upward planar embedding, or computes the feasible sets $\mathcal{F}_{\rho_1}, \ldots, \mathcal{F}_{\rho_m}$.*

5   Single-Connected Graphs

In this section, we extend Lemma 10 from the biconnected case to arbitrary partial 2-trees. To this end, we obtain a general lemma that allows us to test upward planarity of digraphs from the feasible sets of biconnected components.

Lemma 11 ($\star$). *Let G be an n-vertex digraph. Let $B_1, \ldots, B_t$ be the maximal biconnected components of G. For $i \in [t]$, let the edges of B_i be $e^i_1, \ldots, e^i_{m_i}$, and the respective Q-nodes in the SPQR-tree of B_i be $\rho^i_1, \ldots, \rho^i_{m_i}$. There is an algorithm that, given G and the feasible sets $\mathcal{F}_{\rho^i_j}$ for each $i \in [t]$ and $j \in [m_i]$, in time $\mathcal{O}(n^2)$ correctly decides whether G admits an upward planar embedding.*

Note that Lemma 11 holds for all digraphs, not only partial 2-trees. In fact, it generalizes [5, Section 5], where an analogous statement has been shown for all expanded graphs. Our main result follows from Lemmas 11 and 10.

Theorem 2 ($\star$). *Let G be an n-vertex directed partial 2-tree. It is possible to determine whether G admits an upward planar embedding in time $\mathcal{O}(n^2)$.*

Hence, all that remains now is to prove Lemma 11. To give an intuition of the proof, we start by guessing the root of the block-cut tree of G, which corresponds to a biconnected component that is assumed to see the outer face in the desired upward planar embedding of G. The core of the proof is the following lemma, which states that leaf components can be disregarded as long as certain simple conditions on their parent cut-vertex are met.

Lemma 12 (⋆). *Consider a rooted block-cut tree of a digraph G, its cut vertex v that is adjacent to leaf blocks $B_1,...,B_\ell$, and the parent block P. Denote by G^P the subgraph $G\left[\left(V(G) \setminus \bigcup_{i \in [\ell]} B_i\right) \cup \{v\}\right]$. Any upward planar embedding of G^P in which the root block is adjacent to the outer face, can be extended to an embedding of G with the same property if the following conditions hold:*

1. *Each B_i has an upward planar embedding with v on the outer face f_i.*
2. *If v is a non-switch vertex in P, each B_i has an upward planar embedding with v on f_i where the angle at v in f_i is not small.*
3. *If there is $j \in [\ell]$ such that v is a non-switch vertex in B_j, and all upward planar embeddings of B_j with v on f_j have a small angle at v in f_j, then for all $i \in [\ell]$ s.t. $i \neq j$ and v is a non-switch vertex in B_i, B_i has an upward planar embedding with v on f_i where the angle at v in f_i is flat.*

Moreover, if G admits an upward planar embedding in which the root block is adjacent to the outer face, the conditions above are necessarily satisfied.

The proof of Lemma 12 essentially boils down to a case distinction on how the leaf blocks are attached; the cases that need to be considered are intuitively illustrated in Fig. 4. With this, we finally have all the components necessary to prove Theorem 2. Intuitively, the algorithm proceeds in a leaf-to-root fashion along the block-cut tree, and at each point it checks whether the conditions of Lemma 12 are satisfied. If they are, the algorithm removes the respective leaf components and proceeds upwards, while otherwise we reject the instance.

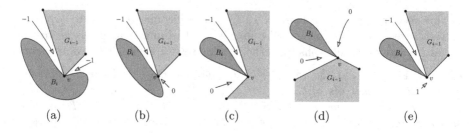

| (a) | (b) | (c) | (d) | (e) |

Fig. 4. Illustrations for the proof of Lemma 12.

6   Concluding Remarks

We have provided an $\mathcal{O}(n^2)$-time algorithm for testing the upward planarity of n-vertex directed partial 2-trees, substantially improving on the state of the art [10]. There are several major obstacles to overcome for improving this runtime to linear; hence, it would be worth investigating whether the quadratic bound is tight. Another interesting direction for future work is to see whether our new techniques can be used to obtain quadratic algorithms for related problems, such as computing orthogonal drawings with the minimum number of bends [7,9].

Acknowledgments. Ganian and Simonov acknowledge support by the Austrian Science Foundation (FWF Project Y1329). Di Giacomo and Frati acknowledge support by MIUR (PRIN project 20174LF3T8).

References

1. Bertolazzi, P., Di Battista, G., Liotta, G., Mannino, C.: Upward drawings of triconnected digraphs. Algorithmica **12**(6), 476–497 (1994)
2. Bertolazzi, P., Di Battista, G., Mannino, C., Tamassia, R.: Optimal upward planarity testing of single-source digraphs. SIAM J. Comput. **27**(1), 132–169 (1998)
3. Brückner, G., Himmel, M., Rutter, I.: An SPQR-tree-like embedding representation for upward planarity. In: Archambault, D., Tóth, C.D. (eds.) GD 2019. LNCS, vol. 11904, pp. 517–531. Springer, Cham (2019). https://doi.org/10.1007/978-3-030-35802-0_39
4. Chan, H.: A parameterized algorithm for upward planarity testing. In: Albers, S., Radzik, T. (eds.) ESA 2004. LNCS, vol. 3221, pp. 157–168. Springer, Heidelberg (2004). https://doi.org/10.1007/978-3-540-30140-0_16
5. Chaplick, S., Di Giacomo, E., Frati, F., Ganian, R., Raftopoulou, C.N., Simonov, K.: Parameterized algorithms for upward planarity. In: Goaoc, X., Kerber, M. (eds.) 38th International Symposium on Computational Geometry (SoCG 2022). LIPIcs, vol. 224, pp. 26:1–26:16. Schloss Dagstuhl - Leibniz-Zentrum für Informatik (2022)
6. Chaplick, S., Di Giacomo, E., Frati, F., Ganian, R., Raftopoulou, C.N., Simonov, K.: Testing upward planarity of partial 2-trees. CoRR abs/2208.12548 (2022)
7. Di Battista, G., Liotta, G., Vargiu, F.: Spirality and optimal orthogonal drawings. SIAM J. Comput. **27**(6), 1764–1811 (1998)
8. Di Battista, G., Tamassia, R.: On-line planarity testing. SIAM J. Comput. **25**(5), 956–997 (1996)
9. Di Giacomo, E., Liotta, G., Montecchiani, F.: Orthogonal planarity testing of bounded treewidth graphs. J. Comput. Syst. Sci. **125**, 129–148 (2022)
10. Didimo, W., Giordano, F., Liotta, G.: Upward spirality and upward planarity testing. SIAM J. Discret. Math. **23**(4), 1842–1899 (2009)
11. Didimo, W., Liotta, G., Ortali, G., Patrignani, M.: Optimal orthogonal drawings of planar 3-graphs in linear time. In: Chawla, S. (ed.) SODA 2020. pp. 806–825. SIAM (2020)
12. Frati, F.: Planar rectilinear drawings of outerplanar graphs in linear time. Comput. Geom. **103**, 101854 (2022)
13. Garg, A., Tamassia, R.: On the computational complexity of upward and rectilinear planarity testing. SIAM J. Comput. **31**(2), 601–625 (2001)
14. Gutwenger, C., Mutzel, P.: A linear time implementation of SPQR-trees. In: Marks, J. (ed.) GD 2000. LNCS, vol. 1984, pp. 77–90. Springer, Heidelberg (2001). https://doi.org/10.1007/3-540-44541-2_8
15. Healy, P., Lynch, K.: Two fixed-parameter tractable algorithms for testing upward planarity. Int. J. Found. Comput. Sci. **17**(5), 1095–1114 (2006)
16. Hutton, M.D., Lubiw, A.: Upward planar drawing of single source acyclic digraphs. In: Aggarwal, A. (ed.) 2nd Annual ACM/SIGACT-SIAM Symposium on Discrete Algorithms. SODA 1991, pp. 203–211. ACM/SIAM (1991)
17. Hutton, M.D., Lubiw, A.: Upward planar drawing of single-source acyclic digraphs. SIAM J. Comput. **25**(2), 291–311 (1996)
18. Papakostas, A.: Upward planarity testing of outerplanar dags (extended abstract). In: Tamassia, R., Tollis, I.G. (eds.) GD 1994. LNCS, vol. 894, pp. 298–306. Springer, Heidelberg (1995). https://doi.org/10.1007/3-540-58950-3_385

Computing a Feedback Arc Set Using PageRank

Vasileios Geladaris[(✉)], Panagiotis Lionakis, and Ioannis G. Tollis

Computer Science Department, University of Crete, Heraklion, Greece
{csd3926,lionakis,tollis}@csd.uoc.gr

Abstract. We present a new heuristic algorithm for computing a minimum Feedback Arc Set in directed graphs. The new technique produces solutions that are better than the ones produced by the best previously known heuristics, often reducing the FAS size by more than 50%. It is based on computing the PageRank score of the nodes of the directed line graph of the input directed graph. Although the time required by our heuristic is heavily influenced by the size of the produced line graph, our experimental results show that it runs very fast even for very large graphs used in graph drawing.

Keywords: Feedback arc set · Hierarchical graph drawing · PageRank · Line graph

1 Introduction

In a directed graph, G, a *feedback arc set* (*FAS*) is a set of edges whose removal leave G acyclic. The minimum FAS problem is important for visualizing directed graphs in hierarchical style [7]. In fact, the first step of both known frameworks for hierarchical graph drawing is to compute a minimum FAS [13,18]. Unfortunately, computing a minimum FAS is NP-hard and thus many heuristics have been presented in order to find a reasonably good solution. In this paper we present a new heuristic that uses a different approach and produces FAS that contain about half the number of edges of the best known heuristics. However, it requires superlinear time, and hence it may not be suitable for very large graphs. Finding a minimum FAS has many additional applications beyond Graph Drawing, including misinformation removal, label propagation, and many application domains motivated by Social Network Analysis [6,9,16].

A feedback arc set of a directed graph $G = (V, E)$ is a subset of edges F of E such that removing the edges in F from E leaves G acyclic (no directed cycles). In other words, a FAS contains at least one edge from each cycle of G. In hierarchical drawing algorithms the edges in a FAS are not removed, but instead their direction is inverted. Following the terminology of [7], a set of edges whose reversal makes the digraph acyclic is called a feedback set (FS). Notice that a FAS is not always a FS. However, it is easy to see that every minimal cardinality

© The Author(s), under exclusive license to Springer Nature Switzerland AG 2023
P. Angelini and R. von Hanxleden (Eds.): GD 2022, LNCS 13764, pp. 188–200, 2023.
https://doi.org/10.1007/978-3-031-22203-0_14

FAS is also a FS. Hence it follows that the minimum FS problem is as hard as the well studied minimum FAS problem which is known to be NP-hard [10,11]. Clearly, any heuristic for solving the minimum FAS problem can be applied for solving the minimum FS problem, as discussed in [7,12].

There have been many heuristics for solving the FAS problem due to the multitude of its applications. Two of the most important heuristics/techniques are due to Eades, Lin & Smyth [8] and Brandenburg & Hanauer [4]. The first is a greedy heuristic, that will be called *GreedyFAS*, whereas the second presents a set of heuristics based on sorting. Simpson, Srinivasan & Thomo published an experimental study for the FAS problem on very large graphs at web-scale (also called *webgraphs*) [17]. They implemented and compared many FAS heuristics. According to their study, the aforementioned are the most efficient heuristics, but only GreedyFAS is suitable to run on their extra large webgraphs.

In this paper we present a new heuristic algorithm for computing a minimum FAS in directed graphs. The new technique produces solutions that are better than the ones produced by the best previous heuristics, sometimes even reducing the FAS size by more than 50%. It is based on computing the PageRank score of the nodes of a graph related to the input graph, and runs rather fast for graphs up to 4,000 nodes. However, it is slower than GreedyFAS for webgraphs.

2   Existing Algorithms

In this section we summarize and give a brief description of two important heuristics that currently give the best results for the FAS problem, according to the new experimental study of Simpson, Srinivasan & Thomo [17]. They implemented and compared many heuristics for FAS, and performed experiments on several large and very large webgraphs. Their results show that two of the known heuristic algorithms give the best results.

The first of the two heuristic algorithms that currently produce the best FAS size is called *GreedyFAS* and it is due to Eades, Lin & Smyth [8]. In [17] two different optimized implementations of GreedyFAS that run in $O(n+m)$ are presented and tested. These are the most efficient implementations in their study and are able to run even for their extra large webgraphs. The second algorithm is *SortFAS* of Brandenburg & Hanauer [4]. According to [17], SortFAS, as proposed runs in $O(n^3)$ time but Simpson et al. present an implementation that runs in $O(n^2)$ time.

We will present experimental results that show that our new heuristic algorithm performs better than both of them in terms of the size of the produced FAS. On the other hand, it takes more time than both of them for large graphs. However, for graphs that are typically used for visualization purposes, the running time is acceptable whereas the produced FAS size is about half.

2.1   GreedyFAS

The GreedyFAS algorithm was introduced by Eades, Lin & Smyth in 1993 [8]. It efficiently calculates an approximation to the FAS problem on a graph G. In order to understand the algorithm, we first discuss the *Linear Arrangement*

Problem (LA), which is an equivalent formulation to the FAS problem. The LA problem produces an ordering of the nodes of a graph G for which the number of arcs pointing backwards is minimum. The set of backwards arcs is a FAS since removing them from G leaves the graph acyclic.

GreedyFAS calculates a feedback arc set of a graph G by first calculating a Linear Arrangement of G. More specifically, in each iteration, the algorithm removes all nodes of G that are sinks followed by all the nodes that are sources. It then removes a node u for which $\delta(u) = d^+(u) - d^-(u)$ is a maximum, where $d^+(u)$ denotes the out-degree of u and $d^-(u)$ denotes the in-degree of u. The algorithm also makes use of two sequences of nodes s_1 and s_2. When any node u is removed from G then it is either prepended to s_2 if it's a sink, or appended to s_1 if it's not. The above steps are repeated until G is left with no nodes, then the sequence $s = s_1 s_2$ is returned as a linear arrangement for which the backward arcs make up a feedback arc set. For more details see [7,12]. Using the implementations of [17], GreedyFAS runs very fast, in $O(n + m)$ time, and is suitable for their extra large webgraphs. The pseudocode for GreedyFAS, as described in [7] and [17], is presented in Algorithm 1.

Algorithm 1. GreedyFAS

Input: Directed graph $G = (V, E)$
Output: Linear Arrangement A
$s_1 \leftarrow \emptyset$, $s_2 \leftarrow \emptyset$
while $G \neq \emptyset$ **do**
    while G contains a sink **do**
        choose a sink u
        $s_2 \leftarrow u s_2$
        $G \leftarrow G \backslash u$
    while G contains a source **do**
        choose a source u
        $s_1 \leftarrow s_1 u$
        $G \leftarrow G \backslash u$
    choose a node u for which $\delta(u)$ is a maximum
    $s_1 \leftarrow s_1 u$
    $G \leftarrow G \backslash u$
return $s = s_1 s_2$

2.2 SortFAS

The SortFAS algorithm was introduced in 2011 by Brandenburg & Hanauer [4]. The algorithm is an extension of the KwikSortFAS heuristic by Ailon et al. [1], which is an approximation algorithm for the FAS problem on tournaments. With SortFAS, Brandenburg & Hanauer extended the above heuristic to work for general directed graphs. It uses the underlying idea that the nodes of a graph can be sorted into a desirable Linear Arrangement based on the number of back arcs induced.

In the case of SortFAS, the nodes are processed in order of their ordering $(v_1...v_n)$. The algorithm goes through n iterations. In the i-th iteration, node v_i is inserted into the linear arrangement in the best position based on the first $i-1$ nodes which are already placed. The best position is the one with the least number of back arcs induced by v_i. In case of a tie the leftmost position is taken. Using the implementation of [17], SortFAS runs in $O(n^2)$ time. The pseudocode for SortFAS, as described in [17], is presented in Algorithm 2.

Algorithm 2. SortFAS

Input: Linear arrangement A
for each node v in A **do**
    $val \leftarrow 0$, $min \leftarrow 0$, $loc \leftarrow$ position of v
    for each position j from $loc - 1$ down to $-$ **do**
        $w \leftarrow$ node at position j
        if arc (v, w) exists **then**
            $val \leftarrow val - 1$
        else if arc (w, v) exists **then**
            $val \leftarrow val + 1$
        if $val \leq min$ **then**
            $min \leftarrow val, loc \leftarrow j$
    insert v at position loc

---

3   Our Proposed Approach

Our approach is based on running the well known PageRank algorithm [5,14] on the directed line digraph of the original directed graph. The *line graph* of an undirected graph G is another graph $L(G)$ that is constructed as follows: each edge in G corresponds to a node in $L(G)$ and for every two edges in G that are adjacent to a node v an edge is placed in $L(G)$ between the corresponding nodes. Clearly, the number of nodes of a line graph is m and the number of edges is proportional to the sum of squares of the degrees of the nodes in G, see [15]. If G is a directed graph, its *directed line graph* (or *line digraph*) $L(G)$ has one node for each edge of G. Two nodes representing directed edges incident upon v in G (one incoming into v, and one outgoing from v), called $L(u, v)$, and $L(v, w)$, are connected by a directed edge from $L(u, v)$ to $L(v, w)$ in $L(G)$. In other words, every edge in $L(G)$ represents a directed path in G of length two. Similarly, the number of nodes of a line digraph is m and the number of edges is proportional to $\sum_{u \in V}[d^+(u) \times d^-(u)]$. Hence, the size of $L(G)$ is $O(m + \sum_{u \in V}[d^+(u) \times d^-(u)])$.
    Given a digraph $G = (V, E)$ our approach is to compute its line digraph, $L(G)$, run a number of iterations of PageRank on $L(G)$ and remove the node of highest PageRank in $L(G)$. Our experimental results indicate that PageRank values converge reasonably well within five iterations.
    A digraph G is *strongly connected* if for every pair of vertices of G there is a cycle that contains them. If G is not strongly connected, it can be decomposed

into its *strongly connected components (SCC)* in linear time [19]. An SCC of G is a subgraph that is strongly connected, and is maximal, in the sense that no additional edges or vertices of G can be included in the subgraph without breaking its property of being strongly connected. If each SCC is contracted to a single vertex, the resulting graph is a directed acyclic graph (DAG). It follows that feedback arcs can exist only within some (SCC) of G. Hence we can apply this approach inside each SCC, using their corresponding line digraph, and remove the appropriate edges from each SCC. This approach will avoid performing several useless computations and thus reduce the running time of the algorithm.

3.1   Line Graph

In order to obtain the line digraph of G, we use a DFS-based approach. First, for each edge (u, v) of G, we create a node (u, v) in $L(G)$ and then run the following recursive procedure. For a node v, we mark it as visited and iterate through each one of its outgoing edges. For each outgoing edge (v, u) of v, we add an edge in $L(G)$ from the *prev* $L(G)$ node that was processed before the procedure's call to the node (v, u). Afterwards we call the same procedure for u if it's not visited with (v, u) as *prev*. If u is visited we add an edge from (v, u) to each one of $L(G)$'s nodes corresponding from u. Since this technique is based on DFS, the running time is $O(n + m + |L(G)|)$. The pseudocode for computing a line digraph is presented in Algorithm 3.

Algorithm 3. LineDigraph

Input: Digraph $G = (V, E)$
Output: Line Digraph $L(G)$ of G
Create a line digraph $L(G)$ with every edge of G as a node
$v \leftarrow$ random node of G

    procedure GETLINEGRAPH$(G, L(G), v, prev)$
        mark v as *visited*
        for each edge $e = (v, u)$ outgoing of v **do**
            $z \leftarrow$ node of $L(G)$ representing e
            create an edge in $L(G)$ from *prev* to z ▷ Given that *prev* is not nill
            if u is not *visited* **then**
                GetLineGraph$(G, L(G), u, z)$
            else
                for each node k in $L(G)$ that originates from u **do**
                    create an edge in $L(G)$ from z to k

3.2   PageRank

PageRank was first introduced by Brin & Page in 1998 [5,14]. It was developed in order to determine a measure of importance of web pages in a hyperlinked network of web pages. The basic idea is that PageRank will assign a score of importance to every node (web page) in the network. The underlying assumption is that important nodes are those that receive many "recommendations" (in-links) from other important nodes (web pages). In other words, it is a link analysis algorithm that assigns numerical scores to the nodes of a graph in order to measure the importance of each node in the graph. PageRank works by counting the number and quality/importance of edges pointing to a node and then estimate the importance of that node. We use a similar approach in order to determine the importance of edges in a directed graph. The underlying assumption of our technique is that the number of cycles that contain a specific edge e will be reflected in PageRank score of e. Thus the removal of edges with high PageRank score is likely to break the most cycles in the graph.

Given a graph with n nodes and m edges, PageRank starts by assigning an initial score of $1/n$ to all the nodes of a graph. Then for a predefined number of iterations each node divides its current score equally amongst its outgoing edges and then passes these values to the nodes it is pointing to. If a node has no outgoing links then it keeps its score to itself. Afterwards, each node updates its new score to be the sum of the incoming values. It is obvious that after enough iterations all PageRank values will inevitably gather in the sinks of the graph. In use cases where that is a problem a damping factor is used, where each node gets a percentage of its designated score and the rest gets passed to all other nodes of the graph. For our use case we have no need for this damping factor as we want the scores of the nodes to truly reflect their importance. The number of iterations depends on the size and structure of a graph. We found that for small and medium graphs, which is the case in the scenario for graph visualization, about five iterations were enough for the scores of the nodes to converge. Depending on the implementation, PageRank can run in $O(k(n+m))$ time, where k is the number of iterations. The pseudocode for PageRank is presented in Algorithm 4.

Algorithm 4. PageRank

Input: Digraph $G = (V, E)$, number of iterations k
Output: PageRank scores of G
for each node v in G **do**
    $PR(v) \leftarrow \frac{1}{|V|}$
for k iterations **do**
    for each node v in G **do**
        $PR(v) \leftarrow \sum_{u \in in(v)} \frac{PR_{old}(u)}{|out(u)|}$
return PR

3.3   PageRankFAS

Our proposed algorithm is based on the concepts of PageRank and Line Digraphs. The idea behind *PageRankFAS* is that we can score the edges of

G based on their involvement in cycles: For each strongly connected component $(s_1, s_2, ..., s_j)$ of G, it computes the line digraph $L(s_i)$ of the i-th strongly connected component, to transform edges to nodes; next it runs the PageRank algorithm on $L(s_i)$ to obtain a score for each edge of s_i in G.

We observed that the nodes of the line digraphs with the highest PageRank score correspond to edges that are involved in the most cycles of G. We also observed that the nodes of the line digraphs with lower score correspond to edges of G with low involvement in cycles. Using this knowledge, we run PageRankFAS for a number of iterations. In each iteration, we use PageRank to calculate the node scores of each $L(s_i)$ and remove the node(s) with the highest PageRank score, also removing the corresponding edge(s) from G. We repeat this process until G becomes acyclic. The pseudocode is presented in Algorithm 5.

Algorithm 5. PageRankFAS

    Input: Digraph $G = (V, E)$
    Output: Feedback Arc Set of G
    $fas \leftarrow \emptyset$
    while G has cycles **do**
        Let $(s_1, s_2, ..., s_j)$ be the strongly connected components of G
        for each strongly connected component s_i **do**
            Create a line digraph $L(s_i)$ with every edge of s_i as a node
            $v \leftarrow$ random node of s_i
            GetLineGraph$(s_i, L(s_i), v, nill)$
            PageRank$(L(s_i))$
            $u \leftarrow$ node of $L(s_i)$ with highest PageRank value
            $e \leftarrow$ edge of G corresponding to u
            Add e to fas
            Remove e from G
    return fas

---

4   Experiments and Discussion

Here we report the experimental results and describe some details of our setup. All of our algorithms are implemented in Java 8 using the WebGraph framework [2,3] and tested on a single machine with Apple's M1 processor, 8GB of RAM and running macOS Monterey 12.

Datasets: In order to evaluate our proposed heuristic algorithm we used four different datasets:

1. Randomly generated graphs with 100, 200, 400, 1000, 2000, 4000 nodes and an average out-degree of 1.5, 3 and 5 each.
2. Three directed graphs from the datasets in graphdrawing.org, suitably modified in order to contain cycles (since the originals are DAGs).

3. Randomly generated graphs with 50, 100 and 150 nodes and average out-degrees of 1.5, 3, 5, 8, 10 and 15 each.
4. Two webgraphs from the Laboratory of Web Algorithmics[1], also used in [17].

We randomly generate a total of 36 graphs using a predefined number of nodes, average out-degree and back edge percentage, and we repeat the process 10 times. By construction, this model has the advantage that we know in advance an upper bound to the FAS size, since the number of randomly created back edges divided by the total number of edges, is an upper-bound to the size of a minimum FAS. Finally, in order to avoid having abrupt results due to randomness, for each case we run the three algorithms on 10 created graphs and report the average numbers. This smooths out several points in our curves.

4.1   FAS with Respect to the Number of Nodes

The first set of experiments gives us an idea of how PageRankFAS performs on graphs, with varying number of nodes in comparison to the other two algorithms. It is noteworthy that in most cases the FAS found by PageRankFAS is less than 50% of the FAS found by GreedyFAS and SortFAS. As a matter of fact, for large visualization graphs with 4,000 nodes and 12,000 edges the reduction in the FAS size is almost 55% with respect to the FAS produced by GreedyFAS. The execution time taken by PageRankFAS is less than one second for graphs up to 1,000 nodes, which is similar to the time of the other two heuristics. For the larger graphs, even up to 4,000 nodes the time required is less than 8 s, whereas, the other heuristics run in about 1–2 s. The results of this experiment are shown in Fig. 1. It is interesting to note that the performance of SortFAS is better than the performance of GreedyFAS as the graphs become denser, and in fact, SortFAS actually out-performs GreedyFAS when the graphs have an average out-degree 5 and above, see Fig. 1(c).

4.2   FAS with Respect to the Number of Back Edges

The second type of experiments make use of three graphs from graphdrawing.org. Since these graphs are directed acyclic, we randomly added back edges in different percentages of the total number of edges. We did this in a controlled manner in order to know in advance an upper bound of FAS. PageRankFAS gave by far the best FAS results and GreedyFAS also produced FAS with sizes mostly below 10%. SortFAS was not competitive in this dataset. The results are shown in Fig. 2. The execution time taken by PageRankFAS is well below 0.15 of a second for all graphs, which is similar to the other two heuristics.

---

[1] https://law.di.unimi.it/datasets.php.

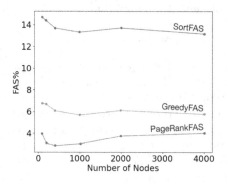

(a) Graphs with average out-degree 1.5

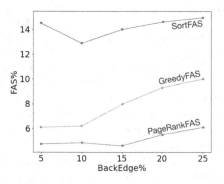

(a) Graph with 50 nodes and 75 edges before modification

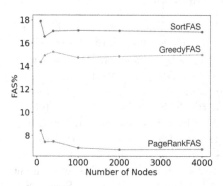

(b) Graphs with average out-degree 3

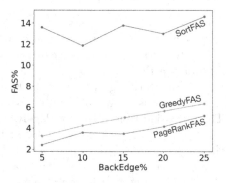

(b) Graph with 75 nodes and 86 edges before modification

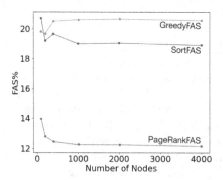

(c) Graphs with average out-degree 5

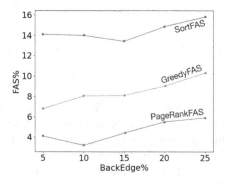

(c) Graph with 99 nodes and 154 edges before modification

Fig. 1. FAS percentage for graphs with increasing number of nodes and three different average out-degrees.

Fig. 2. FAS percentage for 3 types of graphs from graphdrawing.org and for various numbers of back edges.

4.3   FAS with Respect to the Average Out-Degree

Motivated by the results shown in Fig. 1(c) we decided to investigate the correlation between the density of a graph and its potential FAS percentage. In this experiment, we created 18 different graphs, six of them with 50 nodes, six with 100 nodes and six with 150 nodes as follows: For each node size (i.e., 50, 100, 150) six graphs with average out-degrees 1.5, 3, 5, 8, 10 and 15. Again, as with our previous experiments, the results reported here are the averages of 10 runs in order to compensate for the randomness of each graph and to get smoother curves. The results of this experiment are shown in Fig. 3.

The results of PageRankFAS are consistently better than the results of GreedyFAS and SortFAS for all graphs. The results of GreedyFAS and SortFAS are very close to each other, for the graphs with 50 nodes. Notice however that, SortFAS outperforms GreedyFAS when the number of nodes exceeds 100 and the average out-degree exceeds five. This is aligned with the results shown in Fig. 1(c). Furthermore, as expected, when the average out-degree increases the FAS size clearly increases. Consequently, all techniques seem to converge at higher percentages of FAS size. Again, PageRankFAS runs in a small fraction of a second for all graphs, which is similar to the running times of the other two heuristics.

4.4   PageRankFAS on Webgraphs

The experiments reported in [17] use large and extra large benchmark webgraphs. Their smaller benchmarks are *wordassociation-2011* (with 10,617 nodes, 72,172 edges, which implies an average degree 6.80) and *enron* (with 69,244 nodes, 276,143 edges, which implies an average degree 3.86).

The authors report that the sizes of a FAS found by GreedyFAS and SortFAS for wordassociation-2011 are 18.89% and 20.17%, respectively [17]. We ran PageRankFAS for wordassociation-2011 and obtained a FAS of size 14.85%. Similarly, for webgraph enron they report a FAS of 12.54% and 14.16% respectively. We ran PageRankFAS on webgraph enron and obtained a FAS of size 11.05%. The results are shown in Fig. 4. As expected, and consistent with our experimental observations of the previous subsections, the FAS size of the denser webgraph (wordassociation-2011) is larger than the FAS size of the sparser graph (enron), as computed by all heuristics.

Unfortunately, the required execution time of our algorithm does not allow us to test it on the larger webgraphs used in [17]. However, it is interesting that there exists a FAS of smaller size for these large graphs, which, to the best of our knowledge, was not known before.

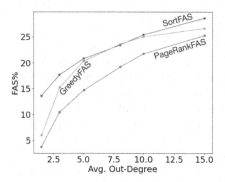

(a) Graphs with 50 nodes

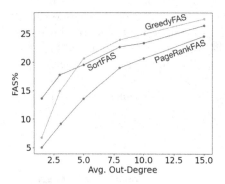

(b) Graphs with 100 nodes

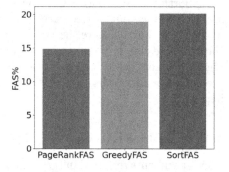

(a) wordassociation-2011

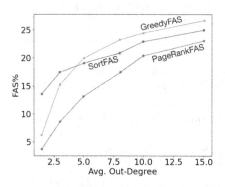

(c) Graphs with 150 nodes

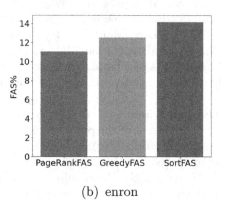

(b) enron

Fig. 3. FAS percentage depending on the average out-degree of three different types of graphs.

Fig. 4. FAS percentage on two web-graphs.

5   Conclusions

We presented a heuristic algorithm for computing a FAS of minimum size based on PageRank. Our experimental results show that the size of a FAS computed by our heuristic algorithm is typically about 50% smaller than the sizes obtained by the best previous heuristics. Our algorithm is more time consuming than the best previous heuristics, but it's running time is reasonable for graphs up to 4,000 nodes. For smaller graphs, up to 1,000 nodes, the execution time is well below one second, which is similar to the running times of the other two heuristics. Therefore, this is acceptable for graph drawing applications. An interesting side result is that we found out that the FAS-size of two large graphs is significantly less than it was known before. Since it is NP-hard to compute the minimum FAS, the optimum solution for these webgraphs is unknown. Hence, we do not know how close our solutions are to the optimum. It would be interesting to investigate techniques to speedup PageRankFAS in order to make it more applicable to larger webgraphs.

References

1. Ailon, N., Charikar, M., Newman, A.: Aggregating inconsistent information: ranking and clustering. Journal of the ACM (JACM) **55**(5), 1–27 (2008). https://doi.org/10.1145/1411509.1411513
2. Boldi, P., Rosa, M., Santini, M., Vigna, S.: Layered label propagation: A multiresolution coordinate-free ordering for compressing social networks. In: Srinivasan, S., Ramamritham, K., Kumar, A., Ravindra, M.P., Bertino, E., Kumar, R. (eds.) Proceedings of the 20th international conference on World Wide Web. pp. 587–596. ACM Press (2011)
3. Boldi, P., Vigna, S.: The WebGraph framework I: Compression techniques. In: Proc. of the Thirteenth International World Wide Web Conference (WWW 2004). pp. 595–601. ACM Press, Manhattan, USA (2004)
4. Brandenburg, F.J., Hanauer, K.: Sorting heuristics for the feedback arc set problem. In: Technical Report MIP-1104. University of Passau Germany (2011)
5. Brin, S., Page, L.: The anatomy of a large-scale hypertextual web search engine. Comput. Networks **30**(1–7), 107–117 (1998)
6. Budak, C., Agrawal, D., El Abbadi, A.: Limiting the spread of misinformation in social networks. In: Proceedings of the 20th International Conference on World Wide Web, pp. 665–674 (2011). https://doi.org/10.1145/1963405.1963499
7. Di Battista, G., Eades, P., Tamassia, R., Tollis, I.G.: Graph Drawing, vol. 357. Prentice Hall, Upper Saddle River (1999)
8. Eades, P., Lin, X., Smyth, W.F.: A fast and effective heuristic for the feedback arc set problem. Inf. Process. Lett. **47**(6), 319–323 (1993)
9. He, X., Song, G., Chen, W., Jiang, Q.: Influence blocking maximization in social networks under the competitive linear threshold model. In: Proceedings of the 2012 SIAM International Conference on Data Mining, pp. 463–474. SIAM (2012)
10. Johnson, D.S.: The np-completeness column: an ongoing gulde. J. Algorithms **3**(4), 381–395 (1982)

11. Karp, R.M.: Reducibility among combinatorial problems. In: Jünger, M., Liebling, T.M., Naddef, D., Nemhauser, G.L., Pulleyblank, W.R., Reinelt, G., Rinaldi, G., Wolsey, L.A. (eds.) 50 Years of Integer Programming 1958-2008, pp. 219–241. Springer, Heidelberg (2010). https://doi.org/10.1007/978-3-540-68279-0_8

12. Nikolov, N.S., Healy, P.: Hierarchical drawing algorithms. In: Handbook of Graph Drawing and Visualization, ed. Roberto Tamassia, pp. 409–453. CRC Press (2014)

13. Ortali, G., Tollis, I.G.: A new framework for hierarchical drawings. J. Graph Algorithms Appl. **23**(3), 553–578 (2019). https://doi.org/10.7155/jgaa.00502

14. Page, L., Brin, S., Motwani, R., Winograd, T.: The pagerank citation ranking: Bringing order to the web. Tech. rep, Stanford InfoLab (1999)

15. Pemmaraju, S., Skiena, S.: Computational Discrete Mathematics: Combinatorics and Graph Theory with Mathematica ®. Cambridge University Press (1990)

16. Simpson, M., Srinivasan, V., Thomo, A.: Clearing contamination in large networks. IEEE Trans. Knowl. Data Eng. **28**(6), 1435–1448 (2016). https://doi.org/10.1109/TKDE.2016.2525993

17. Simpson, M., Srinivasan, V., Thomo, A.: Efficient computation of feedback arc set at web-scale. Proc. VLDB Endowment **10**(3), 133–144 (2016). https://doi.org/10.14778/3021924.3021930

18. Sugiyama, K., Tagawa, S., Toda, M.: Methods for visual understanding of hierarchical system structures. IEEE Trans. Syst. Man Cybern. **11**(2), 109–125 (1981). https://doi.org/10.1109/TSMC.1981.4308636

19. Tarjan, R.: Depth-first search and linear graph algorithms. SIAM J. Comput. **1**(2), 146–160 (1972)

st-Orientations with Few Transitive Edges

Carla Binucci[1]([✉])[iD], Walter Didimo[1][iD], and Maurizio Patrignani[2][iD]

[1] Universià degli Studi di Perugia, Perugia, Italy
{carla.binucci,walter.didimo}@unipg.it
[2] Roma Tre University, Rome, Italy
maurizio.patrignani@uniroma3.it

Abstract. The problem of orienting the edges of an undirected graph such that the resulting digraph is acyclic and has a single source s and a single sink t has a long tradition in graph theory and is central to many graph drawing algorithms. Such an orientation is called an *st*-orientation. We address the problem of computing *st*-orientations of undirected graphs with the minimum number of transitive edges. We prove that the problem is NP-hard in the general case. For planar graphs we describe an ILP model that is fast in practice. We experimentally show that optimum solutions dramatically reduce the number of transitive edges with respect to unconstrained *st*-orientations computed via classical *st*-numbering algorithms. Moreover, focusing on popular graph drawing algorithms that apply an *st*-orientation as a preliminary step, we show that reducing the number of transitive edges leads to drawings that are much more compact.

1 Introduction

The problem of orienting the edges of an undirected graph in such a way that the resulting digraph satisfies specific properties has a long tradition in graph theory and represents a preliminary step of several graph drawing algorithms. For example, Eulerian orientations require that each vertex gets equal in-degree and out-degree; they are used to compute 3D orthogonal graph drawings [17] and right-angle-crossing drawings [2]. Acyclic orientations require that the resulting digraph does not contain directed cycles (i.e., it is a DAG); they can be used as a preliminary step to compute hierarchical and upward drawings that nicely represent an undirected graph, or a partially directed graph, so that all its edges monotonically flow in the same direction [4,5,15,18,22,24].

Specific types of acyclic orientations that are central to many graph algorithms and applications are the so called *st-orientations*, also known as *bipolar orientations* [33], whose resulting digraphs have a single source s and a single sink t. It is well known that an undirected graph G with prescribed vertices s

Work partially supported by: (i) MIUR, grant 20174LF3T8 AHeAD: efficient Algorithms for HArnessing networked Data", (ii) Dipartimento di Ingegneria, Universita degli Studi di Perugia, grant RICBA21LG: Algoritmi, modelli e sistemi per la rappresentazione visuale di reti.

P. Angelini and R. von Hanxleden (Eds.): GD 2022, LNCS 13764, pp. 201–216, 2023.
https://doi.org/10.1007/978-3-031-22203-0_15

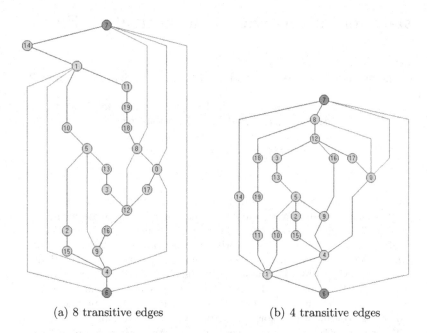

(a) 8 transitive edges   (b) 4 transitive edges

Fig. 1. Two polyline drawings of the same plane graph, computed using two different *st*-orientations, with $s = 6$ and $t = 7$; transitive edges are in red. (a) An unconstrained *st*-orientation with 8 transitive edges, computed through an *st*-numbering; (b) An *st*-orientation with the minimum number (four) of transitive edges; the resulting drawing is more compact and has shorter edges. (Color figure online)

and t admits an *st*-orientation if and only if G with the addition of the edge (s,t) (if not already present) is biconnected. The digraph resulting from an *st*-orientation is also called an *st-graph*. An *st*-orientation can be computed in linear time via an *st*-numbering (or *st*-ordering) of the vertices of G [7,20], by orienting each edge from the end-vertex with smaller number to the end-vertex with larger number [7]. In particular, if G is planar, a *planar st-orientation* of G additionally requires that s and t belong to the external face in some planar embedding of the graph. Planar *st*-orientations were originally introduced in the context of an early planarity testing algorithm [27], and are largely used in graph drawing to compute different types of layouts, including visibility representations, polyline drawings, dominance drawings, and orthogonal drawings (refer to [10,26]). Planar *st*-orientations and related graph layout algorithms are at the heart of several graph drawing libraries and software (see, e.g., [8,9,25,35]). Algorithms that compute *st*-orientations with specific characteristics (such as bounds on the length of the longest path) are also proposed and experimented in the context of visibility and orthogonal drawings [30,31].

Our paper focuses on the computation of *st*-orientations with a specific property, namely we address the following problem: "Given an undirected graph G

and two prescribed vertices s and t for which $G \cup (s, t)$ is biconnected, compute an *st*-orientation of G such that the resulting *st*-graph G' has the minimum number of transitive edges (possibly none)". We recall that an edge (u, v) of a digraph G' is *transitive* if there exists a directed path from u to v in $G' \setminus (u, v)$. An *st*-orientation is *non-transitive* if the resulting digraph has no transitive edges; *st*-graphs with no transitive edges are also known as *transitively reduced st-graphs* [10, 19], *bipolar posets* [23], or *Hasse diagrams of lattices* [11, 32]. The problem we study, besides being of theoretical interest, has several practical motivations in graph drawing. We mention some of them:

- Planar *st*-oriented graphs without transitive edges admit compact dominance drawings with straight-line edges, a type of upward drawings that can be computed in linear time with very simple algorithms [12]; when a transitive edge is present, one can temporarily subdivide it with a dummy vertex, which will correspond to an edge bend in the final layout. Hence, having few transitive edges helps to reduce bends in a dominance drawing.
- As previously mentioned, many layout algorithms for undirected planar graphs rely on a preliminary computation of an *st*-orientation of the input graph. We preliminary observed that reducing the number of transitive edges in such an orientation has typically a positive impact on the readability of the layout. Indeed, transitive edges often result in long curves; avoiding them produces faces where the lengths of the left and right paths are more balanced and leads to more compact drawings (see Fig. 1).
- Algorithms for computing upward confluent drawings of transitively reduced DAGs are studied in [19]. Confluent drawings exploit edge bundling to create "planar" layouts of non-planar graphs, without introducing ambiguity [14]. These algorithms can be applied to draw undirected graphs that have been previously *st*-oriented without transitive edges when possible.

We also mention algorithms that compute two-page book embeddings of two-terminal series-parallel digraphs, which either assume the absence of transitive edges [1] or which are easier to implement if transitive edges are not present [13].

Contribution. In this paper we first prove that deciding whether a graph admits an *st*-orientation without transitive edges is NP-complete. This is in contrast with the tractability of a problem that is at the opposite of ours, namely, deciding whether an undirected graph has an orientation such that the resulting digraph is its own transitive closure; this problem can be solved in linear time [28].

From a practical point of view, we provide an Integer Linear Programming (ILP) model for planar graphs, whose solution is an *st*-orientation with the minimum number of transitive edges. In our setting, s and t are two prescribed vertices that belong to the same face of the input graph in at least one of its planar embeddings. We prove that the ILP model works very fast in practice. Popular solvers such as CPLEX can find a solution in few seconds for graphs up to 1000 vertices and the resulting *st*-orientations save on average 35% of transitive edges (with improvements larger than 80% on some instances) with respect to applying classical unconstrained *st*-orientation algorithms. Moreover, focusing on popular graph drawing algorithms that apply an *st*-orientation as a preliminary step, we show that reducing the number of transitive edges leads to drawings that are much more compact.

For space restrictions, some details are omitted. Full proofs and additional material can be found in [6].

2   NP-Completeness of the General Problem

We prove that given an undirected graph $G = (V, E)$ and two vertices $s, t \in V$, it is NP-complete to decide whether there exists a non-transitive *st*-orientation of G. We call this problem NON-TRANSITIVE ST-ORIENTATION (NTO). To prove the hardness of NTO we describe a reduction from the NP-complete problem NOT-ALL-EQUAL 3SAT (NAE3SAT) [34], where one has a collection of clauses, each composed of three literals out of a set X of Boolean variables, and is asked to determine whether there exists a truth assignment to the variables in X so that each clause has at least one `true` and one `false` literal.

Starting from a NAE3SAT instance φ, we construct an instance $I_\varphi = \langle G, s, t \rangle$ of NTO such that I_φ is a yes instance of NAE3SAT if and only if φ is a yes instance of NTO. Instance I_φ has one variable gadget V_x for each Boolean variable x and one clause gadget C_c for each clause c of φ. By means of a split gadget, the truth value encoded by each variable gadget V_x is transferred to all the clause gadgets containing either the direct literal x or its negation $\bar{x}$. Observe that the NAE3SAT instance is in general not "planar", in the sense that if you construct a graph where each variable x and each clause c is a vertex and there is an edge between x and c if and only if a literal of x belongs to c, then such a graph would be non-planar. The NAE3SAT problem on planar instances is, in fact, polynomial [29]. Hence, G has to be assumed non-planar as well.

The main ingredient of the reduction is the *fork gadget* (Fig. 2), for which the following lemma holds (the proof is in [6]).

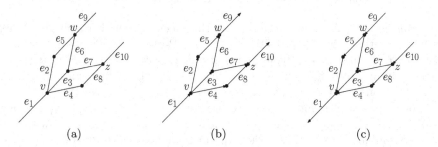

Fig. 2. (a) The fork gadget. (b)–(c) The two possible orientations of the fork gadget in a non-transitive st-orientation of the whole graph.

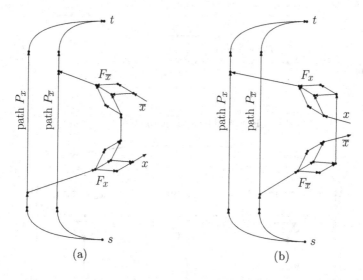

Fig. 3. The variable gadget V_x and its **true** (a) and **false** (b) orientations.

Lemma 1. *Let G be an undirected graph containing a fork gadget F that does not contain the vertices s or t. In any non-transitive st-orientation of G, the edges e_9 and e_{10} of F are oriented either both exiting F or both entering F. They are oriented exiting F if and only if edge e_1 is oriented entering F.*

For each Boolean variable x of ϕ we construct a *variable gadget V_x* by suitably combining two fork gadgets, denoted F_x and $F_{\bar{x}}$, as follows (see Fig. 3). We introduce two paths P_x and $P_{\bar{x}}$ of length four from s to t. The edge e_1 of F_x (of $F_{\bar{x}}$, respectively) is attached to the middle vertex of path P_x (of path $P_{\bar{x}}$, respectively). Edge e_{10} of $F_{\bar{x}}$ is identified with edge e_9 of F_x. The two edges e_9 of $F_{\bar{x}}$ and e_{10} of F_x are denoted $\bar{x}$ and x, respectively. We have the following lemma (see [6] for the proof).

Lemma 2. *Let G be an undirected graph containing a variable gadget V_x. In any non-transitive st-orientation of G the two edges of V_x denoted x and $\bar{x}$ are one entering and one exiting V_x or vice versa.*

By virtue of Lemma 2 we associate the **true** value of variable x with the orientation of V_x where edge x is oriented exiting and edge $\bar{x}$ is oriented entering V_x (see Fig. 3(a)). We call such an orientation the *true orientation of V_x*. Analogously, we associate the **false** value of variable x with the orientation of V_x where edge x is oriented entering and edge $\bar{x}$ is oriented exiting V_x (see Fig. 3(b)). Observe that edge x (edge $\bar{x}$, respectively) is oriented exiting V_x when the literal x (the literal $\bar{x}$, respectively) is **true**. Otherwise edge x (edge $\bar{x}$, respectively) is oriented entering V_x.

The *split gadget S_k* is composed of a chain of $k - 1$ fork gadgets $F_1, F_2, \ldots F_{k-1}$, where, for $i = 1, 2, \ldots, k - 2$, the edge e_9 of F_i is identified

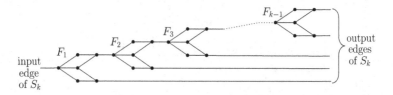

Fig. 4. The split gadget S_k.

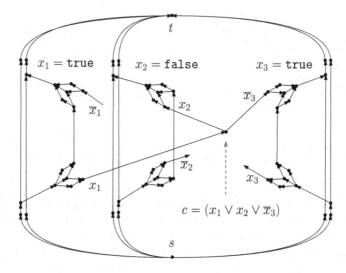

Fig. 5. The clause gadget C_c for clause $c = (x_1 \vee x_2 \vee \overline{x}_3)$. The configurations of the three variable gadgets correspond to the truth values $x_1 = \text{true}$, $x_2 = \text{false}$, and $x_3 = \text{true}$. The clause is satisfied because the first literal x is **true** and the second and third literals x_2 and $\overline{x}_3$ are **false**.

with the edge e_1 of F_{i+1}. We call *input edge of S_k* the edge denoted e_1 of F_1. Also, we call *output edges of S_k* the $k-1$ edges denoted e_{10} of the fork gadgets $F_1, F_2, \ldots F_{k-1}$ and the edge e_9 of F_{k-1} (see Fig. 4). The next lemma is immediate and we omit the proof.

Lemma 3. *Let G be an undirected graph containing a split gadget S_k that does not contain the vertices s or t. In any non-transitive st-orientation of G, the k output edges of S_k are all oriented exiting S_k if the input edge of S_k is oriented entering S_k. Otherwise, if the input edge of S_k is oriented exiting S_k the ouput edges of S_k are all oriented entering S_k.*

If the directed literal x (negated literal $\overline{x}$, respectively) occurs in k clauses, we attach the edge denoted x (denoted $\overline{x}$, respectively) of V_x to a split gadget S_x, and use the k output edges of S_x to carry the truth value of x (of $\overline{x}$, respectively) to the k clauses. The *clause gadget C_c* for a clause $c = (l_1 \vee l_2 \vee l_3)$ is simply a vertex v_c that is incident to three edges encoding the truth values of the three literals l_1, l_2, and l_3 (see Fig. 5). We prove the following.

Theorem 1. NTO *is NP-complete.*

Sketch of proof: The reduction from an instance φ of NAE3SAT to an instance I_φ described above is performed in time linear in the size of φ. Also, I_φ is positive if and only if φ is positive. Indeed, in any non-transitive *st*-orientation of G each vertex v_c of a clause gadget C_c has at least one incoming and one outgoing edge, as well as in any truth assignment that satisfies φ each clause c has at least one **true** and one **false** literal. Finally, NTO is trivially in NP, as one can non-deterministically explore all possible orientations of the graph. □

The analogous problem where the source and the target vertices of G are not prescribed but can be freely choosen is also NP-complete (see [6]).

3 ILP Model for Planar Graphs

Let G be a planar graph with two prescribed vertices s and t, such that $G \cup (s,t)$ is biconnected and such that G admits a planar embedding with s and t on the external face. In this section we describe how to compute an *st*-orientation of G with the minimum number of transitive edges by solving an ILP model.

Suppose that G' is the plane *st*-graph resulting from a planar *st*-orientation of G, along with a planar embedding where s and t are on the external face. It is well known (see, e.g., [10]) that for each vertex $v \neq s, t$ in G', all incoming edges of v (as well as all outgoing edges of v) appear consecutively around v. Thus, the circular list of edges incident to v can be partitioned into two linear lists, one containing the incoming edges of v and the other containing the outgoing edges of v. Also, the boundary of each internal face f of G' consists of two edge-disjoint directed paths, called the *left path* and the *right path* of f, sharing the same end-vertices (i.e., the same source and the same destination). It can be easily verified that an edge e of G' is transitive if and only if it coincides with either the left path or the right path of some face of G' (see also Claim 2 in [23]). Note that, since the transitivity of e does not depend on the specific planar embedding of G', the aforementioned property for e holds for every planar embedding of G'. Due to this observation, in order to compute a planar *st*-orientation of G with the minimum number of transitive edges, we can focus on any arbitrarily chosen planar embedding of G with s and t on the external face.

Let e_1 and e_2 be two consecutive edges encountered moving clockwise along the boundary of a face f, and let v be the vertex of f shared by e_1 and e_2. The triple (e_1, v, e_2) is an *angle of G at v in f*. Denote by $\deg(f)$ the number of angles in f and by $\deg(v)$ the number of angles at v. As it was proved in [16], all planar *st*-orientations of the plane graph G can be characterized in terms of labelings of the angles of G. Namely, each planar *st*-orientation of G has a one-to-one correspondence with an angle labeling, called an *st-labeling* of G, that satisfies the following properties:

(L1) Each angle is labeled either S (small) or F (flat), except the angles at s and at t in the external face, which are not labeled;

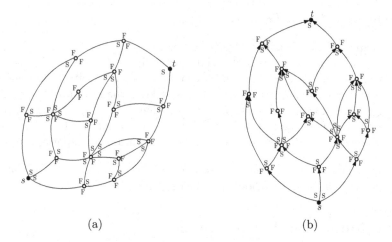

Fig. 6. (a) An st-labeling of a plane graph G with prescribed nodes s and t. (b) The corresponding st-orientation of G.

(L2) Each internal face f has 2 angles labeled S and $\deg(f) - 2$ angles labeled F;
(L3) For each vertex $v \neq s, t$ there are $\deg(v) - 2$ angles at v labeled S and 2 angles at v labeled F;
(L4) All angles at s and t in their incident internal faces are labeled S.

Given an st-labeling of G, the corresponding st-orientation of G is such that for each vertex $v \neq s, t$, the two F angles at v separate the list of incoming edges of v to the list of outgoing edges of v, while the two S angles in a face f separate the left and the right path of f. See Fig. 6 for an illustration. The st-orientation can be constructed from the st-labeling in linear time by a breadth-first-search of G that starts from s, makes all edges of s outgoing, and progressively orients the remaining edges of G according to the angle labels.

Thanks to the characterization above, an edge $e = (u, v)$ of the st-graph resulting from an st-orientation is transitive if and only if in the corresponding st-labeling the angle at u and the angle at v in one of the two faces incident to e (possibly in both faces) are labeled S. Based on this, we present an ILP model that describes the possible st-labelings of G (for any arbitrary planar embedding of G with s and t on the external face) and that minimizes the number of transitive edges. The model aims to assign angle labels that satisfy Properties (L1)–(L4) and counts pairs of consecutive S labels that occur in the circular list of angles in an internal face; additional constraints are needed to avoid that a transitive edge is counted twice when it coincides with both the left and the right path of its two incident faces. The model, which uses a number of variables and constraints that is linear in the size of G, is as follows.

Sets. Denote by V, E, and F the sets of vertices, edges, and faces of G, respectively. Also let $F_{\text{int}} \subset F$ be the set of internal faces of G. For each face $f \in F$, let $V(f)$ and $E(f)$ be the set of vertices and the set of edges incident to f, respectively. For each vertex $v \in V$, let $F(v)$ be the set of faces incident to v and let $F_{\text{int}}(v)$ be the set of internal faces incident to v. For each edge $e \in E$, let $F(e)$ be the set consisting of the two faces incident to e.

Variables. We define a binary variable x_{vf} for each vertex $v \in V \setminus \{s, t\}$ and for each face $f \in F(v)$. Also, we define the binary variables x_{sf} (resp. x_{tf}) for each face $f \in F_{\text{int}}(s)$ (resp. $f \in F_{\text{int}}(t)$). If $x_{vf} = 1$ (resp. $x_{vf} = 0$) we assign an S label (resp. an F label) to the angle at v in f.

For each internal face $f \in F_{\text{int}}$ and for each edge $(u, v) \in E(f)$, we define a binary variable y_{uvf}. An assignment $y_{uvf} = 1$ indicates that both the angles at u and at v in f are labeled S, that is, $x_{uf} = 1$ and $x_{vf} = 1$. As a consequence, if $y_{uvf} = 1$ edge (u, v) is transitive. Note however that the sum of all y_{uvf} does not always correspond to the number of transitive edges; indeed, if f and g are the two internal faces incident to edge (u, v), it may happen that both y_{uvf} and y_{uvg} are set to one, thus counting (u, v) as transitive twice. To count the number of transitive edges without repetitions, we introduce another binary variable z_{uv}, for each edge $(u, v) \in E$, such that $z_{uv} = 1$ if and only if (u, v) is transitive.

Objective Function and Constraints. The objective function and the set of constraints are described by the formulas (1)–(8). The objective is to minimize the total number of transitive edges, i.e., the sum of the variables z_{uv}. Constraints 2 and 3 guarantee Properties (L2) and (L3) of the *st*-labeling, respectively, while Constraints 4 and 5 guarantee Property (L4). Constraints 6 relate the values of the variables y_{uvf} to the values of x_{uf} and x_{vf}. Namely, they guarantee that $y_{uvf} = 1$ if and only if both x_{uf} and x_{vf} are set to 1. Constraints 7 relate the values of the variables z_{uv} to those of the variables y_{uvf}; they guarantee that an edge (u, v) is counted as transitive (i.e., $z_{uv} = 1$) if and only if in at least one of the two faces f incident to (u, v) both the angle at u and the angle at v are labeled S. Finally, we explicitly require that x_{uv} and y_{uv} are binary variables, while we only require that each z_{uv} is a non-negative integer; this helps to speed-up the solver and, along with the objective function, is enough to guarantee that each z_{uv} takes value 0 or 1.

$$\min \sum_{(u,v) \in E} z_{uv} \tag{1}$$

$$\sum_{v \in V(f)} x_{vf} = 2 \quad \forall f \in F_{\text{int}} \tag{2}$$

$$\sum_{f \in F(v)} x_{vf} = \deg(v) - 2 \quad \forall v \in V \setminus \{s, t\} \tag{3}$$

$$x_{sf} = 1 \quad \forall f \in F_{\text{int}} \cap F(s) \tag{4}$$

$$x_{tf} = 1 \quad \forall f \in F_{\text{int}} \cap F(t) \tag{5}$$

$$x_{uf} + x_{vf} \leq y_{uvf} + 1 \quad \forall f \in F_{\text{int}} \quad \forall (u, v) \in E(f) \tag{6}$$

$$z_{uv} \geq y_{uvf} \quad \forall e = (u, v) \in E \quad \forall f \in F(e) \tag{7}$$

$$x_{vf} \in \{0, 1\} \quad y_{uvf} \in \{0, 1\} \quad z_{uv} \in \mathbb{N} \tag{8}$$

4   Experimental Analysis

We evaluated the ILP model with the solver IBM ILOG CPLEX 20.1.0.0 (using the default setting), running on a laptop with Microsoft Windows 11 v.10.0.22000 OS, Intel Core i7-8750H 2.20 GHz CPU, and 16 GB RAM.

Instances. The experiments have been executed on a large benchmark of instances, each instance consisting of a plane biconnected graph and two vertices s and t on the external face. These graphs are randomly generated with the same approach used in previous experiments in graph drawing (see, e.g., [3]). Namely, for a given integer $n > 0$, we generate a plane graph with n vertices starting from a triangle and executing a sequence of steps, each step preserving biconnectivity and planarity. At each step the procedure randomly performs one of the two following operations: (i) an Insert-Edge operation, which splits a face by adding a new edge, or (ii) an Insert-Vertex operation, which subdivides an existing edge with a new vertex. The Insert-Vertex operation is performed with a prescribed probability p_{iv} (which is a parameter of the generation process), while the Insert-Edge operation is performed with probability $1 - p_{iv}$. For each operation, the elements (faces, vertices, or edges) involved are randomly selected with uniform probability distribution. To avoid multiple edges, if an Insert-Edge operation selects two end-vertices that are already connected by an edge, we discard the selection and repeat the step. Once the plane graph is generated, we randomly select two vertices s and t on its external face, again with uniform probability distribution. We generated a sample of 10 instances for each pair (n, p_{iv}), with $n \in \{10, 20, \ldots, 90, 100, 200, \ldots, 900, 1000\}$ and $p_{iv} \in \{0.2, 0.4, 0.5, 0.6, 0.8\}$, for a total of 950 graphs. Note that, higher values of p_{iv} lead to sparser graphs.

See [6] for a table that reports for each sample the average, the minimum, and the maximum density (number of edges divided by the number of vertices) of the graphs in that sample, together with the standard deviation. On average, for $p_{iv} = 0.8$ we have graphs with density of 1.23 (close to the density of a tree), for $p_{iv} = 0.5$ we have graphs with density of 1.76, and for $p_{iv} = 0.2$ we have graphs with density 2.53 (close to the density of maximal planar graphs).

Experimental Goals. We have three main experimental goals: (G1) Evaluate the efficiency of our approach, i.e., the running time required by our ILP model; (G2) Evaluate the percentage of transitive edges in the solutions of the ILP model and how many transitive edges are saved w.r.t. applying a classical linear-time algorithm that computes an unconstrained st-orientation of the graph [21]; (G3) Evaluate the impact of minimizing the number of transitive edges on the area (i.e. the area of the minimum bounding box) of polyline drawings constructed with algorithms that compute an st-orientation as a preliminary step.

About (G1), we refer to the algorithm that solves the ILP model as OPTST. About (G2) and (G3) we used implementations available in the GDToolkit library [9] for the following algorithms: (a) A linear-time algorithm that computes an unconstrained st-orientation of the graph based on the classical st-numbering algorithm by Even and Tarjan [21]. We refer to this algorithm as

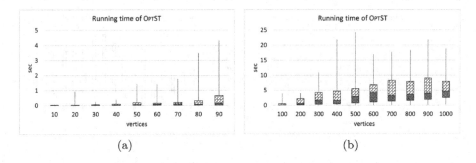

Fig. 7. Box-plots of the running time of OPTST.

HEURST. (*b*) A linear-time algorithm that first computes a visibility representation of an undirected planar graph based on a given *st*-orientation of the graph, and then computes from this representation a planar polyline drawing [11]. We call DRAWHEURST and DRAWOPTST the applications of this drawing algorithm to the *st*-graphs obtained by HEURST and of OPTST, respectively.

Experimental Results. About (G1), Fig. 7 reports the running time (in seconds) of OPTST, i.e., the time needed by CPLEX to solve our ILP model. To make the charts more readable we split the results into two sets, one for the instances with number of vertices up to 90 and the other for the larger instances. OPTST is rather fast: 75% of the instances with up to 90 vertices is solved in less than one second and all these instances are solved in less than five seconds. For the larger instances (with up to 1000 vertices), 75% of the instances are solved in less than 10 s and all instances are solved in less than 25 s. These results clearly indicate that our ILP model can be successfully used in several application contexts that manage graphs with up to thousand vertices.

About (G2), Fig. 8 shows the reduction (in percentage) of the number of transitive edges in the solutions of OPTST with respect to the solutions of HEURST. More precisely, Fig. 8(a) reports values averaged over all instances with the same number of vertices; Fig. 8(b), Fig. 8(c), and Fig. 8(d) report the same data, partitioning the instances by different values of p_{iv}, namely 0.8 (the sparsest instances), 0.4–0.6 (instances of medium density), and 0.2 (the densest instances). For each instance, denoted by trOpt and trHeur the number of transitive edges of the solutions computed by OPTST and HEURST, respectively, the reduction percentage equals the value $\left(\frac{\text{trHeur}-\text{trOpt}}{\max\{1,\text{trHeur}\}} \times 100 \right)$. Over all instances, the average reduction is about 35%; it grows above 60% on the larger graphs if we restrict to the sparsest instances (with improvements larger than 80% on some graphs), while it is below 30% for the densest instances, due to the presence of many 3-cycles, for which a transitive edge cannot be avoided.

About (G3), Fig. 9 shows the percentage of instances for which DRAWOPTST produces drawings that are better than those produced by DRAWHEURST in terms of area requirement (the label "better" of the legend). It can be seen that DRAWOPTST computes more compact drawings for the majority of the

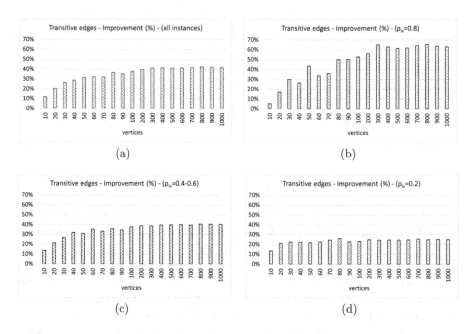

Fig. 8. Improvement (%) in the number of transitive edges.

instances. In particular, it is interesting to observe that this is most often the case even for the densest instances (i.e., those for $p_{iv} = 0.2$), for which we have previously seen that the average reduction of transitive edges is less evident. For those instances for which DRAWOPTST computes more compact drawings than DRAWHEURST, Fig. 10 reports the average percentage of improvement in terms of area requirement (i.e., the percentage of area reduction). The values are mostly between 30% and 50%. To complement this data, Fig. 11 reports the trend of the improvement (reduction) in terms of drawing area with respect to the reduction of the transitive edges (discretized in four intervals). For the instances with $p_{iv} = 0.8$ and $p_{iv} = 0.2$, the correlation between these two measures is quite evident. For the instances of medium density ($p_{iv} \in \{0.4, 0.5, 0.6\}$), the highest values of improvement in terms of area requirement are observed for reductions of transitive edges between 22% and 66%. Drawings of our instances computed by DRAWHEURST and DRAWOPTST are reported in [6].

5   Final Remarks and Open Problems

We addressed the problem of computing st-orientations with the minimum number of transitive edges. This problem has practical applications in graph drawing, as finding an st-orientation is at the heart of several graph drawing algorithms. Although st-orientations without transitive edges have been studied from a combinatorial perspective [23], there is a lack of practical algorithms, and the com-

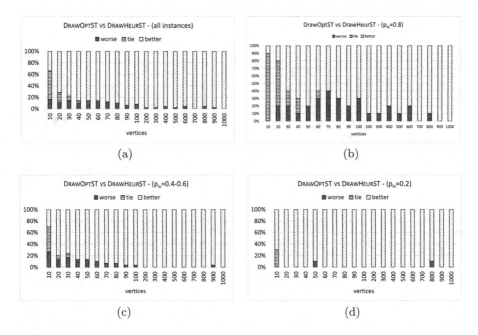

Fig. 9. Instances for which DRAWOPTST produces drawings that are more compact than DRAWHEURST (label "better").

plexity of deciding whether a graph can be oriented to become an *st*-graph without transitive edges seems not to have been previously addressed.

We proved that this problem is NP-hard in general and we described an ILP model for planar graphs based on characterizing planar *st*-graphs without transitive edges in terms of a constrained labeling of the vertex angles inside its faces. An extensive experimental analysis on a large set of instances shows that our model is fast in practice, taking few seconds for graphs of thousand vertices. It saves on average 35% of transitive edges w.r.t. a classical algorithm that computes an unconstrained *st*-orientation. We also showed that for classical layout algorithms that compute polyline drawings of planar graphs through an *st*-orientation, minimizing the number of transitive edges yields more compact drawings most of the time.

We suggest two future research directions: (*i*) It remains open to establish the time complexity of the problem for planar graphs. Are there polynomial-time algorithms that compute *st*-orientations with the minimum number of transitive edges for all planar graphs or for specific subfamilies of planar graphs? (*ii*) One can extend the experimental analysis to real-world graphs and design fast heuristics, which can be compared to the optimal algorithm.

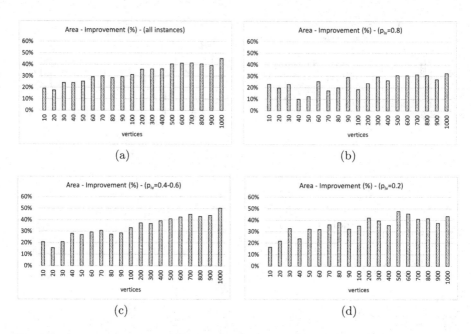

Fig. 10. Area improvement (%) of DrawOptST w.r.t. DrawHeurST, for the instances where DrawOptST is "better" (i.e., the "better" instances in Fig. 9).

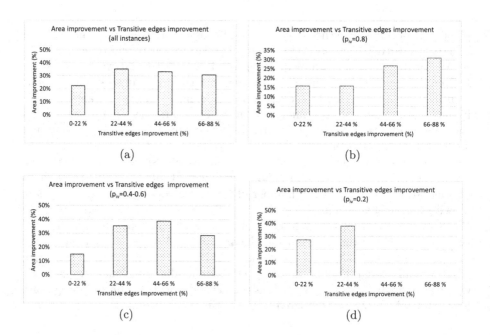

Fig. 11. Correlation between the improvement (reduction) in terms of drawing area and in terms of transitive edges improvement.

References

1. Alzohairi, M., Rival, I.: Series-parallel planar ordered sets have pagenumber two. In: North, S. (ed.) GD 1996. LNCS, vol. 1190, pp. 11–24. Springer, Heidelberg (1997). https://doi.org/10.1007/3-540-62495-3_34
2. Angelini, P., Cittadini, L., Didimo, W., Frati, F., Di Battista, G., Kaufmann, M., Symvonis, A.: On the perspectives opened by right angle crossing drawings. J. Graph Algorithms Appl. **15**(1), 53–78 (2011)
3. Bertolazzi, P., Di Battista, G., Didimo, W.: Computing orthogonal drawings with the minimum number of bends. IEEE Trans. Computers **49**(8), 826–840 (2000)
4. Binucci, C., Didimo, W.: Computing quasi-upward planar drawings of mixed graphs. Comput. J. **59**(1), 133–150 (2016)
5. Binucci, C., Didimo, W., Patrignani, M.: Upward and quasi-upward planarity testing of embedded mixed graphs. Theor. Comput. Sci. **526**, 75–89 (2014)
6. Binucci, C., Didimo, W., Patrignani, M.: *st*-orientations with few transitive edges. CoRR 2208.11414 (2022). http://arxiv.org/abs/2208.11414
7. Brandes, U.: Eager *st*-Ordering. In: Möhring, R., Raman, R. (eds.) ESA 2002. LNCS, vol. 2461, pp. 247–256. Springer, Heidelberg (2002). https://doi.org/10.1007/3-540-45749-6_25
8. Chimani, M., Gutwenger, C., Jünger, M., Klau, G.W., Klein, K., Mutzel, P.: The open graph drawing framework (OGDF). In: Handbook of Graph Drawing and Visualization, pp. 543–569. Chapman and Hall/CRC (2013)
9. Di Battista, G., Didimo, W.: Gdtoolkit. In: Handbook of Graph Drawing and Visualization, pp. 571–597. Chapman and Hall/CRC (2013)
10. Di Battista, G., Eades, P., Tamassia, R., Tollis, I.G.: Graph Drawing: Algorithms for the Visualization of Graphs. Prentice-Hall (1999)
11. Di Battista, G., Tamassia, R.: Algorithms for plane representations of acyclic digraphs. Theor. Comput. Sci. **61**, 175–198 (1988)
12. Battista, G.D., Tamassia, R., Tollis, I.G.: Area requirement and symmetry display of planar upward drawings. Discrete Comput. Geomet. **7**(4), 381–401 (1992). https://doi.org/10.1007/BF02187850
13. Di Giacomo, E., Didimo, W., Liotta, G., Wismath, S.K.: Book embeddability of series-parallel digraphs. Algorithmica **45**(4), 531–547 (2006)
14. Dickerson, M., Eppstein, D., Goodrich, M.T., Meng, J.Y.: Confluent drawings: visualizing non-planar diagrams in a planar way. J. Graph Algorithms Appl. **9**(1), 31–52 (2005)
15. Didimo, W.: Upward graph drawing. In: Encyclopedia of Algorithms, pp. 2308–2312 (2016)
16. Didimo, W., Pizzonia, M.: Upward embeddings and orientations of undirected planar graphs. J. Graph Algorithms Appl. **7**(2), 221–241 (2003)
17. Eades, P., Symvonis, A., Whitesides, S.: Three-dimensional orthogonal graph drawing algorithms. Discret. Appl. Math. **103**(1–3), 55–87 (2000)
18. Eiglsperger, M., Kaufmann, M., Eppinger, F.: An approach for mixed upward planarization. J. Graph Algorithms Appl. **7**(2), 203–220 (2003)
19. Eppstein, D., Simons, J.A.: Confluent Hasse diagrams. J. Graph Algorithms Appl. **17**(7), 689–710 (2013)
20. Even, S., Tarjan, R.E.: Computing an *st*-numbering. Theor. Comput. Sci. **2**(3), 339–344 (1976)
21. Even, S., Tarjan, R.E.: Corrigendum: Computing an *st*-numbering. TCS **2**(1976), 339–344. Theor. Comput. Sci. **4**(1), 123 (1977)

22. Frati, F., Kaufmann, M., Pach, J., Tóth, C.D., Wood, D.R.: On the upward planarity of mixed plane graphs. J. Graph Algorithms Appl. **18**(2), 253–279 (2014)

23. Fusy, É., Narmanli, E., Schaeffer, G.: On the enumeration of plane bipolar posets and transversal structures. CoRR abs/2105.06955 (2021)

24. Healy, P., Nikolov, N.S.: Hierarchical drawing algorithms. In: Handbook of Graph Drawing and Visualization, pp. 409–453. Chapman and Hall/CRC (2013)

25. Jünger, M., Mutzel, P. (eds.): Graph Drawing Software. Springer (2004)

26. Kaufmann, M., Wagner, D. (eds.): Drawing Graphs, Methods and Models (the book grow out of a Dagstuhl Seminar, April 1999), Lecture Notes in Computer Science, vol. 2025. Springer (2001)

27. Lempel, A., Even, S., Cederbaum, I.: An algorithm for planarity testing of graphs. In: Theory of Graphs: International Symposium (Rome 1966), pp. 215–232. Gordon and Breach, New York (1967)

28. McConnell, R.M., Spinrad, J.P.: Modular decomposition and transitive orientation. Discret. Math. **201**(1–3), 189–241 (1999)

29. Moret, B.M.E.: Planar NAE3SAT is in P. SIGACT News **19**(2), 51–54 (1988)

30. Papamanthou, C., Tollis, I.G.: Algorithms for computing a parameterized st-orientation. Theor. Comput. Sci. **408**(2–3), 224–240 (2008)

31. Papamanthou, C., Tollis, I.G.: Applications of parameterized st-orientations. J. Graph Algorithms Appl. **14**(2), 337–365 (2010)

32. Platt, C.: Planar lattices and planar graphs. J. Comb. Theory Ser. B **21**(1), 30–39 (1976)

33. Rosenstiehl, P., Tarjan, R.E.: Rectilinear planar layouts and bipolar orientations of planar graphs. Discrete & Computational Geometry **1**(4), 343–353 (1986). https://doi.org/10.1007/BF02187706

34. Schaefer, T.J.: The complexity of satisfiability problems. In: Proceedings of the 10th Annual ACM Symposium on Theory of Computing, pp. 216–226 (1978)

35. Wiese, R., Eiglsperger, M., Kaufmann, M.: yFiles - visualization and automatic layout of graphs. In: Graph Drawing Software, pp. 173–191. Springer (2004)

Beyond Planarity

Quasiplanar Graphs, String Graphs, and the Erdős-Gallai Problem

Jacob Fox[1], János Pach[2], and Andrew Suk[3(✉)]

[1] Stanford University, Stanford, CA, USA
jacobfox@stanford.edu
[2] Rényi Institute, Budapest and IST Austria, Budapest, Hungary
pach@cims.nyu.edu
[3] University of California at San Diego, La Jolla, CA, USA
asuk@ucsd.edu

Abstract. An *r-quasiplanar graph* is a graph drawn in the plane with no r pairwise crossing edges. Let $s \geq 3$ be an integer and $r = 2^s$. We prove that there is a constant C such that every r-quasiplanar graph with $n \geq r$ vertices has at most $n \left(Cs^{-1} \log n \right)^{2s-4}$ edges.

A graph whose vertices are continuous curves in the plane, two being connected by an edge if and only if they intersect, is called a *string graph*. We show that for every $\epsilon > 0$, there exists $\delta > 0$ such that every string graph with n vertices, whose chromatic number is at least n^ϵ contains a clique of size at least n^δ. A clique of this size or a coloring using fewer than n^ϵ colors can be found by a polynomial time algorithm in terms of the size of the geometric representation of the set of strings.

In the process, we use, generalize, and strengthen previous results of Lee, Tomon, and others. All of our theorems are related to geometric variants of the following classical graph-theoretic problem of Erdős, Gallai, and Rogers. Given a K_r-free graph on n vertices and an integer $s < r$, at least how many vertices can we find such that the subgraph induced by them is K_s-free?

Keywords: Quasi-planar graphs · String graphs · Graph coloring

1 Introduction

A *topological graph* is a graph drawn in the plane with points as vertices and edges as continuous curves connecting some pairs of vertices. The curves are allowed to cross, but they may not pass through vertices other than their endpoints. If the edges are drawn as straight-line segments, then the graph is *geometric*. If no r edges in a topological graph G are pairwise crossing, then G is called *r-quasiplanar*.

J. Fox—Supported by a Packard Fellowship and by NSF award DMS-1855635.
J. Pach—Supported by NKFIH grants K-131529, Austrian Science Fund Z 342-N31, and ERC Advanced Grant "GeoScape."
A. Suk—Supported by an NSF CAREER award and NSF award DMS-1952786.

P. Angelini and R. von Hanxleden (Eds.): GD 2022, LNCS 13764, pp. 219–231, 2023.
https://doi.org/10.1007/978-3-031-22203-0_16

The following is a longstanding unsolved problem in the theory of topological graphs; see, *e.g.,* [5].

Conjecture 1. The number of edges of every r-quasiplanar graph of n vertices is $O_r(n)$.

Conjecture 1 has been proved for $r \leq 4$. See [1,3,4].

The *intersection graph* of a family of geometric objects, $\mathcal{S}$, is a graph with vertex set $\mathcal{S}$, in which two vertices are joined by an edge if and only if their intersection is nonempty. If S consists of continuous curves (or line segments) in the plane, then their intersection graph is called a *string graph* (resp., a *segment graph*).

A natural approach to prove Conjecture 1 is the following. Removing a small disc around every vertex of an r-quasiplanar graph G, we are left with a family of continuous curves $\mathcal{S}$ in the plane, no r of which are pairwise crossing. These curves define a K_r-free string graph H. Suppose that the chromatic number of H satisfies $\chi(H) \leq f(r)$. Then each color class corresponds to the edges of a planar subgraph of G. Thus, the size of each color class is at most $3n - 6$, provided that $n \geq 3$. This would immediately imply that every r-quasiplanar graph with n vertices has at most $(3n - 6)f(r) = O_r(n)$ edges, as required.

Surprisingly, this approach is not viable. In 2014, Pawlik, Kozik, Krawczyk, Lasoń, Micek, Trotter, and Walczak [28] represented a class of K_3-free graphs originally constructed by Burling [6] as segment graphs whose chromatic numbers can be arbitrarily large. Shortly after, Walczak [35] strengthened this result by proving that there are K_3-free segment graphs on n vertices in which every independent set is of size $O(\frac{n}{\log \log n})$.

Using the same approach, in order to prove Conjecture 1 for some r, it would be sufficient to show that there is a constant $g(r)$ with the property that the vertex set of every K_r-free string graph can be colored by $g(r)$ colors such that each (string) graph induced by one of the color classes is K_4-free. Indeed, the result of Ackerman [1] cited above implies that the number of edges in each color class is $O(n)$. The first question to answer is the following.

Problem 1. Fix an integer $r \geq 4$. Is it true that every K_r-free segment graph on n vertices has an induced subgraph on $\Omega_r(n)$ vertices which is K_{r-1}-free?

Building upon the work of McGuinness [26], Suk [32] showed that every K_r-free *segment* graph on n vertices has a K_{r-1}-free induced subgraph with at least $\Omega_r(\frac{n}{\log n})$ vertices. (See also [29,30].) For *string* graphs, in general, until now the best known result, due to Fox and Pach [17], was weaker: they could only guarantee the existence of an independent set and, hence, a K_{r-1}-free induced subgraph, of size at least $\frac{n}{(\log n)^{O(\log r)}}$.

In a different range, where r grows polynomially in n, Tomon [33] solved a longstanding open problem by showing that there is a constant $c' > 0$ such that every string graph on n vertices has a clique or an independent set of size $n^{c'}$.

Our next theorem slightly strengthens the result of Fox and Pach [17].

Theorem 1. *Let s be a positive integer and $r = 2^s$. Every K_r-free string graph on $n \geq r$ vertices has an independent set of size at least $n(cs/\log n)^{2s-2}$, where $c > 0$ is an absolute constant.*

At the beginning of Sect. 4, we show how to deduce from Theorem 1 the following strengthening of Tomon's above mentioned theorem [33].

Corollary 1. *For any $\epsilon > 0$, there is $\delta > 0$ such that every string graph G on n vertices has a clique of size at least n^δ or its chromatic number is at most n^ϵ. (In the latter case, obviously, G has an independent set of size at least $n^{1-\epsilon}$.)*

Theorem 1 guarantees the existence of a large *independent* set in a K_r-free string graph G. If, in the spirit of Problem 1, we want to find only a large K_{r-1}-*free* induced subgraph in G, we can do better.

Theorem 2. *For any $n \geq r \geq 3$, every K_r-free string graph with n vertices has a K_{r-1}-free induced subgraph with at least $c\frac{n}{\log^2 n}$ vertices, where $c > 0$ is an absolute constant.*

At the expense of another logarithmic factor, we can also find an induced subgraph with no clique of size $\lceil r/2 \rceil$.

Theorem 3. *For any $n \geq r \geq 3$, every K_r-free string graph with n vertices has a $K_{\lceil r/2 \rceil}$-free induced subgraph with at least $c\frac{n}{\log^3 n}$ vertices, where $c > 0$ is an absolute constant.*

Now we return to the original motivation behind our present note: to estimate from above the number of edges of an r-quasiplanar *topological* graph of n vertices. As mentioned before, for $r \leq 4$, Conjecture 1 is true. For any $r \geq 5$, the best previously known upper bounds were $n(\log n)^{O(\log r)}$ and $O(n(\log n)^{4r-16})$, established in [17] and [27], respectively. For *geometric* graphs, for any fixed $r \geq 5$, Valtr [34] obtained the upper bound $O(n \log n)$. See [2] for a survey.

Using the result of Ackerman [1] as the base case of an induction argument, and exploiting several properties of string graphs established by Lee [24], Tomon [33], and Fox and Pach [16,17], we will deduce the following improved upper bound for the number of edges of r-quasiplanar topological graphs.

Theorem 4. *Let $s \geq 3$ be an integer and $r = 2^s$. Then every r-quasiplanar graph with $n \geq r$ vertices has at most $n(\frac{c \log n}{s})^{2s-4}$ edges, where $c > 0$ is an absolute constant.*

Setting $s = 3$, for instance, we obtain that every 8-quasiplanar topological graph on n vertices has $O(n(\log n)^2)$ edges, which is better than the previously known bound of $O(n(\log n)^{16})$ [1,27]. For $r = \delta \log n$, Theorem 4 immediately implies the following.

Corollary 2. *For any $\epsilon > 0$ there is $\delta > 0$ such that every topological graph with n vertices and at least $3n^{1+\epsilon}$ edges has n^δ pairwise crossing edges.*

The factor 3 in front of the term $n^{1+\epsilon}$ guarantees that the graph is not planar. Otherwise, we could not even guarantee that there is one crossing pair of edges.

In the special case where the strings are allowed to cross only a bounded number of times, some results very similar to Theorems 1 and 4 were established in [15].

Theorems 1, 2, 3, and Corollary 1 guarantee the existence of an independent set or a K_p-free induced subgraph for some $p > 2$ in a string graph satisfying certain conditions. All of these sets and subgraphs can be found by *efficient polynomial time algorithms* in terms of the size of a geometric representation of the underlying string graph. For example, the proof of Corollary 1 yields the following algorithmic result.

Proposition 1. *For any $\epsilon > 0$ there is $\delta > 0$ with the following property. Given a representation of a string graph on n vertices as an intersection graph of strings, there is a polynomial time algorithm which either properly colors the vertices with n^ϵ colors or finds a clique of size n^δ.*

Erdős and Gallai [11] raised the following question. Given a pair of integers, $2 \leq p < r$, how large of a K_p-free induced subgraph must be contained in every K_r-free graph of n vertices? For $p = 2$, we obtain Ramsey's problem: how large of an independent set must be contained in every K_r-free graph of n vertices. The special case $p = r - 1$ was considered by Erdős and Rogers [12]. These problems have since been extensively studied. For many striking results, see, *e.g.,* [9,10,19,22,23,31,36]. Apart from our last two results listed in the introduction, all statements in this paper can be regarded as geometric variants of the Erdős-Gallai-Rogers problem for string graphs.

The rest of this note is organized as follows. In Sect. 2, we apply the analogues of the separator theorem and the Kővári-Sós-Turán theorem for string graphs [17,24] to establish Theorems 2 and 3. In Sect. 3, we present a simple technical lemma (Lemma 3) and some of its consequences needed for the proof of Theorems 1 and 4. The proofs of these two theorems and Corollary 1 are given in Sect. 4. The last section contains some concluding remarks.

Throughout this paper, log always stands for the binary logarithm. The letters c and C appearing in different theorems denote unrelated positive constants. Whenever they are not important, we will simply omit floor and ceiling signs.

2   Separators—Proofs of Theorems 2 and 3

In this section, we prove Theorems 2 and 3. We need the separator theorem for string graphs, due to Lee [24]. A *separator* in a graph $G = (V, E)$ is a subset S of the vertex set V such that no connected component of $G \setminus S$ has more than $\frac{2}{3}|V|$ vertices. Equivalently, S is a separator of G if there is a partition $V = S \cup V_1 \cup V_2$ with $|V_1|, |V_2| \leq \frac{2}{3}|V|$ such that no vertex in V_1 is adjacent to any vertex in V_2.

Lemma 1 ([24]). *Every string graph with m edges has a separator of size at most $c_1\sqrt{m}$, where c_1 is an absolute constant.*

We now prove the following theorem which immediately implies Theorem 2. Let us remark that the *neighborhood* of vertex v does not include v.

Theorem 5. *There is an absolute constant $c > 0$ with the following property. Every string graph G on n vertices contains an induced subgraph G' on $c\frac{n}{\log^2 n}$ vertices whose every connected component is contained in the neighborhood of a vertex or is an isolated vertex.*

Proof. Let $c > 0$ be a sufficiently small constant to be specified later. We proceed by induction on n. The base case when $n = 1$ is trivial. For the inductive step, assume that the statement holds for all $n' < n$. Let $G = (V, E)$ be an n-vertex string graph.

If G contains a vertex v of degree at least $cn/\log^2 n$, then we are done by setting G' to be the neighborhood of v. Otherwise, we know that there are at most $cn^2/\log^2 n$ edges in G. By Lemma 1, G has a separator $S \subset V$ of size at most $c_1\sqrt{cn}/\log n$, where c_1 is the absolute constant from Lemma 1. Hence, there is a partition $V = S \cup V_1 \cup V_2$ with $|V_1|, |V_2| \leq \frac{2}{3}|V|$ such that no vertex in V_1 is adjacent to any vertex in V_2, and $|S| \leq c_1\sqrt{cn}/\log n$. By applying induction on V_1 and V_2 and setting $c < \left(\frac{\log 3/2}{c_1}\right)^2$, we obtain an induced subgraph G' on at least

$$c\frac{|V_1|}{\log^2 |V_1|} + c\frac{|V_2|}{\log^2 |V_2|} \geq c\frac{|V_1| + |V_2|}{\log^2 (2n/3)} \geq c\frac{n - \frac{c_1\sqrt{cn}}{\log n}}{\log^2 (2n/3)}$$

$$= c\frac{n}{\log^2 n} \cdot \frac{1 - \frac{c_1\sqrt{c}}{\log n}}{\left(1 - \frac{\log(3/2)}{\log n}\right)^2} \geq c\frac{n}{\log^2 n}$$

vertices such that each component of G' is contained in the neighborhood of a vertex or is an isolated vertex.

To see that Theorem 5 implies Theorem 2, it is sufficient to notice that if G' has a clique of size $r - 1$, then G has a clique of size r.

The proof of Theorem 3 is very similar to that of Theorem 2. Here, we need the following analogue of the Kővári-Sós-Turán theorem, which can also be deduced from Lemma 1 (see Conjecture 3.3 in [17]).

Lemma 2 ([17,24]).
Every $K_{t,t}$-free string graph on n vertices has at most $c_2 t(\log t)n$ edges, where c_2 is an absolute constant.

Proof of Theorem 3. Let $c > 0$ be a sufficiently small constant to be determined later. We proceed by induction on n. The base case $n = 1$ is trivial. For the inductive step, assume that the statement holds for all $n' < n$. Let $G = (V, E)$ be a K_r-free string graph on n vertices, and let c_1 and c_2 be the constants from Lemmas 1 and 2, respectively.

If G has at least $cc_2 \frac{n^2}{\log^2 n}$ edges, then, by Lemma 2, G contains a complete bipartite graph $K_{t,t}$, where $t \geq c \frac{n}{\log^3 n}$. Since G is K_r-free, one of these parts must be $K_{\lceil r/2 \rceil}$-free, and we are done.

Otherwise, if G has fewer than $cc_2 \frac{n^2}{\log^2 n}$ edges, then, by Lemma 1, there is a partition $V = S \cup V_1 \cup V_2$ with $|V_1|, |V_2| \leq \frac{2}{3}|V|$ such that no vertex in V_1 is adjacent to any vertex in V_2, and $|S| \leq c_1 \sqrt{cc_2} n / \log n$. Applying the induction hypothesis to V_1 and V_2, and setting $c < \frac{\log^2(3/2)}{c_2 c_1^2}$, we obtain a $K_{\lceil r/2 \rceil}$-free induced subgraph $G' \subseteq G$ with at least

$$c \frac{|V_1|}{\log^3 |V_1|} + c \frac{|V_2|}{\log^3 |V_2|} \geq c \frac{|V_1| + |V_2|}{\log^3 (2n/3)} \geq c \frac{n - \frac{c_1 \sqrt{cc_2} n}{\log n}}{\log^3 (2n/3)}$$

$$= c \frac{n}{\log^3 n} \cdot \frac{1 - \frac{c_1 \sqrt{cc_2}}{\log n}}{\left(1 - \frac{\log(3/2)}{\log n}\right)^3} \geq c \frac{n}{\log^3 n}$$

vertices. $\square$

3   A Technical Lemma for String Graphs

The *average degree* in a graph $G = (V, E)$ is $d = \frac{2|E|}{|V|}$. The *edge density* of G is defined as $\frac{|E|}{\binom{|V|}{2}} = \frac{d}{|V|-1}$. We say that a graph is *dense* if its edge density is larger than some positive constant (which we will conveniently specify for our purposes).

Using Lee's separator theorem for string graphs (Lemma 1), it is easy to deduce the following technical lemma which states that every string graph G has a dense induced subgraph G' whose average degree is not much smaller than the average degree in G.

Lemma 3. *For any $\epsilon > 0$, there is $C = C(\epsilon)$ with the following property. Every string graph $G = (V, E)$ with average degree $d = 2|E|/|V|$ has an induced subgraph $G[V']$ with average degree $d' \geq (1 - \epsilon)d$ and $|V'| \leq Cd'$.*

Proof. Let $G = (V, E)$ be a string graph with average degree d. We recursively define a nested sequence of induced subgraphs $G_0 \supset G_1 \supset \cdots$.

We begin with $G_0 = G$, and let $V_0 = V$, $E_0 = E$ and $d_0 = d$. After obtaining $G_i = (V_i, E_i)$ with $E_i = E(G[V_i])$ and with average degree $d_i = 2|E_i|/|V_i|$, we show that G_i is the desired induced subgraph if $d_i \geq |V_i|/C$. Otherwise if

$d_i < |V_i|/C$, we have $|E_i| \leq |V_i|^2/(2C)$, and by Lemma 1, there is a partition $V_i = U_0 \cup U_1 \cup U_2$ with $|U_1|, |U_2| \leq 2|V_i|/3$,

$$|U_0| \leq c_1\sqrt{|E_i|} \leq c_1\frac{|V_i|}{\sqrt{2C}} \leq |V_i|/12,$$

and there are no edges from U_1 to U_2. The last inequality above follows from the fact that $C \geq (12c_1)^2$. We can assume this as we can choose C as large as we want.

We take G_{i+1} to be the induced subgraph on whichever of $G[U_1 \cup U_0]$ and $G[U_2 \cup U_0]$ has larger average degree. As all edges of G_i are in at least one of these two induced subgraphs and $|U_1 \cup U_0| + |U_2 \cup U_0| = |V_i| + |U_0|$, the average degree of G_{i+1} satisfies

$$d_{i+1} \geq d_i\frac{|V_i|}{|V_i| + |U_0|} = d_i\frac{1}{1 + |U_0|/|V_i|}$$

$$\geq d_i\frac{1}{1 + c_1\sqrt{|E_i|}/|V_i|} \geq d_i\frac{1}{1 + c_1\sqrt{d_i/(2|V_i|)}}.$$

As $d_i < |V_i|/C$ and C can be chosen sufficiently large, the above inequality implies that $d_{i+1} \geq \frac{9}{10}d_i$. The inequality $|U_0| \leq |V_i|/12$ implies that $|V_{i+1}| \leq \frac{3}{4}|V_i|$. These two inequalities together imply that $d_{i+1}/|V_{i+1}| \geq \frac{6}{5}d_i/|V_i|$. It follows from the inequality above that

$$d_{i+1} = d\prod_{j=0}^{i} d_{j+1}/d_j \geq d\prod_{j=0}^{i} \frac{1}{1 + c_1\sqrt{d_j/(2|V_j|)}} \geq de^{-\Sigma_{j=0}^{i} c_1\sqrt{d_j/(2|V_j|)}},$$

where the last inequality uses that $\frac{1}{1+x} \geq e^{-x}$ for any $x > 0$. The sum in the exponent is dominated by a geometric series with common ratio $\sqrt{6/5} > 1$, and its largest summand is at most $c_1(1/(2C))^{1/2}$, as $d_i \leq |V_i|/C$. Hence, the sum in the exponent is $O(C^{-1/2})$. Taking C large enough, we have that $d_{i+1} \geq (1 - \epsilon)d$ for every i for which d_{i+1} is defined. (We can choose $C = O(1/\epsilon^2)$ to satisfy this.) Further, as $|V_{i+1}| \leq \frac{3}{4}|V_i| < |V_i|$ for every i for which V_{i+1} is defined, after at most $|V|$ iterations, the above process will terminate with the desired induced subgraph G_i.

The main result of [16] is that every dense string graph contains a dense spanning subgraph which is an incomparability graph. Applying Lemma 3 to this spanning subgraph with $\epsilon = 1/2$, we obtain the following corollary.

Corollary 3. *There is a constant $c > 0$ with the following property. Every string graph with n vertices and m edges has a subgraph with at least $c\frac{m}{n}$ vertices which is an incomparability graph with edge density at least c.*

Given a graph $G = (V, E)$ and two disjoint subsets of vertices $X, Y \in V$, we say that X is *complete* to Y if $xy \in E$ for all $x \in X$ and $y \in Y$.

The next lemma can be deduced by combining Corollary 3 above with Lemmas 6 and 7 of Tomon [33].

Lemma 4. *There is a constant $c > 0$ with the following property. If $G = (V, E)$ is a string graph with n vertices and at least αn^2 edges, for some $\alpha > 0$, then there are disjoint vertex subsets $X_1, \ldots, X_t \subset V$ for some $t \geq 2$ such that*

1. *X_i is complete to X_j for all $i \neq j$, and*
2. *$|X_i| \geq c\alpha \frac{n}{t^2}$ for every i.*

4   Back to Quasiplanar Graphs—Proofs of Theorems 1 and 4

Before turning to the proof of Theorem 1, we show how it implies Corollary 1.

Proof of Corollary 1. The most natural technique for properly coloring a graph is by successively extracting maximum independent sets from it. Using this greedy method and the bound in Theorem 1, for $r = 2^s$, we obtain a proper coloring of any K_r-free string graph on n vertices with at most $(\frac{\log n}{cs})^{2s-2} \log n$ colors. Indeed, each time we extract a maximum independent set, the fraction of remaining vertices is at most $1 - \alpha$ with $\alpha = (\frac{cs}{\log n})^{2s-2}$. As $1 - \alpha < e^{-\alpha}$, after at most $\frac{\log n}{\alpha}$ iterations, no vertex remains.

For a given $\epsilon > 0$, choose a sufficiently small $\delta > 0$ be such that

1. $2\delta \log \frac{1}{c\delta} < \frac{\epsilon}{2}$ and
2. $\log n < n^{\epsilon/2}$ provided that $n^\delta \geq 2$.

Consider any K_{n^δ}-free string graph G on n vertices. If $n^\delta < 2$, then G has no edges and, hence, its chromatic number is $1 \leq n^\epsilon$. Otherwise, substituting $s = \delta \log n$, Theorem 1 yields that the chromatic number of G is at most

$$n^{2\delta \log \frac{1}{c\delta}} \log n < n^\epsilon.$$

$\square$

Now we turn to the proof of Theorem 1 which gives, for $r = 2^s$, a lower bound on the independence number of a K_r-free string graph on n vertices.

Proof of Theorem 1. Our proof is by double induction on s and n. Throughout we let $r = 2^s$. The base cases are when $s = 1$ (in which case we get an independent set of size n) or $n = r$ (in which case we get an independent set of size 1) and are trivial. The induction hypothesis is that the theorem holds for all $s' < s$ and all n', and for $s' = s$ and all $n' < n$. Note that we may assume that $r \leq n/4$, as otherwise the theorem holds. Let $\alpha = c'(\frac{s}{\log n})^2$, where $c' > 0$ is a sufficiently small absolute constant. Let G be a K_r-free string graph on n vertices.

If G has at most αn^2 edges, applying Lemma 1, there is a vertex partition $V = V_0 \cup V_1 \cup V_2$ with $|V_0| \leq c_1 \alpha^{1/2} n$, $|V_1|, |V_2| \leq 2n/3$, and there are no edges from V_1 to V_2. Note that $|V_0| \leq n/12$ so $|V_1|, |V_2| \geq n/4$. We obtain a large independent set in G by taking the union of large independent sets in V_1 and V_2. Using the induction hypothesis applied to $G[V_1]$ and $G[V_2]$, we obtain an independent set in G of order at least

$$|V_1| \left(\frac{cs}{\log |V_1|} \right)^{2s-2} + |V_2| \left(\frac{cs}{\log |V_2|} \right)^{2s-2} \geq (|V_1| + |V_2|) \left(\frac{cs}{\log(2n/3)} \right)^{2s-2}.$$

Note that $|V_1| + |V_2| = n - |V_0| \geq n(1 - c_1 c'^{1/2} \cdot \frac{s}{\log n})$. We also have

$$(\log(2n/3))^{2s-2} = (\log n)^{2s-2} \left(1 - \frac{\log(3/2)}{\log n} \right)^{2s-2} \leq (\log n)^{2s-2} \left(1 - \frac{s}{2\log n} \right).$$

Substituting in these estimates and using $c' > 0$ is sufficiently small, we obtain an independent set of the desired size.

Suppose next that G has at least αn^2 edges. By Lemma 4, there is an integer $t \geq 2$ and disjoint vertex subsets $X_1, \ldots, X_t$ such that X_i is complete to X_j for all $i \neq j$ and $|X_i| \geq 4c'' \alpha n/t^2$ for $i = 1, \ldots, t$ where $0 < 4c'' < 1$ is an absolute constant. Losing a factor at most 2 in the number of sets X_i, we may assume $t = 2^p$ for a positive integer $p < s$, which implies $|X_i| \geq c'' \alpha n/t^2$. As G is K_{2^s}-free, one of these sets X_i induces a subgraph which is $K_{2^{s-p}}$-free. Let $n_0 = |X_i|$. Applying the induction hypothesis to $G[X_i]$, we obtain an independent set of size at least

$$n_0 \left(\frac{c(s-p)}{\log n_0} \right)^{2(s-p)-2} \geq c'' c' \left(\frac{s}{\log n} \right)^2 n 2^{-2p} \left(\frac{c(s-p)}{\log n_0} \right)^{2(s-p)-2}$$

$$\geq n \left(\frac{cs}{\log n} \right)^{2s-2}.$$

The last inequality holds, because after substituting $\log n_0 \leq \log n$, the ratio of the right-hand side and the expression in the middle reduces to

$$\frac{(2c)^{2p}}{c'' c'} \left(\frac{s}{\log n} \right)^{2(p-1)} \left(1 + \frac{p}{s-p} \right)^{2(s-p)-2} \leq \frac{(2ec)^{2p}}{c'' c'} \left(\frac{s}{\log n} \right)^{2(p-1)} \leq 1.$$

At the first inequality, we used $1 + x \leq e^x$ with $x = \frac{p}{s-p}$. As for the second inequality, we know that $s \leq \log n$, and we are free to choose the constant $c > 0$ as small as we wish (for instance, $c = c'' c'/30$ will suffice). This completes the proof. $\qquad\square$

A careful inspection of the proof of Theorem 1 shows that it recursively constructs an independent set of the desired size in a K_r-free string graph in polynomial time in terms of the size of the geometric representation of the set of strings. Indeed, the proof itself is essentially algorithmic. In the first case, when the string graph is relatively sparse, we apply Lee's separator theorem for string graphs, and take the union of large independent sets from the string graph of the two remaining large vertex subsets after deleting the small separator. In the second case, when the string graph is relatively dense, we apply Lemma 4 to get in the string graph a complete multipartite subgraph with large parts, and we can find a large independent set in one of the parts. However, this does require checking that results from several earlier papers each yield desirable structures in string graphs and incomparability graphs in polynomial time. These results include Lee's separator theorem for string graphs [24], the Fox-Pach result that every dense string graph contains a dense spanning subgraph which is an incomparability graph [16], and some extremal results of Tomon [33] for incomparability graphs.

A set of vertices $X \subseteq V$ in a graph $G = (V, E)$ is said to be *r-independent* if it does not induce a clique of size r, that is, if $G[X]$ is K_r-free. In particular, a 2-independent set is simply an independent set. Note that the proof of Theorem 1 carries through to the following generalization concerning the Erdős-Gallai problem for string graphs.

Theorem 6. *Let s, q be positive integers with $s > q$. Every K_{2^s}-free string graph G on $n \geq 2^s$ vertices contains a 2^q-independent set of size at least*

$$\min \left(\left(\frac{c(s+1-q)}{\log n} \right)^{2s-2q} n, \left(\frac{c(s+1-q)}{2^s \log n} \right)^2 n \right),$$

where $c > 0$ is an absolute constant.

Proof. (Sketch) We follow the proof of Theorem 1, making minor modifications. The proof is by double induction on s and n, with the base cases $s = q$ or $n = 2^s$ being trivial. We let $\alpha = c' \left(\frac{s+1-q}{\log n} \right)^2$. As in the proof of Theorem 1, if G has at most αn^2 edges, we apply the string graph separator lemma (Lemma 1). We delete the separator, use the induction hypothesis on the resulting components, and take the union of the 2^q-independent sets in the components to get a 2^q-independent set of the desired size in G.

So, we may assume G has more than αn^2 edges. By Lemma 4, there is an integer $t \geq 2$ and disjoint vertex subsets $X_1, \ldots, X_t$ such that X_i is complete to X_j for all $i \neq j$ and $|X_i| \geq 4c'' \alpha n / t^2$ for $i = 1, \ldots, t$, where $c'' > 0$ is an absolute constant. Losing a factor at most 2 in the number of sets X_i, we may assume that $t = 2^p$ for a positive integer $p < s$, which implies $|X_i| \geq c'' \alpha n / t^2$. As G is K_{2^s}-free, at least one of the sets X_i induces a subgraph which is $K_{2^{s-p}}$-free.

The proof now splits into two cases, depending on whether $s - p > q$ or not. If $s - p > q$, the rest of the proof goes through as in the proof of Theorem 1. If $s - p \leq q$, then X_i is the desired 2^q-independent set. Indeed, we have

$$|X_i| \geq c'' \alpha n / t^2 \geq c'' \alpha n / 2^{2s} = c'' c' \left(\frac{s+1-q}{2^s \log n} \right)^2 n \geq \left(\frac{c(s+1-q)}{2^s \log n} \right)^2 n,$$

for a sufficiently small absolute constant $c > 0$, as desired. $\qquad\blacksquare$

We complete the section by proving Theorem 4, which gives an upper bound on the number of edges of a r-quasiplanar graph with n vertices for $r = 2^s$ a perfect power of two.

Proof of Theorem 4. Let $s \geq 3$ be an integer and $r = 2^s$. We have to prove that every r-quasiplanar graph on $n \geq r$ vertices has at most $n(C\frac{\log n}{s})^{2s-4}$ edges, where C is an absolute constant.

For any r-quasiplanar graph $G = (V, E)$ on n vertices, delete a small disk around each vertex and consider the string graph whose vertex set consists of the (truncated) curves in E. As G is r-quasiplanar, the resulting string graph is K_{2^s}-free.

Applying Theorem 6 with $q = 2$, we obtain a subset $E' \subset E$ with

$$|E'| \geq |E| \left(\frac{c(s-1)}{\log |E|} \right)^{2s-4} \geq |E| \left(\frac{c(s-1)}{2 \log n} \right)^{2s-4},$$

for some absolute constant $c > 0$ such that $G' = (V, E')$ is 4-quasiplanar. According to Ackerman's result [1], every 4-quasiplanar graph on n vertices has at most a linear number of edges in n, that is, we have $|E'| \leq C'n$ for a suitable constant $C' > 0$. Putting these two bounds together, we get the desired upper bound

$$|E| \leq C'n \left(\frac{2 \log n}{c(s-1)} \right)^{2s-4} \leq n \left(C\frac{\log n}{s} \right)^{2s-4},$$

provided that C is sufficiently large.

$\square$

5   Concluding Remarks

A. A family of graphs $\mathcal{G}$ is said to be *hereditary* if for any $G \in \mathcal{G}$, all induced subgraphs of G also belong to $\mathcal{G}$. Obviously, the family of string graphs is hereditary.

The proof of Lemma 3 only uses the fact that there is a separator theorem for string graphs. A careful inspection of the proof shows that the same result holds, instead of string graphs, for any hereditary family of graphs $\mathcal{G}$ such that every $G = (V, E) \in \mathcal{G}$ has a separator of size $O(|E|^\alpha |V|^{1-2\alpha})$, for a suitable constant $\alpha = \alpha(\mathcal{G}) > 0$.

Similar techniques were used in [13–15,25]. Our Lemma 3 enables us to simplify some of the proofs in these papers.

B. Circle graphs are intersection graphs of chords of a circle. Gyárfás [18] proved that every circle graph with clique number r has chromatic number at most $O(r^2 4^r)$. Kostochka [20], and Kostochka and Kratochvíl [21] improved this bound to $O(r^2 2^r)$ and $O(2^r)$ respectively. Recently, a breakthrough was made by Davies and McCarty [7], who obtained the upper bound $O(r^2)$. Shortly after this, Davies [8] further improved this bound to $O(r \log r)$, which is asymptotically best possible due to a construction of Kostochka [20]. By taking the union of the $r - 2$ largest color classes in a proper coloring with the minimum number of colors, Davies' result implies that every circle graph on n vertices with clique number r contains an induced subgraph on $\Omega(n/\log r)$ vertices that is K_{r-1}-free. We conjecture that this "naive" bound can be improved as follows.

Conjecture 2. Every K_r-free circle graph on n vertices contains an induced subgraph on $\Omega(n)$ vertices which is K_{r-1}-free.

Acknowledgement. We are grateful to Zach Hunter for carefully reading our manuscript and pointing out several mistakes.

References

1. Ackerman, E.: On the maximum number of edges in topological graphs with no four pairwise crossing edges. Discrete Comput. Geom. **41**, 365–375 (2009)
2. Ackerman, E.: Quasi-planar graphs. In: Hong, S.-H., Tokuyama, T. (eds.) Beyond Planar Graphs, pp. 31–45. Springer, Singapore (2020). https://doi.org/10.1007/978-981-15-6533-5_3
3. Ackerman, E., Tardos, G.: On the maximum number of edges in quasi-planar graphs. J. Comb. Theory, Ser. A **114**(3), 563–571 (2007)
4. Agarwal, P., Aronov, B., Pach, J., Pollack, R., Sharir, M.: Quasi-planar graphs have a linear number of edges. Combinatorica **17**, 1–9 (1997)
5. Brass, P., Moser, W.O.J., Pach, J.: Research problems in discrete geometry. Springer (2005)
6. Burling, J.: On coloring problems of families of polytopes. (Ph.D. thesis), University of Colorado, Boulder 1(5) (1965)
7. Davies, J., McCarty, R.: Circle graphs are quadratically χ-bounded. Bull. London Math. Soc. **53**, 673–679 (2021)
8. Davis, J.: Improved bounds for colouring circle graphs. Preprint, arXiv:2107.03585 (2021)
9. Dudek, A., Retter, T., Rödl, V.: On generalized ramsey numbers of erdos and rogers. J. Combin. Theory Ser. B **109**, 213–227 (2014)
10. Dudek, A., Rödl, V.: On k_s-free subgraphs in k_{s+k}-free graphs and vertex folkman numbers. Combinatorica **31**, 39–53 (2011)
11. Erdős, P., Gallai, T.: On the minimal number of vertices representing the edges of a graph. Magyar Tud. Akad. Mat. Kutató Int. Közl. **6**, 181–203 (1961)
12. Erdős, P., Rogers, C.A.: The construction of certain graphs. Canad. J. Math. **14**, 702–707 (1962)
13. Fox, J., Pach, J.: Separator theorems and turán-type results for planar intersection graphs. Adv. Math. **219**, 1070–1080 (2008)
14. Fox, J., Pach, J.: A separator theorem for string graphs and its applications. Combin. Probab. Comput. **19**, 371–390 (2010)
15. Fox, J., Pach, J.: Coloring k_k-free intersection graphs of geometric objects in the plane. European J. Combin. **33**, 853–866 (2012)
16. Fox, J., Pach, J.: String graphs and incomparability graphs. Adv. Math. **230**, 1381–1401 (2012)
17. Fox, J., Pach, J.: Applications of a new separator theorem for string graphs. Combin. Probab. Comput. **23**, 66–74 (2014)
18. Gyárfás, A.: On the chromatic number of multiple interval graphs and overlap graphs. Discrete Math. **55**, 161–166 (1985)
19. Janzer, O., Gowers, W.T.: Improved bounds for the erdos-rogers function. Adv. Comb. **3**, 1–27 (2020)
20. Kostochka, A.: O verkhnikh otsenkakh khromaticheskogo chisla grafov (on upper bounds for the chromatic number of graphs). In: Dementyev, V.T. (ed.) Modeli i metody optimizacii, Akad. Nauk SSSR SO 10, pp. 204–226 (1988)

21. Kostochka, A., Kratochvíl, J.: Covering and coloring polygon-circle graphs. Discrete Math. **163**, 299–305 (1997)
22. Krivelevich, M.: k_s-free graphs without large k_r-free subgraphs. Combin. Probab. Comput. **3**, 349–354 (1994)
23. Krivelevich, M.: Bounding ramsey numbers through large deviation inequalities. Random Struct. Algorithms **7**, 145–155 (1995)
24. Lee, J.: Separators in region intersection graphs. ITCS **7**, 1:1–1:8 (2017). Full version: arXiv:1608.01612
25. Lipton, R., Rose, D., Tarjan, R.: Generalized nested dissections. SIAM J. Numer. Anal. **16**, 346–358 (1979)
26. McGuinness, S.: Colouring arcwise connected sets in the plane, i. Graphs Combin. **16**, 429–439 (2000)
27. Pach, J., Radoičić, R., Tóth, G.: Relaxing planarity for topological graphs. In: Akiyama, J., Kano, M. (eds.) JCDCG 2002. LNCS, vol. 2866, pp. 221–232. Springer, Heidelberg (2003). https://doi.org/10.1007/978-3-540-44400-8_24
28. Pawlik, A., Kozik, J., Krawczyk, T., Lasoń, M., Micek, P., Trotter, W., Walczak, B.: Triangle-free intersection graphs of line segments with large chromatic number. J. Combin. Theory Ser. B **105**, 6–10 (2014)
29. Rok, A., Walczak, B.: Coloring curves that cross a fixed curve. Discrete Comput. Geom. **61**, 830–851 (2019)
30. Rok, A., Walczak, B.: Outerstring graphs are χ-bounded. SIAM J. Discrete Math. **13**, 2181–2199 (2019)
31. Sudakov, B.: Large k_r-free subgraphs in k_s-free graphs and some other ramsey-type problems. Random Struct. Algorithms **26**, 253–265 (2005)
32. Suk, A.: Coloring intersection graphs of x-monotone curves in the plane. Combinatorica **34**, 487–505 (2014)
33. Tomon, I.: String graphs have the erdos-hajnal property. arXiv preprint. arXiv:2002.10350 (2020)
34. Valtr, P.: On geometric graphs with no k pairwise parallel edges. Discrete Comput. Geom. **19**, 461–469 (1998)
35. Walczak, B.: Triangle-free geometric intersection graphs with no large independent sets. Discrete Comput. Geom. **53**, 221–225 (2015)
36. Wolfovitz, G.: k_4-free graphs without large induced triangle-free subgraphs. Combinatorica **33**, 623–631 (2013)

Planarizing Graphs and Their Drawings by Vertex Splitting

Martin Nöllenburg[1], Manuel Sorge[1], Soeren Terziadis[1],
Anaïs Villedieu[1(✉)], Hsiang-Yun Wu[2,3], and Jules Wulms[1]

[1] Algorithms and Complexity Group, TU Wien, Vienna, Austria
{noellenburg,manuel.sorge,sterziadis,avilledieu,jwulms}@ac.tuwien.ac.at
[2] Research Unit of Computer Graphics, TU Wien, Vienna, Austria
hsiang.yun.wu@acm.org
[3] St. Pölten University of Applied Sciences, Sankt Pölten, Austria

Abstract. The splitting number of a graph $G = (V, E)$ is the minimum number of vertex splits required to turn G into a planar graph, where a vertex split removes a vertex $v \in V$, introduces two new vertices v_1, v_2, and distributes the edges formerly incident to v among v_1, v_2. The splitting number problem is known to be NP-complete for abstract graphs and we provide a non-uniform fixed-parameter tractable (FPT) algorithm for this problem. We then shift focus to the splitting number of a given topological graph drawing in $\mathbb{R}^2$, where the new vertices resulting from vertex splits must be re-embedded into the existing drawing of the remaining graph. We show NP-completeness of this *embedded* splitting number problem, even for its two subproblems of (1) selecting a minimum subset of vertices to split and (2) for re-embedding a minimum number of copies of a given set of vertices. For the latter problem we present an FPT algorithm parameterized by the number of vertex splits. This algorithm reduces to a bounded outerplanarity case and uses an intricate dynamic program on a sphere-cut decomposition.

Keywords: Vertex splitting · Planarization · Parameterized complexity

1 Introduction

While planar graphs admit compact and naturally crossing-free drawings, computing good layouts of large and dense non-planar graphs remains a challenging task, mainly due to the visual clutter caused by large numbers of edge crossings.

Manuel Sorge is supported by the Alexander von Humboldt Foundation, Anaïs Villedieu is supported by the Austrian Science Fund (FWF) under grant P31119, Jules Wulms is partially supported by the Austrian Science Fund (FWF) under grant P31119 and partially by the Vienna Science and Technology Fund (WWTF) under grant ICT19-035. Colored versions of figures can be found in the online version of this paper.

P. Angelini and R. von Hanxleden (Eds.): GD 2022, LNCS 13764, pp. 232–246, 2023.
https://doi.org/10.1007/978-3-031-22203-0_17

However, graphs in many applications are typically non-planar and hence several methods have been proposed to simplify their drawings and minimize crossings, both from a practical point of view [25,27] and a theoretical one [28,36]. Drawing algorithms often focus on reducing the number of visible crossings [33] or improving crossing angles [34], aiming to achieve similar beneficial readability properties as in crossing-free drawings of planar graphs.

One way of turning a non-planar graph into a planar one while retaining the entire graph and not deleting any of its vertices or edges, is to apply a sequence of *vertex splitting* operations, a technique which has been studied in theory [9,12,24,28], but which is also used in practice, e.g., by biologists and social scientists [17,18,31,37,38]. For a given graph $G = (V, E)$ and a vertex $v \in V$, a *vertex split* of v removes v from G and instead adds two non-adjacent copies v_1, v_2 such that the edges formerly incident to v are distributed among v_1 and v_2. Similarly, a *k-split* of v for $k \geq 2$ creates k copies $v_1, \ldots, v_k$, among which the edges formerly incident to v are distributed. On the one hand, splitting a vertex can resolve some of the crossings of its incident edges, but on the other hand the number of objects in the drawing to keep track of increases. Therefore, we aim to minimize the number of splits needed to obtain a planar graph, which is known as the *splitting number* of the graph. Computing it is NP-hard [13], but it is known for some graph classes including complete and complete bipartite graphs [15,16,19]. A related concept is the folded covering number [24] or equivalently the planar split thickness [12] of a graph G, which is the minimum k such that G can be decomposed into at most k planar subgraphs by applying a k-split to each vertex of G at most once. Eppstein et al. [12] showed that deciding whether a graph has split thickness k is NP-complete, even for $k = 2$, but can be approximated within a constant factor and is fixed-parameter tractable (FPT) for graphs of bounded treewidth.

While previous work considered vertex splitting in the context of abstract graphs, our focus in this paper is on vertex splitting for non-planar, topological graph drawings in $\mathbb{R}^2$. In this case we want to improve the given input drawing by applying changes to a minimum number of split vertices, which can be freely re-embedded, while the non-split vertices must remain at their original positions in order to maintain layout stability [30], see Fig. 1.

The underlying algorithmic problem for vertex splitting in drawings of graphs is two-fold: firstly, a suitable (minimum) subset of vertices to be split must be selected, and secondly the split copies of these vertices must be re-embedded in a crossing-free way together with a partition of the original edges of each split vertex into a subset for each of its copies.

The former problem is closely related to the NP-complete problem VERTEX PLANARIZATION, where we want to decide whether a given graph can be made planar by deleting at most k vertices, and to related problems of hitting graph minors by vertex deletions. Both are very well-studied in the parameterized complexity realm [20–22,35]. For example, it follows from results of Robertson and Seymour [35] that VERTEX PLANARIZATION can be solved in cubic time for

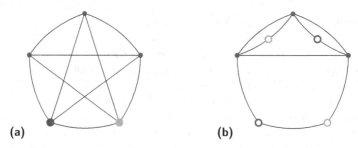

Fig. 1. Vertex splitting in a drawing of K_5. The red and orange disks in **(a)** are split once into the red and orange circles in **(b)**. Note that an abstract K_5 without drawing has splitting number 1. (Color figure online)

fixed k and a series of papers [20,21,29] improved the dependency on the input size to linear and the dependency on k to $2^{O(k \log k)}$.

The latter re-embedding problem is related to drawing extension problems, where a subgraph is drawn and the missing vertices and edges must be inserted in a (near-)planar way into this drawing [3,4,6,7,10,11]. In these works, however, the incident edges of each vertex are given, while we still need to distribute them among the copies. Furthermore, as we show in Sect. 4, it generalizes natural problems on covering vertices by faces in planar graphs [1,2,5,23].

Contributions. In this paper we extend the investigation of the splitting number problem and its complexity from abstract graphs to graphs with a given (non-planar) topological drawing. In Sect. 3.1, we first show that the original splitting number problem is non-uniformly FPT when parameterized by the number of split operations using known results on minor-closed graph classes. We then describe a polynomial-time algorithm for minimizing crossings in a given drawing when re-embedding the copies of a single vertex, split at most k times for a fixed integer k, in Sect. 3.2.

For the remainder of the paper we shift our focus to two basic subproblems of vertex splitting in topological graph drawings. We distinguish the *candidate selection step*, where we want to compute a set of vertices that requires the minimum number of splits to obtain planarity, and the *re-embedding step*, that asks where each copy should be put back into the drawing and with which neighborhood. We prove in Sect. 4 that both problems are NP-complete, using a reduction from vertex cover in planar cubic graphs for the candidate selection problem and showing that the re-embedding problem generalizes the NP-complete FACE COVER problem. Finally, in Sect. 5 we present an FPT algorithm for the re-embedding problem parameterized by the number of splits. Given a partial planar drawing and a set of vertices to split and re-embed, the algorithm first reduces the instance to a bounded-outerplanarity case and then applies dynamic programming on the decomposition tree of a sphere-cut decomposition of the remaining partial drawing. We note that our reduction for showing NP-hardness of the re-embedding problem is indeed a parameterized reduction from FACE

COVER parameterized by the solution size to the re-embedding problem parameterized by the number of allowed splits. FACE COVER is known to be FPT in this case [1], and hence our FPT algorithm is a generalization of that result.

Due to space constraints, missing proofs and details are found in the full version [32].

2 Preliminaries

Let $G = (V, E)$ be a simple graph with vertex set $V(G) = V$ and edge set $E(G) = E$. For a subset $V' \subset V$, $G[V']$ denotes the subgraph of G induced by V'. The neighborhood of a vertex $v \in V(G)$ is defined as $N_G(v)$. If G is clear from the context, we omit the subscript G. A *split* operation applied to a vertex v results in a graph $G' = (V', E')$ where $V' = V \setminus \{v\} \cup \{\dot{v}^{(1)}, \dot{v}^{(2)}\}$ and E' is obtained from E by distributing the edges incident to v among $\dot{v}^{(1)}, \dot{v}^{(2)}$ such that $N_G(v) = N_{G'}(\dot{v}^{(1)}) \cup N_{G'}(\dot{v}^{(2)})$ (copies are written with a dot for clarity). Splits with $N(\dot{v}^{(1)}) = N(v)$ and $N(\dot{v}^{(2)}) = \emptyset$ (equivalent to moving v to $\dot{v}^{(1)}$), or with $N(\dot{v}^{(1)}) \cap N(\dot{v}^{(2)}) \neq \emptyset$ (which is never beneficial, but can simplify proofs) are allowed. The vertices $\dot{v}^{(1)}, \dot{v}^{(2)}$ are called *split vertices* or *copies* of v. If a copy $\dot{v}$ of a vertex v is split again, then any copy of $\dot{v}$ is also called a copy of the original vertex v and we use the notation $\dot{v}^{(i)}$ for $i = 1, 2, \ldots$ to denote the different copies of v.

Problem 1 (SPLITTING NUMBER). Given a graph $G = (V, E)$ and an integer k, can G be transformed into a planar graph G' by applying at most k splits to G?

SPLITTING NUMBER is NP-complete, even for cubic graphs [13]. We extend the notion of vertex splitting to drawings of graphs. Let G be a graph and let Γ be a *topological drawing* of G, which maps each vertex to a point in $\mathbb{R}^2$ and each edge to a simple curve connecting the points corresponding to the incident vertices of that edge. We still refer to the points and curves as vertices and edges, respectively, in such a drawing. Furthermore, we assume Γ is a *simple* drawing, meaning no two edges intersect more than once, no three edges intersect in one point (except common endpoints), and adjacent edges do not cross.

Problem 2 (EMBEDDED SPLITTING NUMBER). Given a graph $G = (V, E)$ with a simple topological drawing Γ and an integer k, can G be transformed into a graph G' by applying at most k splits to G such that G' has a planar drawing that coincides with Γ on $G[V(G) \cap V(G')]$?

Problem 2 includes two interesting subproblems, namely an embedded vertex deletion problem (which corresponds to selecting candidates for splitting) and a subsequent re-embedding problem, both defined below.

Problem 3 (EMBEDDED VERTEX DELETION). Given a graph $G = (V, E)$ with a simple topological drawing Γ and an integer k, can we find a set $S \subset V$ of at most k vertices such that the drawing Γ restricted to $G[V \setminus S]$ is planar?

Problem 3 is closely related to the NP-complete problem VERTEX SPLIT-TING [20, 21, 26, 29], yet it deals with deleting vertices from an arbitrary given drawing of a graph with crossings. One can easily see that Problem 3 is FPT, using a bounded search tree approach, where for up to k times we select a remaining crossing and branch over the four possibilities of deleting a vertex incident to the crossing edges. The vertices split in a solution of Problem 2 necessarily are a solution to Problem 3; otherwise some crossings would remain in Γ after splitting and re-embedding. However, a set corresponding to a minimum-split solution of Problem 2 is not necessarily a minimum cardinality vertex deletion set as vertices can be split multiple times. Moreover, an optimal solution to Problem 2 may also split vertices that are not incident to any crossed edge and thus do not belong to an inclusion-minimal vertex deletion set. We note here that a solution to Problem 3 solves a problem variation where rather than minimizing the number of splits required to reach planarity, we instead minimize the number of split vertices: Splitting each vertex in an inclusion-minimal vertex deletion set its degree many times trivially results in a planar graph.

In the re-embedding problem, a graph drawing and a set of candidate vertices to be split are given. The task is to decide how many times to split each candidate vertex, where to re-embed each copy, and to which neighbors of the original candidate vertex to connect each copy.

Problem 4 (SPLIT SET RE-EMBEDDING). Given a graph $G = (V, E)$, a candidate set $S \subset V$ such that $G[V \setminus S]$ is planar, a simple planar topological drawing Γ of $G[V \setminus S]$, and an integer $k \geq |S|$, can we perform in G at most k splits to the vertices in S, where each vertex in S is split at least once, such that the resulting graph has a planar drawing that coincides with Γ on $G[V \setminus S]$?

We note that if no splits were allowed ($k = 0$) then Problem 4 would reduce to a partial planar drawing extension problem asking to re-embed each vertex of set S at a new position without splitting, which can be solved in linear time [3].

3   Algorithms for (Embedded) Splitting Number

SPLITTING NUMBER is known to be NP-complete in non-embedded graphs [13]. In Sect. 3.1, we show that it is FPT when parameterized by the number of allowed split operations. Indeed, we will show something more general, namely, that we can replace planar graphs by any class of graphs that is closed under taking minors and still get an FPT algorithm. Essentially we will show that the class of graphs that can be made planar by at most k splitting operations is closed under taking minors and then apply a result of Robertson and Seymour that asserts that membership in such a class can be checked efficiently [8].

For vertex splitting in graph drawings, we consider in Sect. 3.2 the restricted problem to split a single vertex. We show that selecting such a vertex and re-embedding at most k copies of it, while minimizing the number of crossings, can be done in polynomial time for constant k. For details see the full paper [32].

3.1 A Non-uniform Algorithm for Splitting Number

We use the following terminology. Let G be a graph. A *minor* of G is a graph H obtained from a subgraph of G by a series of edge contractions. *Contracting* an edge uv means to remove u and v from the graph, and to add a vertex that is adjacent to all previous neighbors of u and v. A graph class Π is *minor closed* if for every graph $G \in \Pi$ and each minor H of G we have $H \in \Pi$. Let Π be a graph class and $k \in \mathbb{N}$. We define the graph class Π_k to contain each graph G such that a graph in Π can be obtained from G by at most k vertex splits.

Theorem 1. *For a minor-closed graph class Π and $k \in \mathbb{N}$, Π_k is minor closed.*

The proof is given in the full paper [32] and essentially shows that whenever we have a graph $G \in \Pi_k$ and a minor H of G, then we can retrace vertex splits in H analogous to the splits that show that G is in Π_k. By results of Robertson and Seymour we obtain (again see [32]):

Proposition 1. *Let Π be a minor-closed graph class. There is a function $f : \mathbb{N} \to \mathbb{N}$ such that for every $k \in \mathbb{N}$ there is an algorithm running in $f(k) \cdot n^3$ time that, given a graph G with n vertices, correctly determines whether $G \in \Pi_k$.*

Since the class of planar graphs is minor closed, we obtain the following.

Corollary 1. SPLITTING NUMBER *is non-uniformly fixed-parameter tractable[1] with respect to the number of allowed vertex splits.*

3.2 Optimally Splitting a Single Vertex in a Graph Drawing

Let Γ be a drawing of a graph $G = (V, E)$ and let $v \in V$ be the single vertex to be split k times. Chimani et al. [6] showed that inserting a single star into an embedded graph while minimizing the number of crossings can be solved in polynomial time by considering shortest paths between faces in the dual graph, whose length correspond to the edges crossed in the primal. We build on their algorithm by computing the shortest paths in the dual of the planarized sub-drawing of Γ for $G[V \setminus \{v\}]$ between all faces incident to $N(v)$ and all possible faces for re-inserting the copies of v as the center of a star. We branch over all combinations of k faces to embed the copies, compute the nearest copy for each neighbor and select the combination that minimizes the number of crossings.

Theorem 2. *Given a drawing Γ of a graph G, a vertex $v \in V(G)$, and an integer k, we can split v into k copies such that the remaining number of crossings is minimized in time $O((|F| + |\mathcal{E}|) \cdot |N(v)| \cdot |F|^k)$, where F and $\mathcal{E}$ are respectively the sets of faces and edges of the planarization of Γ.*

---

[1] A parameterized problem is non-uniformly fixed-parameter tractable if there is a function $f : \mathbb{N} \to \mathbb{N}$ and a constant c such that for every parameter value k there is an algorithm that decides the problem and runs in $f(k) \cdot n^c$ time on inputs with parameter value k and length n.

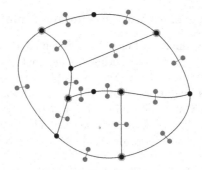

Fig. 2. The drawing Γ in black, and the vertices and edges added to obtain Γ' in blue. The vertex cover highlighted in orange corresponds to the deletion set. (Color figure online)

4 NP-Completeness of Subproblems

While it is known that SPLITTING NUMBER is NP-complete [13], in the correctness proof of the reduction Faria et al. [13] assume that it is permissible to draw all vertices, split or not, at new positions as there is no initial drawing to be preserved. The reduction thus does not seem to easily extend to EMBEDDED SPLITTING NUMBER. Here we show the NP-completeness of each of its two subproblems.

Theorem 3. EMBEDDED VERTEX DELETION *is NP-complete.*

Proof. We reduce from the NP-complete VERTEX COVER problem in planar graphs [14], where given a planar graph $G = (V, E)$ and an integer k, the task is to decide if there is a subset $V' \subseteq V$ with $|V'| \leq k$ such that each edge $e \in E$ has an endpoint in V'. Given the planar graph G from such a VERTEX COVER instance and an arbitrary plane drawing Γ of G we construct an instance of EMBEDDED VERTEX DELETION as follows. We create a drawing Γ' by drawing a *crossing edge* e' across each edge e of Γ such that e' is orthogonal to e and has a small enough positive length such that e' intersects only e and no other crossing edge or edge in Γ, see Fig. 2. Drawing Γ' can be computed in polynomial time.

Let C be a vertex cover of G with $|C| = k$. We claim that C is also a deletion set that solves EMBEDDED VERTEX DELETION for Γ'. We remove the vertices in C from Γ', with their incident edges. By definition of a vertex cover, this removes all the edges of G from Γ'. The remaining edges in Γ' are the crossing edges and they form together a (disconnected) planar drawing which shows that C is a solution of EMBEDDED VERTEX DELETION for Γ'.

Let D be a deletion set of Γ' such that $|D| = k$. We find a vertex cover of size at most k for G in the following manner. Assume that D contains a vertex w that is an endpoint of a crossing edge e that crosses the edge (u, v) of G. Since w has degree one, deleting it only resolves the crossing between e and (u, v), thus we can replace w in D by u (or v) and resolve the same crossing as well as all the crossings induced by the edges incident to u (or v). Thus we can find a deletion set D' of size smaller or equal to k that contains only vertices in G and removing this deletion set from Γ' removes only edges from G. Since every edge of G is crossed in Γ', every edge of G must have an incident vertex in D', thus D' is a vertex cover for G.

Containment in NP is easy to see. Given a deletion set D, we only need to verify that Γ' is planar after deleting D and its incident edges.

Next, we prove that also the re-embedding subproblem itself is NP-complete, by showing that FACE COVER is a special case of the re-embedding problem. The problem FACE COVER is defined as follows. Given a planar graph $G = (V, E)$, a subset $D \subseteq V$, and an integer k, can G be embedded in the plane, such that at most k faces are required to cover all the vertices in D? FACE COVER is NP-complete, even when G has a unique planar embedding [5].

Theorem 4. SPLIT SET RE-EMBEDDING *is NP-complete.*

Proof. We give a parameterized reduction from FACE COVER (with unique planar embedding) parameterized by the solution size to the re-embedding problem parameterized by the number of allowed splits. We first create a graph $G' = (V', E')$, with a new vertex v, vertex set $V' = V \cup \{v\}$ and $E' = E \cup \{dv \mid d \in D\}$. Then we compute a planar drawing Γ of G corresponding to the unique embedding of G. Finally, we define the candidate set $S = \{v\}$ and allow for $k - 1$ splits in order to create up to k copies of v. Then G', Γ, S, and $k - 1$ form an instance I of SPLIT SET RE-EMBEDDING.

In a solution of I, every vertex in D is incident to a face in Γ, in which a copy of v was placed. Therefore, selecting these at most k faces in Γ gives a solution for the FACE COVER instance. Conversely, given a solution for the FACE COVER instance, we know that every vertex in D is incident to at least one of the at most k faces. Therefore, placing a copy of v in every face of the FACE COVER solution yields a re-embedding of at most k copies of v, each of which can realize all its edges to neighbors on the boundary of the face without crossings.

Finally, a planar embedding of the graph can be represented combinatorially in polynomial space. We can also verify in polynomial time that this embedding is planar and exactly the right connections are realized, for NP-containment.

5   Split Set Re-embedding Is Fixed-Parameter Tractable

In this section we show that SPLIT SET RE-EMBEDDING (SSRE) can be solved by an FPT-algorithm, with the number k of splits as a parameter. We provide an overview of the involved techniques and algorithms in this section and refer to [32] for the full technical details.

Preparation. First, from the given set S of $s = |S|$ candidate vertices (disks in Fig. 3(a)) we choose how many copies of each candidate we will insert back into the graph; these copies form a set S_λ. Vertices with neighbors in S are called *pistils* (squares in Fig. 3), and faces incident to pistils are called *petals*. The

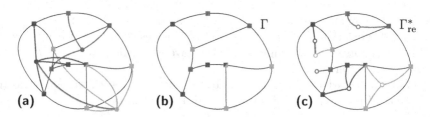

Fig. 3. (a) An example graph G, (b) a planar drawing Γ of G where S has been removed, and (c) a solution drawing Γ^*_{re}. Pistils are squares, copies are circles and vertices in S are disks.

copies in S_λ must be made adjacent to the corresponding pistils. Since vertices in S can be pistils, we also determine which of their copies in S_λ are connected by edges. For both of these choices, the number of copies per candidate and the edges between copies, we branch over all options. In each of the $\mathcal{O}(2^k k^4)$ resulting branches we apply dynamic programming to solve SSRE.

Second, we prepare the drawing Γ for the dynamic programming. A face f is not necessarily involved in a solution, e.g., if it is not a petal: a copy embedded in f either has no neighbors and can be re-embedded in any face, or its neighbors are not incident with f, and this embedding induces a crossing. Therefore we remove all vertices not incident to petals, which actually results in a drawing of a $10k$-outerplanar graph. We show this in the following way. We find for each vertex a *face path* to the outerface, which is a path alternating between incident vertices and faces. Each face visited in that path is either a petal or adjacent to a petal (from the above reduction rule). Thus, each face in the path might either have a copy embedded in it in the solution, or a face up to two hops away has a copy embedded in it. If we label each face by a closest copy with respect to the length of the face path, then one can show that no label can appear more than five times on any shortest face path to the outerface. Since there are at most $2k$ different copies (labels), we can bound the maximum distance to the outerface by $10k$. Because of the $10k$-outerplanarity, we can now compute a special branch decomposition of our adapted drawing Γ in polynomial time, a so called *sphere-cut decomposition* of branchwidth $20k$. A sphere-cut decomposition (λ, T) is a tree T and a bijection λ that maps the edges of Γ to the leaves of T (see Fig. 4).

Each edge e of T splits T into two subtrees, which induces via λ a bipartition (A_e, B_e) of $E(G)$ into two subgraphs. We define a vertex set $\text{mid}(e)$, which contains all vertices that are incident to an edge in both A_e and B_e. Additionally, e also corresponds to a curve, called a *noose* $\eta(e)$, that intersects Γ only in $\text{mid}(e)$. We define a root for T, and we say that for edge e, the subtree further from the root corresponds to the drawing *inside* the noose $\eta(e)$. Having a root for T allows us to solve SSRE on increasingly larger subgraphs in a structured way, starting with the leaves of T (the edges of Γ) and continuing bottom-up to the root (the complete drawing Γ).

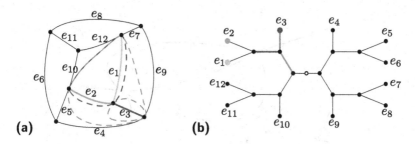

Fig. 4. (a) A graph and **(b)** its sphere-cut decomposition. Each labeled leaf corresponds to the same labeled edge of the graph. The middle set of each colored edge in the tree corresponds to the vertices of the corresponding colored dashed noose in the graph. (Color figure online)

Initialization. During the dynamic programming, we want to determine whether there is a partial solution for the subgraphs of Γ that we encounter when traversing T. For one such subgraph we describe such a *partial solution* with a tuple $(S'_\lambda, (N_{\dot{v}})_{\dot{v} \in S'_\lambda}, \Gamma')$. In this tuple, the set $S'_\lambda \subseteq S_\lambda$ corresponds to the embedded copies, $(N_{\dot{v}})$ are their respective neighborhoods in Γ, and Γ' is the resulting drawing. However, during the dynamic programming, we need more information to determine whether a partial solution exists. For example, for the faces completely inside the noose η enclosing Γ' (*processed* faces) a solution must already be found, while faces intersected by η (*current* faces) still need to be considered.

We store this information in a *signature*, which is a tuple $(S_{\mathrm{in}}, C_{\mathrm{out}}, M, N_\eta)$. An example of a signature is visualized in Fig. 5. The set $S_{\mathrm{in}}(t) \subseteq S_\lambda$ corresponds to the copies embedded in processed faces, and N_η contains a set $X(p)$ for each pistil p on the noose, such that $X(p)$ describes which neighbors of p are still missing in the partial solution. The set C_{out} corresponds to a set of planar graphs describing the combinatorial embedding of copies in current faces. One such graph C_f (Fig. 6) associated with a current face f consists of a cycle, whose vertices represent the pistils of f, and of copies embedded inside the cycle. Since f is current, not all of its pistils are necessarily inside the noose, and M describes which section of the cycle, and hence which pistils, should be used.

Saving a single local optimal partial solution, one that uses the smallest number of copies, for a given noose is not sufficient. This sub-solution may result in a no-instance when considering the rest of the graph outside the noose. We therefore keep track of all signatures that lead to partial solutions, which we call *valid* signatures. These signatures allow us to realize the required neighborhoods for pistils inside the noose with a crossing-free drawing. The number of distinct signatures $N_s(k)$ depends on the number k of splits and we prove an upper bound of $2^{O(k^2)}$ by counting all options for each element of a signature tuple. Since the number of signatures is bounded by a function of our parameter k, we can safely enumerate all signatures. We then determine which signatures are valid for each noose in T. The number $N_s(k)$ of signatures will be part of the leading term in the total running time.

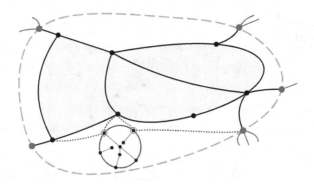

Fig. 5. The information stored in a signature of the partial solution inside the orange dashed noose: copies in S_{in} are used in the grey faces, blue noose vertices who have missing neighbors (outgoing edges outside the noose) are stored in N_η, in red an example of a nesting graph for a face traversed by the noose, with four dotted edges connecting to the cycle and the vertices described by M in green. (Color figure online)

Lemma 1. *The number $N_s(k)$ of possible signatures is upper bounded by $2^{O(k^2)}$.*

Dynamic Programming. Finally, we give an overview of how the valid signatures are found. In each branch, we perform bottom-up dynamic programming on T. We want to find a valid signature at the root node of T, and we start from the leaves of T. Each leaf corresponds to an edge (u_1, u_2) of the input graph G, for which we consider all enumerated signatures and check if a signature is valid and thus corresponds to a partial solution. Such a partial solution should cover all missing neighbors of u_1 and u_2 not in $N_\eta = \{X(u_1), X(u_2)\}$, using for each incident face f the subgraph of $C_f \in C_{\text{out}}$ as specified by M.

For internal nodes of T we merge some pairs of valid child signatures corresponding to two nooses η_1 and η_2. We merge if the partial solutions corresponding to the child signatures can together form a partial solution for the union of the graphs inside η_1 and η_2. The signature of this merged partial solution is hence valid for the internal node when (1) faces not shared between the nooses do not have copies in common, (2) shared faces use identical nesting graphs and (3) use disjoint subgraphs of those nesting graphs to cover pistils, and (4) noose vertices have exactly a prescribed set of missing neighbors. Thus we can find valid signatures for all nodes of T and notably for its root. If we find a valid signature for the root, a partial solution $(S'_\lambda, (N_{\dot{v}})_{\dot{v} \in S'_\lambda}, \Gamma')$ must exist. In Γ' all pistils are covered and it is planar, as the nesting graphs are planar and they represent a combinatorial embedding of copies that together cover all pistils. It is possible that certain split vertices are in no nesting graph, and hence $S_\lambda \setminus S'_\lambda \neq \emptyset$. We verify that the remaining copies that are pistils in $S_\lambda \setminus S'_\lambda$ induce a planar graph, which allows us to embed them in a face of Γ' to obtain the final drawing. The running time for every node of T is polynomial in $N_s(k)$, thus, over all created branches SPLIT SET RE-EMBEDDING is solved in $2^{O(k^2)} \cdot n^{O(1)}$ time.

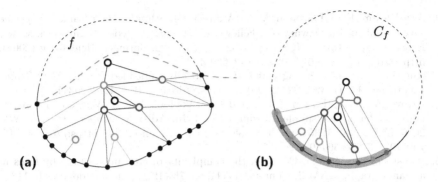

Fig. 6. (a) A face f and copies inside the orange noose, and (b) the corresponding nesting graph C_f with the interval described by M highlighted in grey. The two light blue vertices represent two different copies of the same removed vertex. Copies in C_f have no edges to copies in other nesting graphs. (Color figure online)

Theorem 5. SPLIT SET RE-EMBEDDING *can be solved in* $2^{O(k^2)} \cdot n^{O(1)}$ *time, using at most k splits on a topological drawing Γ of input graph G with n vertices.*

6   Conclusions

We have introduced the embedded splitting number problem. However, fixed-parameter tractability is only established for the SPLIT SET RE-EMBEDDING subproblem. The main open problem is to investigate the parameterized complexity of EMBEDDED SPLITTING NUMBER. A trivial XP-algorithm for EMBEDDED SPLITTING NUMBER can provide appropriate inputs to SPLIT SET RE-EMBEDDING as follows: check for any subset of up to k vertices whether removing those vertices results in a planar input drawing, and branch on all such subsets.

Many variations of embedded splitting number are interesting for future work. For example, rather than aiming for planarity, we can utilize vertex splitting for crossing minimization. Other possible extensions can adapt the splitting operation, for example, the split operation allows both creating an additional copy of a vertex and re-embedding it, and the cost of these two parts can differ: simply re-embedding a vertex can be a cheaper operation.

Acknowledgments. We would like to thank an anonymous reviewer for their input to simplify the proof of Theorem 3.

References

1. Abu-Khzam, F.N., Fernau, H., Langston, M.A.: A bounded search tree algorithm for parameterized face cover. J. Discrete Algorithms **6**(4), 541–552 (2008). https://doi.org/10.1016/j.jda.2008.07.004

2. Abu-Khzam, F.N., Langston, M.A.: A direct algorithm for the parameterized face cover problem. In: Downey, R., Fellows, M., Dehne, F. (eds.) Parameterized and Exact Computation. LNCS, vol. 3162, pp. 213–222. Springer, Heidelberg (2004). https://doi.org/10.1007/978-3-540-28639-4_19

3. Angelini, P., et al.: Testing planarity of partially embedded graphs. ACM Trans. Algorithms 11(4), 32:1–32:42 (2015). https://doi.org/10.1145/2629341

4. Arroyo, A., Klute, F., Parada, I., Seidel, R., Vogtenhuber, B., Wiedera, T.: Inserting one edge into a simple drawing is hard. In: Adler, I., Müller, H. (eds.) WG 2020. LNCS, vol. 12301, pp. 325–338. Springer, Cham (2020). https://doi.org/10.1007/978-3-030-60440-0_26

5. Bienstock, D., Monma, C.L.: On the complexity of covering vertices by faces in a planar graph. SIAM J. Comput. 17(1), 53–76 (1988). https://doi.org/10.1137/0217004

6. Chimani, M., Gutwenger, C., Mutzel, P., Wolf, C.: Inserting a vertex into a planar graph. In: Mathieu, C. (ed.) Proceedings of 20th Symposium on Discrete Algorithms (SODA), pp. 375–383. SIAM (2009). https://doi.org/10.1137/1.9781611973068.42

7. Chimani, M., Hlinený, P.: Inserting multiple edges into a planar graph. In: Fekete, S.P., Lubiw, A. (eds.) Proceedings of 32nd International Symposium on Computational Geometry (SoCG). LIPIcs, vol. 51, pp. 30:1–30:15 (2016). https://doi.org/10.4230/LIPIcs.SoCG.2016.30

8. Cygan, M., Fomin, F.V., Kowalik, Ł., Lokshtanov, D., Marx, D., Pilipczuk, M., Pilipczuk, M., Saurabh, S.: Parameterized Algorithms. Springer, Cham (2015). https://doi.org/10.1007/978-3-319-21275-3

9. Eades, P., de Mendonça N, C.F.X.: Vertex splitting and tension-free layout. In: Brandenburg, F.J. (ed.) GD 1995. LNCS, vol. 1027, pp. 202–211. Springer, Heidelberg (1996). https://doi.org/10.1007/BFb0021804

10. Eiben, E., Ganian, R., Hamm, T., Klute, F., Nöllenburg, M.: Extending nearly complete 1-planar drawings in polynomial time. In: Esparza, J., Král', D. (eds.) Proceedings of 45th International Symposium on Mathematical Foundations of Computer Science (MFCS). LIPIcs, vol. 170, pp. 31:1–31:16. Schloss Dagstuhl - Leibniz-Zentrum für Informatik (2020). https://doi.org/10.4230/LIPIcs.MFCS.2020.31

11. Eiben, E., Ganian, R., Hamm, T., Klute, F., Nöllenburg, M.: Extending partial 1-planar drawings. In: Czumaj, A., Dawar, A., Merelli, E. (eds.) Proceedings of 47th International Colloquium on Automata, Languages, and Programming (ICALP). LIPIcs, vol. 168, pp. 43:1–43:19. Schloss Dagstuhl-Leibniz-Zentrum für Informatik (2020). https://doi.org/10.4230/LIPIcs.ICALP.2020.43

12. Eppstein, D., Kindermann, P., Kobourov, S., Liotta, G., Lubiw, A., Maignan, A., Mondal, D., Vosoughpour, H., Whitesides, S., Wismath, S.: On the planar split thickness of graphs. Algorithmica 80(3), 977–994 (2017). https://doi.org/10.1007/s00453-017-0328-y

13. Faria, L., de Figueiredo, C.M.H., de Mendonça N., C.F.X.: Splitting number is NP-complete. Discrete Appl. Math. 108(1), 65–83 (2001). https://doi.org/10.1016/S0166-218X(00)00220-1

14. Garey, M.R., Johnson, D.S.: The rectilinear steiner tree problem is NP-complete. J. SIAM Appl. Math. 32(4), 826–834 (1977). https://doi.org/10.1137/0132071

15. Hartsfield, N.: The toroidal splitting number of the complete graph K_n. Discret. Math. 62(1), 35–47 (1986). https://doi.org/10.1016/0012-365X(86)90039-7

16. Hartsfield, N., Jackson, B., Ringel, G.: The splitting number of the complete graph. Graphs Comb. 1(1), 311–329 (1985). https://doi.org/10.1007/BF02582960

17. Henry, N., Bezerianos, A., Fekete, J.: Improving the readability of clustered social networks using node duplication. IEEE Trans. Visual Comput. Graph. **14**(6), 1317–1324 (2008). https://doi.org/10.1109/TVCG.2008.141

18. Henry Riche, N., Dwyer, T.: Untangling euler diagrams. IEEE Trans. Visual Comput. Graph. **16**(6), 1090–1099 (2010). https://doi.org/10.1109/TVCG.2010.210

19. Jackson, B., Ringel, G.: The splitting number of complete bipartite graphs. Arch. Math. **42**(2), 178–184 (1984). https://doi.org/10.1007/BF01772941

20. Jansen, B.M.P., Lokshtanov, D., Saurabh, S.: A near-optimal planarization algorithm. In: Chekuri, C. (ed.) Proceedings of 2014 Annual ACM-SIAM Symposium on Discrete Algorithms (SODA), pp. 1802–1811. Proceedings, Society for Industrial and Applied Mathematics (2013). https://doi.org/10.1137/1.9781611973402.130

21. Kawarabayashi, K.i.: Planarity allowing few error vertices in linear time. In: Proceedings of 50th Annual IEEE Symposium on Foundations of Computer Science (FOCS), pp. 639–648 (2009). https://doi.org/10.1109/FOCS.2009.45

22. Kim, E.J., Langer, A., Paul, C., Reidl, F., Rossmanith, P., Sau, I., Sikdar, S.: Linear kernels and single-exponential algorithms via protrusion decompositions. ACM Trans. Algorithms **12**(2), 21:1–21:41 (2015). https://doi.org/10.1145/2797140

23. Kloks, T., Lee, C.M., Liu, J.: New algorithms for k-Face Cover, k-feedback vertex set, and k-disjoint cycles on plane and planar graphs. In: Goos, G., Hartmanis, J., van Leeuwen, J., Kučera, L. (eds.) Graph-Theoretic Concepts in Computer Science. LNCS, vol. 2573, pp. 282–295. Springer, Heidelberg (2002). https://doi.org/10.1007/3-540-36379-3_25

24. Knauer, K.B., Ueckerdt, T.: Three ways to cover a graph. Discret. Math. **339**(2), 745–758 (2016). https://doi.org/10.1016/j.disc.2015.10.023

25. von Landesberger, T., Kuijper, A., Schreck, T., Kohlhammer, J., van Wijk, J.J., Fekete, J.D., Fellner, D.W.: Visual analysis of large graphs: State-of-the-art and future research challenges. Comput. Graph. Forum **30**(6), 1719–1749 (2011). https://doi.org/10.1111/j.1467-8659.2011.01898.x

26. Lewis, J.M., Yannakakis, M.: The node-deletion problem for hereditary properties is NP-complete. J. Comput. Syst. Sci. **20**(2), 219–230 (1980). https://doi.org/10.1016/0022-0000(80)90060-4

27. Lhuillier, A., Hurter, C., Telea, A.C.: State of the art in edge and trail bundling techniques. Comput. Graph. Forum **36**(3), 619–645 (2017). https://doi.org/10.1111/cgf.13213

28. Liebers, A.: Planarizing graphs - a survey and annotated bibliography. J. Graph Algorithms Appl. **5**(1), 1–74 (2001). https://doi.org/10.7155/jgaa.00032

29. Marx, D., Schlotter, I.: Obtaining a planar graph by vertex deletion. Algorithmica **62**(3–4), 807–822 (2012). https://doi.org/10.1007/s00453-010-9484-z

30. Misue, K., Eades, P., Lai, W., Sugiyama, K.: Layout adjustment and the mental map. J. Vis. Lang. Comput. **6**(2), 183–210 (1995). https://doi.org/10.1006/jvlc.1995.1010

31. Nielsen, S.S., Ostaszewski, M., McGee, F., Hoksza, D., Zorzan, S.: Machine learning to support the presentation of complex pathway graphs. IEEE/ACM Trans. Comput. Biol. Bioinf. **18**(3), 1130–1141 (2019). https://doi.org/10.1109/TCBB.2019.2938501

32. Nöllenburg, M., Sorge, M., Terziadis, S., Villedieu, A., Wu, H., Wulms, J.: Planarizing graphs and their drawings by vertex splitting. CoRR abs/2202.12293 (2022). https://arxiv.org/abs/2202.12293

33. Nöllenburg, M.: Crossing layout in non-planar graph drawings. In: Hong, S.-H., Tokuyama, T. (eds.) Beyond Planar Graphs, pp. 187–209. Springer, Singapore (2020). https://doi.org/10.1007/978-981-15-6533-5_11

34. Okamoto, Y.: Angular resolutions: around vertices and crossings. In: Hong, S.-H., Tokuyama, T. (eds.) Beyond Planar Graphs, pp. 171–186. Springer, Singapore (2020). https://doi.org/10.1007/978-981-15-6533-5_10

35. Robertson, N., Seymour, P.D.: Graph minors. XIII. the disjoint paths problem. J. Comb. Theory Ser. B **63**(1), 65–110 (1995). https://doi.org/10.1006/jctb.1995.1006

36. Schaefer, M.: Crossing Numbers of Graphs. CRC Press (2018)

37. Wu, H.Y., Nöllenburg, M., Sousa, F.L., Viola, I.: Metabopolis: scalable network layout for biological pathway diagrams in urban map style. BMC Bioinform. **20**(1), 1–20 (2019). https://doi.org/10.1186/s12859-019-2779-4

38. Wu, H.Y., Nöllenburg, M., Viola, I.: Multi-level area balancing of clustered graphs. IEEE Trans. Visualization Comput. Graph., 1–15 (2020). https://doi.org/10.1109/TVCG.2020.3038154

The Thickness of Fan-Planar Graphs is At Most Three

Otfried Cheong[1], Maximilian Pfister[2], and Lena Schlipf[2(✉)]

[1] Institut für Informatik, Universität Bayreuth, Bayreuth, Germany
otfried.cheong@uni-bayreuth.de
[2] Wilhelm-Schickard-Institut für Informatik, Universität Tübingen,
Tübingen, Germany
{maximilian.pfister,lena.schlipf}@uni-tuebingen.de

Abstract. We prove that in any strongly fan-planar drawing of a graph $\mathcal{G}$ the edges can be colored with at most three colors, such that no two edges of the same color cross. This implies that the thickness of strongly fan-planar graphs is at most three. If $\mathcal{G}$ is bipartite, then two colors suffice to color the edges in this way.

Keywords: Thickness · Fan-planarity · Beyond planarity

1 Introduction

In order to visualize non-planar graphs, the research field of *graph drawing beyond planarity* emerged as a generalization of drawing planar graphs by allowing certain edge-crossing patterns. Among the most popular classes of beyond-planar graph drawings are k-planar drawings, where every edge can have at most k crossings, k-quasiplanar drawings, which do not contain k mutually crossing edges, *RAC*-drawings, where the edges are straight-line segments that can only cross at right angles, fan-crossing-free drawings, where an edge is crossed by independent edges, and fan-planar drawings, where no edge is crossed by independent edges (see below for the full definition). Many more details and further classes can be found in the recent survey [9] and recent book [10] about beyond-planarity.

Thickness. A different notion of non-planarity was introduced by Tutte in 1963 [19]: the *thickness* of a graph is the minimum number of planar graphs into which the edges of the graph can be partitioned. Subsequent research derived tight bounds on the thickness of complete and complete bipartite graphs [2,4]. For general graphs, however, it turns out to be hard to determine their thickness—even deciding if the thickness of a given graph is two is already NP-hard [15]. More details about thickness and related concepts can be found in [16].

The first connection between thickness and beyond-planarity was laid in 1973, when Kainen [11] studied the relationship between the thickness of a graph and

Research of Schlipf was supported by the Ministry of Science, Research and the Arts Baden-Württemberg (Germany).

k-planarity. He observed that a k-planar graph $\mathcal{G}$ has thickness at most $k + 1$. This follows from the fact that $\mathcal{G}$ admits by definition a drawing where every edge crosses at most k other edges, hence the edge intersection graph (refer to Sect. 2 for a formal definition) has maximum degree k and admits a vertex coloring using at most $k + 1$ colors. For other beyond-planar graph classes, however, the thickness is not so closely related to a coloring of the edge intersection graph. For 3-quasiplanar graphs, for instance, there exist 3-quasi-planar drawings whose edge intersection graph cannot be colored with a bounded number of colors [18], yet the thickness of 3-quasi-planar graphs is at most seven.

The most general tool to determine the thickness of a graph $\mathcal{G}$ is to compute its *arboricity*, which is the minimum number of forests into which the edges of $\mathcal{G}$ can be partitioned. Since a forest is planar, the arboricity is an upper bound on the thickness, and since any planar graph has arboricity at most three, the arboricity of $\mathcal{G}$ is at most three times its thickness. By the Nash-Williams theorem [17], the arboricity of $\mathcal{G}$ is $a(\mathcal{G}) = \max_S \frac{m_S}{n_S - 1}$, where S ranges over all subgraphs of $\mathcal{G}$ with m_S edges and n_S vertices. For typical beyond-planar graph classes, a subgraph of a graph is contained inside the same class, and so a linear bound on the number of edges of graphs in the class implies a constant bound on the thickness. This insight actually improves the result of Kainen for large enough k, as k-planar graphs have at most $3.81\sqrt{k}n$ edges [1], and therefore the thickness of k-planar graphs is $\mathcal{O}(\sqrt{k})$.

Fan-Planar Graphs. We consider the class of *fan-planar* graphs. This class was introduced in 2014 by Kaufmann and Ueckerdt [12] as graphs that admit a *fan-planar drawing*, which they define by the requirement that for each edge e, the edges crossing e have a common endpoint on the same side of e. This can be formulated equivalently by two forbidden patterns, patterns (I) and (II) in Fig. 1: one is the configuration where e is crossed by two independent edges, the other where e is crossed by incident edges with the common endpoint on different sides of e. Fan-planarity has evolved into a popular subject of study with many related publications, a good overview of which can be found in the recent survey article by Bekos and Grilli [6].

And now it get's complicated: the recent journal version [13] of the 2014 paper [12] gives a more restricted definition of fan-planarity, where also pattern (III) of Fig. 1 is forbidden. The restriction is necessary to allow the proof for the bound $5n - 10$ on the number of edges of a fan-planar graph to go through. Patterns (II) and (III) can formally be defined as follows: if one removes the edge e and the two edges intersecting e from the plane, we are left with two connected regions called cells, a bounded cell and an unbounded cell. In pattern (II), one endpoint of e lies in the bounded cell, in pattern (III), both endpoints lie in the bounded cell. Both patterns are forbidden, so both endpoints of e must lie in the unbounded cell.

To distinguish this definition from the "classic" definition, let's call a fan-planar drawing without pattern (III) a *strongly fan-planar* drawing, and a graph with such a drawing a *strongly fan-planar* graph. Note that—quite unusual for a topological drawing style—strongly fan-planar drawings are only defined in the

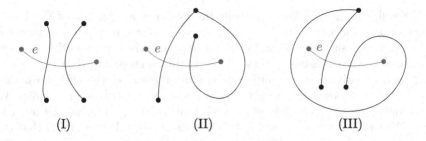

Fig. 1. The three forbidden configurations in strongly fan-planar drawings.

plane, the definition does not work for drawings on the sphere! Clearly, strongly fan-planar graphs are also fan-planar (in the "classic" sense), so most of the literature applies to this restricted class as well.

Our Contribution. We establish the first result on the thickness of a beyond-planar graph class that is stronger than bounds based only on the density or arboricity of the class. We show that the edges of a simple, strongly fan-planar drawing of a graph G can be colored with three colors, such that no two edges of the same color cross, implying that G has thickness at most three. We also show that if G is bipartite, then two colors suffice for the same coloring, and therefore bipartite strongly fan-planar graphs have thickness two (and this bound is tight unless G is planar). For comparison, the upper bound implied by density (and therefore arboricity) of strongly fan-planar graphs is five [13] (and four for bipartite strongly fan-planar graphs [3]). Our proof relies on a complete characterization of chordless cycles in the intersection graph, which should be of independent interest for the study of fan-planar drawings.

2   Preliminaries

Throughout the paper, we assume that any graph G and its corresponding drawing is *simple*, that is, G has no self-loops or multiple edges, adjacent edges cannot cross, any two edges are allowed to cross at most once, and crossing points of distinct pairs of edges do not coincide. We will refer to all of these as property (S). (Non-simple fan-planar graphs where discussed by Klemz et al. [14].) The *intersection graph* $\mathcal{I}$ of a drawing Γ of graph G has a vertex for every edge of G, and two vertices are connected in $\mathcal{I}$ if the corresponding edges cross in Γ.

Let G be a strongly fan-planar graph with a fixed, strongly fan-planar drawing Γ. (Throughout the proofs, we will simply say "fan-planar" with the understanding that we require strong fan-planarity.) This means that none of the forbidden configurations shown in Fig. 1 occur. For ease of reference, we will refer to the three forbidden patterns as properties (I), (II), and (III). If an edge e is crossed by more than one edge, all edges crossing e share a common endpoint—we call this point the *anchor* of e. Throughout the paper, we will use the letter v to denote the anchor of the edge denoted by e, so that v_i is always the anchor of e_i.

We will discuss *chordless* cycles in the intersection graph $\mathcal{I}$ of $\mathcal{G}$, that is, cycles without diagonals. In $\mathcal{G}$, a chordless cycle C corresponds to a sequence of edges $e_1, \ldots, e_k$, such that e_i and e_{i+1} intersect, but there are no other intersections between the edges of C—we will refer to this as property (M). (Throughout the paper, all arithmetic on indices is modulo the size of the cycle.) Fixing a chordless cycle C, we define x_i as the intersection point between e_{i-1} and e_i. We let a_i and b_i be the endpoints of e_i, such that $a_i x_i x_{i+1} b_i$ appear in this order on e_i. We will call a_i the *source*, b_i the *target* of e_i—but keep in mind that this orientation is only defined with respect to C. Let $\hat{e}_i$ be the oriented segment of e_i from x_i to x_{i+1}—we will call $\hat{e}_i$ the *base* of e_i. If we concatenate the bases $\hat{e}_i$ in order, we obtain a closed loop that we call $\mathcal{L}$. Since by (M) two bases $\hat{e}_i$, $\hat{e}_j$ do not intersect, $\mathcal{L}$ is a Jordan curve and partitions the plane into two regions. Since $\mathcal{G}$ is fan-planar, edges e_i and e_{i+2} share an endpoint, namely the anchor v_{i+1} of e_{i+1}. We will use $\mathcal{G}_C$ to denote the subgraph of $\mathcal{G}$ consisting of only the edges of C, with the same embedding.

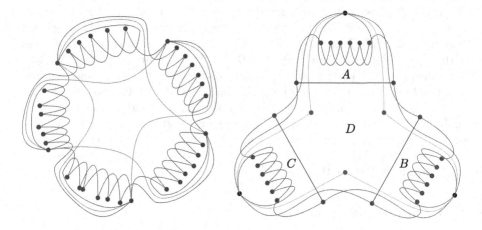

Fig. 2. Chordless cycles can intersect and share edges.

3   Characterizing Chordless Cycles

In this section we characterize chordless cycles in $\mathcal{I}$. In the next section we then study how chordless cycles can interact, and we will be able to break all odd cycles simultaneously by coloring a carefully chosen set of edges with one color. Figure 2 shows two examples of strongly fan-planar graphs whose intersection graphs have several chordless cycles that cross and share edges. The graph on the left has 32 distinct chordless cycles, as for each of the five red edges we can instead traverse the blue edges "behind." In the graph on the right, the boundaries of the faces labeled A, B, C are loops of chordless cycles of length 11.

There is a chordless cycle of length 9 surrounding face D, there is a chordless cycle of length 30 surrounding all four faces, and there are three chordless cycles of length 23 surrounding ABD, BCD, and CAD, respectively.

It is easy to see using (I) and (S) that $\mathcal{I}$ cannot have cycles of length three, while cycles of length four have a unique shape, see Fig. 6(a). It remains to study the structure of chordless cycles of length at least five.

We will call an edge e_i of a chordless cycle *canonical* if the anchor v_i of e_i is the target of e_{i-1} and the source of e_{i+1}, that is, if $b_{i-1} = v_i = a_{i+1}$. If, in addition, no other edge of C is incident to v_i—that is, if v_i has degree two in $\mathcal{G}_C$—then e_i is *strictly canonical*.

Figure 3 shows a sequence of canonical edges. Note that some of the endpoints of the edges of this sequence can coincide, see Fig. 4. For a canonical edge e_i, we

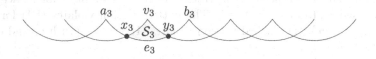

$$a_3 \qquad v_3 \qquad b_3$$
$$x_3 \quad \mathcal{S}_3 \quad y_3$$
$$e_3$$

Fig. 3. A sequence of canonical edges.

will call the "triangle" $\mathcal{S}_i$ with corners x_i, x_{i+1}, v_i and bounded by the edges e_{i-1}, e_i, and e_{i+1} the *spike* of e_i, see the shaded region in Fig. 3.

Lemma 1. *Let e_1, e_2, e_3, e_4, e_5 be five consecutive canonical edges of a chordless cycle. Then the anchors of the five edges are distinct.*

Proof. Since v_2 is an endpoint of e_1, $v_2 \neq v_1$ holds. By construction, v_1 and v_3 are the endpoints of e_2, so $v_3 \neq v_1$. If $v_1 = v_4$, then e_2 and e_3 share an endpoint and intersect, a contradiction to (S). Finally, if $v_1 = v_5$, then both e_2 and e_4 connect v_1 and v_5 and are therefore identical. Using analogous arguments one establishes the remaining inequalities. $\qquad\square$

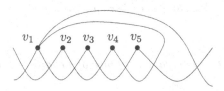

$$v_1 \qquad v_2 \quad v_3 \quad v_4 \quad v_5$$

Fig. 4. The anchors of canonical edges can coincide, but only after at least four distinct anchors.

Lemma 2. *Let $C = (e_1, \ldots e_k)$ be a chordless cycle in $\mathcal{I}$ with $k \geq 5$. Then, $b_{i-1} \neq b_{i+1}$ and $a_{i-1} \neq a_{i+1}$ for $1 \leq i \leq k$.*

Proof. Suppose for a contradiction that there is an index i such that $b_{i-1} = b_{i+1}$. By renumbering we can assume that $b_1 = v_2 = b_3$. This implies that x_4 lies on the segment $x_3 v_2$, and x_1 lies on the segment $a_1 x_2$, see Fig. 5. Consider e_4, which contains x_4 and is incident to v_3. Since by (M) e_4 does not cross e_1, we have $v_3 \neq a_2$ as otherwise e_3 forms forbidden configuration (II) with e_2 and e_4. Consider the region $\mathcal{R}$ bounded by $v_2 x_2 x_3$. By (M) the two boundary segments $v_2 x_2$ and $x_2 x_3$ do not cross any edge of C, and the segment $x_3 v_2$ only crosses e_4. It follows that e_4 is incident to b_2 and its other endpoint lies in $\mathcal{R}$. If x_5 lies inside $\mathcal{R}$, then the closed curve $\mathcal{L}$ has to intersect the boundary of $\mathcal{R}$ to return to x_1, a contradiction. So x_5 lies on $x_4 b_2$, and therefore $b_4 = b_2$. The same argument now implies that $b_5 = b_3 = b_1$, $b_6 = b_4 = b_2$, and so on. The final edge e_k must contain x_1, so it cannot be incident to b_1 by (S), and so k is even, e_k is incident to b_2, and x_k lies on the edge e_{k-1} incident to b_1. Since x_1 lies on e_k between x_k and $b_k = b_2$, there are two possibilities for drawing e_k, shown dashed and dotted in Fig. 5. The dotted version violates (III) for e_{k-1} (with e_{k-2} and e_k), the dashed version violates (III) for e_k (with e_1 and e_{k-1}), a contradiction.

Assume now that $a_{i-1} = a_{i+1}$ for some i, and let $C' = (e_k, \ldots, e_1)$ be the chordless cycle obtained by reversing C. Reversing the direction flips a_i and b_i for each edge, so C' now has an index j such that $b_{j-1} = b_{j+1}$, and we already showed this cannot be the case.　□

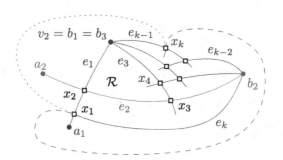

Fig. 5. The cycle cannot be closed.

Corollary 1. *Let e_i be a non-canonical edge of a chordless cycle C. Then $a_{i-1} = v_i = b_{i+1}$.*

Proof. The anchor v_i is a common endpoint of e_{i-1} and e_{i+1}. This cannot be b_{i-1}, because e_i is not canonical and $b_{i-1} \neq b_{i+1}$ by Lemma 2. So $v_i = a_{i-1}$, and since $a_{i-1} \neq a_{i+1}$ by Lemma 2 again, we have $v_i = a_{i-1} = b_{i+1}$.　□

We say that the chordless cycle C is *fully canonical* if all its edges are canonical. Such cycles can be created for any $k \geq 5$ by taking the corners of a regular k-gon and connecting every other corner with an edge, see Fig. 6(b). Note that

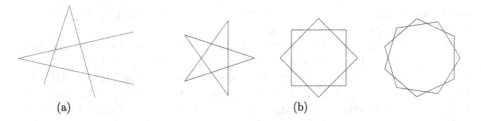

Fig. 6. (a) The only possible cycle of length four, and (b) fully canonical cycles for $k = 5, 8, 11$.

such a fully canonical cycle corresponds to a single closed trail of length k in $\mathcal{G}$ for odd k, but to two closed trails of length $k/2$ for even k. These closed trails in $\mathcal{G}$ are not necessarily cycles, as the anchors of the edges of a fully canonical cycle can coincide, see e.g. Fig. 2(left).

Lemma 3. *A chordless cycle of length at least five has at least four consecutive canonical edges.*

Proof. Let $C = (e_1, \ldots e_k)$ be a chordless cycle of length $k \geq 5$ such that $e_1, \ldots, e_m$ is a longest consecutive sequence of canonical edges in C, and assume $m < 4$. We distinguish four different cases according to the value of m.

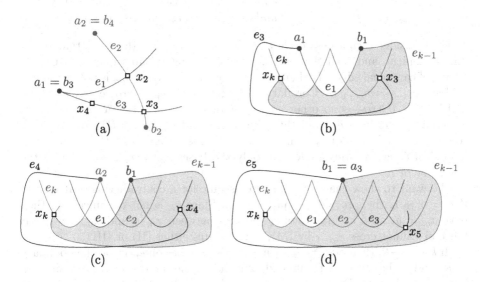

Fig. 7. (a) $m = 0$, (b) $m = 1$, (c) $m = 2$ and (d) $m = 3$.

If $m = 0$, that is, C does not have any canonical edge: By Corollary 1, $a_1 = v_2 = b_3$ and $a_2 = v_3 = b_4$. This implies that x_4 lies on the segment $b_3 x_3$, and e_4 must connect $b_4 = a_2$ with x_4 without intersecting either e_1 or e_2, creating forbidden configuration (II), see Fig. 7(a).

If $m = 1$: Since e_2 and e_k are not canonical, by Corollary 1 $a_1 = b_3$ and $b_1 = a_{k-1}$. By (M), e_3 does not cross e_k, and e_2 does not cross e_{k-1}. Therefore, $x_k \in e_{k-1}$ and $x_3 \in e_3$ implies that we have the situation of Fig. 7(b) (by (II) and (III)). But then e_3 and e_{k-1} cross, so we must have $k = 5$. Since e_4 crosses e_3 and e_5, e_3 and e_5 must have a common endpoint. By (S), e_5 cannot be incident to a_1, and e_3 cannot reach either endpoint of e_5 without introducing a crossing that contradicts (S).

If $m = 2$: Since e_3 and e_k are not canonical, by Corollary 1 $a_2 = b_4$ and $b_1 = a_{k-1}$. Since e_4 and e_k have a common endpoint, they cannot cross, so $k > 5$. Also e_{k-1} and e_3 do not cross. Therefore $x_k \in e_{k-1}$ and $x_4 \in e_4$ implies that we have the situation of Fig. 7(c). So e_4 and e_{k-1} cross and we have $k = 6$. However, e_4, e_5, e_6 (and also e_3, e_4, e_5) form forbidden configuration (II).

If $m = 3$: Since e_4 and e_k are not canonical, by Corollary 1 $a_3 = b_5$ and $b_1 = a_{k-1}$, so, since e_2 is canonical, we have $a_{k-1} = b_1 = a_3 = b_5$. Since e_5 and e_{k-1} share this endpoint, they cannot cross, so $k \neq 5$. If $k = 6$, then $a_5 = b_5$, a contradiction, so we have $k \geq 7$. It follows that e_5 does not cross $\{e_1, e_2, e_3, e_k\}$, and e_{k-1} does not cross $\{e_1, e_2, e_3, e_4\}$. Therefore $x_k \in e_{k-1}$ and $x_5 \in e_5$ implies that we have the situation of Fig. 7(d). But this requires e_5 and e_{k-1} to cross, a contradiction to the above. □

Theorem 1. *If a chordless cycle of length $k \geq 5$ is not fully canonical, then $k \geq 9$, edges $e_1, \ldots, e_{k-1}$ are canonical, anchors $v_2 = v_{k-2}$ coincide so that $b_1 = a_3 = b_{k-3} = a_{k-1}$, and b_{k-1} and a_1 are vertices of degree one in $\mathcal{G}_C$.*

Proof. Consider a chordless cycle $C = (e_1, \ldots, e_k)$ of length $k \geq 5$ that is not fully canonical, and such that $e_1, \ldots, e_{m-1}$ is a longest sub-sequence of C consisting of canonical edges. By Lemma 3 we have $m \geq 5$. Since e_m is not canonical, we have $a_{m-1} = b_{m+1}$ by Corollary 1. By definition, e_{m+1} crosses e_m in x_{m+1}, and by (M) it crosses no other edge e_i with $i \neq m + 2$. Let $\mathcal{R}$ be the region enclosed by the base $\hat{e}_m$ and the edges e_{m-1} and e_{m+1}. By (II) and (III), $\mathcal{R}$ contains no endpoint of e_m, and so we have the situation of Fig. 8(a). The loop $\mathcal{L}$ of C lies entirely in $\mathcal{R}$, and only the bases $\hat{e}_{m-1}$, $\hat{e}_m$, $\hat{e}_{m+1}$ lie on the boundary of $\mathcal{R}$.

We want to show that $k = m$, so we assume for a contradiction that $k > m$. If $k = m + 1$, then $e_k = e_{m+1}$ in Fig. 8(a). Since e_k is not canonical, we then have $a_m = b_1$, and e_1 intersects $e_k = e_{m+1}$ between x_{m+1} and $b_{m+1} = a_{m-1}$. Since e_1 cannot cross e_{m-1} by (M), that violates either (II) or (III).

If $k = m + 2$, then $e_{k-1} = e_{m+1}$ in Fig. 8(a), x_k lies on e_{m+1} on the boundary of $\mathcal{R}$, and $\hat{e}_k$ lies (except for its endpoint x_k) in the interior of $\mathcal{R}$. But since e_k is not canonical, we have $a_{m+1} = b_1$, and since e_1 does not cross the boundary of $\mathcal{R}$, it cannot contain x_1 in the interior of $\mathcal{R}$.

We assume next that $k > m + 2$, which implies that $\hat{e}_k$ lies in the interior of $\mathcal{R}$. Since e_k is not canonical, we have $a_{k-1} = b_1$. Symmetrically to the argument above, the edge e_{k-1} must be such that the region $\mathcal{R}'$ formed by $\hat{e}_k$, e_{k-1}, and e_1 contains no endpoint of e_k, so we are in the situation of Fig. 8(b). Again, the

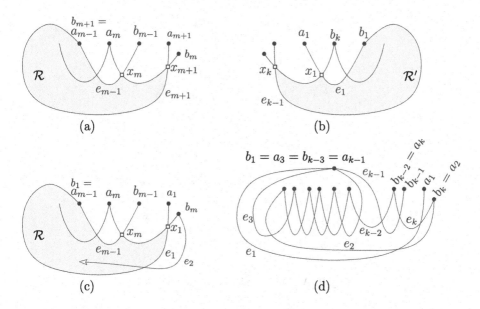

Fig. 8. Proof of Theorem 1.

loop $\mathcal{L}$ lies in $\mathcal{R}'$, with only $\hat{e}_{k-1}, \hat{e}_k, \hat{e}_1$ on the boundary of $\mathcal{R}'$. In particular, $\hat{e}_m$ lies in the interior of $\mathcal{R}'$.

Since $\hat{e}_m$ is in the interior of $\mathcal{R}'$ but on the boundary of $\mathcal{R}$, while $\hat{e}_k$ is in the interior of $\mathcal{R}$ but on the boundary of $\mathcal{R}'$, the two regions cannot be nested, and their boundaries must intersect. The boundary of $\mathcal{R}$ consists of $e_{m-1}, \hat{e}_m$, and e_{m+1}, the boundary of $\mathcal{R}'$ consists of $e_{k-1}, \hat{e}_k$, and e_1, so by (M) the only possible edge crossing occurs when $k = m + 3$, so that e_{m+1} and e_{k-1} can cross. Since two intersecting closed curves must intersect an even number of times, we must in addition have that the vertices a_{m-1} of $\mathcal{R}$ and b_1 of $\mathcal{R}'$ coincide, but then e_{m+1} and e_{k-1} have a common endpoint and cannot cross at all by (S).

It follows that our assumption that $k > m$ is false, and so $k = m$. Relabeling the edges in Fig. 8(a) we obtain Fig. 8(c). Since e_1 is canonical, we have $b_m = a_2$, so e_2 starts in b_m and enters $\mathcal{R}$ through e_1, see Fig. 8(c). Since e_2 cannot cross either e_{m-1} or e_m, its other endpoint b_2 lies in the interior of $\mathcal{R}$. Now it remains to observe that since $e_2, \ldots, e_{k-2}$ are canonical, we end up with the situation shown in Fig. 8(d).

Since $v_2 = b_1 = a_{k-1} = v_{k-2}$, Lemma 1 implies that $k - 2 \geq 7$, so $k \geq 9$. □

In Fig. 9 we draw again the non-canonical chordless cycle of Fig. 8(d), showing the symmetry in the characterization. The reader may enjoy determining which of the chordless cycles in Fig. 2 are fully canonical, and which edges are the non-canonical ones of the other cycles.

It is easy to see that when k is odd, then the graph $\mathcal{G}$ contains a closed trail of length $k - 2$ consisting of all edges of C except for e_{k-1} and e_1, namely the closed trail $e_2 e_4 e_6 \ldots e_{k-3} e_3 e_5 \ldots e_{k-2} e_k$. In Fig. 8(d) we have $k = 11$, so there

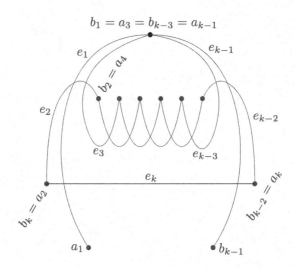

Fig. 9. A non-canonical chordless cycle.

is a closed trail of length 9. Figure 9 shows the smallest possible non-canonical chordless cycle: here $k = 9$, and so there is a closed trail of length 7. We obtain the following corollary:

Corollary 2. *The edges of a strongly fan-planar drawing of a bipartite graph $\mathcal{G}$ can be colored using two colors such that no edges of the same color cross. As a consequence, a bipartite, strongly fan-planar graph has thickness at most two.*

Proof. We show that $\mathcal{I}$ is bipartite. Assume otherwise: then $\mathcal{I}$ has an odd cycle, which contains a chordless odd cycle C of length k. If C is fully canonical, then the edges of C form an odd cycle of length k in $\mathcal{G}$, so $\mathcal{G}$ is not bipartite. If C is not fully canonical, then Theorem 1 implies that $\mathcal{G}$ has an odd cycle of length $k - 2$, again a contradiction. □

The bound on the thickness is tight: every bipartite, strongly fan-planar graph that is not planar has thickness exactly two, an example being $K_{3,3}$.

In the remainder of this section, we study the regions induced by the loop $\mathcal{L}$ and the spikes $\mathcal{S}_i$ in a little more detail, in particular how they can be crossed by *other* edges, that is, edges not part of C. The edges of C intersect only in the corners of the loop $\mathcal{L}$. The loop $\mathcal{L}$ partitions the plane into two regions. If C is fully canonical, then one of these regions is empty in $\mathcal{G}_C$—this could be either the bounded or the unbounded region delimited by $\mathcal{L}$. If C has a non-canonical edge, then from Theorem 1 it follows that the *bounded* region delimited by $\mathcal{L}$ is empty in $\mathcal{G}_C$ (here it plays a role that property (III) cannot be used for drawings on a sphere). In both cases, we will denote the region delimited by $\mathcal{L}$ that is empty in $\mathcal{G}_C$ by $\mathcal{L}$ as well. Adjacent to $\mathcal{L}$ is, for each base $\hat{e}_i$ of a canonical edge e_i, the spike $\mathcal{S}_i$, which is itself an empty region in $\mathcal{G}_C$.

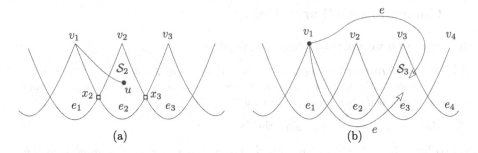

Fig. 10. (a) u must lie in $\mathcal{S}_2$ or $e = e_2$, and (b) e starting in v_1 cannot enter $\mathcal{S}_3$.

Lemma 4. *Let e_1, e_2, e_3 be three consecutive canonical edges of a chordless cycle. Let $e = (v_1, u)$ be an edge that crosses e_1 in the relative interior of the segment $x_2 v_2$. Then u is contained in $\mathcal{S}_2$.*

Proof. Traversing e from v_1 to u, e enters $\mathcal{S}_2$ by crossing the segment $x_2 v_2$, see Fig. 10(a). Since e cannot cross e_2 and cannot cross e_1 again by (S), e could leave $\mathcal{S}_2$ only through the segment $v_2 x_3$, by crossing e_3. But then v_3 would have to be an endpoint of e. By Lemma 1, $v_3 \neq v_1$, so $u = v_3$, but then $e = e_2$, a contradiction since e_2 does not intersect the relative interior of $x_2 v_2$.   □

Lemma 5. *Let e_1, e_2, e_3, e_4 be four consecutive canonical edges of a chordless cycle. Let $e = (v_1, u)$ be an edge incident to v_1. Then e does not enter the interior of $\mathcal{S}_3$.*

Proof. By (S) e cannot cross e_2, so to enter $\mathcal{S}_3$, it would have to cross either e_3 or e_4, see Fig. 10(b). If e crosses e_3, then it must be incident to v_3, and since $v_3 \neq v_1$, that means $e = (v_1, v_3) = e_2$, which does not intersect the interior of $\mathcal{S}_3$. If e crosses e_4, then it must be incident to v_4, and since $v_4 \neq v_1$, that means $e = (v_1, v_4)$. But v_4 lies outside $\mathcal{S}_3$, so e would have to cross e_4 again to reach v_4, a contradiction to (S).   □

Lemma 6. *Let C be a chordless cycle of length at least five, and let e be an edge not part of C such that e intersects the loop $\mathcal{L}$ of C. Then e starts in the anchor v_i of a canonical edge e_i of C, passes through the spike $\mathcal{S}_i$, crosses the base $\hat{e}_i$, and either (1) ends in $\mathcal{L}$; or (2) crosses the base $\hat{e}_j$ of another canonical edge e_j of C, then passes through $\mathcal{S}_j$ and terminates in v_j; or (3) crosses the base of the non-canonical edge e_j of C, never enters $\mathcal{L}$ again, and terminates in a vertex that is not a vertex of C. The second and third case can only happen if e_i and e_j share an endpoint. In particular, (2) cannot happen when e_{i-1} and e_{i+1} are strictly canonical, and (3) implies that $i \in \{j - 2, j + 2\}$.*

The proof can be found in the full version [8].

4 Coloring with Three Colors

In this section we will show our main theorem:

Theorem 2. *In every strongly fan-planar drawing of a graph G there is a set S of edges such that (1) S is independent in $\mathcal{I}$, that is, no two edges in S cross; and (2) every odd cycle in $\mathcal{I}$ contains an edge in S.*

The theorem immediately implies the following:

Corollary 3. *The edges of a strongly fan-planar drawing of a graph G can be colored using three colors such that no two edges of the same color cross. As a consequence, a strongly fan-planar graph has thickness at most three.*

Proof. Pick the set S of edges according to Theorem 2 and color them with the first color. Then $\mathcal{I} \setminus S$ contains no odd cycle and is therefore bipartite, and can be colored with the remaining two colors.   □

We construct the set S for the proof of Theorem 2 using the following lemma.

Lemma 7. *Let C be a chordless cycle of length at least five. Then C contains an edge e_i such that*

- $e_{i-2}, e_{i-1}, e_i, e_{i+1}, e_{i+2}$ *are all canonical in C;*
- e_{i-1}, e_i, e_{i+1} *are all strictly canonical in C.*

Proof. If C is fully canonical and no two anchors coincide, we can pick any edge of C as e_i and are done. Otherwise, we pick a sequence of canonical edges $e_1, \ldots, e_m$ in C such that $v_1 = v_m$ and such that the spikes $\mathcal{S}_2, \ldots, \mathcal{S}_{m-1}$ are contained in the region bounded by $\mathcal{L} \cup \mathcal{S}_1 \cup \mathcal{S}_m$. (When C is not fully canonical, then its sub-sequence $e_2, \ldots, e_{k-2}$ has this property by Theorem 1.)

Since spikes cannot intersect (except for touching at their anchors), the coinciding anchors form a bracket structure. Pick any innermost interval $e_j, \ldots, e_\ell$ such that $v_j = v_\ell$. Then the anchors $v_j, \ldots, v_{\ell-1}$ are all distinct. By Lemma 1, $\ell - j \geq 5$. We can thus pick $i = j + 2$ to satisfy the requirements of the lemma.   □

For each chordless odd cycle C of $\mathcal{I}$, we pick an edge e as in Lemma 7 and call it the *ground edge* of C. Our proof of Theorem 2 relies on the following key lemma:

Lemma 8. *Let C and C' be two chordless odd cycles with ground edges e and e'. If e and e' cross, then e is part of C' and e' is part of C.*

The proof can be found in the full version [8]. We can now prove our main theorem.

Proof (of Theorem 2). Starting with $S = \emptyset$, we consider all chordless odd cycles C in $\mathcal{I}$ one by one. If S does not already contain an edge of C, we add the ground edge of C to S.

We claim that the resulting set S satisfies the conditions of the theorem. Indeed, by construction S contains an edge of every chordless odd cycle. Since every odd cycle contains a chordless odd cycle as a subset, condition (2) holds. Assume now for a contradiction that there are two edges $e, e' \in S$ such that e and e' cross. Let C and C' be the chordless odd cycles that caused e and e' to be added to S, and assume w.l.o.g. that e was added to S before e'. By Lemma 8, e is an edge of C'—but that means that when considering C', no edge has been added to S, a contradiction. $\qquad\square$

5   Open Problems

We conclude with some open problems:

1. Do these results hold for fan-planar graphs in the "classic" definition?
2. Are there actually strongly fan-planar graphs that have thickness three?
3. Are there 2-planar graphs of thickness three?
4. What is the thickness of the optimal 2-planar graphs? (These graphs have been fully characterized [7].) It can be shown that these graph admit an edge decomposition into a 1-planar graph and a bounded-degree planar graph [5]. If the thickness of this graph is actually three, this would answer both previous questions in the affirmative as the optimal 2-planar graphs are also fan-planar.
5. If there is no 2-planar graph of thickness three, what is the smallest k such that there exists a k-planar graph of thickness three? Note that K_9 is 4-planar and requires thickness three, hence $k \in \{2, 3, 4\}$.
6. Are there strongly fan-planar graphs $\mathcal{G}$ such that *every* fan-planar drawing of $\mathcal{G}$ requires three colors for the edges? In other words, are there strongly fan-planar graphs where odd cycles in the intersection graph of its drawing are unavoidable?

References

1. Ackerman, E.: On topological graphs with at most four crossings per edge. Comput. Geom. **85** (2019)
2. Alekseev, V.B., Gončakov, V.S.: The thickness of an arbitrary complete graph. Math. USSR-Sbornik **30**(2), 187–202 (1976). https://doi.org/10.1070/sm1976v030n02abeh002267
3. Angelini, P., Bekos, M.A., Kaufmann, M., Pfister, M., Ueckerdt, T.: Beyond-Planarity: Turán-type results for non-planar bipartite graphs. In: 29th International Symposium on Algorithms and Computation (ISAAC 2018), vol. 123, pp. 28:1–28:13. LIPIcs (2018). https://doi.org/10.4230/LIPIcs.ISAAC.2018.28
4. Beineke, L.W., Harary, F., Moon, J.W.: On the thickness of the complete bipartite graph. Math. Proc. Cambridge Philos. Soc. **60**(1), 01–05 (1964). https://doi.org/10.1017/S0305004100037385
5. Bekos, M.A., Giacomo, E.D., Didimo, W., Liotta, G., Montecchiani, F., Raftopoulou, C.N.: Edge partitions of optimal 2-plane and 3-plane graphs. Discret. Math. **342**(4), 1038–1047 (2019). https://doi.org/10.1016/j.disc.2018.12.002
6. Bekos, M.A., Grilli, L.: Fan-planar graphs. In: Hong and Tokuyama [10], chap. 8, pp. 131–148. https://doi.org/10.1007/978-981-15-6533-5_8

7. Bekos, M.A., Kaufmann, M., Raftopoulou, C.N.: On optimal 2- and 3-planar graphs. In: 33rd International Symposium on Computational Geometry (SoCG 2017), vol. 77, pp. 16:1–16:16. LIPIcs (2017). https://doi.org/10.4230/LIPIcs.SoCG.2017.16

8. Cheong, O., Pfister, M., Schlipf, L.: The thickness of fan-planar graphs is at most three. CoRR abs/2208.12324 (2022). https://arxiv.org/abs/2208.12324

9. Didimo, W., Liotta, G., Montecchiani, F.: A survey on graph drawing beyond planarity. ACM Comput. Surv. **52**(1) (2019). https://doi.org/10.1145/3301281

10. Hong, S.-H., Tokuyama, T. (eds.): Beyond Planar Graphs. Springer, Singapore (2020). https://doi.org/10.1007/978-981-15-6533-5

11. Kainen, P.: Thickness and coarseness of graphs. Abhandlungen aus dem Mathematischen Seminar der Universität Hamburg 39, 88–95 (sept 1973). https://doi.org/10.1007/BF02992822

12. Kaufmann, M., Ueckerdt, T.: The density of fan-planar graphs. CoRR abs/1403.6184v1 (2014). http://arxiv.org/abs/1403.6184v1

13. Kaufmann, M., Ueckerdt, T.: The density of fan-planar graphs. Electron. J. Comb. **29**(1) (2022). https://doi.org/10.37236/10521

14. Klemz, B., Knorr, K., Reddy, M.M., Schröder, F.: Simplifying non-simple fan-planar drawings. In: Purchase, H.C., Rutter, I. (eds.) GD 2021. LNCS, vol. 12868, pp. 57–71. Springer, Cham (2021). https://doi.org/10.1007/978-3-030-92931-2_4

15. Mansfield, A.: Determining the thickness of graphs is NP-hard. Math. Proc. Cambridge Philos. Soc. **93**(1), 9–23 (1983). https://doi.org/10.1017/S030500410006028X

16. Mutzel, P., Odenthal, T., Scharbrodt, M.: The thickness of graphs: a survey. Graphs Comb. **14** (2000). https://doi.org/10.1007/PL00007219

17. Nash-Williams, C.S.A.: Edge-disjoint spanning trees of finite graphs. J. Lond. Math. Soc. **s1-36**(1), 445–450 (1961). https://doi.org/10.1112/jlms/s1-36.1.445

18. Pawlik, A.: Triangle-free intersection graphs of line segments with large chromatic number. J. Comb. Theory Ser. B **105**, 6–10 (2014)

19. Tutte, W.T.: The thickness of a graph. In: Indagationes Mathematicae (Proceedings), vol. 66, pp. 567–577. Elsevier (1963)

An FPT Algorithm for Bipartite Vertex Splitting

Reyan Ahmed[1]([✉]), Stephen Kobourov[2][ID], and Myroslav Kryven[2]

[1] Colgate University, Hamilton, USA
aboureyanahmed@arizona.edu
[2] University of Arizona, Tucson, USA
{kobourov,myroslav}@arizona.edu

Abstract. Bipartite graphs model the relationship between two disjoint sets of objects. They have a wide range of applications and are often visualized as 2-layered drawings, where each set of objects is visualized as vertices (points) on one of two parallel horizontal lines and the relationships are represented by (usually straight-line) edges between the corresponding vertices. One of the common objectives in such drawings is to minimize the number of crossings. This, in general, is an NP-hard problem and may still result in drawings with so many crossings that they affect the readability of the drawing. We consider a recent approach to remove crossings in such visualizations by splitting vertices, where the goal is to find the minimum number of vertices to be split to obtain a planar drawing. We show that determining whether a planar 2-layered drawing exists after splitting at most k vertices is fixed parameter tractable in k.

Keywords: Fixed parameter tractability · Graph drawing · Vertex splitting

1 Introduction

Bipartite graphs are used in many applications to study complex systems and their dynamics [20]. We can visualize a bipartite graph $G = (T \cup B, E)$ as a 2-layered drawing where vertices in T are placed (at integer coordinates) along the horizontal line defined by $y = 1$ and vertices in B along the line below (at integer coordinates) defined by $y = 0$.

A common optimization goal in graph drawing is to minimize the number of crossings. Deciding whether a planar 2-layered drawing exists for a given graph can be done in linear time, although most graphs, including sparse ones such as cycles and binary trees, do not admit planar 2-layered drawings [6]. The problem of minimizing the number of crossings in 2-layered layouts is NP-hard, even if the maximum degree of the graph is at most four [16], or if the permutation of vertices is fixed on one of the layers [6]. The latter variant of the problem is known as One-Sided Crossing Minimization (OSCM). The minimum number of crossings in a 2-layered drawing can be approximated within a factor of 1.47 and $1.3 + 12/(\delta - 4)$, where δ is the minimum degree, given that

P. Angelini and R. von Hanxleden (Eds.): GD 2022, LNCS 13764, pp. 261–268, 2023.
https://doi.org/10.1007/978-3-031-22203-0_19

$\delta > 4$ [17]. Dujmović et al. [5] gave a fixed-parameter tractable (FPT) algorithm with runtime $O(1.62^k \cdot n^2)$, which was later improved to $O(1.4656^k + kn^2)$ [4]. Fernau et al. [10] reduced this problem to weighted FAST (feedback arc sets in tournaments) obtaining a subexponential time algorithm that runs in time $2^{O(\sqrt{k}\log k)} + n^{O(1)}$. Finally Kobayashi and Tamaki [14] gave a straightforward dynamic programming algorithm on an interval graph associated with each OSCM instance with runtime $2^{O(\sqrt{k}\log k)} + n^{O(1)}$. They also showed that the exponent $O(\sqrt{k})$ in their bound is asymptotically optimal under the Exponential Time Hypothesis (ETH) [11], a well-known complexity assumption which states that, for each $k \geq 3$, there is a positive constant c_k such that k-SAT cannot be solved in $O(2^{c_k n})$ time where n is the number of variables.

Minimizing the number of crossings in 2-layer drawings may still result in visually complex drawings from a practical point of view [12]. Hence, we study vertex splitting [7,8,13,15] which aims to construct planar drawings, and thus, avoid crossings altogether. In the *split* operation for a vertex u we delete u from G, add two new copies u_1 and u_2, and distribute the edges originally incident to u between the two new vertices u_1 and u_2. There are two main variations of the objective in vertex splitting: minimizing the number of split operations (or *splits*) and minimizing the number of *split vertices* (each vertex can be split arbitrary many times) to obtain a planar drawing of G. Minimizing the number of splits is NP-hard even for cubic graphs [9]. Nickel et al. [18] extend the investigation of the problem and its complexity from abstract graphs to drawings of graphs where splits are performed on an underlying drawing.

Vertex splitting in bipartite graphs with 2-layered drawings has not received much attention [2]. In several applications, such as visualizing graphs defined on anatomical structures and cell types in the human body [19], the two vertex sets of G play different roles and vertex splitting is allowed only on one side of the layout. This has motivated the interest in splitting the vertices in only one vertex partition of the bipartite graph. It has been shown that minimizing splits in this setting is NP-hard for an arbitrary bipartite graph [3].

The other variant – minimizing the number of split vertices – has been recently considered and was shown to be NP-hard [1]. On the positive side, we show that the problem is FPT parameterized by the natural parameter, that is, the number of split vertices.

Problem (Crossing Removal with k Split Vertices – CRSV(k)). *Let $G = (T \cup B, E)$ be a bipartite graph. Decide whether there is a planar 2-layer drawing of G after splitting at most k vertices of B.*

In the next section we prove the following theorem.

Theorem 1. *Given a bipartite graph $G = (T \cup B, E)$, the CRSV(k) problem can be decided in time $2^{O(k^6)} \cdot m$, where m is the number of edges of G.*

We prove Theorem 1 using *kernelization*, one of the standard techniques for designing FPT algorithms. The goal of kernelization is to reduce the input instance to its computationally hard part on which a slower exact algorithm can be applied. If the size of the reduced instance is bounded by a function of the parameter, the problem can be solved by brute force on the reduced instance yielding FPT runtime. Our reduction consists of two parts.

In the first part we identify and remove vertices that necessarily belong to the solution (Step 1. below) and remove redundant vertices of the input graph G; Step 2. below. Then we show that there is a solution for the reduced graph G_1'' if and only if there is a solution for the original graph G; see Claim 1. Then we prove two structural properties about the degrees of the vertices of G_1''; see Lemmas 1 and 2. These two properties allow us to bound the size of the "essential" part (called the *core*) of the reduced graph G_1''; see Lemma 3.

In the second part of the reduction we remove more redundant vertices of G_1'' and identify and remove the vertices that necessarily belong to the solution. Then we show that the resulting reduced graph G_2' has size bounded by a polynomial function of the parameter; see Lemma 4. Finally, we show that there is a solution for G_2' if and only if there is a solution for G_1''; see Claim 2. The proof is concluded by applying an exact algorithm to the graph G_2'.

2   Proof of Theorem 1

Let $G = (T \cup B, E)$ be a bipartite graph and k be the number of vertices that we are allowed to split.

First Reduction Rule: Before we describe our first reduction rule, we make a useful observation.

Observation 1. *If a vertex $v \in B$ has at least three neighbours of degree at least two, it must be split in any planar 2-layered drawing of G; see Fig. 1a.*

Let B_{tr} be the set of such vertices of degree 3 or more in B (as described in Observation 1). The first reduction rule consists of two steps described below.

1. We initialize our solution set S with the vertices in B_{tr}, that is, $S := B_{\mathrm{tr}}$ and remove them from the graph G. Let the resulting graph be $G_1' = (T_1' \cup B_1', E_1')$ and $k_1' = k - |B_{\mathrm{tr}}|$; note that $T_1' = T_1$.
2. Let $T_s \subset T_1'$ be the set of vertices v such that $\deg(v) = 1$ and $\deg(u) \geq 3$, where u is the unique neighbor of v in G_1'. Similarly, let $B_s \subset B_1'$ be the set of vertices v such that $\deg(v) = 1$ and $\deg(u) \geq 3$, where u is the unique neighbor of v in G_1'. We remove the vertices T_s and B_s from the graph G_1'. Let the resulting graph be $G_1'' = (T_1'' \cup B_1'', E_1'')$.

Let us now show the following.

Claim 1. *The graph G is a* YES *instance for* CRSV (k) *if and only if G_1'' is a* YES *instance for* CRSV (k_1').

Proof. We first argue the "only if" direction: consider a planar 2-layered drawing of G with at most k vertices split. According to Observation 1, each vertex in B_{tr} is split, moreover, none of the vertices in B_s are split because each of them

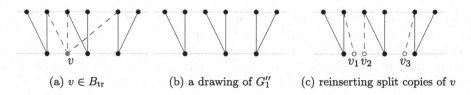

(a) $v \in B_{\mathrm{tr}}$ (b) a drawing of G_1'' (c) reinserting split copies of v

Fig. 1. Reinserting split copies $v_1, v_2, \ldots, v_{\deg(v)}$ of $v \in B_{\mathrm{tr}}$ into a planar 2-layered drawing of G_1'' to get a planar 2-layered drawing of G.

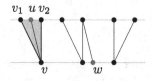

Fig. 2. Reinserting $w \in B_s$ and $u \in T_s$ into a 2-layered drawing of G_1'' to obtain a 2-layered drawing of G. A safe wedge $v_1 v v_2$ is filled green. (Color figure online)

has degree one. Therefore, there are at most $k - |B_{\mathrm{tr}}|$ vertices in $B \setminus (B_{\mathrm{tr}} \bigcup B_s)$ that are split. Because $B_1'' = B \setminus (B_{\mathrm{tr}} \bigcup B_s)$ and $k_1' = k - |B_{\mathrm{tr}}|$ there exists a planar 2-layered drawing of G_1'' with at most k_1' vertices split.

For the "if" direction, consider a planar 2-layered drawing of G_1'' with at most k_1' vertices split. Note that after applying Step 1. and Step 2. the vertices in B_1'' have degree at most two. Thus for each vertex $v \in B_{\mathrm{tr}}$ we can reinsert its split copies $v_1, v_2, \ldots, v_{\deg(v)}$ (each reinserted vertex has degree one) without crossings; see Fig. 1. For the same reason we can reinsert the vertices in B_s of degree one removed at Step 2.; see Fig. 2. To see that we can reinsert each vertex $u \in T_s$ of degree one removed at Step 2. observe that we always connect it to a vertex $v \in B_1''$ of degree at least two, therefore, in any planar 2-layered drawing of G_1'' there is always a *safe wedge* formed by two edges vv_1 and vv_2 where we can fit in the edge vu without causing any crossings; see Fig. 2.   □

Now we state two observations about the degrees of the vertices of the graph G_1''.

Lemma 1. *For each vertex $v \in T_1''$ it holds that $\deg(v) \leq k_1' + 2$ if there exists a planar 2-layered drawing of G_1'' with at most k_1' split vertices.*

Proof. Consider for contradiction that there is a vertex $v \in T_1''$ that has $\deg(v) = k_1' + 3$; see Fig. 3. According to Step 2. v does not have any neighbors of degree one in B_1'', therefore, to obtain a planar 2-layered drawing of G_1'' all but two neighbors of v must be split, that is, $k_1' + 1$ vertices must be split; contradiction.   □

To make our second observation let T_1'', $_{\deg(v) \geq 3}$ be the set of all the vertices of degree at least three in T_1''.

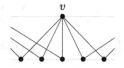

Fig. 3. All but two neighbors of v must be split to obtain a planar 2-layered drawing of G_1'' because each neighbour of v has degree at least two.

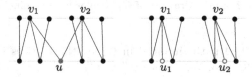

Fig. 4. For each vertex in $v \in T_{1, \deg(v) \geq 3}''$ at least one of its neighbors $u \in B_1''$ must be split to obtain a planar 2-layered drawing of G_1'' because each of the neighbours of v has degree at least two. Splitting u can resolve crossings for at most two vertices $v_1, v_2 \in T_{1, \deg(v) \geq 3}''$ because it has degree at most two.

Lemma 2. *It holds that* $\left| T_{1, \deg(v) \geq 3}'' \right| \leq 2k_1'$ *if there exists a planar 2-layered drawing of G_1'' with at most k_1' split vertices.*

Proof. Observe that according to Step 2. no vertex in $T_{1, \deg(v) \geq 3}''$ has any neighbors of degree one in B_1''. This implies that for each vertex v in $T_{1, \deg(v) \geq 3}''$ at least one of its neighbors $u \in B_1''$ must be split to obtain a planar 2-layered drawing of G_1''; see Fig. 4. But the degree of u is at most two, therefore, splitting u can resolve crossings for at most two vertices $v_1, v_2 \in T_{1, \deg(v) \geq 3}''$. Thus, if $|T_{1, \deg(v) \geq 3}''| > 2k_1'$, more than k_1' vertices in B_1'' must be split to obtain a planar 2-layered drawing of G_1''; contradiction. $\qquad\square$

For a subset of vertices W let $N(W)$ denote the set of neighbors of W. From Lemma 1 and 2 we obtain the following.

Lemma 3. *The graph induced by the vertices $T_{1, \deg(v) \geq 3}'' \bigcup N(T_{1, \deg(v) \geq 3}'')$ has at most $2k_1'(k_1' + 2)$ vertices if there exists a planar 2-layered drawing of G_1'' with at most k_1' split vertices.*

Let $C = T_{1, \deg(v) \geq 3}'' \bigcup N(T_{1, \deg(v) \geq 3}'')$ and call the graph induced by the vertices in C the *core* of G_1''. Now we can proceed to the second reduction rule.

Second Reduction Rule: Observe that all the vertices in $(B_1'' \bigcup T_1'') \setminus C$ have degree at most two, and therefore, induce paths or cycles in G_1''. Since the cycles are not connected to the core in G_1'' (because their vertices have degree at most two in G_1'') we can remove them and handle separately. We need to account for one split vertex per each such cycle. Let $\mathcal{E}$ be the set of these cycles and let $k_2' = k_1' - |\mathcal{E}|$. In addition, let Z be the set of vertices that we split in these cycles, $\mathcal{S} := \mathcal{S} \bigcup Z$.

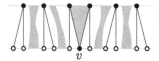   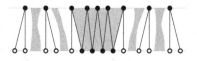

(a) p (black) in G_2', safe wedge (green)   (b) reinserting the missing part of p

Fig. 5. Reinserting the missing part of the path $p \in \mathcal{P}$ into a planar 2-layered drawing of G_2' to get a planar 2-layered drawing of G_1''. (Color figure online)

Let $\mathcal{P}$ be the set of paths induced in G_1'' by the vertices in $(B_1'' \bigcup T_1'') \setminus C$ of length at least $2k_2' + 5$. We reduce G_1'' to $G_2' = (T_2' \cup B_2', E_2')$ by *shortening* each path $p \in \mathcal{P}$ (that is, iteratively removing one of the middle vertices of p from T_1'' and identifying its two neighbours in B_1'') until p has at most $2k_2' + 5$ vertices. Because during shortening step the length of p decreases by two, after the shortening process p will still have at least $2k_2' + 3$ vertices.

Claim 2. *The graph G_1'' is a* YES *instance for* CRSV(k_1') *if and only if G_2' is a* YES *instance for* CRSV(k_2').

Proof. In one direction the claim is obvious, because shortening paths in a planar 2-layered drawing of the graph G_1'' does not cause any crossings.

For the other direction, consider a planar 2-layered drawing of the graph G_2'. To obtain from it a planar 2-layered drawing of the graph G_1'' we need to: (1) reinsert each of the cycles in $\mathcal{E}$ that we have removed from G_1'' to obtain G_2', and (2) reinsert back the missing parts of the paths of $\mathcal{P}$, which are made up of the vertices from $(B_1'' \bigcup T_1'') \setminus C$. Because the cycles in $\mathcal{E}$ are disconnected from G_1'' we can reinsert them anywhere in the drawing wherever there is space with one split vertex in Z.

Let us now argue why we can reinsert the missing vertices from $(B_1'' \bigcup T_1'') \setminus C$ into the paths in $\mathcal{P}$; we will refer to Fig. 5 for illustration. Because for any such path $p \in \mathcal{P}$ the length of p is at least $2k_2' + 3$ there must be at least one vertex v in B_2' that was not split in a planar 2-layered drawing of G_2' (see Fig. 5a), as otherwise a planar 2-layered drawing of G_2' cannot be constructed with at most k_2' splits. Therefore, there must be a safe wedge formed by the unsplit vertex v and the two edges of the path p incident to v providing space to reinsert the missing vertices without causing any crossings; see Fig. 5b.   □

Lemma 4. *The graph G_2' has at most $O(k^6)$ vertices.*

Proof. According to Lemma 3 the core C has at most $2k_1'(k_1' + 2)$ vertices and according to Lemma 1 the highest degree of each vertex in C is at most $k_1' + 2$. Therefore, there can be at most $\binom{2k_1'(k_1'+2)}{2}(k_1' + 2)$ many paths induced by the vertices in $(B_2' \bigcup T_2') \setminus C$. Moreover, after applying the second reduction rule each such path has at most $2k_2' + 5$ vertices. Thus the total number of vertices in G_2' is at most $\binom{2k_1'(k_1'+2)}{2}(k_1' + 2)(2k_2' + 5) \in O(k^6)$.   □

Finally we decide $\text{CRSV}(k_2')$ for G_2' by brute force. More precisely, we check all subsets X of B_2' such that $|X| \leq k_2' \leq k$. For each vertex v in X we check all ways to partition its incident edges (at most $k + 2$) into non-empty subsets, this represents splitting of v. The number of such partitions is bounded by the Bell number of order $k + 2$, which in turn is bounded by $(k + 2)!$. Then we run a linear time algorithm to check whether a planar 2-layered drawing of the resulting graph exists. This can be done in time $2^{O(k^6)}(k!)^{O(k)} \cdot m \subset 2^{O(k^6)} \cdot m$, where m is the number of edges of G. If G_2' is a YES instance for $\text{CRSV}(k_2')$ with the subset of split vertices X, we update our solution set $\mathcal{S} := \mathcal{S} \bigcup X$ and return it. It is worth noting that the kernelization itself can be done in time $O(m)$ since we process each vertex in constant time given that we know its degree. Thus, the kernelization does not affect the total asymptotic runtime of the algorithm.

3   Conclusion and Open Problems

We presented an FPT algorithm for the $\text{CRSV}(k)$ problem parameterized by k. Improving the runtime is needed for this algorithm to be useful in practice, as the constants are very large. Another natural direction is to look for an FPT algorithm for the other variant of the problem, that is, minimizing the number of splits, which was recently shown to be NP-hard [1].

Problem. Crossing Removal with k Splits – CRS(k)). *Let $G = (T \cup B, E)$ be a bipartite graph. Decide whether there is a planar 2-layer drawing of G after applying at most k splits to the vertices in B.*

Is there an FPT algorithm for the $\text{CRS}(k)$ problem parameterized by k? It is not clear how to adjust the algorithm in Theorem 1 as it splits every vertex in B_{tr} as many times as its degree, and thus, the number of splits is not bounded by a function of the parameter k.

References

1. Ahmed, R., et al.: Splitting Vertices in 2-Layer Graph Drawings (2022, unpublished manuscript)
2. Börner, K., Kobourov, S.: Multi-level graph representation for big data arising in science mapping (Dagstuhl Seminar 21152). Dagstuhl Rep. **11**(3), 1–15 (2021). https://doi.org/10.4230/DagRep.11.3.1, https://drops.dagstuhl.de/opus/volltexte/2021/14688
3. Chaudhary, A., Chen, D.Z., Hu, X.S., Niemier, M.T., Ravichandran, R., Whitton, K.: Fabricatable interconnect and molecular QCA circuits. IEEE Trans. Comput. Aided Des. Integr. Circuits Syst. **26**(11), 1978–1991 (2007)
4. Dujmović, V., Fernau, H., Kaufmann, M.: Fixed parameter algorithms for one-sided crossing minimization revisited. J. Discrete Algorithms **6**(2), 313–323 (2008)
5. Dujmović, V., Whitesides, S.: An efficient fixed parameter tractable algorithm for 1-sided crossing minimization. Algorithmica **40**(1), 15–31 (2004)
6. Eades, P., McKay, B.D., Wormald, N.C.: On an edge crossing problem. In: Proceedings of 9th Australian Computer Science Conference, vol. 327, p. 334 (1986)

7. Eades, P., de Mendonça N, C.F.X.: Vertex-splitting and tension-free layout. In: Brandenburg, F.J. (ed.) GD 1995. LNCS, vol. 1027, pp. 202–211. Springer, Heidelberg (1996). https://doi.org/10.1007/BFb0021804
8. Eppstein, D., et al.: On the planar split thickness of graphs. Algorithmica **80**, 977–994 (2018)
9. Faria, L., de Figueiredo, C.M.H., Mendonça, C.F.X.: Splitting number is NP-complete. DAM **108**(1), 65–83 (2001)
10. Fernau, H., Fomin, F.V., Lokshtanov, D., Mnich, M., Philip, G., Saurabh, S.: Ranking and drawing in subexponential time. In: Iliopoulos, C.S., Smyth, W.F. (eds.) IWOCA 2010. LNCS, vol. 6460, pp. 337–348. Springer, Heidelberg (2011). https://doi.org/10.1007/978-3-642-19222-7_34
11. Impagliazzo, R., Paturi, R.: On the complexity of k-sat. J. Comput. Syst. Sci. **62**(2), 367–375 (2001)
12. Jünger, M., Mutzel, P.: 2-layer straightline crossing minimization: performance of exact and heuristic algorithms. JGAA **1**(1), 1–25 (1997)
13. Knauer, K., Ueckerdt, T.: Three ways to cover a graph. DM **339**(2), 745–758 (2016)
14. Kobayashi, Y., Tamaki, H.: A fast and simple subexponential fixed parameter algorithm for one-sided crossing minimization. In: Epstein, L., Ferragina, P. (eds.) ESA 2012. LNCS, vol. 7501, pp. 683–694. Springer, Heidelberg (2012). https://doi.org/10.1007/978-3-642-33090-2_59
15. Liebers, A.: Planarizing graphs - a survey and annotated bibliography. JGAA **5**(1), 1–74 (2001)
16. Muñoz, X., Unger, W., Vrt'o, I.: One sided crossing minimization is NP-hard for sparse graphs. In: Mutzel, P., Jünger, M., Leipert, S. (eds.) GD 2001. LNCS, vol. 2265, pp. 115–123. Springer, Heidelberg (2002). https://doi.org/10.1007/3-540-45848-4_10
17. Nagamochi, H.: An improved approximation to the one-sided bilayer drawing. In: Liotta, G. (ed.) GD 2003. LNCS, vol. 2912, pp. 406–418. Springer, Heidelberg (2004). https://doi.org/10.1007/978-3-540-24595-7_38
18. Nickel, S., Nöllenburg, M., Sorge, M., Villedieu, A., Wu, H.Y., Wulms, J.: Planarizing graphs and their drawings by vertex splitting (2022). https://doi.org/10.48550/ARXIV.2202.12293, https://arxiv.org/abs/2202.12293
19. Paul, H., Börner, K., Herr II, B.W., Quardokus, E.M.: ASCT+B REPORTER (2022). https://hubmapconsortium.github.io/ccf-asct-reporter/. Accessed 06 June 2022
20. Pavlopoulos, G.A., Kontou, P.I., Pavlopoulou, A., Bouyioukos, C., Markou, E., Bagos, P.G.: Bipartite graphs in systems biology and medicine: a survey of methods and applications. GigaScience **7**(4), giy014 (2018)

Dynamic Graph Visualization

On Time and Space: An Experimental Study on Graph Structural and Temporal Encodings

Velitchko Filipov$^{(\boxtimes)}$ ⓘ, Alessio Arleo ⓘ, Markus Bögl ⓘ, and Silvia Miksch ⓘ

TU Wien, Vienna, Austria
{velitchko.filipov,alessio.arleo,markus.bogl,silvia.miksch}@tuwien.ac.at

Abstract. Dynamic networks reflect temporal changes occurring to the graph's structure and are used to model a wide variety of problems in many application fields. We investigate the design space of dynamic graph visualization along two major dimensions: the network structural and temporal representation. Significant research has been conducted evaluating the benefits and drawbacks of different structural representations for static graphs, however, few extend this comparison to a dynamic network setting. We conduct a study where we assess the participants' response times, accuracy, and preferences for different combinations of the graph's structural and temporal representations on typical dynamic network exploration tasks, with and without support of common interaction methods. Our results suggest that matrices provide better support for tasks on lower-level entities and basic interactions require longer response times while increasing accuracy. Node-link with auto animation proved to be the quickest and most accurate combination overall, while animation with playback control the most preferred temporal encoding.

Keywords: User study · Evaluation · Time-oriented data · Graphs and networks

1 Introduction

The increased availability of time-dependent datasets contributed to the rise of research interest toward dynamic network visualization, nowadays considered a mature and thriving research field [12]. Kerracher et al. [41] define a two-dimensional design space for dynamic network visualization: *structural representation* (how the graph's topology is represented) and *temporal encoding* (how time and, consequently, the graph temporal dynamics are illustrated). This two-dimensional design space is expressive enough to characterize the majority of existing dynamic network visualization approaches.

There is extensive literature on studies designed to evaluate different graph representations for typical exploration tasks on static networks. Similar studies have been conducted for dynamic approaches, however mostly focused on node-link diagrams coupled with different temporal encodings (see Sect. 2). This also comes as a consequence of the limited number of dynamic network visualization

P. Angelini and R. von Hanxleden (Eds.): GD 2022, LNCS 13764, pp. 271–288, 2023.
https://doi.org/10.1007/978-3-031-22203-0_20

approaches that have matrices as their base graph representation [12] (see, e.g., [39,53]). Therefore, empirical evidence about the performance and preference for different dynamic network visualization approaches in our design space is still scattered between different studies, experimental settings, procedures, and tasks. Similarly, existing user studies in this context incorporate simple interaction methods to support the network exploration (see, e.g., [6,46]), however, their effect on the participants' experience has not been fully investigated, thus motivating the need for broader and rigorous experimentation.

Our Contribution. In this paper we design, conduct, and discuss the results of an experimental study aimed at assessing and comparing different dynamic visualization approaches centered around combinations of graph representation (node-link and adjacency matrix), temporal encoding (juxtaposition, superimposition, animation with playback control, and auto animation), and interaction support for offline dynamic graph visualization. We conduct a statistical analysis of the study results and condense our findings in a concise discussion meant to support further studies and design of dynamic network visualization techniques.

2 Related Work

We outline recent related studies conducted along the two dimensions of the design space introduced by Kerracher et al. [41], focusing on user studies.

Structural Representations. In graph drawing literature, several studies assess the readability, task performance, and effects of aesthetic criteria on human cognition of different graph structural encodings (e.g., [14,25,33,45,46, 48,49,51]). Ghoniem et al. [33] evaluate, in a controlled experiment, the readability of graphs when represented as node-link diagrams compared to adjacency matrices on generic graph tasks. Their findings suggest that the ability of either visualization to support typical exploration tasks depends on the size and density of the network; the authors concluded how matrix-based techniques were underexploited, despite their proved potential with larger and denser networks. Okoe et al. [45,46] conduct further comparative evaluations between node-link and matrix representations on a large scale ($\sim$ 800 participants). Their results show that node-link diagrams better support memorability and connectivity tasks. Matrices have quicker and more accurate results for tasks that involve finding common neighbors and group tasks (i.e., involving clusters). Concurrently, Ren et al. [51] conduct a large scale study ($\sim$ 600 participants) comparing the readability of node-link diagrams against two different sorting variants of matrix representations on small to medium social networks ($\sim$ 50 nodes). Their findings do not differ significantly from the ones by Okoe et al. [46], suggesting that node-link provided a better implicit understanding of the network, with lower response times and higher accuracy than matrices. However, the gap between the two tended to reduce as the size of the graph increased.

Temporal Encodings. One of the most studied problems concerning dynamic network visualization, is the ability of participants to retain a "mental map" of the graph while investigating its evolution [4–7,50]. Archambault and Purchase investigate the effect of drawing stability on the node-link graph representation coupled with animation and small multiples [6,7]. Drawing stability proved to have a positive effect on task performance, with animation able to improve over timeline in low stability scenarios. Ghani et al. [32] investigate the perception of different visual graph metrics on animated node-link diagrams. Results suggest that animation speed and target separation have the most impact on performance for event sequencing tasks. Linhares et al. [42] compare four different approaches for visualization of dynamic networks, namely the `Massive Sequence View` [26] (timeline-based), the `Temporal Activity Map` [43], and animated node-link and matrix diagrams. While all techniques reached satisfactory results, the animated node-link was the favorite choice of the participants. Even though matrix-based approaches are included in this study, it does not exhaustively cover all the possible combinations of our design space. Filipov et al. [30] conduct an exploratory study comparing different combinations of structural and temporal representations. The results suggest that tasks with matrices were completed quicker and more accurately, the participants preferred matrices with superimposition, and juxtaposition was among the least preferred approaches. However, these results require further formal investigation. Overall, related literature shows that the perception of different temporal encodings has been mainly investigated on node-link diagrams, with few papers focusing on the other combinations of structural and temporal encodings. In this sense, our paper constitutes an effort in understanding whether the differences between node-link and matrix representations still hold in a dynamic scenario, what is the efficacy of the temporal representations, and how effective (and how important) is it to include interactions when designing such approaches.

3   Dynamic Graph Visualization Design Space

We refer to a dynamic graph Γ as a sequence of individual graphs each one representing its state at a specific point in time: $\Gamma = (G_1, G_2, ..., G_k)$; we denote the individual G_x as a dynamic graph *timeslice*. We now briefly describe the different structural and temporal encodings, along with the interactions included in the scope of our experiment, detailing their implementation.

3.1   Network Structural Encoding

The structural dimension focuses on the challenges of laying out a graph to visually present the relationships between elements in an understandable, accurate, and usable manner [41]. **Node-Link (NL)** diagrams present the relational structure of the graph using lines to connect the entities that are depicted using circles, whose coordinates on the plane are computed using specialized algorithms. In our study, we compute the NL layouts using the force-directed implementation of

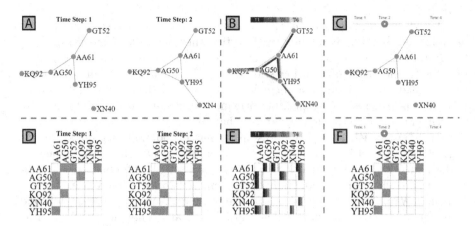

Fig. 1. Network structural and temporal encodings: Juxtaposition (A,D), Superimposition(B,E), and Animation with Playback Controls (C,F)

d3js [19] considering all the timeslices simultaneously instead of on a per-frame basis. This process of *aggregation* [22] is simple to implement and provides a stable layout throughout the sequence of timeslices, at the expense of the quality of individual layouts. We refer to the following for a broader discussion on dynamic network layout algorithms [8,11,12,21,23,24,27,28,55]. **Adjacency Matrices (M)** visualize the network as an $n \times n$ table. A non-zero value in the cell indicates the presence of an edge between the nodes identified by the corresponding row and column. In our study we order the rows and columns alphabetically according to the node's label. More advanced reordering methods exist [13], however, matrix reordering is still under-investigated in a dynamic context and we decided to exclude this aspect from the study design.

3.2   Network Temporal Encoding

In dynamic networks the temporal dimension plays an important role in the analysis process and requires special attention to enable effective exploration and better understand the behavior of the network [44]. **Superimposition (SI)** encodes the temporal dimension of the network in the same screen space by overlaying the timeslices (see, e.g., [21,28]) or making use of explicit encoding (see., e.g., [34,40]). In our study we represent the temporal information in SI using colorblind-friendly color palettes [31]. In NL, we generate multiple parallel edges between the nodes, one for each timeslice where the edge is present, and color-code them individually. In M we subdivide each cell uniformly into rectangles, each representing the existence of that edge during that timeslice, which are colored similarly (see Fig. 1 B-E). **Juxtaposition (JP)** represents the graph's temporal dynamics as distinct layouts, each with dedicated screen space, similar to the small multiples approach by Tufte [56] (see Fig. 1A,D). In our study we generate one diagram per timeslice and arrange them adjacent to each other. **Animation with Playback Control (ANC)** uses a time slider to control the state of the animation and move to any of the available timeslices in no particular order (see Fig. 1C,F). This enables for a more fine and controlled exploration and analysis compared to animation, where speed and time progression is typically fixed. **Auto Animation (AN)** depicts the change of the graph over time as smooth transitions between subsequent timeslices. Differently from ANC, with AN it is not possible to skip forward or navigate backward in time and it automatically goes over each of the timeslices in a sequence.

3.3 Interactions

The interactions we implement are meant to support the network exploration. The following apply regardless of the temporal encoding: (i) zooming and panning (both for M and NL); (ii) hovering over a M cell highlights its corresponding row and column; (iii) in NL, nodes can be moved by dragging in order to declutter some denser areas of the drawing. Moreover, for AN only and regardless of the structural representation, the time between consecutive timeslices can be increased (7 sec maximum) or decreased (1 sec minimum). This selection should not favor any specific combination of structural and temporal encoding techniques over the others. Zooming, panning, and node rearrangement are commonly available in graph exploration software, like *Gephi* [10]. M mouse-over was also used by Okoe et al. [46]. AN speed could also be manipulated in the study by Archambault and Purchase [5].

4 Study Design

In this section we present the study design, including our tasks, research hypotheses, stimuli, and study procedure.

4.1 Tasks and Research Hypotheses

Tasks. The tasks used in our experiment are available in Table 1.

Table 1. The test questions (trials), per task (rows) and entity type (columns).

T	Low-level	High-level
T1	At which time step is the relationship between {source} and {target} introduced for the first time?	At which time step does the clique between {nodes} appear for the first time?
T2	Sum up the changes (additions and removals) of {node}s degree across all time steps.	Calculate the change of the cliques size between {nodes} across all time steps.
T3	At which time step does the node {node} have its highest degree?	Consider the set of nodes {nodes}. Find the size of the largest maximal clique across all the time steps between the given nodes.

Rationale. We picked one task for each category of temporal feature in the taxonomy proposed by Ahn et al. [2], namely, *Individual Temporal Features* (**T1**), *Rate of changes* (**T2**), and *Shape of changes* (**T3**). We selected the most common tasks referenced in the taxonomy and included in our experiment these tasks for both low- (nodes and links) and higher-level (cliques) entities.

Table 2. The research hypotheses that were evaluated in our experiments.

H	Research Hypothesis
H1	Matrices have lower response times and higher accuracy for all tasks compared to node-link diagrams, regardless of the temporal encoding.
H2	From all temporal encoding techniques, superimposition has the lowest response times and highest accuracy, regardless of the structural representation.
H3	Providing interaction techniques increases the response times but not the accuracy.
H4	Matrices have lower response times and higher accuracy for tasks on low-level entities and node-link diagrams have lower response times and higher accuracy for tasks on higher-level entities, regardless of the temporal encoding.
H5	The combination of matrices with superimposition results in the lowest response times and highest accuracy compared to other combinations of network structural and temporal encoding.

Hypotheses. We base our research hypotheses on the proposed tasks and we report them in Table 2.

Rationale. Hypotheses **H1**, **H2**, and **H5** are derived from the results of a previous exploratory study [30] (see also Sect. 2). While the focus of this experiment is centered around the *visual* encoding combinations within our design space, **H3** is intended to investigate the effects of common interactions techniques in this context. We argue that they might increase the response times over visual inspection alone, but without significant impact on accuracy. In **H4** we conjecture that following the evolution of a cluster or clique is more difficult with M compared to NL, as the participant must track several elements at once. We assume this would be easier to achieve with NL as the nodes are drawn closer together.

4.2   Experiment Setting

Stimuli. We generated 24 different scale-free random [15] graphs ($35 \leq |V| \leq 45, 46 \leq |E| \leq 71$) with the *NetworkX* python library [35, 36]. We chose this category of networks as they resemble real-world data examples of scientific interest (e.g., the world-wide-web, authors' co-citation networks [3]). We augmented each graph with 4 timeslices by randomly deleting edges from the original input graph to simulate temporal dynamics (at each subsequent timeslice the edges were added back and a new set was selected for removal). Finally, we split the datasets into two different types: 12 graphs with cliques and 12 without. Cliques were artificially introduced in the graphs by choosing 5 random nodes which were fully connected in one or more of the graph timeslices. The size of the graphs is comparable with the majority of empirical studies on graph visualization [51, 60].

Trials. Each of the tasks is applied to all combinations of structural and temporal encodings of interest in our study (see Sect. 3) resulting in 48 unique trials: $3(task\ types) \times 2(entity\ types) \times 2(network\ encodings) \times 4(temporal\ encodings)$. The order of the trials during the study is randomized in order to mitigate learning effects. The participants take part in the online experiment by completing the trials prepared using SurveyJS [1].

Study Design. Our experiment follows a between-subject arrangement: all participants complete the same entire set of 48 trials on the same graphs, but are exposed to one of two conditions, either *without* (Group A) or *with* (Group B) the support of the interactions discussed in Sect. 3.3. Participants are assigned to one of the two groups when they first access the online experiment, with a 75% probability of being assigned to Group B. As only one hypothesis (**H3**) deals with the group subdivision, we design the experiment to have a higher number of participants with interaction support. For each trial we ask the participant to provide a confidence score of their answer using a 5 point Likert scale (1 least confident - 5 most confident). At the end of the experiment, the participants express their thoughts in text about the encoding combinations they encountered and rank them on a 5 point Likert scale (1 least preferred - 5 most preferred).

Participants. For our study, we enrolled students part of a graduate course on information visualization design. To ensure that participants had a sufficient level of knowledge on the topic, we gave an introductory lecture about the visualizations and the experiment modalities. Participation was optional and its performance did not impact the final grade of the students. The online setting was necessary to guarantee a sufficient number of participants, while ensuring a safe social-distancing protocol. However, this also meant giving up control on the experiment environment.

5 Study Results

We received a total of 76 submissions from as many participants, of which we removed 8 that were trying to game the experiment. This resulted in a final set of 68 valid submissions that were used as the basis of our analysis. Further details can be found in the full version [29].

5.1 Analysis Approach

For each question of our study, we collected the participants' answers, their corresponding response times, and confidence values. We ignored the group subdivision (Group A and B) for hypotheses which did not focus on the presence of interactions (all except **H3**, see Sect. 4.1), as ANOVA tables do not show a statistically significant interaction effect between the independent variables for **H1**, **H2**, **H4**, **H5** (for more information we refer to [29]).

We conduct our analysis as follows, supported by Python libraries for statistical analysis [37,57,58]. We consider the structural and temporal encoding, the task type, entity type, and the groups (Group A and B) as *independent variables*, the response times and accuracy are taken as *dependent variables*. As the group subdivision is not even (25–75), we choose methods that are robust against these unbalanced designs [9,18,38,59]. For each of the hypotheses, we grouped the data according to the hypothesis and visually inspected response times and accuracy (number of correct answers ÷ total number of answers). To remove outliers from the data before the analysis we employed the inter-quantile range (IQR) [52]. We set the IQR lower ($q_1 - 1.5 \cdot$ IQR) and upper ($q_2 + 1.5 \cdot$ IQR) bounds at $q_1 = 0.25$ and $q_2 = 0.75$ as the outlier cut-off boundaries. This resulted in 116 trials (or 3.43%) being detected as outliers and omitted from the analysis.

Table 3. The results of the statistical test (p-values) for each hypothesis. We mark the cells with * if $p < 0.05$, ** if $p < 0.01$, *** if $p < 0.001$. If multiple comparisons are performed, [b] indicates the Bonferroni correction [17].

Hypothesis	Groups	MWU	T-Test	Binomial
H1	*(NL T1) vs (M T1)*	**0.0104*[b]**	**<0.001***[b]**	**0.0013*[b]**
	(NL T2) vs (M T2)	0.1579	**<0.001***[b]**	0.9313
	(NL T3) vs (M T3)	**<0.001***[b]**	**<0.001***[b]**	**0.0022**[b]**
H2	*(SI) vs (JP)*	**<0.001***[b]**	0.1065	0.166
	(SI) vs (ANC)	0.8662	0.1429	0.0883
	(SI) vs (AN)	0.2766	0.7751	**<0.001***[b]**
H3	*(Grp A) vs (Grp B)*	**<0.001*****	**<0.001*****	**<0.001*****
H4	*(M Low) vs (NL Low)*	**<0.001***[b]**	0.1392	**<0.001***[b]**
	(M High) vs (NL High)	**<0.001***[b]**	**<0.001***[b]**	0.4321
H5	*(M+SI) vs (M+JP)*	**0.0056****	0.2567	0.2424
	(M+SI) vs (M+ANC)	0.6301	0.2989	0.0261
	(M+SI) vs (M+AN)	0.2766	0.6328	0.0646
	(M+SI) vs (NL+SI)	**0.0038**[b]**	**<0.001***[b]**	0.449
	(M+SI) vs (NL+JP)	**<0.001***[b]**	**<0.001***[b]**	0.1389
	(M+SI) vs (NL+ANC)	0.0088	**<0.001***[b]**	**<0.001***[b]**
	(M+SI) vs (NL+AN)	0.0331	**<0.001***[b]**	**<0.001***[b]**

The task response times in our experiment are not normally distributed. To mitigate this, we perform a Box-Cox transformation [20]. Visual inspection of the quantile-quantile (Q-Q) plots confirmed a normal distribution of the transformed data. This allows us to run parametric tests, specifically, ANOVA (see [29] for further information about the ANOVA tables) and T-tests [9,18,38,59]. The standard ANOVA and T-tests are robust against such skewed distributions [16,47,54], therefore, we rely on them for our analysis as they both have more statistical power than non-parametric tests and detect significant effects if they truly exist. In presence of statistically significant difference (p-value < 0.05), we check, with T- and Mann-Whitney-U (MWU) tests, whether the significance held and visually explored the corresponding box plots to come to a conclusion. To evaluate our hypotheses on accuracy, we also perform Binomial tests to detect statistical significance between the distributions.

5.2 Quantitative Results

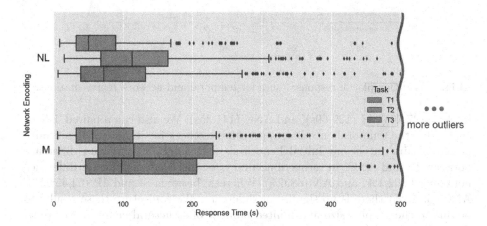

Fig. 2. H1: Box plot of response times for NL and M per task.

H1. We presume based on previous work [30] that M would perform better overall compared to NL for all tasks. Figure 2 depicts differences in response times between M and NL diagrams per task type. The results (see Table 3) indicate that NL is generally faster and more accurate than M. However, when looking at their differences per task we discover for **T1** that NL is significantly faster than M (NL: 73.49s, M: 97.93s), whereas M proves to be more accurate (NL: 74.9%, M: 80.7%). For **T2** the T-Test detects a significant difference in response times between NL and M (NL: 133.41s, M: 194.20s), however, in terms of accuracy they both perform similarly (NL: 52.5%, M: 52.7%). For **T3** NL representations significantly outperform M in terms of response times (NL: 107.32s, M: 175.92s) as well as accuracy (NL: 65.7%, M: 59.4%). Summarizing, the results suggest NL to generally have the lowest response times and higher accuracy compared to M for the proposed tasks. Thus, our results do not support H1.

H2. We assume SI to have the lowest response times and highest accuracy out of all the temporal encoding techniques. In our analysis, however, we do not detect any statistical significance in the comparisons shown in Table 3, with the only exception being JP, which has considerably lower response times than SI (see Fig. 3). Concerning response times, JP has the lowest (118.32s), followed

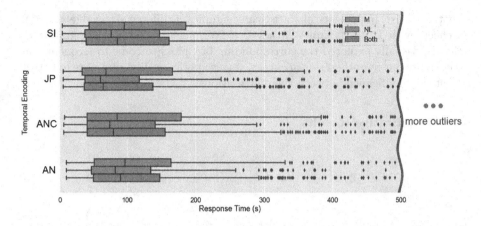

Fig. 3. H2: Box plot of response times for temporal and network representations.

by AN (127.76s), SI (129.69s), and ANC (141.35s). We also run a paired T-Test comparing the temporal encoding approaches to check for statistical significance between pairs out of our initial hypothesis and detect a significant difference between JP and ANC. In terms of accuracy, we discover a significant difference between SI (62.1%) and AN (68.6%). Whereas, between SI and JP (64.45%) or ANC (59.13%) there is no significant difference. We conjecture these results to be due to the graph's size and limited number of structural changes over time, that might favor AN as it is possible for participants to follow all changes the during animation. Our analysis shows no evidence to support H2.

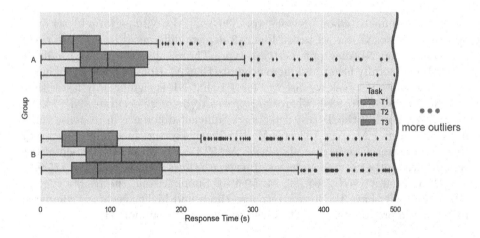

Fig. 4. H3: Box plot of response times for interaction groups per task.

H3. We conjecture that providing interactions influences the response times but not the accuracy. Our tests detect a significant difference (see Table 3) in the response times between group A (no interactions; 114.76s) and group B (interactions; 163.83s). As we initially assume, the group with interactions is much slower in completing tasks than the group with no interactions (see Fig. 4), however, the difference in accuracy is unexpected. The group with interactions is significantly more accurate than the one without (group A: 58%, group B: 65%). This suggests that interactions indeed increase response times, but at the same time provide the participants with a much better understanding of the visualized graphs and corresponding network dynamics regardless of the temporal encoding, therefore, leading to more accurate responses. The analysis shows that our results support H3 in terms of response times, but not accuracy.

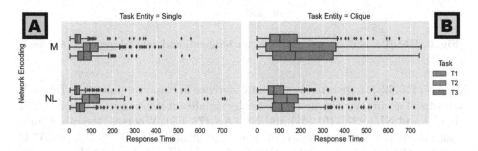

Fig. 5. H4: Box plot of response times for (A) single entities and (B) cliques.

H4. We formulate this hypothesis to evaluate whether the response times and accuracy of M and NL representations is affected by the type of target entity in a dynamic context (*low-level* - individual nodes and links; or *higher-level* - cliques), regardless of the temporal representation. For low-level entities, we do not detect any significant differences of the response times between network representations (see Table 3), both NL and M diagrams perform similarly with no clear winner. The results (see Fig. 5) for tasks on low-level entities indicate that M has lower response times (NL: 97.08s , M: 90.24s), whereas for higher-level entities NL has significantly lower response times (NL: 146.66s, M: 245.2s). However, in terms of accuracy M is significantly better than NL for lower-level entities (NL: 82.1%, M: 86.4%). For the higher-level entities, NL and M representations perform quite similarly in terms of accuracy (NL: 42.1%, M: 41.3%) Based on these findings, the results suggest that H4 is partially supported.

H5. Finally, we want to assess the response times and accuracy for all possible combinations of network structural and temporal encodings. Our assumption is that M representations with SI temporal encoding have the lowest response times and highest accuracy. We compare M+SI to all other combinations of network structural and temporal encodings (see Fig. 6). The results of the statistical tests yield significant differences in response times when comparing M+SI (154.53s) with M+JP (140.13s), NL+SI (105.25s), NL+JP (99.54s), NL+AN (108.8s), and NL+ANC (110.97s). Between M+SI (154.53s) and M+ANC (168.87s) and

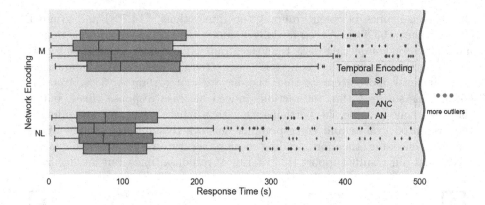

Fig. 6. H5: Box plot of response times for temporal and network representations.

M+AN (160.62s) there is no significant difference in response times (see Table 3). In terms of accuracy we detect statistically significant differences between M+SI (61.1%) and NL+ANC (51.8%) and NL+AN (71.4%). Whereas, the other combinations do not differ enough to warrant significance: M+JP (64%), M+ANC (66.4%), M+AN (65.5%), NL+JP (64.6%), and NL+SI (62.9%). From these results, the most balanced combination in terms of response times and accuracy is NL+AN followed by NL+JP. Therefore, we find no evidence supporting H5.

5.3 Qualitative Results

We collect the participants' ratings per combination of network structural and temporal encoding along with textual feedback pertaining to their preferences and experience during the experiment (see Fig. 7). There are no major differences in the preferences between the SI and JP encodings; ANC is the most preferred temporal encoding when coupled with a NL base representation. The NL representation is generally the most preferred approach, regardless of the temporal encoding. In terms of the participants confidence, we observe that most participants seemed to be fairly confident in their answers across all approaches (see Fig. 8). Most notably, the participants were most confident with NL+JP, followed by M+ANC, M+JP, and NL+ANC. There is general consensus that NL+SI was a very cluttered combination, whereas for M it performed a lot better and was easier to understand (*"SI was really confusing for some of the NL tasks but really useful for many of the M tasks"*). This is presumably due to the clutter generated by parallel edges crossings that occur in NL diagrams, which does not affect M. As in previous studies [30], the feedback on JP outlines that it requires participants to split their attention between multiple views in order to compare the temporal information. The ANC approach was preferred by the study participants for its flexibility due to the additional controls (i.e., time slider). AN was not considered to be a very good temporal encoding technique with the feedback being consistent across structural representations. Some

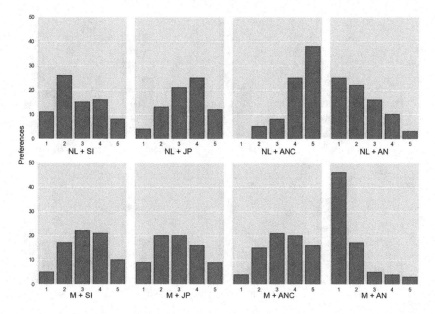

Fig. 7. Preferences per network and temporal encoding on a Likert scale (1–5).

participants commented that they needed to *"screenshot every timestamp to look at the different connections between the nodes"* and wait to watch the whole animation from the beginning. NL+AN, therefore, appears to be the least practical of the approaches, however, it also provides the best results. We conjecture this to be due to the size of the graphs and the amount of structural changes occurring. M+AN is the lowest rated by the participants. The general consensus for AN is that it was difficult to keep track of the changes occurring between the nodes, requiring the viewer to memorize node positions and labels incurring a high cognitive effort to complete the tasks. Despite the aforementioned drawbacks, AN scales better to a larger amount of timeslices compared to SI and JP. Finally, the group with interactions had a better experience overall compared to the group without. The majority of the members of this group explicitly requested interactions to be implemented, supporting our findings concerning H3.

5.4 Limitations

This experiment's limitations open potential future research directions. First, the **size** of the graph was not considered. We chose small graphs as stimuli for this study, both in the amount of nodes/links and number of timeslices. M scales better to larger graphs than NL, while AN and ANC support a greater number of timeslices compared to SI and JP. Future studies on dynamic network visualizations might provide evidence on the scalability of the different potential combinations. Second, we chose simple, custom implementations for our structural and temporal encodings, disregarding more advanced solutions in literature

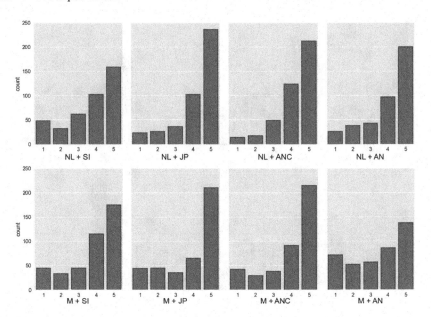

Fig. 8. Confidence per network and temporal encoding on a Likert scale (1–5).

(see Sect. 3). While this was done with the intention of testing the fundamental principles of the techniques in our design space, evaluating more sophisticated approaches might have significantly impacted the results. Finally, we focus on a selection of tasks from a taxonomy on network evolution [2], other graph-based taxonomies could present relevent benchmarks for the proposed techniques.

6   Conclusion

In this paper we presented an experimental study assessing the response times, accuracy, and preferences of participants on different combinations of network structural and temporal encodings, with and without interaction support for the network exploration. Overall, the participants expressed a preference for NL over M, specifically preferring the ANC temporal encoding over the other options, despite AN being more accurate and having lower response times. We also note that our results suggest that the use of M as base representation proved to be more accurate for tasks on low-level entities and counting across different temporal representations. The results of our experiment also suggest a significant effect of interactions on participants' performance. Therefore, as directions for future work, we consider evaluating in more detail the influence that interactions have on accuracy and response times for dynamic network visualization, also considering the potential influence of the graph size on the perception of different combinations of network and temporal encodings.

Acknowledgements. This work was conducted within the framework of the project KnoVA (P31419-N31) and ArtVis (P35767) funded by the Austrian Science Fund (FWF).

References

1. Surveyjs - survey and form javascript libraries. https://surveyjs.io/. Accessed 09 Sept 2022
2. Ahn, J.W., Plaisant, C., Shneiderman, B.: A task taxonomy for network evolution analysis. IEEE Trans. Vis. Comput. Graphics **20**(3), 365–376 (2014)
3. Albert, R., Barabási, A.L.: Statistical mechanics of complex networks. Rev. Mod. Phys. **74**(1), 47 (2002)
4. Archambault, D., Purchase, H., Pinaud, B.: Animation, small multiples, and the effect of mental map preservation in dynamic graphs. IEEE Trans. Vis. Comput. Graphics **17**, 539-552 (2011)
5. Archambault, D., Purchase, H.C.: Mental map preservation helps user orientation in dynamic graphs. In: Didimo, W., Patrignani, M. (eds.) GD 2012. LNCS, vol. 7704, pp. 475–486. Springer, Heidelberg (2013). https://doi.org/10.1007/978-3-642-36763-2_42
6. Archambault, D., Purchase, H.C.: The map in the mental map: experimental results in dynamic graph drawing. Int. J. Hum Comput Stud. **71**(11), 1044–1055 (2013)
7. Archambault, D., Purchase, H.C.: Can animation support the visualisation of dynamic graphs? Inf. Sci. **330**, 495–509 (2016)
8. Arleo, A., Miksch, S., Archambault, D.: Event-based dynamic graph drawing without the agonizing pain. Computer Graphics Forum. **41**, 226–244 (2022). Wiley Online Library (2022)
9. Backhaus, K., Erichson, B., Weiber, R.: Fortgeschrittene Multivariate Analysemethoden. Springer, Heidelberg (2015). https://doi.org/10.1007/978-3-662-46087-0
10. Bastian, M., Heymann, S., Jacomy, M.: Gephi: an open source software for exploring and manipulating networks. In: Proceedings of the International AAAI Conference on Web and Social Media, vol. 3, pp. 361–362 (2009)
11. Baur, M., et al.: Visone software for visual social network analysis. In: Mutzel, P., Jünger, M., Leipert, S. (eds.) GD 2001. LNCS, vol. 2265, pp. 463–464. Springer, Heidelberg (2002). https://doi.org/10.1007/3-540-45848-4_47
12. Beck, F., Burch, M., Diehl, S., Weiskopf, D.: A taxonomy and survey of dynamic graph visualization. Comput. Graphics Forum **36**(1), 133–159 (2017)
13. Behrisch, M., Bach, B., Henry Riche, N., Schreck, T., Fekete, J.D.: Matrix reordering methods for table and network visualization. Comput. Graphics Forum **35**, 693–716 (2016). Wiley Online Library (2016)
14. Bennett, C., Ryall, J., Spalteholz, L., Gooch, A.: The aesthetics of graph visualization. In: Proceedings of the Eurographics Conference on Computational Aesthetics in Graphics, Visualization and Imaging, pp. 57–64. The Eurographics Association (2007)
15. Bollobás, B., Borgs, C., Chayes, J.T., Riordan, O.: Directed scale-free graphs. SODA **3**, 132–139 (2003)
16. Boneau, C.A.: The effects of violations of assumptions underlying the t test. Psychol. Bull. **57**(1), 49 (1960)

17. Bonferroni, C.: Teoria statistica delle classi e calcolo delle probabilita. Pubblicazioni del R Istituto Superiore di Scienze Economiche e Commericiali di Firenze **8**, 3–62 (1936)
18. Bortz, J.: Statistik: Für Sozialwissenschaftler. Springer-Verlag (2013). https://doi.org/10.1007/978-3-642-12770-0
19. Bostock, M., Ogievetsky, V., Heer, J.: D3 data-driven documents. IEEE Trans. Vis. Comput. Graphics **17**(12), 2301–2309 (2011)
20. Box, G.E., Cox, D.R.: An analysis of transformations. J. Roy. Stat. Soc.: Ser. B (Methodol.) **26**(2), 211–243 (1964)
21. Brandes, U., Corman, S.R.: Visual unrolling of network evolution and the analysis of dynamic discourse. Inf. Vis. **2**(1), 40–50 (2003)
22. Brandes, U., Mader, M.: A quantitative comparison of stress-minimization approaches for offline dynamic graph drawing. In: van Kreveld, M., Speckmann, B. (eds.) GD 2011. LNCS, vol. 7034, pp. 99–110. Springer, Heidelberg (2012). https://doi.org/10.1007/978-3-642-25878-7_11
23. Collberg, C., Kobourov, S., Nagra, J., Pitts, J., Wampler, K.: A system for graph-based visualization of the evolution of software. In: Proceedings of the 2003 ACM symposium on Software visualization, pp. 77-ff (2003)
24. Crnovrsanin, T., Chu, J., Ma, K.L.: An incremental layout method for visualizing online dynamic graphs. J. Graph Algorithms Appl. **21**(1), 55–80 (2017)
25. Di Giacomo, E., Didimo, W., Montecchiani, F., Tappini, A.: A user study on hybrid graph visualizations. In: Purchase, H.C., Rutter, I. (eds.) GD 2021. LNCS, vol. 12868, pp. 21–38. Springer, Cham (2021). https://doi.org/10.1007/978-3-030-92931-2_2
26. van den Elzen, S., Holten, D., Blaas, J., van Wijk, J.J.: Dynamic network visualization withextended massive sequence views. IEEE Trans. Visual Comput. Graphics **20**(8), 1087–1099 (2013)
27. Erten, C., Harding, P.J., Kobourov, S.G., Wampler, K., Yee, G.: Graph animations with evolving layouts. In: Liotta, G. (ed.) GD 2003. LNCS, vol. 2912, pp. 98–110. Springer, Heidelberg (2004). https://doi.org/10.1007/978-3-540-24595-7_9
28. Erten, C., Harding, P.J., Kobourov, S.G., Wampler, K., Yee, G.: Exploring the computing literature using temporal graph visualization. In: Visualization and Data Analysis 2004, vol. 5295, pp. 45–56 SPIE (2004)
29. Filipov, V., Arleo, A., Bögl, M., Miksch, S.: On time and space: an experimental study on graph structural and temporal encodings (2022). arxiv:1048550/ARXIV.2208.13716. https://arxiv.org/abs/2208.13716
30. Filipov, V., Arleo, A., Miksch, S.: Exploratory user study on graph temporal encodings. In: 2021 IEEE 14th Pacific Visualization Symposium (PacificVis), pp. 131–135. IEEE (2021)
31. Garnier, S., Ross, N., Rudis, R., Camargo, A.P., Sciaini, M., Scherer, C.: viridis - Colorblind-Friendly Color Maps for R, r package version 0.6.2
32. Ghani, S., Elmqvist, N., Yi, J.S.: Perception of animated node-link diagrams for dynamic graphs. Comput. Graphics Forum **31**, 1205–1214 (2012)
33. Ghoniem, M., Fekete, J., Castagliola, P.: A comparison of the readability of graphs using node-link and matrix-based representations. In: IEEE Symposium on Information Visualization, pp. 17–24 (2004)
34. Gleicher, M., Albers, D., Walker, R., Jusufi, I., Hansen, C.D., Roberts, J.C.: Visual comparison for information visualization. Inf. Vis. SAGE **10**(4), 289–309 (2011)
35. Hagberg, A., Conway, D.: NetworkX: network analysis with python (2020)

36. Hagberg, A., Swart, P., S Chult, D.: Exploring network structure, dynamics, and function using networkx. In: Tech. rep., Los Alamos National Lab. (LANL), Los Alamos, NM (United States) (2008)
37. Harris, C.R., et al.: Array programming with NumPy. Nature **585**(7825), 357–362 (2020)
38. Hedderich, J., Sachs, L.: Angewandte Statistik. Springer (2016). https://doi.org/10.1007/978-3-662-05749-0
39. Henry, N., Fekete, J.-D.: MatLink: enhanced matrix visualization for analyzing social networks. In: Baranauskas, C., Palanque, P., Abascal, J., Barbosa, S.D.J. (eds.) INTERACT 2007. LNCS, vol. 4663, pp. 288–302. Springer, Heidelberg (2007). https://doi.org/10.1007/978-3-540-74800-7_24
40. Javed, W., Elmqvist, N.: Exploring the design space of composite visualization. In: IEEE Pacific Visualization Symposium, pp. 1–8 (2012)
41. Kerracher, N., Kennedy, J., Chalmers, K.: The design space of temporal graph visualisation. In: Elmqvist, N., Hlawitschka, M., Kennedy, J. (eds.) EuroVis - Short Papers. The Eurographics Association (2014)
42. Linhares, C.D.G., Ponciano, J.R., Paiva, J.G.S., Travençolo, B.A.N., Rocha, L.E.C.: A comparative analysis for visualizing the temporal evolution of contact networks: a user study. J. Visual. **24**(5), 1011–1031 (2021). https://doi.org/10.1007/s12650-021-00759-x
43. Linhares, C.D., Travençolo, B.A., Paiva, J.G.S., Rocha, L.E.: Dynetvis: a system for visualization of dynamic networks. In: Proceedings of the Symposium on Applied Computing, pp. 187–194 (2017)
44. Miksch, S., Aigner, W.: A matter of time: applying a data-users-tasks design triangle to visual analytics of time-oriented data. Comput. Graphics **38**, 286–290 (2014)
45. Okoe, M., Jianu, R., Kobourov, S.: Node-link or adjacency matrices: old question, new insights. IEEE Trans. Visual. Comput. Graphics **25**(10), 2940–2952 (2019)
46. Okoe, M., Jianu, R., Kobourov, S.: Revisited experimental comparison of node-link and matrix representations. In: Frati, F., Ma, K.-L. (eds.) GD 2017. LNCS, vol. 10692, pp. 287–302. Springer, Cham (2018). https://doi.org/10.1007/978-3-319-73915-1_23
47. Posten, H.O.: Robustness of the two-sample t-test. In: Robustness of statistical methods and nonparametric statistics, pp. 92–99 Springer (1984). https://doi.org/10.1007/978-94-009-6528-7_36
48. Purchase, H., Carrington, D., Allder, J.A.: Empirical evaluation of aesthetics-based graph layout. Empirical Softw. Eng. **7**, 233–255 (2002)
49. Purchase, H.C.: The effects of graph layout. In: Proceedings of the Australasian Conference on Computer Human Interaction, p. 80 IEEE Computer Society (1998)
50. Purchase, H.C., Hoggan, E., Görg, C.: How important is the mental map? – an empirical investigation of a dynamic graph layout algorithm. In: Kaufmann, M., Wagner, D. (eds.) GD 2006. LNCS, vol. 4372, pp. 184–195. Springer, Heidelberg (2007). https://doi.org/10.1007/978-3-540-70904-6_19
51. Ren, D., et al.: Understanding node-link and matrix visualizations of networks: a large-scale online experiment. Netw. Sci. **7**(2), 242–264 (2019)
52. Rousseeuw, P.J., Croux, C.: Alternatives to the median absolute deviation. J. Am. Stat. Assoc. **88**(424), 1273–1283 (1993)
53. Rufiange, S., Melançon, G.: Animatrix: a matrix-based visualization of software evolution. In: 2014 second IEEE Working Conference on Software Visualization, pp. 137–146. IEEE (2014)

54. Schminder, E., Ziegler, M., Danay, E., Beyer, L., Bühner, M.: Is it really robust? reinvestigating the robustness of anova against violations of the normal distribution. Method.- Eur. Res. J. Methods Behav. Soc. Sci. **6**(4), 147–151 (2010)
55. Simonetto, P., Archambault, D., Kobourov, S.: Event-based dynamic graph visualisation. IEEE Trans. Visual Comput. Graphics **26**(7), 2373–2386 (2018)
56. Tufte, E.R.: The visual display of quantitative information. Quantitative Information, p. 13 (1983)
57. Vallat, R.: Pingouin: statistics in python. J. Open Source Softw. **3**, 1026 (2018)
58. Virtanen, P., et al.: SciPy 1.0 contributors: sciPy 1.0: fundamental algorithms for scientific computing in python. Nat. Methods **17**, 261–272 (2020)
59. Weiß, C.: Basiswissen Medizinische Statistik. S, Springer, Heidelberg (2019). https://doi.org/10.1007/978-3-662-56588-9
60. Yoghourdjian, V., et al.: Exploring the limits of complexity: a survey of empirical studies on graph visualisation. Vis. Inform. **2**(4), 264–282 (2018)

Small Point-Sets Supporting Graph Stories

Giuseppe Di Battista[1] , Walter Didimo[2] , Luca Grilli[2] ,
Fabrizio Grosso[1](✉) , Giacomo Ortali[2] , Maurizio Patrignani[1] ,
and Alessandra Tappini[2]

[1] Roma Tre University, Rome, Italy
{giuseppe.dibattista,fabrizio.grosso,maurizio.patrignani}@uniroma3.it
[2] University of Perugia, Perugia, Italy
{walter.didimo,luca.grilli,giacomo.ortali,alessandra.tappini}@unipg.it

Abstract. In a graph story the vertices enter a graph one at a time and each vertex persists in the graph for a fixed amount of time ω, called viewing window. At any time, the user can see only the drawing of the graph induced by the vertices in the viewing window and this determines a sequence of drawings. For readability, we require that all the drawings of the sequence are planar. For preserving the user's mental map we require that when a vertex or an edge is drawn, it has the same drawing for its entire life. We study the problem of drawing the entire sequence by mapping the vertices only to $\omega + k$ given points, where k is as small as possible. We show that: (i) The problem does not depend on the specific set of points but only on its size; (ii) the problem is NP-hard and is FPT when parameterized by $\omega + k$; (iii) there are families of graph stories that can be drawn with $k = 0$ for any ω, while for $k = 0$ and small values of ω there are families of graph stories that can be drawn and others that cannot; (iv) there are families of graph stories that cannot be drawn for any fixed k and families of graph stories that require at least a certain k.

Keywords: Dynamic graphs · Planar graphs · Time and graph drawing

1 Introduction

In this paper we address "graph stories", a model introduced by Borrazzo et al. in [5] as a framework for exploring temporal data. In a *graph story* the vertices enter a graph one at a time and persist in the graph for a fixed amount of time ω, called the *size of the viewing window*. At any time, the user can see only the drawing of the graph induced by the vertices in the viewing window and

This work was partially supported by: (i) MIUR, grant 20174LF3T8 "AHeAD: efficient Algorithms for HArnessing networked Data"; (ii) Dipartimento di Ingegneria - Università degli Studi di Perugia, grants RICBA20EDG: "Algoritmi e modelli per la rappresentazione visuale di reti" and RICBA21LG: "Algoritmi, modelli e sistemi per la rappresentazione visuale di reti".

P. Angelini and R. von Hanxleden (Eds.): GD 2022, LNCS 13764, pp. 289–303, 2023.
https://doi.org/10.1007/978-3-031-22203-0_21

this determines a sequence of drawings. For readability, all the drawings of the sequence are required to be planar. For preserving the user's mental map, when an edge is drawn it has the same drawing for its entire life. Also, in order to limit the constraints, we allow the edges to be represented by Jordan arcs.

Graph stories are related to a rich body of literature devoted to the visualization of dynamic graphs (surveys can be found in [2,17]). One of the main classification criteria of dynamic graph problems is whether the story is entirely known in advance (*off-line model*) or not (*on-line model*). In this respect, our contribution falls in the off-line model. A third intermediate category (*look-ahead model*) is when a small chunk of the incoming events is known in advance to the drawing algorithms. The events are also a classification criterion, as they may refer to vertices, edges, or both. Finally, further constraints may regard the timings of the events, the more common being that they occur one at a time at regular intervals and that the incoming objects have a fixed lifetime as in the case of graph stories. In some cases, the order of the events is constrained to correspond to a specific kind of visit of the graph.

Several results focus on dynamic trees. In [3], it is shown how to draw a tree in $O(\omega^3)$ area where the model is on-line, the incoming objects are edges that arrive in the order of a Eulerian tour of the tree and whose straight-line drawing persists for a fixed lifetime ω. In [10], a small look ahead on the sequence of vertices is used in order to add one vertex at a time to the current drawing of an infinite tree, balancing the readability of the drawings with respect to the difference between consecutive drawings. In [23], a sequence of trees (their union, though, may be an arbitrary graph) is completely known in advance. Vertices and edges can move during the animation and can have arbitrary lifetime. The purpose is to pursue aesthetic criteria commonly adopted for tree drawings [20].

Only a few results regard more complex families of graphs. For instance, in [14], a stream of edges enter the drawing and never leave it, forming an outerplanar graph that has to be drawn according to an on-line model, moving the previously drawn vertices by a polylogarithmic distance. In [8] the drawings of several families of graphs are updated as vertices and edges enter and leave the current graph according to the on-line model.

More feebly related to our setting is the literature about dynamic planarity [12,13,15,19,21], where the model is on-line and the planar embedding of the graph is allowed to change. When the embedding has to be preserved, instead, planarly adding a stream of edges with a fixed lifetime is NP-complete even for the off-line model [9]. Also, related somehow to dynamic graph drawing is geometric simultaneous embedding [4,6], which can be used to model temporal graphs.

Coming more properly to the graph story model, Borrazzo et al. [5] address the setting where all the drawings of the story are straight-line and planar, and where vertices do not change their position once drawn. It is shown that graph stories of paths and trees can be drawn on a $2\omega \times 2\omega$ and on an $(8\omega+1) \times (8\omega+1)$ grid, respectively. Further, there exist graph stories of planar graphs that cannot be drawn straight-line within an area that is only a function of ω.

Contribution. We study the problem of drawing a graph story by mapping the vertices only to $\omega + k$ given points, where k is as small as possible. We call this a *realizability problem*. Our contribution is as follows. In Sect. 2 we show that the realizability of graph stories is a topological problem, as it does not depend on the specific set of points but only on its size. We also give a characterization of realizable graph stories based on the concept of "compatible embeddings". In Sect. 3 the realizability problem is proven to be NP-complete, even for any given constant k, and to belong to FPT when parameterized by the size $\omega + k$ of the point set. In Sect. 4 we study the realizability of graph stories with $k = 0$, which we call *minimal*. In particular, we show that: (*i*) Every minimal graph story of an outerplanar graph is realizable; (*ii*) for every $\omega \geq 5$ there exist minimal graph stories of series-parallel graphs that are not realizable; (*iii*) all minimal graph stories with $\omega \leq 5$ whose graph does not contain K_5 are realizable if we are allowed to redraw at most one edge at each vertex arrival; and (*iv*) minimal graph stories with $\omega \leq 5$ are always realizable for planar triconnected cubic graphs. Finally, in Sect. 5 we show that there are families of graph stories that are not realizable for any fixed k and families of graph stories that, to be realizable, require at least a certain value for k.

Some proofs have been sketched or omitted and can be found in the full version [11].

Preliminaries. A *drawing* Γ of a graph $G = (V, E)$ maps each vertex of V to a distinct point of the plane and each edge of E to a Jordan arc connecting its end-vertices; Γ is *planar* if no two edges intersect except at common endpoints. A planar drawing Γ of G subdivides the plane into connected regions called *faces*, and the set of circular orders of the edges incident to each vertex is called a *rotation system*. The unbounded face of Γ is the *external face*. Walking on the (not necessarily connected) border of a face f of Γ so to keep f to the left determines a set, called the *boundary of f*, of circular lists of alternating vertices and edges. Each list describes a (not necessarily simple) cycle, which can also consist of an isolated vertex: Each edge of G occurs either once in exactly two circular lists of different face boundaries or twice in the circular list of one face boundary.

Two planar drawings of G are *equivalent* if they have the same rotation system, face boundaries, and external face. An equivalence class of planar drawings of G is a *planar embedding of G*. Note that, if G is connected then each face boundary consists of exactly one circular list; in this case an embedding of G is fully specified by its rotation system and by its external face. If G is equipped with a planar embedding ϕ, it is a *plane* graph; a planar drawing Γ of G is *embedding-preserving* if $\Gamma \in \phi$. If G' is a subgraph of G and Γ' is the restriction of Γ to G', the planar embedding ϕ' of Γ' is the *restriction of ϕ to G'*.

2 Graph Stories

Definition 1. A *graph story* is a tuple $\mathcal{S} = (G, \omega, k, \tau)$ where: *(i)* $G = (V, E)$ is an n-vertex graph; *(ii)* $\omega \leq n$ is a positive integer, called the *size of the viewing*

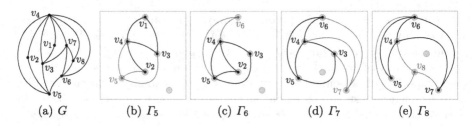

(a) G (b) Γ_5 (c) Γ_6 (d) Γ_7 (e) Γ_8

Fig. 1. A realization of a graph story $\mathcal{S} = (G, 5, 1, \tau)$ on a set P of 6 points. The points of P are yellow disks. For each Γ_i ($5 \le i \le 8$), vertex v_i and its incident edges are red. (Color figure online)

window; *(iii)* k is a non-negative integer, called the *number of extra points*; and *(iv)* $\tau = \langle v_1, v_2, \ldots, v_n \rangle$ is a linear ordering of the vertices of G (i.e., $v_i \in V$ is the vertex at position i according to τ).

Let $G_i = (V_i, E_i)$ denote the subgraph of G induced by all vertices v_j such that $\max\{1, i - \omega + 1\} \le j \le i$. Observe that, if $i \le \omega$ then G_i consists of the i vertices $\{v_1, v_2, \ldots, v_i\}$; otherwise G_i consists of the ω vertices $\{v_{i-\omega+1}, v_{i-\omega+2}, \ldots, v_i\}$. In other words, G_i is the subgraph induced by v_i and by the (up to) $\omega - 1$ vertices of G that precede v_i in τ. For each i, we say that v_i *enters* the viewing window at time i, and for each $i \in \{\omega + 1, \ldots, n\}$, we say that $v_{i-\omega}$ *leaves* the viewing window at time i.

Definition 2. *A* realization *of a graph story* $\mathcal{S} = (G, \omega, k, \tau)$ *on a set* P *of* $\omega + k$ *points is a sequence of drawings* $\mathcal{R} = \langle \Gamma_1, \Gamma_2, \ldots, \Gamma_n \rangle$ *with the following two properties: (R1)* Γ_i ($1 \le i \le n$) *is a planar drawing of* G_i, *where distinct vertices of* V_i *are mapped to distinct points of* P; *(R2) the restrictions of* Γ_{i-1} *and of* Γ_i ($2 \le i \le n$) *to their common subgraph* $G_{i-1} \cap G_i$ *are identical.*

Figure 1 shows a realization of a graph story $\mathcal{S} = (G, 5, 1, \tau)$ on a set of 6 points. A graph story $\mathcal{S}$ is *realizable* if there exists a set P of $\omega + k$ points such that $\mathcal{S}$ admits a realization on P. Since the planarity of all graphs G_i is necessary for realizability, from now on we consider graph stories that satisfy this requirement.

Remark 1 (Edge Visibility). We assume that G only consists of *visible edges*, i.e., edges (v_i, v_j) such that $|i - j| < \omega$. Indeed if $|i - j| \ge \omega$, (v_i, v_j) can be ignored, as it never appears in a realization. Our assumption has two implications: *(i)* G has vertex-degree at most $2\omega - 2$ (every G_i has vertex-degree at most $\omega - 1$); and *(ii)* G has *bandwidth* at most $\omega - 1$ and hence *pathwidth* at most $\omega - 1$ [16] (the set of bags of this decomposition is $\{V_1, V_2, \ldots, V_n\}$).

Remark 2 (Minimality). Clearly, if a graph story $\mathcal{S} = (G, \omega, k, \tau)$ is realizable, every other story $\mathcal{S}' = (G, \omega, k', \tau)$ with $k' > k$ is realizable too. Hence, a natural scenario is when the number of extra points k is zero. We call such a story *minimal* and we denote it as $\mathcal{S} = (G, \omega, \tau)$. For a minimal graph story, Property R2 of Definition 2 implies that each vertex v_i with $\omega + 1 \le i \le n$ is mapped to the same point as $v_{i-\omega}$, thus the mapping of the whole realization is fully determined by the mapping of Γ_ω (i.e., of the first ω drawings of the realization).

2.1   Geometry and Topology of Graph Stories

The following lemma shows that the realizability problem is in essence more a topological problem than a geometric problem.

Lemma 1. *A graph story* $S = (G, \omega, k, \tau)$ *is realizable on a set of points* P, *with* $|P| = \omega + k$, *if and only if it is realizable on any set of points* P' *with* $|P'| = |P|$.

Proof (sketch). Let $\mathcal{R}$ be a realization of S on P. Starting from $\mathcal{R}$, we construct a realization $\mathcal{R}'$ of S on a given arbitrary set of points P'. Let $\rho(\cdot)$ be a function that for each edge e of G gives the Jordan arc $\rho(e)$ used by $\mathcal{R}$ to represent e and let J be the codomain of ρ, i.e., the set (without repetitions) of Jordan arcs used by $\mathcal{R}$. Without loss of generality, we may assume that any two Jordan arcs c and c' of J have a finite intersection. This can be obtained by perturbing c or c'.

Starting from P and J, we construct a multigraph $\mathcal{M}$ that has a vertex w_i for each point $p_i \in P$, with $i = 1, 2, \ldots, \omega + k$, and an edge (w_i, w_j) for each Jordan arc $c \in J$ with endpoints p_i and p_j. Observe that the Jordan arcs in J also provide a (non-planar) drawing $\Gamma(\mathcal{M})$ of $\mathcal{M}$. We planarize $\mathcal{M}$ by replacing crossings with dummy vertices. Further, we subdivide multiple edges of $\mathcal{M}$ in order to obtain a plane graph $\mathcal{G}$. By exploiting one of the algorithms described in [1,18], we can draw $\mathcal{G}$ while preserving its planar embedding on the set of points P' plus an arbitrary set of additional points to host the planarization and subdivision dummy vertices, obtaining $\Gamma(\mathcal{G})$.

Observe that a vertex v of G corresponds to a point of P, which is associated with a vertex of $\mathcal{G}$ drawn on a point of P'. Also, an edge e of G corresponds to a Jordan arc $\rho(e)$ in J, which is a simple path π in $\mathcal{G}$. Hence, we define a function $\rho'(e)$ that gives, for each edge e of G, a Jordan arc that is the concatenation of the curves used in $\Gamma(\mathcal{G})$ to draw the path π. Finally, the Jordan arcs $\rho(e)$ and $\rho(e')$ of two edges e and e' of G cross if and only if $\rho'(e)$ and $\rho'(e')$ cross.

The other direction of the proof is obvious. $\square$

It is natural to ask whether for every realizable graph story where G is planar, there exists a planar embedding of G such that each drawing of the realization preserves this embedding. We formalize this concept and show that this is not always the case. Let S be a story whose graph G is planar. A *supporting embedding* for S is a planar embedding ϕ of G such that S admits a realization $\langle \Gamma_1, \ldots, \Gamma_n \rangle$ where the embedding of Γ_i is the restriction of ϕ to G_i $(i = 1, \ldots, n)$.

Lemma 2. *There exists a minimal graph story* $S = (G, \omega, \tau)$ *such that: (i) G is planar; (ii) S is realizable; and (iii) S does not admit a supporting embedding.*

Proof (sketch). We produce a minimal graph story $S = (G, \omega, \tau)$ such that G admits a single planar embedding ϕ (up to a flip and up to the choice of the external face) and such that in any realization of S there is at least one embedding ϕ_i of G_i that is not the restriction of ϕ to G_i. In this story $\omega = 8$, G is the graph in Fig. 2, and τ is given by the indices of the vertices of G. $\square$

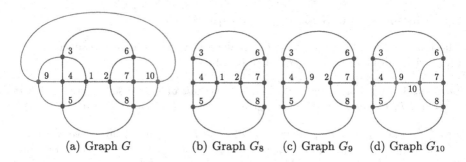

(a) Graph G (b) Graph G_8 (c) Graph G_9 (d) Graph G_{10}

Fig. 2. An illustration for Lemma 2.

2.2 Characterizing Realizable Graph Stories

We now give a characterization of a realizable graph story $\mathcal{S} = (G, \omega, k, \tau)$ in terms of a sequence of "compatible embeddings". To this aim, we give a generalization of the definition of planar embedding that associates with each face a weight representing how many of k notable points are inside such a face.

A *face-k-weighted planar embedding* ϕ *of a planar graph* H is a planar embedding of H together with a non-negative integer, called *weight*, for each face of ϕ such that the sum of all weights is k. The *removal of a vertex* v from a face-k-weighted planar embedding ϕ of H produces a face-$(k + 1)$-weighted planar embedding ϕ^{-v} of $H \setminus v$ such that the planar embedding of ϕ^{-v} is the restriction of the planar embedding of ϕ to $H \setminus v$ and the weights of the faces are changed as follows: (i) all the faces in common between ϕ and ϕ^{-v} have the same weight in ϕ^{-v} as in ϕ, and (ii) the new face of ϕ^{-v} resulting by the removal of v has a weight that is one plus the sum of the weights of the faces of ϕ incident to v.

Let $\mathcal{S}$ be a graph story and let ϕ_i be a face-k-weighted planar embedding of G_i, for $i \in \{\omega, \dots, n\}$. Two face-$k$-weighted planar embeddings ϕ_{i-1} and ϕ_i, with $i = \omega + 1, \dots, n$, are *compatible* if removing $v_{i-\omega}$ from ϕ_{i-1} produces the same face-$(k+1)$-weighted planar embedding of $G_{i-1} \cap G_i$ as removing v_i from ϕ_i.

Lemma 3. *A graph story* $\mathcal{S} = (G, \omega, k, \tau)$ *is realizable if and only if there exists a sequence* $\langle \phi_\omega, \phi_{\omega+1}, \dots, \phi_n \rangle$ *of face-k-weighted planar embeddings for the graphs* $\langle G_\omega, G_{\omega+1}, \dots, G_n \rangle$, *such that* ϕ_{i-1} *and* ϕ_i *are compatible* $(\omega+1 \leq i \leq n)$.

Proof (sketch). We prove here only one direction. Suppose there exists a sequence of face-k-weighted planar embeddings $\langle \phi_\omega, \phi_{\omega+1}, \dots, \phi_n \rangle$ such that any two consecutive face-k-weighted planar embeddings are compatible. Let Γ_ω be any planar drawing of G_ω and let P be the set of points of Γ_ω corresponding to the vertices of G_ω plus k unused points arbitrarily distributed inside the faces of Γ_ω, according to the weights of ϕ_ω. For each $i = 1, \dots, \omega - 1$, define Γ_i as the restriction of Γ_ω to G_i. For each $i = \omega + 1, \dots, n$, by the compatibility of ϕ_i with ϕ_{i-1}, the removal of vertex $v_{i-\omega}$ from Γ_i yields a drawing $\Gamma^\cap$ of $G^\cap = G_{i-1} \cap G_i$ that has the same face-$(k+1)$-weighted embedding $\phi^\cap = \phi_{i-1} \setminus v_{i-\omega} = \phi_i \setminus v_i$ of $G^\cap$. We construct Γ_i from $\Gamma^\cap$ by inserting v_i inside the face of $\Gamma^\cap$ corresponding to the face of $\phi^\cap$ generated by the removal of v_i from ϕ_i. Also, we planarly insert each edge connecting v_i to each of its neighbors according to ϕ_i, without changing the starting drawing and by leaving on each generated face f the number of unused points that corresponds to the weight of f in ϕ_i (e.g., using the technique in [7], where the unused points are regarded as isolated vertices). The sequence $\langle \Gamma_1, \Gamma_2, \ldots, \Gamma_n \rangle$ satisfies Properties R1 and R2, i.e., it is a realization of $\mathcal{S}$.   □

3 Realizability Testing of Graph Stories

We first prove that testing whether a graph story is realizable is NP-hard for any given integer $k \geq 0$ (Theorem 1). Then we prove the that the problem is in FPT when parameterized by $\sigma = \omega + k$ (Theorem 2).

Theorem 1. *For any integer $k \geq 0$, testing the realizability of a graph story $\mathcal{S} = (G, \omega, k, \tau)$ is NP-hard.*

Proof (sketch). We use a reduction from the SUNFLOWER SEFE problem, which is defined as follows. Let $G'_1, G'_2, \ldots, G'_l$ be graphs on the same vertex-set such that each edge in the union of all graphs belongs either to only one of the input graphs or to all the input graphs. SUNFLOWER SEFE asks whether there exists a drawing Γ' of $G'_1 \cup G'_2 \cup \cdots \cup G'_l$ such that two edges cross only if they do not belong to the same graph G'_i. SUNFLOWER SEFE is NP-hard for $l \geq 3$ [22].

Starting from an instance of SUNFLOWER SEFE with $l = 3$, we construct a non-minimal graph story $\mathcal{S} = (G = (V, E), \omega, k, \tau)$ as follows; refer to Fig. 3 for an example with $k = 4$. Let G'_1, G'_2, and G'_3 be the input graphs of SUNFLOWER SEFE with vertex-set V', let E'_i be the set of edges that belong only to graph G'_i ($1 \leq i \leq 3$), and let $E'_\cap$ be the set of edges that belong to all the input graphs. Without loss of generality, we can assume that $|E'_1| \geq |E'_2| \geq |E'_3|$.

We now show how we define sets V and E. For every graph G'_i ($1 \leq i \leq 3$), we subdivide each edge e of E'_i with two vertices d_i^A and d_i^B and we add them to two sets D_i^A and D_i^B, respectively; see $d_{(u,y)}^A$ and $d_{(u,y)}^B$ in Fig. 3(b). We add the three edges obtained by subdividing e to a set E''_i. If needed, we enrich sets D_2^A, D_2^B, D_3^A, and D_3^B with isolated vertices so that all the sets D_i^X have the same cardinality $\tilde{\omega} = |E'_1|$ (note that $\tilde{\omega} = |D_1^A| = |D_1^B|$), with $1 \leq i \leq 3$ and $X \in \{A, B\}$; see the green isolated vertices in Fig. 3(b). Also, we create four sets Δ_j ($1 \leq j \leq 4$) of $\tilde{\omega}$ isolated vertices $\delta_{j,1}, \ldots, \delta_{j,\tilde{\omega}}$; see the purple isolated vertices in Fig. 3(b). We define $V = V' \cup D_1^A \cup D_1^B \cup D_2^A \cup D_2^B \cup D_3^A \cup D_3^B \cup \Delta_1 \cup \Delta_2 \cup \Delta_3 \cup \Delta_4$ and $E = E'_\cap \cup E''_1 \cup E''_2 \cup E''_3$. We define ω as $\omega = |V'| + 6\tilde{\omega}$. Finally, we suitably define τ in such a way that the vertices of the various subsets of V appear in the following order: $\langle D_1^A, \Delta_1, D_2^A, \Delta_2, D_3^A, V', D_1^B, \Delta_3, D_2^B, \Delta_4, D_3^B \rangle$.

$\mathcal{S}$ is constructed in $O(|V'|)$ time and $\omega \in O(|V'|)$. We show in the full version [11] that $\mathcal{S}$ is realizable if and only if $\{G'_1, G'_2, G'_3\}$ is a **yes** instance of SUNFLOWER SEFE. Refer to Fig. 3(c) to (e) for an example.   □

Theorem 2. *Let $\mathcal{S} = (G, \omega, k, \tau)$ be a graph story and let n be the number of vertices of G. There exists an $O(n \cdot 2^{(4\sigma+1)\log_2 \sigma})$-time algorithm that tests whether $\mathcal{S}$ is realizable, where $\sigma = \omega + k$.*

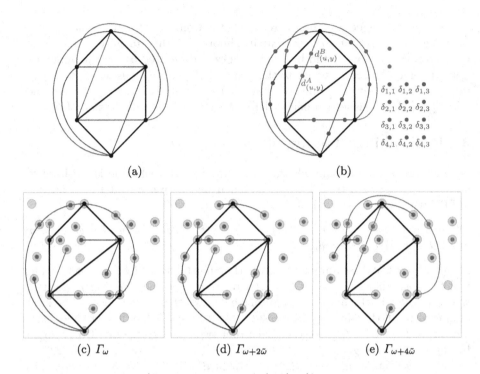

(a)   (b)

(c) Γ_ω (d) $\Gamma_{\omega+2\bar{\omega}}$ (e) $\Gamma_{\omega+4\bar{\omega}}$

Fig. 3. (a) A drawing Γ' of a **yes** instance $\{G'_1, G'_2, G'_3\}$ of SUNFLOWER SEFE. The edges of $E'_\cap$ are black; the edges of E'_1, E'_2, and E'_3 are red, blue, and green, respectively. (b) A drawing of graph G of the story $\mathcal{S} = (G, \omega, 4, \tau)$ constructed from the instance of Fig. 3(a). The vertices of D_1^A (D_2^A, D_3^A, resp.) and D_1^B (D_2^B, D_3^B, resp.) are red (blue, green, resp.); the vertices of Δ_i ($1 \le i \le 4$) are purple. The points of P are represented as yellow disks. (Color figure online)

Proof (sketch). For each subgraph G_i ($i = \omega, \ldots, n$), let $\mathcal{E}_i = \{\phi_i^1, \phi_i^2, \ldots, \phi_i^{s_i}\}$ be the set of all planar face-k-weighted embeddings of G_i. We construct a directed acyclic graph D as follows: (i) For each $\phi_i^j \in \mathcal{E}_i$ ($i = \omega, \ldots, n$ and $j = 1, \ldots, s_i$), D has a node v_i^j corresponding to ϕ_i^j. (ii) For each pair of elements ϕ_i^j and ϕ_{i+1}^r ($\omega \le i \le n-1; 1 \le j \le s_i; 1 \le r \le s_{i+1}$), D contains a directed edge (v_i^j, v_{i+1}^r) if and only if ϕ_i^j and ϕ_{i+1}^r are compatible face-k-weighted embeddings.

Each set $\mathcal{E}_i$, with $i = \omega, \ldots, n$, defines a distinct *layer* of vertices of D, called *layer* i. By construction, each vertex of layer i can only have outgoing edges towards vertices of layer $i + 1$ (if $i < n$) and incoming edges from vertices of layer $i - 1$ (if $i > \omega$). We finally augment D with a dummy source s connected with outgoing edges to all vertices of layer 1 and with a dummy sink t connected with incoming edges to all vertices of layer n. By Lemma 3, $\mathcal{S}$ is realizable if and only if there is a directed path from s to t in D. The time complexity of the algorithm is analyzed in the full version [11]. □

When $k = 0$, we have Corollary 1. Also, Theorems 1 and 2 imply Corollary 2.

Corollary 1. *Let $S = (G, \omega, \tau)$ be a minimal graph story and let n be the number of vertices of G. There exists an $O(n \cdot 2^{(4\omega+1)\log_2 \omega})$-time algorithm that tests whether S is realizable.*

Corollary 2. *For any integer $k \geq 0$, testing the realizability of a graph story $S = (G, \omega, k, \tau)$ is NP-complete.*

4   Minimal Graph Stories

We now turn our attention to minimal graph stories that can be realized for small values of ω. If $\omega \leq 4$ every minimal graph story is easily realizable, independent of G and of τ, and even if G is not a planar graph (just use any predefined planar drawing of the complete graph K_4 as a support for each Γ_i $(i = 1, \ldots, n)$). Establishing which minimal graph stories are realizable when $\omega \geq 5$ is more challenging. We show that every graph story is realizable if G is outerplanar (Theorem 3), while if G is a series-parallel graph this is not always the case, even if $\omega = 5$ (Lemma 4). However, we prove that stories of partial 2-trees (which include series-parallel graphs) are always realizable for $\omega = 5$ if we are allowed to "reroute" at most one edge per time (a formal definition is given later); this result is an implication of a more general result for stories with $\omega = 5$ (Theorem 4). Lemma 4 and Theorem 4 together close the gap on the realizability of minimal graph stories of partial 2-trees when $\omega = 5$. Finally, for $\omega = 5$ we prove that every minimal graph story is realizable if G is a planar triconnected cubic graph (Theorem 5). A graph is *cubic* if all its vertices have degree three.

For a story of an outerplanar graph, we show that any outerplanar embedding is a supporting embedding (see the full version [11] for details).

Theorem 3. *Every minimal graph story $S = (G, \omega, \tau)$ with G outerplanar is realizable. Also, any outerplanar embedding of G is a supporting embedding for S.*

Proof. Let ϕ be any outerplanar embedding of G, and let ϕ_i be the restriction of ϕ to G_i $(1 \leq i \leq n)$. Consider any two consecutive planar embeddings ϕ_{i-1} and ϕ_i, for $\omega + 1 \leq i \leq n$. Since they are restrictions of the same planar embedding of G, then their restrictions to $G_i \cap G_{i-1}$ determine the same set F of faces. Also, both v_i and $v_{i-\omega}$ lie in the plane region corresponding to the external face of F. Hence, ϕ_{i-1} and ϕ_i are compatible and, by Lemma 3, S is realizable. $\square$

Lemma 4. *For any $\omega \geq 5$, there exists a minimal graph story $S = (G, \omega, \tau)$ such that G is a series-parallel graph and S is not realizable.*

Proof (sketch). Consider the story $(G, 5, \tau)$ in Fig. 4(a), where the vertices are labeled with their subscript in the sequence $\tau = \langle v_1, v_2, \ldots, v_8 \rangle$. Graph G_5 admits one of the four embeddings in Figs. 4(b) to (e). Observe that, in all four cases either cycle $3, 4, 5$ separates 6 from 7 in G_7 (Figs. 5(c) and (n)), or cycle $4, 5, 6$ separates 7 from 8 in G_8 (Figs. 5(g) and (k)). See the full version [11] for $\omega > 5$. $\square$

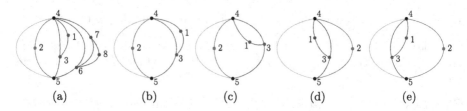

Fig. 4. (a) A minimal graph story of a series-parallel graph that is not realizable. (b),(c),(d),(e) the four combinatorial embeddings of G_5.

Since Property R2 of Definition 2 is a strict requirement, one can think of relaxing it by allowing a partial change of the drawing of $G_{i-1} \cap G_i$ when vertex v_i enters the viewing window. Let Γ be a planar drawing of G, (u, v) be an edge of G incident to two distinct faces f and f' of Γ, and p be a point of the plane inside face f; see Fig. 6(a). *Rerouting* (u, v) *with respect to* p consists of planarly redrawing (u, v) such that u and v keep their positions and p lies inside f'; see Fig. 6(b). The obtained drawing has the same planar embedding as Γ.

An *h-reroute realization* of $S = (G, \omega, k, \tau)$ on a set P of $\omega + k$ points ($h \geq 0$) is a sequence $\langle \Gamma_1, \Gamma_2, \ldots, \Gamma_n \rangle$ satisfying Property R1 of Definition 2 and such that the restriction of Γ_i to $G_{i-1} \cap G_i$ ($2 \leq i \leq n$) is obtained from the restriction of Γ_{i-1} to $G_{i-1} \cap G_i$ by rerouting at most h distinct edges with respect to h points of P. S is *h-reroute realizable* if it has an h-reroute realization on a set of $\omega + k$ points.

The next theorem characterizes the set of graph stories $S = (G, 5, \tau)$ that are 1-reroute realizable. It properly includes those stories whose G is planar.

Theorem 4. *Every minimal graph story* $S = (G, 5, \tau)$ *is 1-reroute realizable if and only if G does not contain K_5.*

Proof (sketch). We only sketch here the proof of one of the two directions. Suppose that G does not contain K_5. Let Γ_4 be a planar drawing of G_4 on P. Let p be the point of P to which no vertex of G_4 is mapped, let f be the face of Γ_4 that contains p, and let $N(v_5)$ be the set of neighbors of v_5 in G_5. If the boundary of f has four vertices, then v_5 can be mapped to p and it can be connected to all its neighbors without creating edge crossings, so to obtain a planar drawing Γ_5 of G_5. If the boundary of f has three vertices, mapping v_5 to p and connecting it to its neighbors may create an edge crossing. To avoid this crossing, it is possible to reroute an edge of the boundary of f with respect to p such that p lies inside a face whose boundary contains all vertices in $N(v_5)$. Such an edge always exists because the faces of Γ_4 are pairwise adjacent. More precisely, if G_4 is not K_4, then there is a face f' of Γ_4 (adjacent to f) that contains all vertices of G_4. If G_4 is K_4, then $|N(v_5)| \leq 3$, as G does not contain K_5. Also, there is a face f' of Γ_4 that contains all vertices of $N(v_5)$. In both cases, we can reroute any edge e shared by f and f' so that p lies inside f'. This procedure can be applied for each pair of graphs G_{i-1} and G_i ($5 < i \leq n$): Γ_i is obtained by mapping v_i to the same point p of P to which v_{i-5} is mapped in Γ_{i-1}, by rerouting at most one edge with respect to p. $\square$

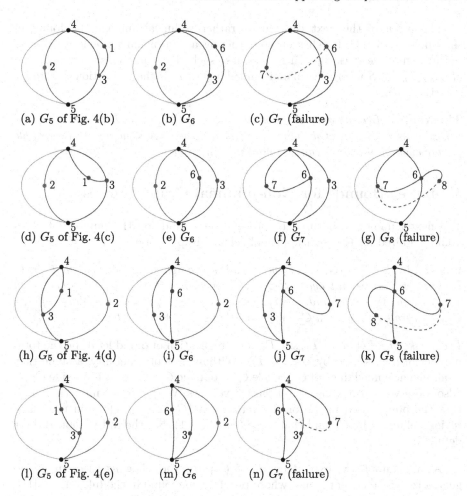

(a) G_5 of Fig. 4(b)   (b) G_6 (c) G_7 (failure)

(d) G_5 of Fig. 4(c)   (e) G_6 (f) G_7 (g) G_8 (failure)

(h) G_5 of Fig. 4(d)   (i) G_6 (j) G_7 (k) G_8 (failure)

(l) G_5 of Fig. 4(e)   (m) G_6 (n) G_7 (failure)

Fig. 5. Tentative realizations of the story of Fig. 4(a) starting from the embeddings of Figs. 4(b) to (e). They all lead to a failure.

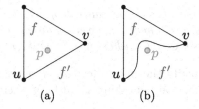

(a)   (b)

Fig. 6. Rerouting edge (u, v) with respect to point p.

The proof of the next theorem is rather technical and can be found in the full version [11]. It relies on constructing a non-planar embedding ϕ of G (with dummy vertices replacing crossings) such that there exists a realization $\langle \Gamma_1, \ldots, \Gamma_n \rangle$ of $\mathcal{S}$ where the planar embedding of Γ_i is the restriction of ϕ to G_i $(i = 1, \ldots, n)$.

Theorem 5. *Every minimal graph story $\mathcal{S} = (G, 5, \tau)$ such that G is an n-vertex planar triconnected cubic graph is realizable. A sequence of compatible planar embeddings for $\mathcal{S}$ can be found in $O(n)$ time.*

5   Lower Bounds for Non-minimal Graph Stories

The next lemma can be used to prove lower bounds on the number of extra points required for the realizability of certain graph stories.

Lemma 5. *Let $\mathcal{S} = (G, \omega, k, \tau)$ be a realizable graph story. Suppose that: (i) G contains vertex-disjoint cycles $C_1, \ldots, C_h$ such that $C_{j-2}, C_{j-1}, C_j \in G_{i_j}$, with $j = 3, \ldots, h$, $i_{j-1} < i_j$ and $i_j - i_{j-1} < \omega$; (ii) in all planar embeddings of G_{i_j}, C_{j-1} separates C_{j-2} from C_j. We have that $\sigma = \omega + k \in \Omega(h)$.*

Proof (sketch). Let $\mathcal{R} = \langle \Gamma_1, \ldots, \Gamma_n \rangle$ be a realization of $\mathcal{S}$ and let σ_i be the total number of points used by $\langle \Gamma_1, \ldots, \Gamma_i \rangle$. Without loss of generality, assume that in all planar embeddings of G_{i_j}, cycle C_j is outside C_{j-1}, which is outside C_{j-2}. Also, observe that a cycle has at least 3 vertices. We prove, by induction on j, that the points used by $\mathcal{R}$ for the vertices in $\{C_1, \ldots, C_{h-1}\}$ lie in the plane region delimited by C_h and that $\sigma_h \geq 9 + 3(h - 1)$. See the full version [11] for details.   □

As an example, we exploit Lemma 5 to prove the following theorem, which generalizes [5, Theorem 1] and whose proof can be found in the full version [11].

Let $n \equiv 0 \mod 3$. An *n-vertex nested triangles graph* G contains the vertices and edges of the 3-cycle $C_i = (v_{i-2}, v_{i-1}, v_i)$, for $i = 3, 6, \ldots, n$, plus the edges (v_i, v_{i+3}), for $i = 1, 2, \ldots, n - 3$. For $n \geq 6$, G is triconnected, thus it has a unique planar embedding (up to the choice of the external face) [24].

Theorem 6. *Let $\mathcal{S} = (G, 9, k, \tau)$ be a realizable graph story such that G is a $3h$-vertex nested triangles graph, where τ is given by the indices of the vertices of G. Any realization of $\mathcal{S}$ has $k \in \Omega(n)$, where $n = 3h$ is the number of vertices of G.*

While Lemma 5 exploits the uniqueness of the embedding of G, the next result provides lower bounds also for graphs that have several planar embeddings.

Theorem 7. *For any $\omega \geq 8$, there exists a graph story $\mathcal{S} = (G, \omega, k, \tau)$ such that G is a series-parallel graph and $\mathcal{S}$ is not realizable for $k < \lfloor \frac{\omega}{2} \rfloor - 3$.*

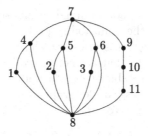

Fig. 7. Illustration for Theorem 7. Case $\omega = 8$.

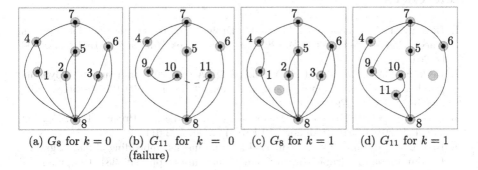

(a) G_8 for $k = 0$ (b) G_{11} for $k = 0$ (c) G_8 for $k = 1$ (d) G_{11} for $k = 1$
                         (failure)

Fig. 8. Illustration for Theroem 7. Case $\omega = 8$. Drawings of G_8 and G_{11}.

Proof (sketch). We prove here the statement for $\omega = 8$, and we show in the full version [11] how to extend the result to any $\omega > 8$. Consider the instance $S = (G, 8, 0, \tau)$ in Fig. 7, where the vertices are labeled with their subscript in the order $\tau = \langle v_1, v_2, \ldots, v_{11} \rangle$. Graph G is a parallel composition of four components, three of which are a series of an edge and a triangle, and the other one is a path of length four.

Observe that, in any planar embedding of $G_\omega = G_8$ at most two among v_1, v_2, and v_3 can be incident to the same face (see Fig. 8(a)). Graph G_{11} contains the paths $(v_7, v_4, v_8), (v_7, v_5, v_8), (v_7, v_6, v_8)$, and $(v_7, v_9, v_{10}, v_{11}, v_8)$. Since v_9, v_{10}, and v_{11} are mapped to the points where v_1, v_2 and v_3 are mapped, respectively, it is not possible to obtain a planar embedding of G_{11} (see Fig. 8(b)). Thus, S does not admit a realization.

To prove that $S = (G, 8, k, \tau)$ is realizable for $k \geq \lfloor \frac{\omega}{2} \rfloor - 3 = 1$, suppose that v_1 and v_2 are drawn on the same face f and there is an extra point p inside f. In this case S is realizable, and G_8 and G_{11} are drawn as in Fig. 8(c) (d). $\square$

6   Final Remarks and Open Problems

We conclude with some open research directions. *(i)* Theorem 1 implies that the realizability testing of graph stories is paraNP-hard when parameterized by k. On the other hand, Theorem 2 proves that the problem is in FPT when parameterized by $\omega + k$. For non-minimal graph stories, it remains open to establish the complexity of the realizability problem when parameterized by ω alone. *(ii)* About minimal graph stories, we showed that for $\omega \geq 5$ there are stories of series-parallel graphs that are not realizable. For $k = 1$, the smaller ω for which we have a non-realizable story of a series-parallel graph is 10. What about the realizability of series-parallel graphs for $k = 1$ and $5 \leq \omega \leq 9$? *(iii)* Finally, is any (minimal) graph story h-reroute realizable for h being a constant or a sublinear function of ω?

References

1. Badent, M., Di Giacomo, E., Liotta, G.: Drawing colored graphs on colored points. Theor. Comput. Sci. **408**(2–3), 129–142 (2008). https://doi.org/10.1016/j.tcs.2008. 08.004
2. Beck, F., Burch, M., Diehl, S., Weiskopf, D.: A taxonomy and survey of dynamic graph visualization. Comput. Graph. Forum **36**(1), 133–159 (2017). https://doi. org/10.1111/cgf.12791
3. Binucci, C., et al.: Drawing trees in a streaming model. Inf. Process. Lett. **112**(11), 418–422 (2012). https://doi.org/10.1016/j.ipl.2012.02.011
4. Bläsius, T., Kobourov, S.G., Rutter, I.: Simultaneous embedding of planar graphs. In: Handbook on Graph Drawing and Visualization, pp. 349–381. Chapman and Hall/CRC (2013)
5. Borrazzo, M., Da Lozzo, G., Di Battista, G., Frati, F., Patrignani, M.: Graph stories in small area. J. Graph Algorithms Appl. **24**(3), 269–292 (2020). https:// doi.org/10.7155/jgaa.00530
6. Braß, P., et al.: On simultaneous planar graph embeddings. Comput. Geom. **36**(2), 117–130 (2007). https://doi.org/10.1016/j.comgeo.2006.05.006
7. Chan, T.M., Frati, F., Gutwenger, C., Lubiw, A., Mutzel, P., Schaefer, M.: Drawing partially embedded and simultaneously planar graphs. J. Graph Algorithms Appl. **19**(2), 681–706 (2015). https://doi.org/10.7155/jgaa.00375
8. Cohen, R.F., Di Battista, G., Tamassia, R., Tollis, I.G.: Dynamic graph drawings: trees, series-parallel digraphs, and planar ST-digraphs. SIAM J. Comput. **24**(5), 970–1001 (1995). https://doi.org/10.1137/S0097539792235724
9. Da Lozzo, G., Rutter, I.: Planarity of streamed graphs. Theor. Comput. Sci. **799**, 1–21 (2019). https://doi.org/10.1016/j.tcs.2019.09.029
10. Demetrescu, C., Di Battista, G., Finocchi, I., Liotta, G., Patrignani, M., Pizzonia, M.: Infinite trees and the future. In: Kratochvíyl, J. (ed.) GD 1999. LNCS, vol. 1731, pp. 379–391. Springer, Heidelberg (1999). https://doi.org/10.1007/3-540-46648-7_39
11. Di Battista, G., et al.: Small point-sets supporting graph stories (2022). https:// doi.org/10.48550/ARXIV.2208.14126
12. Di Battista, G., Tamassia, R.: On-line planarity testing. SIAM J. Comput. **25**(5), 956–997 (1996). https://doi.org/10.1137/S0097539794280736
13. Di Battista, G., Tamassia, R., Vismara, L.: Incremental convex planarity testing. Inf. Comput. **169**(1), 94–126 (2001). https://doi.org/10.1006/inco.2001.3031
14. Goodrich, M.T., Pszona, P.: Streamed graph drawing and the file maintenance problem. In: Wismath, S., Wolff, A. (eds.) GD 2013. LNCS, vol. 8242, pp. 256–267. Springer, Cham (2013). https://doi.org/10.1007/978-3-319-03841-4_23

15. Italiano, G.F.: Fully dynamic planarity testing. In: Encyclopedia of Algorithms, pp. 806–808. Springer (2016). https://doi.org/10.1007/978-1-4939-2864-4_157

16. Kaplan, H., Shamir, R.: Pathwidth, bandwidth, and completion problems to proper interval graphs with small cliques. SIAM J. Comput. **25**(3), 540–561 (1996). https://doi.org/10.1137/S0097539793258143

17. Michail, O.: An introduction to temporal graphs: an algorithmic perspective. Internet Math. **12**(4), 239–280 (2016). https://doi.org/10.1080/15427951.2016.1177801

18. Pach, J., Wenger, R.: Embedding planar graphs at fixed vertex locations. Graphs Comb. **17**(4), 717–728 (2001). https://doi.org/10.1007/PL00007258

19. Poutré, J.A.L.: Alpha-algorithms for incremental planarity testing (preliminary version). In: Leighton, F.T., Goodrich, M.T. (eds.) Proceedings of the Twenty-Sixth Annual ACM Symposium on Theory of Computing, 23–25 May 1994. Montréal, Québec, Canada, pp. 706–715. ACM (1994). https://doi.org/10.1145/195058.195439

20. Reingold, E.M., Tilford, J.S.: Tidier drawings of trees. IEEE Trans. Software Eng. **7**(2), 223–228 (1981). https://doi.org/10.1109/TSE.1981.234519

21. Rextin, A., Healy, P.: A fully dynamic algorithm to test the upward planarity of single-source embedded digraphs. In: Tollis, I.G., Patrignani, M. (eds.) GD 2008. LNCS, vol. 5417, pp. 254–265. Springer, Heidelberg (2009). https://doi.org/10.1007/978-3-642-00219-9_24

22. Schaefer, M.: Toward a theory of planarity: Hanani-Tutte and planarity variants. J. Graph Algorithms Appl. **17**(4), 367–440 (2013). https://doi.org/10.7155/jgaa.00298

23. Skambath, M., Tantau, T.: Offline drawing of dynamic trees: algorithmics and document integration. In: Hu, Y., Nöllenburg, M. (eds.) GD 2016. LNCS, vol. 9801, pp. 572–586. Springer, Cham (2016). https://doi.org/10.1007/978-3-319-50106-2_44

24. Whitney, H.: Congruent graphs and the connectivity of graphs. Am. J. Math. **54**(1), 150–168 (1932). http://www.jstor.org/stable/2371086

On the Complexity of the Storyplan Problem

Carla Binucci[1], Emilio Di Giacomo[1]([⊠]), William J. Lenhart[2],
Giuseppe Liotta[1], Fabrizio Montecchiani[1], Martin Nöllenburg[3],
and Antonios Symvonis[4]

[1] Department of Engineering, University of Perugia, Perugia, Italy
{carla.binucci,emilio.digiacomo,giuseppe.liotta,
fabrizio.montecchiani}@unipg.it
[2] Department of Computer Science, Williams College, Williamstown, USA
wlenhart@williams.edu
[3] Algorithms and Complexity Group, TU Wien, Vienna, Austria
noellenburg@ac.tuwien.ac.at
[4] School of Applied Mathematical and Physical Sciences, NTUA, Athens, Greece
symvonis@math.ntua.gr

Abstract. Motivated by dynamic graph visualization, we study the problem of representing a graph G in the form of a *storyplan*, that is, a sequence of frames with the following properties. Each frame is a planar drawing of the subgraph of G induced by a suitably defined subset of its vertices. Between two consecutive frames, a new vertex appears while some other vertices may disappear, namely those whose incident edges have already been drawn in at least one frame. In a storyplan, each vertex appears and disappears exactly once. For a vertex (edge) visible in a sequence of consecutive frames, the point (curve) representing it does not change throughout the sequence.

Note that the order in which the vertices of G appear in the sequence of frames is a total order. In the STORYPLAN problem, we are given a graph and we want to decide whether there exists a total order of its vertices for which a storyplan exists. We prove that the problem is NP-complete, and complement this hardness with two parameterized algorithms, one in the vertex cover number and one in the feedback edge set number of G. Also, we prove that partial 3-trees always admit a storyplan, which can be computed in linear time. Finally, we show that the problem remains NP-complete in the case in which the total order of the vertices is given as part of the input and we have to choose how to draw the frames.

Keywords: Dynamic graph drawing · NP-hardness · Parameterized analysis · Pathwidth

Research partially supported by MIUR grant 20174LF3T8 *"AHeAD: efficient Algorithms for HArnessing networked Data"*, Progetto RICBA21LG *"Algoritmi, modelli e sistemi per la rappresentazione visuale di reti"*, and the Vienna Science and Technology Fund (WWTF) grant ICT19-035.

P. Angelini and R. von Hanxleden (Eds.): GD 2022, LNCS 13764, pp. 304–318, 2023.
https://doi.org/10.1007/978-3-031-22203-0_22

1   Introduction

Let $G = (V, E)$ be a graph with n vertices. We write $[n]$ as shorthand for the set $\{1, 2, \ldots, n\}$. A *storyplan* $\mathcal{S} = \langle \tau, \{D_i\}_{i \in [n]} \rangle$ of G is a pair defined as follows. The first element is a bijection $\tau : V \to [n]$ that represents a total order of the vertices of G. For a vertex $v \in V$, let $i_v = \tau(v)$ and let $j_v = \max_{u \in N[v]} \tau(u)$, where $N[v]$ is the set containing v and its neighbors. The *lifespan* of v is the interval $[i_v, j_v]$. We say that v *appears* at step i_v, is *visible* at *step* i for each $i \in [i_v, j_v]$, and *disappears* at step $j_v + 1$. Note that a vertex does not disappear until all its neighbors have appeared. The second element of $\mathcal{S}$ is a sequence of drawings $\{D_i\}_{i \in [n]}$, such that: (i) each drawing D_i contains all vertices visible at step i, (ii) each drawing D_i is planar, (iii) the point representing a vertex v is the same over all drawings that contain v (i.e., it does not change during the lifespan of v), and (iv) the curve representing an edge e is the same over all drawings that contain e. We introduce the STORYPLAN problem.

STORYPLAN
Input: Graph $G = (V, E)$
Question: Does G admit a storyplan?

In what follows, each drawing D_i of a storyplan $\mathcal{S}$ is called a *frame* of $\mathcal{S}$. Also, we denote by $|D_i|$ the number of vertices of D_i, while the *width* of $\mathcal{S}$ is $w(\mathcal{S}) = \max_{i \in [n]} |D_i| - 1$ (we subtract one to align the definition with other width parameters). If G admits a storyplan, then the *framewidth* of G, denoted by $\mathrm{fw}(G)$, is the minimum width over all its storyplans; otherwise the framewidth of G is conventionally set to $+\infty$. We will observe that the framewidth of G upper bounds its pathwidth [12], since each frame can be interpreted as a bag of a path decomposition with the addition of conditions (ii)–(iv).

Motivation and Related Work. Testing for the existence of a storyplan of a graph generalizes planarity and it is of theoretical interest as it combines classical width parameters of graphs with topological properties. From a more practical perspective, computing a storyplan (if any) of a graph G is a natural way to gradually visualize G in a story-like or small-multiples fashion, such that each single drawing is planar and the reader's mental map is preserved throughout the sequence of drawings (see, e.g., [8] for a similar approach). More in general, the problem of visualizing graphs that change over time has motivated a notable amount of literature in graph drawing and network visualization (see, e.g., [2,3,6,7,15,16]). While numerous dynamic graph visualization models have been proposed, two works are of particular interest for our research. The first one is the work by Borrazzo et al. [6], in which the following problem is introduced. A *graph story* is formed by a graph G, a total order of its vertices τ, and a positive integer W. The problem is to find a sequence of drawings $\{D_i\}_{i \in [n]}$ in which each D_i contains all vertices v such that $i - W < \tau(v) \leq i$, and the position of a vertex is the same over all drawings it belongs to. Borrazzo et al. prove that any story of a path or a tree can be drawn on a $2W \times 2W$ and on an $(8W + 1) \times (8W + 1)$ grid, respectively, so that all the drawings of the story are straight-line and planar. Note that having a fixed window of size W implies that at most $O(W \cdot n)$ edges of G can be represented, in particular, any edge whose endpoints are at distance larger than W in τ does not appear in any drawing. Having both a fixed order and a fixed lifespan are the key differences with our setting. In particular, unconstrained lifespans allow us to find stories in which all edges are drawn in at least one step, while planarity still guarantees that even large frames are readable. Besides such differences in the models, our focus is on the complexity of the decision problem, rather than on area bounds for specific graph families. The second work is by Da Lozzo and Rutter [7], who introduce *stream planarity*. Given a graph G, a total order τ of the *edges* of G, and a positive integer W, stream planarity asks for a sequence of drawings $\{D_i\}_{i \in [n]}$ in which each D_i contains all edges e such that $i - W < \tau(e) \leq i$, and the subdrawing of the vertices and edges shared by D_i and D_{i-1} is the same in both drawings. Da Lozzo and Rutter prove that there exists a constant value for W for which the stream planarity problem is NP-complete. They also study a variant where a backbone graph is given whose edges must stay in the drawing at each time step; for this variant they prove that the problem is NP-complete for all $W \geq 2$ and can be solved in polynomial time when $W = 1$ or when the backbone graph is biconnected. The difference of stream planarity with our problem, besides the fact that edges are streamed rather than vertices, is again having a fixed order and a fixed lifespan.

Contribution. The main results in this paper can be summarized as follows.

- We show that STORYPLAN is NP-complete (Sect. 3.1). As we reduce from ONE-IN-THREE 3SAT and we blow up the instance by a linear factor, it follows that there is no algorithm that solves STORYPLAN in $2^{o(n)}$ time unless ETH fails. On the other hand, such a lower bound can be complemented with a simple algorithm running in $2^{O(n \log n)}$ time.
- Motivated by the above hardness, we study the parameterized complexity of STORYPLAN and describe two fixed-parameter tractable algorithms. We first show that STORYPLAN belongs to FPT when parameterized by the vertex cover number via the existence of a kernel, whose size is however super-polynomial (Sect. 3.2). We then prove that STORYPLAN parameterized by the feedback edge set number (i.e., the minimum number of edges whose removal makes the graph acyclic) admits a kernel of linear size (Sect. 3.3).
- In parameterized analysis, a central parameter to consider is treewidth. In this direction, finding a parameterized algorithm for STORYPLAN appears to be an elusive task. However, we show that for partial 3-trees, a storyplan always exists and can be computed in linear time (Sect. 3.4).
- Finally, we initiate the study of the complexity of a variant of STORYPLAN in which the total order of the vertices is fixed in advance (but the vertex lifespan remains unconstrained). We prove NP-completeness for this problem via a reduction from SUNFLOWER SEFE [14] (Sect. 4).

Some proofs are omitted and can be found in [4]; the corresponding statements are marked ($\star$).

2   Preliminaries and Basic Results

A *drawing* Γ of a graph $G = (V, E)$ is a mapping of the vertices of V to points in the plane $\mathbb{R}^2$, and of the edges of E to Jordan arcs connecting their corresponding endpoints but not passing through any other vertex. Drawing Γ is *planar* if no edge is crossed. A graph is *planar* if it admits a planar drawing. A planar drawing of a planar graph G subdivides the plane into topologically connected regions, called *faces*. The infinite region is the *outer face*. A *planar embedding* $\mathcal{E}$ of G is an equivalence class of planar drawings that define the same set of faces and the same outer face. For any $V' \subseteq V$, we denote by $G[V']$ the subgraph of G induced by the vertices of V' and by $\Gamma[V']$ the subdrawing of Γ representing $G[V']$.

Connection with Pathwidth. The next properties show some simple connections between storyplans and path decompositions [12].

Theorem 1 ($\star$). *Let $G = (V, E)$ be a graph, then $\mathrm{pw}(G) \leq \mathrm{fw}(G)$. Also, if G is planar then it always admits a storyplan, and in particular $\mathrm{pw}(G) = \mathrm{fw}(G)$.*

Since computing the pathwidth is NP-hard already for planar graphs of bounded degree [11], the next corollary immediately follows from Theorem 1.

Corollary 1. *Computing the framewidth of a graph is NP-hard for planar graphs of bounded degree.*

Analogously, computing the pathwidth of a graph is FPT in the pathwidth [5], hence computing the framewidth of a planar graph is also FPT in the framewidth.

Complete Bipartite Graphs. It is not difficult to verify that if a graph admits a storyplan, then it does not contain K_5 as a subgraph. However, complete bipartite graphs always admit a storyplan and such storyplans have important properties. The next statement plays a central role in most of our proofs.

Lemma 1 ($\star$). *Let $K_{a,b} = (A \cup B, E)$ be a complete bipartite graph with $a = |A|$, $b = |B|$, and $3 \leq b \leq a$. Let $S = \langle \tau, \{D_i\}_{i \in [a+b]} \rangle$ be a storyplan of $K_{a,b}$. Exactly one of A and B is such that all its vertices are visible at some $i \in [a + b]$.*

In view of Lemma 1, we have the following definition.

Definition 1. *For a complete bipartite graph $K_{a,b}$ with $3 \leq b \leq a$ and a storyplan S of $K_{a,b}$, we call* fixed *the partite set of $K_{a,b}$ whose vertices are all visible at some step of S, and* flexible *the other partite set.*

3   Complexity of STORYPLAN

In this section we prove that: STORYPLAN is NP-complete and cannot be solved in $2^{o(n)}$ time unless ETH fails, but there is an algorithm running in $2^{O(n \log n)}$ time (Sect. 3.1); STORYPLAN is in FPT parameterized by vertex cover number or feedback edge set number (Sects. 3.2 and 3.3); graphs of treewidth at most 3 always admit a storyplan, which can be computed in linear time (Sect. 3.4).

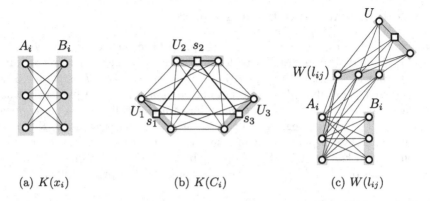

(a) $K(x_i)$ (b) $K(C_i)$ (c) $W(l_{ij})$

Fig. 1. Illustration for the reduction of Theorem 2.

3.1 Hardness

We reduce from ONE-IN-THREE 3SAT, a variant of 3SAT which asks whether there is a satisfying assignment in which *exactly* one literal in each clause is true. Let φ be a 3SAT formula over N variables $\{x_i\}_{i \in [N]}$ and M clauses $\{C_i\}_{i \in [M]}$. We construct an instance of STORYPLAN, i.e., a graph $G = (V, E)$, as follows; refer to Fig. 1 for an illustration.

Variable Gadget. Each variable x_i is represented in G by a copy $K(x_i)$ of $K_{3,3}$ (see Fig. 1(a)). Let A_i and B_i be the two partite sets of $K(x_i)$, which we call the *v-sides* of $K(x_i)$. A true (false) assignment of x_i will correspond to set A_i being flexible (fixed) in a putative storyplan of G (see Definition 1).

Clause Gadget. Consider a copy of $K_{2,2,2} = (U_1 \cup U_2 \cup U_3, F)$. An *extended* $K_{2,2,2}$ is the graph obtained from any such a copy by adding three vertices s_1, s_2, s_3, such that these three vertices are pairwise adjacent, and each s_j is adjacent to both vertices in U_j, for $j \in \{1, 2, 3\}$. In the following, s_1, s_2, s_3 are the *special vertices* of the extended $K_{2,2,2}$, while the other vertices are the *simple vertices*. A clause C_i is represented in G by an extended $K_{2,2,2}$, denoted by $K(C_i)$ (see Fig. 1(b)). In particular, we call each of the three sets of vertices $U_j \cup \{s_j\}$ a *c-side* of $K(C_i)$. The idea is that $K(C_i)$ admits a storyplan if and only if exactly one c-side is flexible (each c-side will be part of a $K_{3,3}$, see the wire gadget below).

Wire Gadget. Refer to Fig. 1(c). Let x_i be a variable having a literal l_{ij} in a clause C_j. Any such variable-clause incidence is represented in G by a set of three vertices, which we call the *w-side* $W(l_{ij})$. All vertices of $W(l_{ij})$ are connected to all vertices of one of the three c-sides of $K(C_j)$, which we call U, such that the graph induced by $W(l_{ij}) \cup U$ in G contains a copy of $K_{3,3}$. Also, each vertex of $W(l_{ij})$ is connected to all vertices of the v-side A_i (B_i) if the literal is positive (negative), such that the graph induced by $W(l_{ij}) \cup A$ ($W(l_{ij}) \cup B$) in G is a copy of $K_{3,3}$. Also, note that each c-side of $K(C_j)$ is adjacent to exactly one w-side.

Lemma 2 (⋆). *If graph G admits a storyplan then φ admits a satisfying assignment with exactly one true literal in each clause.*

Proof (Sketch). Let S be a storyplan of G. For each variable gadget $K(x_i)$ we assign the value *true* to x_i if the v-side A_i is flexible in S. Consider any literal l_{ij} and the wire gadget W_{ij}. If l_{ij} is positive (negative), then A_i (B_i) and W_{ij} form a $K_{3,3}$, hence by Lemma 1 the w-side W_{ij} is fixed (flexible). Analogously, if we consider the clause gadget $K(C_j)$, the c-side connected with W_{ij} is flexible (fixed). Symmetrically, we assign the value *false* to x_i if the v-side B_i is instead flexible in S, and for any positive (negative) literal l_{ij}, the w-side W_{ij} is flexible (fixed), while the corresponding c-side of $K(C_j)$ is fixed (flexible). In other words, the value of x_i propagates consistently throughout all its literals. It remains to prove that, for any clause C_j of $φ$, precisely one literal is true. Namely, we claim that exactly one c-side of $K(C_j)$ is flexible, while the other two are fixed. At high level, we rely on the fact that an extended $K_{2,2,2}$ wants at least two c-sides to be fixed, while the special vertices force at least one c-side to be flexible.

Lemma 3 (⋆). *If the formula φ admits a satisfying assignment with exactly one true literal in each clause, then graph G admits a storyplan.*

Proof (Sketch). Given a satisfying assignment of $φ$ with one true literal per clause, we can compute a storyplan $S = \langle \tau, \{D_i\}_{i \in [n]} \rangle$ of G. In what follows, when the order of a group of vertices is not specified, any relative order is valid.

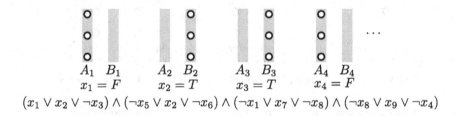

$(x_1 \lor x_2 \lor \neg x_3) \land (\neg x_5 \lor x_2 \lor \neg x_6) \land (\neg x_1 \lor x_7 \lor \neg x_8) \land (\neg x_8 \lor x_9 \lor \neg x_4)$

Fig. 2. Proof of Lemma 3: drawing the vertices of the fixed v-sides.

Consider a single variable gadget $K(x_i)$. If x_i is true in the satisfying assignment, then we let appear the three vertices of the v-side B_i of $K(x_i)$, that is, B_i is the fixed side of $K(x_i)$. If x_i is false, we do the opposite, namely we let appear the three vertices of the v-side A_i of $K(x_i)$. This procedure is repeated for all variables in any order. For ease of presentation, we can imagine that all the drawn v-sides are horizontally aligned, as shown in Fig. 2. Thus, for the variable gadgets, it remains to draw their flexible v-sides.

Consider now a wire gadget $W(l_{ij})$. If x_i is true and l_{ij} is positive, then $W(l_{ij})$ must be fixed because it forms a $K_{3,3}$ with the v-side A_i of $K(x_i)$, which is flexible. Therefore we let appear the three vertices of $W(l_{ij})$. Similarly, if x_i is false and l_{ij} is negative, then $W(l_{ij})$ must be fixed, and we let appear the three vertices of $W(l_{ij})$. Again, this procedure is repeated for all wires in any order.

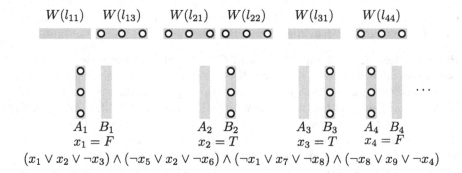

Fig. 3. Proof of Lemma 3: drawing the vertices of the fixed w-sides.

For ease of presentation, we can imagine that all the drawn w-sides are arranged along a horizontal line slightly above the variable gadgets, as shown in Fig. 3. Thus, also for the wire gadgets, it remains to draw the flexible w-sides.

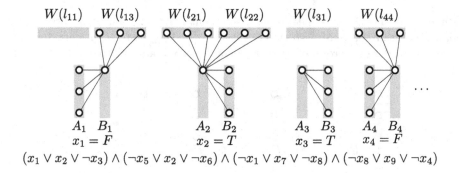

Fig. 4. Proof of Lemma 3: drawing the vertices of the flexible v-sides.

We sketch the remaining part of the proof (see [4] for a full proof). Flexible v-sides can be drawn as in Fig. 4. Figure 5 shows how to draw a clause gadget, ignoring the connections with the linked wire gadgets. Finally, in order to draw the flexible w-sides and their edges, and the edges between the fixed w-sides and the corresponding c-sides, we enclose all the wire and variable gadgets in a face of the current clause gadget where all vertices of the linked c-side are visible.

Theorem 2 (⋆). *The* STORYPLAN *problem is* NP-*hard and it has no* $2^{o(n)}$ *time algorithm unless* ETH *fails.*

The above lower bound for the running time of an algorithm solving STORY-PLAN can be easily complemented with a nearly tight upper bound. The proof of the next theorem also shows that STORYPLAN belongs to NP. Namely, it gives a nondeterministic scheme to generate a set of candidate solutions, and then it

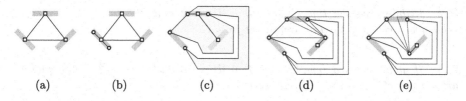

Fig. 5. Proof of Lemma 3: drawing a clause gadget.

shows how to check, in polynomial time, if a candidate solution is valid. However, it intertwines such process in order to obtain a lower time complexity.

Theorem 3 (⋆). *The* STORYPLAN *problem is in* NP. *Also, given an n-vertex graph G, there is an algorithm that solves* STORYPLAN *on G in* $2^{O(n \log n)}$ *time.*

Proof (Sketch). We first guess a total order of the vertices of G. This fixes for each $i \in [n]$ the visible vertices. Next, for each $i \in [n]$, we generate the possible planar embeddings (rather than planar drawings) of the graph induced by the vertices visible at step i, and discard any embedding $\mathcal{E}$ for which there is no planar embedding $\mathcal{E}'$ generated at step $i-1$ (if $i > 1$) such that the restrictions of $\mathcal{E}$ and $\mathcal{E}'$ to the common subgraph coincide. If the algorithm returns at least one planar embedding at step n, there is a sequence of planar embeddings in which common subgraphs share the same embedding, hence G admits a storyplan.

3.2 Parameterization by Vertex Cover Number

A *vertex cover* of a graph $G = (V, E)$ is a set $C \subseteq V$ such that every edge of E is incident to a vertex in C, and the *vertex cover number* of G is the minimum size of a vertex cover of G. We prove the following by means of kernelization.

Theorem 4 (⋆). *Let* $G = (V, E)$ *be a graph with n vertices and vertex cover number* $\kappa = \kappa(G)$. *Deciding whether G admits a storyplan, and computing one if any, can be done in* $O(2^{2^{O(\kappa)}} + n^2)$ *time.*

Algorithm Description. Without loss of generality, we assume that the input graph G does not contain isolated vertices, as they do not affect the existence of a storyplan. Let C be a vertex cover of size $\kappa = \kappa(G)$ of graph G. For $U \subseteq C$, a vertex $v \in V \setminus C$ is of *type* U if $N(v) = U$, where $N(v)$ denotes the set of neighbors of v in G. This defines an equivalence relation on $V \setminus C$ and in particular partitions $V \setminus C$ into at most $\sum_{i=1}^{\kappa} \binom{\kappa}{i} = 2^{\kappa} - 1 < 2^{\kappa}$ distinct types. Denote by V_U the set of vertices of type U. We define three reduction rules.

R.1: *If there exists a type* U *such that* $|U| = 1$, *then pick an arbitrary vertex* $x \in V_U$ *and remove it from G.*

R.2: *If there exists a type U such that $|U| = 2$ and $|V_U| > 1$, then pick an arbitrary vertex $x \in V_U$ and remove it from G.*

R.3: *If there exists a type U such that $|U| \geq 3$ and $|V_U| > 3$, then pick an arbitrary vertex $x \in V_U$ and remove it from G.*

Lemma 4 ($\star$). *Let G' be the graph obtained from G by applying one of the reduction rules **R.1–R.3**. Then G admits a storyplan if and only if G' does.*

Proof (Sketch). For the nontrivial direction, suppose that G' admits a storyplan $\mathcal{S}' = \langle \tau', \{D'\}_{i \in [n']} \rangle$, where $n' = n - 1$. We can distinguish three cases based on the reduction rule applied to G. Here we only prove the simplest of the three cases, namely the case in which **R.1** is applied. See Fig. 6 for an illustration. Let x be the vertex removed from G to obtain G' and let v be its neighbor, whose lifespan according to τ' is $[i_v, j_v]$. We compute τ from τ' by inserting x right after v, thus the lifespan of x in τ is $[i_v + 1, i_v + 1]$. Similarly, we compute $\{D_i\}_{i \in [n]}$ from $\{D'\}_{i \in [n']}$ as follows. For each $i \leq i_v$, we set $D_i = D'_i$. For $i = i_v + 1$, we draw x in D'_{i_v} sufficiently close to v such that edge xv can be drawn as a straight-line segment that does not intersect any other edge. We then set D_i to be equal to the resulting drawing. For each $i > i_v + 1$, we set $D_i = D'_{i-1}$.

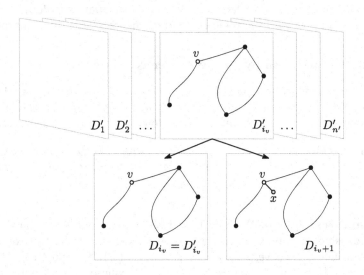

Fig. 6. Illustration for **Case A** of the proof of Lemma 4.

Based on Lemma 4, we can construct an equivalent instance of G of size $O(2^\kappa)$ and use it to conclude the proof of Theorem 4 (see [4]).

3.3   Parameterization by Feedback Edge Set

A *feedback edge set* of a graph $G = (V, E)$ is a set $F \subseteq E$ whose removal results in an acyclic graph, and the *feedback edge set number* of G is the minimum size of a feedback edge set of G. We prove the following.

Theorem 5. *Let G be a graph with n vertices and feedback edge set of size $\psi = \psi(G)$. Deciding whether G admits a storyplan, and computing one if any, can be done in $O(2^{O(\psi \log \psi)} + n^2)$ time.*

Algorithm Description. A *k-chain* of G is a path with $k + 2$ vertices and such that its k inner vertices all have degree two. We define two reduction rules.

R.A: *If there exists a vertex of degree one, then remove it from G .*

R.B: *If there exists a k -chain with $k \geq 3$, then remove its inner vertices from G.*

Based on the above reduction rules we can prove the following.

Lemma 5 ($\star$). STORYPLAN *parameterized by feedback edge set number admits a kernel of linear size.*

To conclude the proof of Theorem 5, observe that computing a linear kernel G^* of G, i.e., applying exhaustively the reduction rules **R.A** and **R.B**, can be done in $O(n + \psi)$ time. Afterwards, following the lines of the proof of Theorem 4, we can brute-force a solution for G^* (if any) in $2^{O(\psi \log \psi)}$ time, and reinsert the missing $O(n)$ vertices each in $O(n)$ time (as detailed in [4, Lemma 9]).

3.4   Partial 3-trees

A *k-tree* has a recursive definition: A complete graph with k vertices is a k-tree; for any k-tree H, the graph obtained from H by adding a new vertex v connected to a clique C of H of size k is a k-tree; C is the *parent clique* of v. A *partial k-tree* is a subgraph of a k-tree and partial k-trees are exactly the graphs of treewidth at most k. Since 2-trees are planar, they admit a storyplan by Theroem 1. We prove that the same holds for partial 3-trees (which may be not planar).

Theorem 6 ($\star$). *Every partial 3-tree G with n vertices admits a storyplan, which can be computed in $O(n)$ time.*

Proof (Sketch). We shall assume that G is a (non-partial) 3-tree. Indeed, if G is a partial 3-tree, a supergraph of G that is a 3-tree always exists by definition. We now construct a specific tree decomposition $\mathcal{T}$ of G that will be used to compute its storyplan; refer to Fig. 7. For a definition of tree decomposition see [13]. The subgraph C_μ induced by the vertices of each bag μ of $\mathcal{T}$ is the *subgraph associated with μ* and it is a 4-clique for each bag μ of T, except for the root ρ of T for which

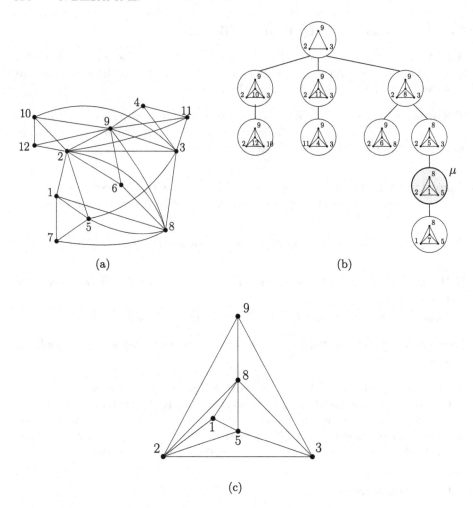

Fig. 7. (a) A 3-tree G. (b) The decomposition tree T of G. (c) The subgraph G_μ of G associated with bag μ, highlighted in (b).

C_ρ is the initial 3-cycle. The subgraph C_μ contains four 3-cliques, three of them are *active* (this means that they can appear in some subgraph C_ν associated with a child ν of μ) and one is *non-active*. The unique 3-clique in C_ρ is active. Each bag μ of T has one child ν for each vertex v whose parent clique is an active 3-clique of μ. The 4-clique C_ν consists of the parent clique C of v, vertex v, and the edges connecting v to C; C is non-active in ν, while the other three 3-cliques are active. For each bag μ distinct from ρ, we denote by v_μ the vertex shared by the three active 3-cliques of C_μ. We say that v_μ is *associated with* μ. One easily verifies that T is a tree decomposition. Also, T has $n - 2$ bags: the root and a bag for each vertex of G that is not in the initial 3-cycle.

We now associate to each bag μ a subgraph G_μ of G. For the root ρ, the subgraph G_ρ is the initial 3-cycle. For a bag μ with parent λ, the subgraph G_μ is obtained from G_λ by connecting v_μ to the vertices of its parent clique.

Property 1. Every graph G_μ is an embedded planar 3-tree such that the active 3-cliques of C_μ are internal faces of G_μ.

The proof of Property 1 is by induction on the length of the path from the root ρ to μ in $\mathcal{T}$. The graph G_ρ consists of a 3-cycle, which is the unique active 3-clique and which is both an internal and an external face. The graph G_μ is obtained by adding the vertex v_μ to G_λ and connecting it to its parent clique C, which is active in C_λ. By induction, C is an internal face of G_λ and therefore, by placing v_μ inside this face, we obtain an embedded planar 3-tree such that the active faces of C_μ are the three faces created by the addition of v_μ inside C.

Let $\mu \neq \rho$ be a bag of $\mathcal{T}$; the next property follows from the definition of $\mathcal{T}$.

Property 2. The neighbors of v_μ distinct from those of its parent clique are all vertices associated with bags of the subtree of $\mathcal{T}$ rooted at μ.

Let $\rho = \mu_1, \mu_2, \ldots, \mu_{n-2}$ be an order of the bags of $\mathcal{T}$ according to a preorder visit of $\mathcal{T}$. To create a storyplan of G, we define an ordering $\tau : v_1, v_2, \ldots, v_n$ of the vertices of G such that v_1, v_2, and v_3 are the vertices of the initial 3-cycle, and each v_i with $i > 3$ is the vertex associated with μ_{i-2}.

Let G_i be the graph induced by the vertices that are visible at step i, for $i \geq 3$; by Property 2 the graph G_i is a subgraph of G_{μ_i} which, by Property 1 is an embedded planar 3-tree such that the three active 3-cliques of C_{μ_i} are faces of G_{μ_i}. To simplify the description we prove that there exists a storyplan $\mathcal{S} = \langle \tau, \{D_i\}_{i \in [n]} \rangle$, where each D_i is a drawing of G_{μ_i}. This implies that there exists a storyplan where each D_i is a drawing of G_i. Let μ_j be the parent of μ_i in $\mathcal{T}$; since the order τ corresponds to a preorder of the bags of $\mathcal{T}$, we have $j < i$. Moreover, all bags μ_k with $j < k < i$, if any, belong to the subtrees of μ_j visited before μ_i and for each such subtree $\mathcal{T}'$ no other bag of $\mathcal{T}'$ exists before μ_j or after μ_i. By Property 2 all the vertices associated with the bags μ_k that belong to $G_{\mu_{i-1}}$ do not have any neighbor after v_{μ_i} and therefore they can be removed. The removal of these vertices transforms $G_{\mu_{i-1}}$ into G_{μ_j} (all the vertices associated with the bags μ_k for $j < k < i$ had been added to G_{μ_j} that had never been changed). By Property 1 the active 3-cliques of G_{μ_j} are faces of G_{μ_j}. It follows that there exists a storyplan $\mathcal{S} = \langle \tau, \{D_i\}_{i \in [n]} \rangle$ whose frames D_i are as follows. D_1 is a planar drawing of a 3-cycle; given D_{i-1} of $G_{\mu_{i-1}}$, a drawing D_i of G_{μ_i} can be computed by removing all vertices associated with the bags μ_k for $j < k < i$, and adding v_{μ_i} inside a face of D_j.

The above storyplan can be computed in $O(n)$ time, see [4].

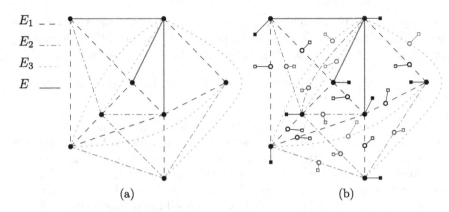

E_1 - -
E_2 - - -
E_3 ....
E ___

(a)   (b)

Fig. 8. Illustration for Theorem 7. (a) An instance G_1, G_2, G_3 of SUNFLOWER SEFE; (b) The instance G constructed from G_1, G_2, G_3; the subdivision vertices are circles with a white fill while the spectators are squares.

4   Complexity with Fixed Order

In this section we study the STORYPLANFIXEDORDER variant of STORYPLAN, defined below. This variant is closer to the setting studied in [6] and it models the case in which the way the graph changes over time is prescribed.

STORYPLANFIXEDORDER
Input: Graph $G = (V, E)$ with n vertices, total order $\tau : V \to [n]$
Question: Does G admit a storyplan $\mathcal{S} = \langle \tau, \{D_i\}_{i \in [n]} \rangle$?

We prove that STORYPLANFIXEDORDER is NP-complete by reducing from SUN-FLOWER SEFE. Let $G_1, \ldots, G_k$ be k graphs on the same set V of vertices. A *simultaneous embedding with fixed edges (SEFE)* of $G_1, \ldots, G_k$ consists of k planar drawings $\Gamma_1, \ldots, \Gamma_k$ of $G_1, \ldots, G_k$, respectively, such that each vertex is mapped to the same point in every drawing and each shared edge is represented by the same simple curve in all drawings sharing it. The SEFE problem asks whether k input graphs on the same set of vertices admit a SEFE, and it is NP-complete even when the pairwise intersection between any two input graphs is the same over all pairs of graphs [1,14]. This variant is called SUNFLOWER SEFE, and the result in [1,14] proves NP-completeness already when $k = 3$.

Construction. Refer to Fig. 8 for an example. Let G_1, G_2, G_3 be an instance of SUNFLOWER SEFE. Let V be the common vertex set of the three graphs, let E be the common edge set, and let E_i be the exclusive edge set of G_i for $i = 1, 2, 3$. We construct an instance $\langle G, \tau \rangle$ of STORYPLANFIXEDORDER as follows. Graph G contains all vertices in V and all edges in E. Also, for each edge $e = uv$ in E_i, it contains a vertex w_e^i, called *a subdivision vertex of E_i*, and the edges uw_e^i and vw_e^i (i.e., it contains the edge e subdivided once). Moreover, for each vertex z, either a vertex in V or a subdivision vertex of an edge, G contains an additional vertex s_z, called the *spectator of z*, and the edge zs_z. To obtain the total order τ we group the vertices of G in a set of blocks $B_1, \ldots, B_8$, and we order the blocks by increasing index, while vertices within the same block can be ordered arbitrarily. We denote by τ_i^- and τ_i^+ the position in τ of the first and of the last vertex of B_i, respectively, for each $i = 1, \ldots, 8$. Block B_1 contains all vertices in V; for $i \in \{2, 4, 6\}$, block B_i contains all subdivision vertices of $E_{\frac{i}{2}}$, while block B_{i+1} contains all spectators of the vertices in B_i; finally B_8 contains all spectators of the vertices in B_1.

Theorem 7 ($\star$). *The* STORYPLANFIXEDORDER *problem is* NP-*complete.*

Proof (Sketch). At a high level, the total order τ is designed to show the three graphs one by one while keeping the common edge set visible. In particular, a spectator vertex s_v forces vertex v to stay visible until s_v appears, while a subdivision vertex w_e^i makes edge e visible only when G_i must be drawn.

5   Discussion and Open Problems

Our work can stimulate further research based on several possible directions.

– It would be interesting to study further parameterizations of STORYPLAN. Is STORYPLAN parameterized by treewidth (pathwidth) in XP? In addition, we note that if the total order is fixed, then an FPT algorithm in the size of the largest frame (or the length of the longest lifespan) readily follows from the proof of Theorem 3.
– Conditions (iii) and (iv) of the definition of a storyplan can be replaced by the existence of a sequence of planar embeddings in which common subgraphs keep the same embedding. This is not true if we study more geometric versions of the problem, in which for instance edges are straight-line segments and/or vertices are restricted on an integer grid of fixed size (as in [6]).
– Condition (ii) of storyplan can be relaxed so to only allow specific crossing patterns [9,10], e.g., right-angle crossings or few crossings per edge.

Acknowledgement. Research in this work started at the Bertinoro Workshop on Graph Drawing 2022.

References

1. Angelini, P., Da Lozzo, G., Neuwirth, D.: Advancements on SEFE and partitioned book embedding problems. Theoret. Comput. Sci. **575**, 71–89 (2015). https://doi.org/10.1016/j.tcs.2014.11.016
2. Beck, F., Burch, M., Diehl, S., Weiskopf, D.: The state of the art in visualizing dynamic graphs. In: Borgo, R., Maciejewski, R., Viola, I. (eds.) EuroVis 2014. Eurographics Association (2014). https://doi.org/10.2312/eurovisstar.20141174
3. Binucci, C., Brandes, U., Di Battista, G., Didimo, W., Gaertler, M., Palladino, P., Patrignani, M., Symvonis, A., Zweig, K.A.: Drawing trees in a streaming model. Inf. Process. Lett. **112**(11), 418–422 (2012). https://doi.org/10.1016/j.ipl.2012.02.011

4. Binucci, C., Di Giacomo, E., Lenhart, W.J., Liotta, G., Montecchiani, F., Nöllenburg, M., Symvonis, A.: On the complexity of the storyplan problem. CoRR abs/2209.00453 (2022). https://arxiv.org/abs/2209.00453

5. Bodlaender, H.L., Kloks, T.: Efficient and constructive algorithms for the pathwidth and treewidth of graphs. J. Algorithms **21**(2), 358–402 (1996). https://doi.org/10.1006/jagm.1996.0049

6. Borrazzo, M., Da Lozzo, G., Di Battista, G., Frati, F., Patrignani, M.: Graph stories in small area. J. Graph Algorithms Appl. **24**(3), 269–292 (2020). https://doi.org/10.7155/jgaa.00530

7. Da Lozzo, G., Rutter, I.: Planarity of streamed graphs. Theoret. Comput. Sci. **799**, 1–21 (2019). https://doi.org/10.1016/j.tcs.2019.09.029

8. Di Giacomo, E., Didimo, W., Liotta, G., Montecchiani, F., Tollis, I.G.: Techniques for edge stratification of complex graph drawings. J. Vis. Lang. Comput. **25**(4), 533–543 (2014). https://doi.org/10.1016/j.jvlc.2014.05.001

9. Didimo, W., Liotta, G., Montecchiani, F.: A survey on graph drawing beyond planarity. ACM Comput. Surv. **52**(1), 1–37 (2019). https://doi.org/10.1145/3301281

10. Hong, S.-H., Tokuyama, T. (eds.): Beyond Planar Graphs. Springer, Singapore (2020). https://doi.org/10.1007/978-981-15-6533-5

11. Monien, B., Sudborough, I.H.: Min Cut is NP-complete for edge weighted trees. Theor. Comput. Sci. **58**, 209–229 (1988). https://doi.org/10.1016/0304-3975(88)90028-X

12. Robertson, N., Seymour, P.D.: Graph minors. I. Excluding a forest. J. Comb. Theory, Ser. B **35**(1), 39–61 (1983). https://doi.org/10.1016/0095-8956(83)90079-5

13. Robertson, N., Seymour, P.: Graph minors. II. Algorithmic aspects of tree-width. J. Algorithms **7**(3), 309–322 (1986). https://doi.org/10.1016/0196-6774(86)90023-4

14. Schaefer, M.: Toward a theory of planarity: Hanani-Tutte and planarity variants. J. Graph Algorithms Appl. **17**(4), 367–440 (2013). https://doi.org/10.7155/jgaa.00298

15. Skambath, M., Tantau, T.: Offline drawing of dynamic trees: algorithmics and document integration. In: Hu, Y., Nöllenburg, M. (eds.) Graph Drawing and Network Visualization (GD'16). LNCS, vol. 9801, pp. 572–586. Springer (2016). https://doi.org/10.1007/978-3-319-50106-2_44

16. Vehlow, C., Beck, F., Weiskopf, D.: Visualizing dynamic hierarchies in graph sequences. IEEE Trans. Vis. Comput. Graph. **22**(10), 2343–2357 (2016). https://doi.org/10.1109/TVCG.2015.2507595

Visualizing Evolving Trees

Kathryn Gray[1], Mingwei Li[2]📧, Reyan Ahmed[3]([✉]) 📧,
and Stephen Kobourov[1]📧

[1] Department of Computer Science, University of Arizona, Tucson, USA
ryngray@arizona.edu, kobourov@cs.arizona.edu
[2] Department of Computer Science, Vanderbilt University, Nashville, USA
mingwei.li@vanderbilt.edu
[3] Department of Computer Science, Colgate University, Hamilton, USA
rahmed1@colgate.edu

Abstract. Evolving trees arise in many real-life scenarios from computer file systems and dynamic call graphs, to fake news propagation and disease spread. Most layout algorithms for static trees do not work well in an evolving setting (e.g., they are not designed to be stable between time steps). Dynamic graph layout algorithms are better suited to this task, although they often introduce unnecessary edge crossings. With this in mind we propose two methods for visualizing evolving trees that guarantee no edge crossings, while optimizing (1) desired edge length realization, (2) layout compactness, and (3) stability. We evaluate the two new methods, along with five prior approaches (three static and two dynamic), on real-world datasets using quantitative metrics: stress, desired edge length realization, layout compactness, stability, and running time. The new methods are fully functional and available on github. (This work was supported in part by NSF grants CCF-1740858, CCF-1712119, and DMS-1839274.)

1   Introduction

Dynamic graph visualization is used in many fields including social networks [28], bibliometric networks [49], software engineering [14], and pandemic modeling [7]; see the survey by Beck *et al.* [10]. Here we focus on a special case, *evolving trees*. In evolving trees the dynamics are captured only by growth (whereas in general dynamic graphs, nodes and edges can also disappear). While this is a significant restriction of the general dynamic graph model, evolving trees are common in many domains including the Tree of Life [37] and the Mathematics Genealogy Graph [36]. An evolving tree can also model disease spread, where nodes correspond to infected individuals and a new node v is added along with an edge to existing node u if u infected v. Visualizing this process can help us see how the infection spreads, the rate of infection, and to identify "super-spreaders."

There are several methods and tools that can be used to visualize evolving trees [2,15,16,39], however, most of them have limitations that can impact their usability in this domain. Some represent nodes only as points ignoring labels [15, 16], which makes them less useful in real-life applications where it is important

P. Angelini and R. von Hanxleden (Eds.): GD 2022, LNCS 13764, pp. 319–335, 2023.
https://doi.org/10.1007/978-3-031-22203-0_23

to see what each node represents. Others utilize the level-by-level approach for drawing hierarchical graphs [27,41], which does not capture the underlying graph structure well. Force-directed algorithms tend to better capture the underlying graph structure [32], although they may introduce unnecessary edge crossings. With this in mind, we propose two methods for drawing crossing-free evolving trees that optimize the following desirable properties:

1. **Desired edge length realization:** The Euclidean distance between two nodes u and v in the layout should realize the corresponding pre-specified edge length $l(u,v)$, or be uniform when no additional information is given. This is important in several domains, e.g., when visualizing phylogenetic trees [6], where the edge length represents evolutionary distance between two species.
2. **Layout compactness:** The drawing area should be proportional to the total area needed for all the labels. A good visualization should have the labeled graph drawn in a compact way [40]. This prevents the trivial solution of scaling the layout until all overlaps and crossings are removed, which can create vast empty spaces in the visualization.
3. **Stability:** Between time steps, nodes should move as little as possible. This helps the viewer maintain a mental map of the graph [38]. If the graph moves around too much, it is difficult to see where new nodes and edges are added and we lose the context of the new node's relation to the rest of the graph.

We propose two force-directed methods that ensure no edge crossings and optimize desired edge length, compactness and stability. Minimizing edge crossings is important in graph readability [43], and since we work with trees, a layout without edge crossings is possible and desirable. We use two trees extracted from Tree of Life [37] and the Mathematics Genealogy [36] projects to demonstrate the new methods and quantitatively evaluate their performance, measuring desired edge length realization, compactness, stability, stress, crossings, and running time. We also evaluate the performance of five earlier methods, showing the two proposed methods perform well overall; see Fig. 1.

2   Related Work

Dynamic graph drawing has a long history [11,47] and two broad categories: offline and online. In the easier offline setting we assume that all the data about the dynamics is known in advance. Algorithms for offline dynamic visualization use different approaches including combining all time-slice instances into a single supergraph [18–20], connecting the same node in consecutive time-slices and optimizing them simultaneously [22–24], providing animation [4], and showing small multiples type visualization [3]. *DynNoSlice* by Simonetto et al. [45] is one of the most recent approaches for this setting and is different from the prior methods as it does not rely on discrete time-slices.

Online dynamic graph drawing deals with the harder problem – when we do not know in advance what changes will occur. One can optimize the current view, given what has happened in the past, but cannot look into the future, as the

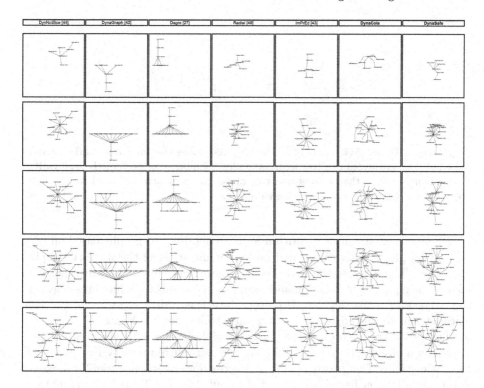

Fig. 1. Layouts from DynNoSlice, DynaGraph, Dagre, Radial, ImPrEd, DynaCola and DynaSafe of the same evolving math genealogy tree; each row adds six new nodes.

information is not available. Cohen *et al.* [15,16] and Workman *et al.* [51] describe algorithms for dynamic drawing of trees that place nodes that are equidistant from the root on the same level (same y coordinate). These algorithms do not take edge lengths into consideration, and the hierarchical nature of the layout can lead to exponential differences between the shortest and longest edges.

DynaDAG is an online graph drawing method for drawing dynamic directed acyclic graphs as hierarchies [41]. This method moves nodes between adjacent ranks based on the median sort. It was not specifically developed[1] for trees and may introduce crossings; see Fig. 2. Other approaches for online dynamic graph drawing have maintained the horizontal and vertical position of nodes [38], used node aging methods [29], and adapted multilevel approaches [17] (using FM^3 [31]). Online approaches have also been implemented on the GPU [25]. However, these methods do not guarantee crossing-free layouts for trees and do not take into account desired edge lengths.

---

[1] We have used the implementation available in the DynaGraph system: https://www.dynagraph.org/.

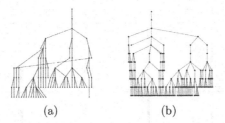

(a)   (b)

Fig. 2. (a) A DynaDAG layout of a tree with 100 nodes that introduces **edge crossings**. (b) A Dagre layout of a tree with 300 nodes, with **large edge length variability**. Both examples show that **nodes are too close to each other** if labeled. The trees are extracted from the math genealogy dataset.

Dagre is a multi-phase algorithm for drawing directed graphs based on [27]. The initial phase finds an optimal rank assignment using the network simplex algorithm. Then it sets the node order within ranks by an iterative heuristic incorporating a weight function and local transpositions to reduce crossings (via the barycenter heuristic) [33]. However, since Dagre draws graphs in a hierarchical structure, the edge lengths may vary arbitrarily; see Fig. 2.

The *radial* layout implemented in yFiles [50] displays each biconnected component in a circular fashion². Radial layouts were introduced by Kar *et al.* [34] for static graphs. Dougrusoz *et al.* [21] described an interactive tool for dynamic graph visualization based on the radial layout. Six and Tollis [46] adapted the radial idea to circular drawings of static biconnected graphs and experimentally showed that their layout has fewer edge crossings. Kaufmann *et al.* [35] extended this model to handle dynamic graphs, providing the basis of the yFiles *radial* implementation. Pavlo *et al.* [42] adapted the idea to make the radial layout computation parallelizable. Bachmaier [5] further improved the radial layout algorithm for static graphs by adapting the hierarchical approach [48] to minimize edge crossings. Radial layout methods have more freedom than the traditional level-by-level tree layout methods. Nevertheless, they are still constrained and can result in unstable visualization for dynamic graph in general and evolving trees in particular; see Fig. 3.

Force-directed algorithms [18–20,44,45] underlie many static and dynamic graph visualization methods. Unlike hierarchical and radial approaches, force-directed algorithms place nodes at arbitrary positions, and so tend to generate more compact layouts that better realize desired edge lengths. By adjusting forces appropriately, one can also generate stable layouts by this approach. In particular, ImPrEd, by Simonetto *et al.* [44], provides a force-directed approach to improve a given initial layout, without introducing new edge crossings. To the best of our knowledge, there are no force-directed methods for evolving trees.

---

² We use this radial layout algorithm later for our experiments.

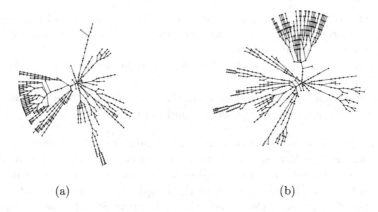

(a)   (b)

Fig. 3. A radial layout of a 400-node evolving tree of life (a) and the layout after adding 100 nodes (b). As most of the growth occurred on the right side, it is easy to see the **instability** – a lot of movement was needed to accommodate it. Also, **nodes are too close to each other** if labeled.

3   Algorithms for Visualizing Evolving Trees

Here, we describe two force-directed algorithms for evolving tree visualization, *DynaCola* and *DynaSafe*, that realize desired edge lengths without creating crossings, and optimize compactness and stability. DynaCola avoids edge crossings by creating and maintaining a "collision region" for each edge. While collision detection/prevention is usually applied to nodes, by carefully applying it to the edges we can prevent all edge crossings. DynaSafe prevents edge crossings with a "safe" coordinate update at every step of the algorithm. Before updating a coordinate, it first checks whether the update will introduce a crossing and then limits the update magnitude to avoid the crossing.

3.1   DynaCola

DynaCola stands for Dynamic Collision, as the algorithm uses the collision forces to prevent edge crossings. This is a force-directed algorithm, augmented with edge-regions used to prevent crossings; the pseudocode of the algorithm can be found in the full version [30]. Recall that we are gradually growing a tree, one node at a time, while maintaining a crossing-free layout and optimizing desirable properties (desired edge lengths, compactness, stability). The DynaCola force-directed algorithm relies on the following forces and is implemented in d3.js [12]:

– A force f_E for each edge, to realize the desired edge length. The strength of this force is proportional to the difference between the edge distance in the layout and the desired edge length.

- A general repulsive force f_R defined for all pairs of nodes and implemented with the Barnes-Hut quad-tree data structure [8]. This helps realize the global structure of the underlying tree.
- A collision force f_C for each edge, described in details below. This force prevents edge crossings.
- A gravitational force f_G that attracts all nodes to the center of mass. This force draws the nodes closer together and improves compactness.

To ensure that no edges cross during an update, we define a collision region around each edge: if any edge/node moves too close to another edge, it will be pushed away. To create a collision region for an edge $e = (u, v)$, we can create collision circles with diameter equal to the length of e for both u and v. Then every point of e will be either inside the collision region of u or v. However, the sum of all collision regions for all nodes will be unnecessarily large and the layout will not be compact. With the help of subdivision nodes along the edges, we can reduce the sum of all collision regions. Let $e = (u, v)$ be an edge in the graph. We use a set of subdivision nodes $V_s = \{v_1, v_2, \cdots, v_k\}$ and replace the edge (u, v) by a set of edges $E_s = \{(u, v_1), (v_1, v_2), \cdots, (v_k, v)\}$. We assign the desired edge length of an edge in E_s equal to $l(u, v)/|E_s|$, where $l(u, v)$ is the desired edge length. In general, the number of subdivision nodes per edge should be a small constant n_s (by default $n_s = 1$), since the complexity of the algorithm increases as n_s increases. Also, note that n_s determines the number of bends per edge (no bends when $n_s = 0$, one bend when $n_s = 1$, and so on). Note that, the collision force does not follow any hard constraint, even after having a collision region edge crossings may happen. If existing edges introduce crossings, then we roll back to previous crossing-free coordinates.

When a new node is added to the tree, a new edge also is added, with one of its endpoints already placed. To place the new node, we randomly sample a set of 100 nearby points at a distance equal to the desired edge length, trying to find a crossing-free position. If we cannot find such a suitable point, we gradually reduce the distance and repeat the search until we find a crossing free position. Once the new node has been placed, we subdivide its adjacent edge as described above.

3.2   DynaSafe

DynaSafe stands for Dynamic Safety, as the algorithm prioritizes safe moves and will not make a move if it introduces an edge crossing, the pseudocode of the algorithm can be found in the full version [30]. DynaSafe is also a force-directed algorithm, however, it differs from DynaCola as it draws straight-line edges (rather than edges with bends). The algorithm utilizes the following forces and is implemented in d3.js [12]

- A force f_E for each edge that is similar to DynaCola.
- A stress-minimizing force f_S on every pair of nodes not connected by an edge, used to improve global structure. The desired distance is the shortest-path

distance between the pair, and the magnitude of the force is proportional to the difference between the realized and desired distance.

- A repulsive force that is similar to DynaCola.
- A gravitational force that is similar to DynaCola.

DynaSafe prevents edge crossings from occuring at any time by updating the coordinates safely: if the proposed new coordinate of a node introduces crossings, we gradually reduce the magnitude of the movement until the crossing is avoided. To place the new node, we randomly sample a set of 100 nearby points to find a crossing-free position for its adjacent edge. If we cannot find a crossing free position using the sample points, we continuously reduce the edge length until we find a crossing free position. Once the node is added, an iteration of force-directed algorithm optimizes the layout (again without introducing crossings).

By the nature of force-directed algorithms, after one phase of force computations each node has a proposed new position. Before moving any node to its proposed new position, we check that the move is "safe," i.e., it does not introduce a crossing. If the movement of a node introduces any crossings, then the magnitude of the move is set to $p\%$ of the original movement. This is repeated (if needed) at most q times, and if the crossing is still unavoidable then the node is not moved in this phase. By default $p = 0.8$ and $q = 12$.

4 Experimental Evaluation

We evaluate DynaCola and DynaSafe, along with five earlier methods: DynNoSlice, DynaGraph, Dagre, Radial, and ImPrEd. We use two evolving trees to visually compare the results, as well quantitatively evaluate the desired properties.

4.1 Datasets

We use two real-world datasets to extract evolving trees for our experiments.

The Tree of Life: captures the evolutionary progression of life on Earth [37]. The underlying data is a tree structure with a natural time component. As a new species evolves, a new node in the tree is added. The edges give the parent-child relation of the nodes, where the parent is the original species, and the child is the new species. We use a subset of this graph with 500 nodes. The maximum node degree of this tree is 5, and the radius is 24.

The Mathematics Genealogy: shows advisor-advisee relationships in the world of mathematics, stretching back to the middle ages [36]. The dataset includes the thesis titles, students, advisors, dates, and number of descendants. The total number of nodes is around 260,000 and is continuously updated. While this data is not quite a tree (or even connected, or planar), we extract a subset to create a tree with 500 nodes. The maximum node degree of this tree is 5 and the radius is 14.

4.2   Evaluation Metrics

We use standard metrics for each of our desired properties: desired edge length preservation, compactness, and stability. Additionally, we compute the stress of the drawing and the number of crossings. This gives a total of five quantitative measures. For each of these measures we define a loss function as follows:

Desired Edge Length (DEL): To measure how close the realized edge lengths are to the desired edge lengths, we find the mean squared error between these two values. Given the desired edge lengths $\{l_{ij} : (i, j) \in E\}$ and coordinates of the nodes X in the computed layout, we evaluate with the following formula:

$$\text{Desired edge length loss} = \sqrt{\frac{1}{|E|} \sum_{(i,j) \in E} \left(\frac{||X_i - X_j|| - l_{ij}}{l_{ij}} \right)^2} \qquad (1)$$

This measures the root mean square of the relative error as in [1], producing a non-negative number, with 0 corresponding to a perfect realization. For DynaCola we subdivide the edges, to compute DEL, we set the length of the subdivided edges such that the summation of the length of the subdivided edges is equal to the length of the original edge.

Compactness: To measure the compactness of each layout, we use the ratio between the drawing area and the sum of the areas for all labels [9]. We assume that a label is at most 16 characters, as we abbreviate longer labels. The sum of the areas for all labels gives the minimum possible area needed to draw all labels without overlaps (ignoring any space needed for edges). The area of the actual drawing is given by the smallest bounding rectangle, once the drawing has been scaled up until there are no overlapping labels. Once we have this scaled drawing, we find the positions of the nodes with the largest and smallest x and y values ($X_{max,0}, X_{min,0}, X_{max,1}$ and $X_{min,1}$). Using these values we calculate the area of the bounding rectangle.

$$\text{Compactness loss} = \frac{(X_{max,0} - X_{min,0})(X_{max,1} - X_{min,1})}{\sum_{v \in V} \text{label\_area}(v)} \qquad (2)$$

This formula produces a non-negative number; the ideal value for this measure is 1 and corresponds to a perfect space utilization.

Stability: To measure stability, we consider how much each of the nodes moved after adding a new node. We then sum the movements of all nodes over all time steps. Since different algorithms use different amounts of drawing areas, we divide the value by the drawing area to normalize the results. This measure is similar to that used in DynNoSlice [45], but since DynNoSlice does not use time slices, it is closer to the measure found in [13]:

$$\text{Stability loss} = \frac{\sum_{v \in V} \sum_{t=1}^{T-1} \|X_v(t+1) - X_v(t)\|}{(X_{max,0} - X_{min,0})(X_{max,1} - X_{min,1})} \qquad (3)$$

Here, T is the maximum time (500 in our two datasets). This formula produces a non-negative number; the ideal value is 0 and corresponds to a perfectly stable layout (no movement of any already placed nodes).

Stress: This measure evaluates the global quality of the layout, looking at the differences between the realized distance between any pair of nodes and the actual distance between them. This measure is used in a variety of graph drawing algorithms [13,26,45]:

$$\text{Stress loss} = \frac{\left(\sum_{i \neq j}\left(D_{i,j} - \|X_i - X_j\|\right)^2\right)^{1/2}}{\sum_{i \neq j} \|X_i - X_j\|} \qquad (4)$$

Here, $D_{i,j}$ is the shortest path distance in the graph. This formula produces a non-negative number; the ideal value is 0 and corresponds to a perfect embedding (that captures all graph distances by the realized Euclidean distances).

Edge Crossings: Finally, we measure the number of edge crossings in each of the outputs. Note that our algorithms DynaSafe and DynaCola enforce "no edge crossings" as a hard constraint. However, DynNoSlice and DynaGraph do not have such a constraint and so can and indeed do, introduce crossings. Therefore we include the number of edge crossings for a complete comparison.

4.3 Experimental Setup

We compare these algorithms to five previous algorithms: DynNoSlice, Dyna-Graph, Dagre, Radial, and ImPrEd. We note that while Dagre, Radial, and ImPrEd are not specifically designed for dynamic graphs, they can be modified for this purpose. Specifically, we can use the layout of a tree at step i to initialize the layout of the tree at step $i + 1$, add the new edge, and update the layout.

We consider the simplest case for the desired edge length by using a uniform length of 100 for all edges. This is a necessary parameter for our algorithms DynaCola and DynaSafe, but only needed in the other four algorithms in order to compute the desired edge length measure. To be able to compare our methods to the other four (that do not take desired edge length into account), we set the desired edge length equal to the average edge length obtained in the layout. We then normalize these values for a fair comparison.

The performance of DynNoSlice depends heavily on two parameters, τ and δ that must be tuned. With the help of the authors, we found $\tau = 16$ and $\delta = 4$ worked well for our 500-node trees. The performance of ImPrEd depends on two parameters: repulsion force and the number of iterations. The default values of repulsion force and the number of iterations are equal to one and 200 respectively. We have used the default values. The larger the number of iterations

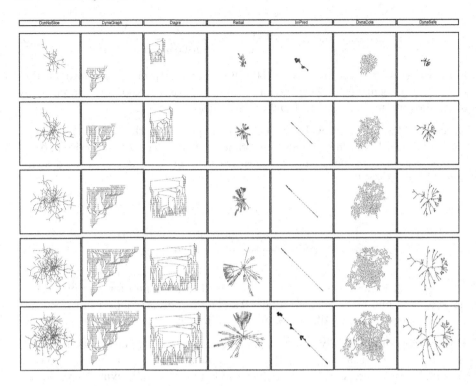

Fig. 4. Layouts obtained by the seven methods for the tree of life dataset.

is, the better the output of ImPrEd is. However, the running time increases as the number of iterations increases. We keep the number of iterations equal to 200 since it already takes more than 4 h to compute the 500-node trees. The performance of DynaCola depends on the number of subdivision nodes n_s; we use $n_s = 1$ for the experiments. For the radial layout algorithm, we have used the default settings in the yFiles [50] implementation. The other algorithms are also used with their default settings. We have implemented our algorithms in d3.js [12]. For other algorithms, we have used the default API. All experiments are conducted in a machine that has macOS 11.3.1 operating system, a 2.3 GHz 8-core Intel core i9 processor, and 32 GB 2667 MHz MHz DDR4 memory.

4.4   Results

Both the visual and quantitative results indicate that the two new methods perform well overall; see Fig. 4.

Desired Edge Lengths: The quantitative results are shown in Table 1. We use green to show the best results and yellow for the second best and indicate that DynNoSlice, DynaCola, and ImPrEd perform well. For the math genealogy 500-node tree, ImPrEd is the best. However, both DynNoSlice and ImPrEd have significantly larger running times (measured in hours, rather than minutes or seconds) as discussed below. Moreover, while ImPrEd does well on the math genealogy graph, it does not do well on the tree of life graph. For the math genealogy 500-node tree, DynaCola is the second best and DynNoSlice is third. For the tree of life dataset, DynaCola is the best, DynNoSlice is the second best and DynaSafe is third. DynaGraph has the worst performance – not surprising given that it is a hierarchical layout, which is forced to use some very long edges near the root.

Table 1. Desired edge lengths of math genealogy tree (MG) and tree of life (TOL).

Nodes	DynNoSlice	DynaGraph	Dagre	Radial	ImPrEd	DynaCola	DynaSafe
100 MG	0.37691	1.95933	0.682568	0.653853	0.219103	0.282521	0.589865
200 MG	0.36552	1.95179	0.679827	0.640628	0.213615	0.270322	0.575430
300 MG	0.35007	1.94213	0.666440	0.63058	0.204821	0.253877	0.564747
400 MG	0.34402	1.93822	0.646479	0.619203	0.193037	0.243184	0.553141
500 MG	0.33377	1.91979	0.639766	0.592694	0.182071	0.237139	0.548756
100 TOL	0.21675	1.28710	0.448483	0.448205	0.45402	0.158071	0.411747
200 TOL	0.21972	1.37271	0.494261	0.460161	0.49120	0.166190	0.443935
300 TOL	0.23986	1.40404	0.510473	0.481034	0.52016	0.176748	0.453332
400 TOL	0.25597	1.45660	0.553543	0.510352	0.59326	0.183856	0.470334
500 TOL	0.26652	1.52650	0.581648	0.530249	0.61093	0.189373	0.485759

Compactness: The quantitative results are shown in Table 2 and indicate that DynNoSlice outperforms the rest of the algorithms. DynaCola is second best and DynaSafe is third. Here, Dagre has the worst performance. Although DynNoSlice performs well, it introduces many edge crossings as discussed later. DynaCola layouts have higher compactness than DynaSafe. The absence of stress-related force allows placing nodes closer even if the graph theoretic distance is higher. Consider a path, the layout will be a straight line if stress is minimized. However, a zig-zag layout will provide better compactness.

Stability: The quantitative results are shown in Table 3 and idicate that DynaCola does best. DynaGraph is second, and DynaSafe is third. The radial layout performs worse in this metric because it rotates the subtrees as more edges are added.

Stress: The quantitative results are shown in Table 4. In general, DynaSafe does much better on this measure than the rest. Again, ImPrEd performs well for the regular-shaped math genealogy tree but does not perform well for the tree

Table 2. Compactness of math genealogy tree (MG) and the tree of life (TOL).

Nodes	DynNoSlice	DynaGraph	Dagre	Radial	ImPrEd	DynaCola	DynaSafe
100 MG	85.60	192.07	219.20	153.53	161.23	124.90	147.80
200 MG	87.94	196.29	224.54	162.80	161.20	130.87	153.31
300 MG	95.24	201.41	225.86	169.34	171.92	137.00	159.87
400 MG	98.89	206.48	227.74	175.07	181.20	145.47	169.68
500 MG	106.82	208.39	236.94	192.53	187.94	149.43	174.93
100 TOL	96.46	196.79	223.43	179.05	179.20	147.82	160.58
200 TOL	100.24	214.29	231.38	183.06	218.29	154.10	169.38
300 TOL	110.56	216.16	239.98	190.82	329.27	157.19	170.04
400 TOL	119.98	233.85	255.76	203.92	416.27	167.52	194.28
500 TOL	126.85	235.72	272.82	214.09	528.01	173.94	196.03

Table 3. Stability of math genealogy tree (MG) and tree of life (TOL).

Nodes	DynNoSlice	DynaGraph	Dagre	Radial	ImPrEd	DynaCola	DynaSafe
100 MG	0.001584	0.001393	0.0016502	0.001998	0.001530	0.001348	0.001459
200 MG	0.000752	0.000447	0.0012497	0.001839	0.000598	0.000264	0.000410
300 MG	0.000577	0.000227	0.0010083	0.001450	0.000437	0.000225	0.000295
400 MG	0.000249	0.000190	0.0009504	0.001203	0.000391	0.000164	0.000216
500 MG	0.000037	0.000014	0.0007591	0.001047	0.000026	0.000011	0.000019
100 TOL	0.000437	0.000139	0.001241	0.003609	0.000491	0.000105	0.000163
200 TOL	0.000323	0.000125	0.001196	0.003408	0.008305	0.000097	0.000146
300 TOL	0.000263	0.000099	0.001163	0.003174	0.005305	0.000072	0.000106
400 TOL	0.000235	0.000073	0.001136	0.002490	0.001937	0.000064	0.000101
500 TOL	0.000199	0.000071	0.000834	0.001941	0.000810	0.000052	0.000101

of life. For the tree of life DynNoSlice is second and DynaCola third. DynaGraph and Dagre perform the worst in this metric due to the limitations inherent in the hierarchical layout. Note that the stress is normalized, so the numbers are comparable.

Edge Crossings: The quantitative results are shown in Table 5. There are five winners here – the five algorithms that prevent any edge crossings: Dyna-Cola, DynaSafe, Radial, Dagre, and ImPrEd. DynNoSlice and DynaGraph do introduce some crossings.

Running time: The Radial layout has the lowest running time, taking 34.93 s and 28.03 s, respectively, to draw the 500-node math genealogy tree and tree of life. On the other end, DynNoSlice is the slowest algorithm, taking more than 6 h to draw the 500-node trees. Both DynNoSlice and ImPrEd take significantly longer running time compared to other algorithms. ImPrEd takes more than four hours to draw the 500-node trees. Our two new methods are not as fast

Table 4. Stress scores of math genealogy tree (MG) and tree of life (TOL).

Nodes	DynNoSlice	DynaGraph	Dagre	Radial	ImPrEd	DynaCola	DynaSafe
100 MG	113.10	150.59	230.75	125.37	89.43	76.51	49.44
200 MG	161.45	186.17	250.98	172.05	120.45	151.74	68.42
300 MG	179.48	264.90	286.77	205.98	148.02	184.47	94.05
400 MG	186.68	292.66	286.83	262.09	173.92	227.98	107.56
500 MG	249.61	393.11	396.72	314.93	203.54	291.39	109.00
100 TOL	136.52	210.06	263.98	192.64	163.02	128.45	59.77
200 TOL	165.75	262.24	325.65	243.59	349.28	201.76	65.10
300 TOL	181.62	305.15	369.57	287.93	427.09	220.63	81.81
400 TOL	254.20	328.59	398.11	317.28	509.32	306.89	93.99
500 TOL	285.19	400.81	461.43	374.02	593.19	351.23	119.77

Table 5. Edge crossings of math genealogy tree (MG) and tree of life (TOL).

Nodes	DynNoSlice	DynaGraph	Dagre	Radial	ImPrEd	DynaCola	DynaSafe
100 MG	43	8	0	0	0	0	0
200 MG	82	11	0	0	0	0	0
300 MG	168	11	0	0	0	0	0
400 MG	217	13	0	0	0	0	0
500 MG	277	13	0	0	0	0	0
100 TOL	21	0	0	0	0	0	0
200 TOL	67	0	0	0	0	0	0
300 TOL	106	0	0	0	0	0	0
400 TOL	176	0	0	0	0	0	0
500 TOL	231	0	0	0	0	0	0

as the Radial algorithm and not as slow as DynNoSlice, taking about 5 min on the 500-node trees. Due to space limitations, we provide more details of running time in the Appendix.

5 Discussion and Limitations

While there are many algorithms and tools for drawing static trees, only a few can handle dynamic trees well. Among those, even fewer takes edge labels into account while also preventing edge crossings. With this in mind, we described two methods that give better, readable layouts for evolving trees. We compared these two algorithms with others that have been set up for dynamic trees. With respect to the criteria that we have put forward, our algorithms match or exceed each of these algorithms. Fully functional prototypes and videos showing them in action are available online https://ryngray.github.io/dynamic-trees/. Source code and all experimental data can be found on github https://github.com/abureyanahmed/evolving_tree.

Naturally, our work comes with several limitations that could be addressed in future work.

Anticipating the Future: Currently, the two new methods, DynaCola and DynaSafe perform well for evolving trees, where growth is the only type of change. A natural question is whether these algorithms can be generalized to the more challenging problems of online dynamic tree visualization. Answering such question may need more precise modeling of the graph dynamics. Even though online dynamic graph drawing assumes no knowledge about the actual changes to the graph in the future, some prior knowledge of the graph may be available or predictable in advance. For example, knowing the expected depth or size of a tree or maximum degree of nodes (e.g., from domain knowledge about the specific type of graph) may help the layout algorithm reserve enough space for growth. In general, we anticipate that if one can model the evolving dynamics of the graph (e.g., probabilistically), incorporating knowledge of such dynamics into the drawing algorithm may help improve the resultant drawing; conversely, carefully defining compatible graph dynamics for a particular drawing algorithm will also allow us to identify the limitations of the given algorithm.

Multi-level Label Display: For simplicity, in this work we assume labels to be always shown in the drawing in a fixed font size. In practice, however, labels may come with different levels of importance and different desired font size. In that case, one might prefer to see only important labels displayed first in a zoomed out view of the graph, and later see more labels when zooming in. Incorporate such multi-level label display into the node placement strategy seems like an interesting and relevant problem.

Finding Desired Properties: We have proposed two different algorithms to solve the same evolving tree visualization problem, and each is associated with different benefits. Finding a continuous *spectrum* of algorithms with tunable parameters to balance the multiple desired properties would provide more flexibility. On the other hand, a careful human-subjects study may also help prioritize existing properties of the drawing, or help identify new desired properties from the specific tasks.

Considering More Dynamic Datasets: The datasets we considered are evolving in nature. For example, in the math genealogy dataset, once an advisee gets related to an advisor, the relationship remains forever. Although we considered only evolving trees, our ideas can be applied to datasets where the elements may get deleted. Applying the algorithms on more dynamic datasets remains future work.

Acknowledgements. We thank the authors of DynNoSlice and DynaGraph for their assistance with running and tuning algorithms. We also thank yFiles whose radial layout implementation we use in the evaluation.

References

1. Ahmed, R., De Luca, F., Devkota, S., Kobourov, S., Li, M.: Graph drawing via gradient descent, $(GD)^2$. In: GD 2020. LNCS, vol. 12590, pp. 3–17. Springer, Cham (2020). https://doi.org/10.1007/978-3-030-68766-3_1
2. Archambault, D., Purchase, H., Pinaud, B.: Animation, small multiples, and the effect of mental map preservation in dynamic graphs. IEEE Trans. Visual Comput. Graphics **17**(4), 539–552 (2010)
3. Bach, B., Henry-Riche, N., Dwyer, T., Madhyastha, T., Fekete, J.D., Grabowski, T.: Small multipiles: piling time to explore temporal patterns in dynamic networks. Comput. Graph. Forum **34**(3), 31–40 (2015)
4. Bach, B., Pietriga, E., Fekete, J.D.: Graphdiaries: animated transitions and temporal navigation for dynamic networks. IEEE Trans. Vis. Comput. Graph. **20**(5), 740–754 (2013)
5. Bachmaier, C.: A radial adaptation of the sugiyama framework for visualizing hierarchical information. IEEE Trans. Vis. Comput. Graph. **13**(3), 583–594 (2007)
6. Bachmaier, C., Brandes, U., Schlieper, B.: Drawing phylogenetic trees. In: International Symposium on Algorithms and Computation, pp. 1110–1121 (2005)
7. Balcan, D., Gonçalves, B., Hu, H., Ramasco, J.J., Colizza, V., Vespignani, A.: Modeling the spatial spread of infectious diseases: the global epidemic and mobility computational model. J. Comput. Sci. **1**(3), 132–145 (2010)
8. Barnes, J., Hut, P.: A hierarchical o(n log n) force-calculation algorithm. Nature **324**(6096), 446–449 (1986)
9. Barth, L., Kobourov, S.G., Pupyrev, S.: Experimental comparison of semantic word clouds. In: Gudmundsson, J., Katajainen, J. (eds.) SEA 2014. LNCS, vol. 8504, pp. 247–258. Springer, Cham (2014). https://doi.org/10.1007/978-3-319-07959-2_21
10. Beck, F., Burch, M., Diehl, S., Weiskopf, D.: The state of the art in visualizing dynamic graphs. In: 16th Eurographics Conference on Visualization, (EuroVis). Eurographics Association (2014)
11. Beck, F., Burch, M., Diehl, S., Weiskopf, D.: A taxonomy and survey of dynamic graph visualization. In: Computer Graphics Forum. vol. 36(1), pp. 133–159. Wiley Online Library (2017)
12. Bostock, M., Ogievetsky, V., Heer, J.: D^3 data-driven documents. IEEE Trans. Vis. Comput. Graph. **17**(12), 2301–2309 (2011)
13. Brandes, U., Mader, M.: A quantitative comparison of stress-minimization approaches for offline dynamic graph drawing. In: van Kreveld, M., Speckmann, B. (eds.) GD 2011. LNCS, vol. 7034, pp. 99–110. Springer, Heidelberg (2012). https://doi.org/10.1007/978-3-642-25878-7_11
14. Burch, M., Müller, C., Reina, G., Schmauder, H., Greis, M., Weiskopf, D.: Visualizing dynamic call graphs. In: Vision, Modeling, and Visualization (VMV), pp. 207–214 (2012)
15. Cohen, R.F., Di Battista, G., Tamassia, R., Tollis, I.G.: Dynamic graph drawings: trees, series-parallel digraphs, and planar ST-digraphs. SIAM J. Comput. **24**(5), 970–1001 (1995)
16. Cohen, R.F., Di Battista, G., Tamassia, R., Tollis, I.G., Bertolazzi, P.: A framework for dynamic graph drawing. In: Proceedings of the Eighth Annual Symposium on Computational Geometry, pp. 261–270 (1992)
17. Crnovrsanin, T., Chu, J., Ma, K.-L.: An incremental layout method for visualizing online dynamic graphs. In: Di Giacomo, E., Lubiw, A. (eds.) GD 2015. LNCS, vol. 9411, pp. 16–29. Springer, Cham (2015). https://doi.org/10.1007/978-3-319-27261-0_2

18. Diehl, S., Görg, C.: Graphs, they are changing. In: 10th International Symposium on Graph Drawing (GD), pp. 23–31 (2002)

19. Diehl, S., Görg, C., Kerren, A.: Foresighted graph layout. Technical Report, University of Saarland (2000)

20. Diehl, S., Görg, C., Kerren, A.: Preserving the mental map using foresighted layout. In: Proceedings of Joint Eurographics - IEEE TCVG Symposium on Visualization (VisSym) (2001)

21. Doğrusöz, U., Madden, B., Madden, P.: Circular layout in the Graph Layout toolkit. In: North, S. (ed.) GD 1996. LNCS, vol. 1190, pp. 92–100. Springer, Heidelberg (1997). https://doi.org/10.1007/3-540-62495-3_40

22. Erten, C., Harding, P.J., Kobourov, S.G., Wampler, K., Yee, G.: GraphAEL: Graph animations with evolving layouts. In: Liotta, G. (ed.) GD 2003. LNCS, vol. 2912, pp. 98–110. Springer, Heidelberg (2004). https://doi.org/10.1007/978-3-540-24595-7_9

23. Erten, C., Kobourov, S.G., Le, V., Navabi, A.: Simultaneous graph drawing: layout algorithms and visualization schemes. In: Liotta, G. (ed.) GD 2003. LNCS, vol. 2912, pp. 437–449. Springer, Heidelberg (2004). https://doi.org/10.1007/978-3-540-24595-7_41

24. Forrester, D., Kobourov, S.G., Navabi, A., Wampler, K., Yee, G.V.: Graphael: a system for generalized force-directed layouts. In: Pach, J. (ed.) GD 2004. LNCS, vol. 3383, pp. 454–464. Springer, Heidelberg (2005). https://doi.org/10.1007/978-3-540-31843-9_47

25. Frishman, Y., Tal, A.: Online dynamic graph drawing. IEEE Trans. Vis. Comput. Graph. 14(4), 727–740 (2008)

26. Gansner, E.R., Koren, Y., North, S.: Graph drawing by stress majorization. In: Pach, J. (ed.) GD 2004. LNCS, vol. 3383, pp. 239–250. Springer, Heidelberg (2005). https://doi.org/10.1007/978-3-540-31843-9_25

27. Gansner, E.R., Koutsofios, E., North, S.C., Vo, K.P.: A technique for drawing directed graphs. IEEE Trans. Software Eng. 19(3), 214–230 (1993)

28. Gilbert, F., Simonetto, P., Zaidi, F., Jourdan, F., Bourqui, R.: Communities and hierarchical structures in dynamic social networks: analysis and visualization. Soc. Netw. Anal. Min. 1(2), 83–95 (2011)

29. Gorochowski, T.E., di Bernardo, M., Grierson, C.S.: Using aging to visually uncover evolutionary processes on networks. IEEE Trans. Vis. Comput. Graph. 18(8), 1343–1352 (2011)

30. Gray, K., Li, M., Ahmed, R., Kobourov, S.: Visualizing evolving trees (2021). https://doi.org/10.48550/ARXIV.2106.08843, https://arxiv.org/abs/2106.08843

31. Hachul, S., Jünger, M.: Drawing large graphs with a potential-field-based multilevel algorithm. In: Pach, J. (ed.) GD 2004. LNCS, vol. 3383, pp. 285–295. Springer, Heidelberg (2005). https://doi.org/10.1007/978-3-540-31843-9_29

32. Hu, Y., Koren, Y.: Extending the spring-electrical model to overcome warping effects. In: 2009 IEEE Pacific Visualization Symposium, pp. 129–136. IEEE (2009)

33. Jünger, M., Mutzel, P.: 2-layer straightline crossing minimization: performance of exact and heuristic algorithms. In: Graph Algorithms and Applications I, pp. 3–27. World Scientific (2002)

34. Kar, G., Madden, B., Gilbert, R.: Heuristic layout algorithms for network management presentation services. IEEE Network 2(6), 29–36 (1988)

35. Kaufmann, M., Wiese, R.: Maintaining the mental map for circular drawings. In: Goodrich, M.T., Kobourov, S.G. (eds.) GD 2002. LNCS, vol. 2528, pp. 12–22. Springer, Heidelberg (2002). https://doi.org/10.1007/3-540-36151-0_2

36. Keller, M.T.: Math genealogy project. https://genealogy.math.ndsu.nodak.edu/
37. Maddison, D., Schulz, K., Lenards, A., Maddison, W.: Tree of life web project. http://tolweb.org/tree/
38. Misue, K., Eades, P., Lai, W., Sugiyama, K.: Layout adjustment and the mental map. J. Vis. Lang. Comput. **6**(2), 183–210 (1995)
39. Moen, S.: Drawing dynamic trees. IEEE Softw. **7**(4), 21–28 (1990)
40. Nguyen, Q.H.: INKA: an ink-based model of graph visualization. CoRR abs/1801.07008 (2018)
41. North, S.C.: Incremental layout in DynaDAG. In: Brandenburg, F.J. (ed.) GD 1995. LNCS, vol. 1027, pp. 409–418. Springer, Heidelberg (1996). https://doi.org/10.1007/BFb0021824
42. Pavlo, A., Homan, C., Schull, J.: A parent-centered radial layout algorithm for interactive graph visualization and animation. arXiv preprint cs/0606007 (2006)
43. Purchase, H.: Which aesthetic has the greatest effect on human understanding? In: DiBattista, G. (ed.) GD 1997. LNCS, vol. 1353, pp. 248–261. Springer, Heidelberg (1997). https://doi.org/10.1007/3-540-63938-1_67
44. Simonetto, P., Archambault, D., Auber, D., Bourqui, R.: Impred: an improved force-directed algorithm that prevents nodes from crossing edges. In: Computer Graphics Forum, vol. 30(3), pp. 1071–1080. Wiley Online Library (2011)
45. Simonetto, P., Archambault, D., Kobourov, S.: Event-based dynamic graph visualisation. IEEE Trans. Vis. Comput. Graph. **26**(7), 2373–2386 (2018)
46. Six, J.M., Tollis, I.G.: A framework for circular drawings of networks. In: Kratochvíyl, J. (ed.) GD 1999. LNCS, vol. 1731, pp. 107–116. Springer, Heidelberg (1999). https://doi.org/10.1007/3-540-46648-7_11
47. Skambath, M., Tantau, T.: Offline drawing of dynamic trees: algorithmics and document integration. CoRR abs/1608.08385 (2016)
48. Sugiyama, K., Tagawa, S., Toda, M.: Methods for visual understanding of hierarchical system structures. IEEE Trans. Syst. Man Cybern. **11**(2), 109–125 (1981)
49. van Eck, N.J., Waltman, L.: Visualizing bibliometric networks. In: Ding, Y., Rousseau, R., Wolfram, D. (eds.) Measuring Scholarly Impact, pp. 285–320. Springer, Cham (2014). https://doi.org/10.1007/978-3-319-10377-8_13
50. Wiese, R., Eiglsperger, M., Kaufmann, M.: yFiles: visualization and automatic layout of graphs. In: Mutzel, P., Jünger, M., Leipert, S. (eds.) GD 2001. LNCS, vol. 2265, pp. 453–454. Springer, Heidelberg (2002). https://doi.org/10.1007/3-540-45848-4_42
51. Workman, D., Bernard, M., Pothoven, S.: An incremental editor for dynamic hierarchical drawing of trees. In: Bubak, M., van Albada, G.D., Sloot, P.M.A., Dongarra, J. (eds.) ICCS 2004. LNCS, vol. 3038, pp. 986–995. Springer, Heidelberg (2004). https://doi.org/10.1007/978-3-540-24688-6_126

Improved Scheduling of Morphing Edge Drawing

Kazuo Misue$^{(\boxtimes)}$

University of Tsukuba, Tsukuba, Japan
misue@cs.tsukuba.ac.jp

Abstract. Morphing edge drawing (MED), a graph drawing technique, is a dynamic extension of partial edge drawing (PED), where partially drawn edges (stubs) are repeatedly stretched and shrunk by morphing. Previous experimental evaluations have shown that the reading time with MED may be shorter than that with PED. The morphing scheduling method limits visual clutter by avoiding crossings between stubs. However, as the number of intersections increases, the overall morphing cycle tends to lengthen in this method, which is likely to have a negative effect on the reading time. In this paper, improved scheduling methods are presented to address this issue. The first method shortens the duration of a single cycle by overlapping a part of the current cycle with the succeeding one. The second method duplicates every morph by the allowable number of times in one cycle. The third method permits a specific number of simultaneous crossings per edge. The effective performances of these methods are demonstrated through experimental evaluations.

Keywords: Graph drawing · Partial edge drawing · Morphing edge drawing · Scheduling of morphing

1 Introduction

Partial edge drawing (PED) is a graph-drawing technique in which the edges are drawn partially to avoid crossings. Morphing edge drawing (MED) is a dynamic graph representation technique in which the stubs (partially drawn edges) are repeatedly stretched and shrunk by morphing [6]. Experiments by Bruckdorfer have suggested that, compared with graph drawings in which edges are drawn as complete line segments, PED may improve the reading accuracy and increase the reading time [2]. An experimental evaluation by Misue & Akasaka showed that MED has the potential to reduce reading time compared to PED [6]. It is possible that, as the stubs change with morphing, less time is needed to guess the erased parts. The scheduling method shown by Misue & Akasaka schedules MED morphing so that stubs do not create new crossings. In other words, in a

Supplementary Information The online version contains supplementary material available at https://doi.org/10.1007/978-3-031-22203-0_24.

situation where two edges intersect, while one stub is stretched, the other must wait as short. Although this type of scheduling maintains the reduction of visual clutter by PED, it forces the morphing cycle to increase as the number of nodes, edges, and intersection points increases. Here, the morphing cycle is the total of the morphing time of all edges. Correspondingly, the latency before morphing used to determine whether two nodes are adjacent to each other may be longer than the time needed for guessing.

To address this issue, three methods to shorten the morphing cycle in MED were developed in this study, as explained below: The first method shortens the duration of a single cycle by initiating the new morphing of some stubs without waiting for all the stubs to regain their shortest states. In fact, in the MED scheduling shown by Misue & Akasaka, the duration of one cycle starts when all the stubs are in their shortest states and eventually ends when they all return to their initial shortest states again. Here, we have exploited the idea that, even if the morphing of some stubs begins before all the stubs return to their shortest states, no crossing may occur, and the duration of a cycle is reduced. In the second method, every morph is duplicated by the allowable number of times in one cycle. In previous MED scheduling, each edge stub is stretched and shrunk only once within each cycle. However, some stubs can be morphed two or more times within one cycle without leading to crossing. Considering this, multiple morphings within a single cycle can shorten the average duration of a morphing cycle. Finally, in the third method, crossings between edges are allowed to occur. Although graph drawings with many crossings are difficult to read, a small number of crossings are considered to have only a limited impact on readability [8,9]. Therefore, we developed a scheduling method that allows up to a certain number of simultaneous crossings per edge.

The contributions of this study can be summarized as follows:

1. Three new scheduling methods were presented to shorten the morphing cycle in MED, and
2. The effectiveness of each scheduling method is demonstrated experimentally.

2   Partial Edge Drawing and Morphing Edge Drawing

A simple undirected graph is denoted by $G = (V, E)$ and the drawing of a graph G is denoted by $\Gamma(G) = (\Gamma(V), \Gamma(E))$. $\Gamma(V) = \{\Gamma(v) | v \in V\}$ and $\Gamma(E) = \{\Gamma(e) | e \in E\}$. Herein, $\Gamma(G)$ is a traditional straight-line drawing, $\Gamma(v)$ of node $v \in V$ is a point located at position p_v, and $\Gamma(e)$ of edge $e \in E$ is a line segment connecting two nodes (points). In other words, it can be expressed as $\Gamma(e) = \{s \cdot p_w + (1 - s) \cdot p_v | s \in [0, 1]\}$, where $e = \{v, w\}$. Drawing $\Gamma(G)$ can be referred to by the retronym *complete edge drawing (CED)* because it completely draws a straight-line segment to represent an edge. The layout of graph G, that is, $\Gamma(G)$, is assumed to be provided in advance within this study. To simplify the description in subsequent sections, Γ is omitted and e is used to replace $\Gamma(e)$ when it is clear from the context that it represents $\Gamma(e)$.

2.1   Partial Edge Drawing

The partial drawing of the edge $e = \{v, w\}$ is represented by the function $\gamma_e :$ $[0,1]^2 \to 2^{\Gamma(e)}$, as shown in Eq. (1).

$$\gamma_e(\alpha, \beta) = \begin{cases} \{s \cdot p_w + (1-s) \cdot p_v | s \in [0, \alpha] \cup [\beta, 1]\} & \text{for } \alpha < \beta \\ \Gamma(e) & \text{for } \alpha \geq \beta. \end{cases} \quad (1)$$

The partial drawing $\gamma_e(\alpha, \beta)$ of edge e is the remainder of the entire $\Gamma(e)$ mapped to the interval $[0,1]$, with the part corresponding to interval (α, β) removed from $\Gamma(e)$. Each of the remaining contiguous parts is called a *stub*. If the part to be deleted is not the end of edge $\Gamma(e)$, that is, if $0 < \alpha$ and $\beta < 1$, two stubs remain at the two nodes incident to the edge e. We refer to them as a *pair of stubs*.

Given α_e and β_e for all edges $e \in E$ and that there exists an edge $e_1 \in E$ such that $\alpha_{e_1} < \beta_{e_1}$, the drawing $\Gamma_{PED}(G) = (\Gamma(V), \Gamma_{PED}(E))$ is called a *PED*, where $\Gamma_{PED}(E) = \{\gamma_e(\alpha_e, \beta_e) | e \in E\}$. When the lengths of a pair of stubs are equal, that is, when there exists a relationship $\alpha_e = 1 - \beta_e$, this is called a *symmetric PED (SPED)*. In this case, the smaller parameter α_e is called the *stub ratio*. If the stub ratios for all edges are the same δ, the drawing is called δ-*symmetric homogeneous PED (δ-SHPED)*.

2.2   Morphing Edge Drawing

Let T be a set of times. A dynamic drawing $\Gamma_{MED}(G) = (\Gamma(V), \Gamma_{MED}(E))$, which is constructed using the morphing function $\mu_e : T \to 2^{\Gamma(e)}$ and defines the partial drawing of edge e at time $t \in T$. It is called the *morphing edge drawing (MED)*, where $\Gamma_{MED}(E) = \{\mu_e | e \in E\}$. Let $\rho_e : T \to [0,1]^2$ be a function that defines the parameters of the partial edge for time $t \in T$. The function μ_e can then be constructed as $\mu_e(t) = \gamma_e(\rho_e(t))$. For all edges $e \in E$, if the function ρ_e satisfies $\rho_e(t) = (\delta_t, 1 - \delta_t)$ (where $0 \leq \delta_t \leq 1/2$) for $\forall t \in T$, then a SPED is obtained at any time. MED constructed based on such function is called a *symmetric MED (SMED)*.

This study focuses on SMED. In one morph, each stub changes from the shortest state (stub-ratio δ) to the longest state (stub-ratio η), remains in the longest state for a certain time and then returns to the shortest state. The range over which a stub stretches and shrinks is called as the *morphing range*. Two paired stubs start and end morphing simultaneously.

3   Related Work

Bruckdorfer et al. have given the formulation of PED [2]. Burch et al. [5] demonstrated the applicability of this approach to directed graphs using tapered links to represent partially drawn edges. Schmauder et al. applied PED to weighted graphs by coloring edges to represent their weights [10]. Information on PED is summarized in the commentary by Nöllenburg [7].

Bruckdorfer et al. conducted experiments comparing CED and 1/4-SHPED with respect to graph-reading performance [3]. Although not statistically significant, the chart visualizing the results of the experiment indicated a slightly more accurate but longer response time for 1/4-SHPED than for CED in terms of the graph reading task. Binucci et al. [1] conducted more detailed evaluation experiments and found that, among the SPEDs, SHPED yielded the best reading accuracy. Burch [4] examined the effects of stub orientation and length on the graph reading accuracy and found that shorter stub lengths tend to result in more misjudgments regarding the target nodes and that stub orientation also affects accuracy.

MED was proposed by Misue & Akasaka [6]. The formalization of the MED is provided herein, and evaluation experiments on the readability of the MED indicate that the MED may be superior to the PED, in terms of the reading time. The formalization presented in Sect. 2 is based on the one proposed by Misue & Akasaka [6].

4   Terminology and Notation

This section describes the terminology and notations used in this paper.

4.1   Set and Set Family

The sets and functions that return a set are capitalized. Let $\#(A)$ denote the number of elements in a finite set A, A^c denote the complement of set A, and 2^A denote the power set of set A. For set A, let $A^{\#k}$ denote the set family created by collecting only all the subsets with k (≥ 1) elements. In other words, $A^{\#k} = \{A' \in 2^A | \#(A') = k\}$.

4.2   Time Periods

Suppose that the time period is a subset of $U = (-\infty, \infty)$ and can be expressed as $P = \bigcup_{i=1}^{\chi}[a_i, b_i)$ ($\chi \geq 0$). In this case, $b_i < a_j$ if $i < j$. In other words, we assumed that the time period can be represented as a union set of noncontiguous half-open intervals. When $\chi = 0$, it is assumed to be empty.

4.3   Intersections and Types of Intersections

It can be assumed that the intersection point of two drawn edges can be represented by a pair of edges because the layout of the graph is assumed to be provided in advance. Therefore, when any two sets of edges cross at the same point, although only one point exists from the geometric perspective, the point is considered as two different intersection points corresponding to the two edge crossings. If edge e_1 crosses another edge e_2 at a point p, then, e_2 is called the opposite edge of e_1 at p and is denoted by e_1/p. In other words, when e_1 and e_2 cross at point p, $e_1/p = e_2$ and $e_2/p = e_1$. Let I denote an entire set of intersection points. Furthermore, let $I(e)$ denote a set of intersection points on edge e.

In MED, crossings between stubs may be unavoidable, as illustrated in Fig. 1. We refer to an intersection point as "e is *always passing*", where a stub of edge e is passing even when the stub is at its shortest state. The intersection points at which both edges always pass are called *always crossing*. The intersection points at which both edges are not always-passing are called *fully avoidable*, whereas intersection points at which only one edge is always passing are called *semi-avoidable*. Fully avoidable and semi-avoidable intersections are collectively called *avoidable*. If semi-avoidable intersections are in the morphing range of a stub, the morphing of the stub will always result in one or more crossings.

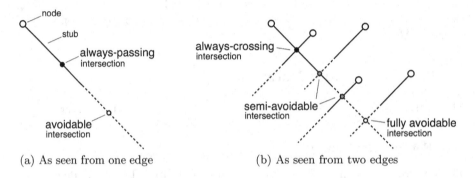

(a) As seen from one edge   (b) As seen from two edges

Fig. 1. Type of intersection points. The solid lines represent the state where the stubs are shortest, and the dashed lines represent the state where the stubs are stretched.

5   Scheduling

The scheduling of a MED defines a morphing function $\mu_e : T \to 2^{\Gamma(e)}$ for all edges e. We assumed that the change in each stub from stretching to shrinking back to the original state with respect to the elapsed time from the start of morphing is already defined by the function $\mu_e^* : \mathbb{R} \to 2^{\Gamma(e)}$. Thus, scheduling implies the determination of the morphing start time for each edge. Once the start time $t_{start}(e)$ has been determined, the morphing function can be defined as $\mu_e(t) = \mu_e^*(t - t_{start}(e))$.

Given a function μ_e^*, we can determine the elapsed time after the start of morphing to each stub state. Let $\tau_{pass}(e, p)$ denote the time it takes for the tip of a stub of edge e to pass the intersection point p for the first time (passing while stretching) after the onset of morphing. Let $\tau_{ret}(e, p)$ denote the time it takes for the tip of the stub of edge e to pass the intersection point p for the second time (passing while shrinking) after the onset of morphing. Let $\tau_{trip}(e)$ denote the time from the start to the end of the morphing of edge e. $\tau_{ret}(e, p)$ and $\tau_{trip}(e)$ include the time when the stub is fully stretched and the morphing is paused. Let C_p^e ($\subseteq U$) denote the time period when the stub of edge e passes through point p on e. If the morphing start time of edge e is $t_{start}(e)$, then $C_p^e = [t_{start}(e) + \tau_{pass}(e,p), t_{start}(e) + \tau_{ret}(e,p))$. Let $C_p^e = \emptyset$ if the morphing start time at edge e is undefined.

In the following sections, we first describe the algorithm proposed by Misue & Akasaka [6] and then extend it in a step-by-step manner to accommodate the overlapping of each cycle, duplicating morphs in one cycle, and allowance of crossings. In this manner, we proceed with the explanation, while extending the functions. Thus, we use the numbered function names like $F^{(1)}$ and $F^{(2)}$. Because the functions with larger numbers are extensions of the smaller-numbered functions, only one function with the largest number needs to be defined for implementation.

5.1 Basic Scheduling Algorithm

Here, all intersection points are assumed to be fully avoidable.

Algorithm 1 shows the algorithm proposed by Misue & Akasaka [6]. Given a set of edges E, this algorithm determines the start time $t_{start}(e)$ of morphing for all edges $e \in E$. Let E be a morphing group consisting of edges whose morphing timings may affect each other. The algorithm sequentially determines the start time of morphing with respect to the edges in set E. Misue & Akasaka [6] sorted the edges in descending order of their lengths and determined the start time of the morphing of each edge in the order. The method examines, for an edge e, the morphing timing of all opposite edges that intersect edge e and have already determined their start time. It then determines the earliest time at which no crossings occur for edge e as the morphing start time for e. Note that the first morphing is assumed to start at time zero ($t = 0$).

Algorithm 1. Scheduling morphing

Input: E – Set of edges in a morphing group
Output: The start time $t_{start}(e)$ of all $e \in E$, and the total morphing time t_{total}
1: **function** SCHEDULEPARALLEL(E)
2:    | **for** e in $sort(E)$ **do**
3:    | | $t_{start}(e) \leftarrow t_{earliest}(P_{fbd}(e))$
4:    | **end for**
5:    | $t_{total} \leftarrow \max_{e \in E}(t_{start}(e) + \tau_{trip}(e))$
6: **end function**

Function $P_{fbd}^{(1)} : E \rightarrow 2^U$ provides the time period during which one or more crossings occur when edge $e \in E$ starts morphing. In other words, $P_{fbd}^{(1)}(e)$ is the forbidden morphing start period used by edge e to avoid crossing with its opposite edges. The definition of function $P_{fbd}^{(1)}$ is expressed in Eqs. (2) and (3).

$$P_{fbd}^{(1)}(e) = P_{crit}^{(1)}(e) \tag{2}$$

$$P_{crit}^{(1)}(e) = \bigcup_{p \in I_s(e)} S_p^e(C_p^{e/p}) \tag{3}$$

Let $I_s(e) = \{p \in I(e) | t_{start}(e/p)$ has been defined$\}$. Function $S_p^e : 2^U \to 2^U$ modifies a given time period by the time needed for the stub of edge e to reach the intersection point p, extend further, and then return to that point. $S_p^e(\emptyset) = \emptyset$, and if $[a, b) \subseteq P$, then $[a - \tau_{ret}(e, p), b - \tau_{pass}(e, p)) \subseteq S_p^e(P)$. $S_p^e(P)$ yields the morphing start period for edge e to pass through point p within the time period P. In other words, if morphing does not start at time $S_p^e(C_p^{e/p})$, edge e can avoid crossing with an opposite edge e/p at the intersection point p. Function $P_{crit}^{(1)} : E \to 2^U$ provides the union of these periods for all intersection points on e. If the morphing start time of edge e is undefined, let $C_p^e = \emptyset$ and $S_p^e(\emptyset) = \emptyset$. Therefore, $I_s(e)$ appearing on the right side of Eq. (3) may be replaced by $I(e)$.

Function $t_{earliest} : 2^U \to U$ yields the minimum non-negative value not included in the time period P. This definition can be expressed as in Eq. (4).

$$t_{earliest}(P) = \min([0, \infty) \cap P^c) \tag{4}$$

One of the outputs from Algorithm 1, t_{total}, is the total morphing time, which denotes the cycle length when morphing is repeated.

6   Overlapping a Part of Each Cycle

Based on the schedule (that has already been determined), let t_{latest} (< 0) be the time before time zero when the morphing of edge e can be started. Shortening the cycle by $t_{total} - (t_{start}(e) + |t_{latest}|)$ will not cause the crossing of e (see Fig. 2). In other words, the period can be shortened to $t_{start}(e) - t_{latest}$ without causing any crossings at edge e. Overall, the graph can shorten the cycle to the maximum value at all edges $e \in E$. Equation (5) shows a shortened cycle length t_{cycle}.

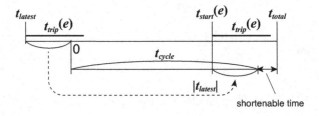

Fig. 2. Schematic of concept for shortening a single cycle

$$t_{cycle} = \max_{e \in E}\{t_{start}(e) - t_{latest}(P_{fbd}^{(2)}(e))\}, \tag{5}$$

where $t_{latest}(P^{(2)}_{fbd}(e))$ provides the latest possible morphing start time before time zero of edge e. Function $P^{(2)}_{fbd}$ is an extension of $P^{(1)}_{fbd}$, and its definition is given in Eqs. (6) and (7).

$$P^{(2)}_{fbd}(e) = P^{(1)}_{crit}(e) \cup P^{(1)}_{self}(e) \tag{6}$$

$$P^{(1)}_{self}(e) = [t_{start}(e) - \tau_{trip}(e), t_{start}(e) + \tau_{trip}(e)) \tag{7}$$

Function $P^{(2)}_{fbd}$ uses $P^{(1)}_{self} : E \to 2^U$ in addition to $P^{(1)}_{crit}$. Function $P^{(1)}_{self}(e)$ yields the time period when morphing is prohibited to start so that it does not overlap with its own morphing. When the start time $t_{start}(e)$ of edge e is undefined, let $P^{(1)}_{self}(e) = \emptyset$. Thus, function $P^{(2)}_{fbd}$ can be used instead of $P^{(1)}_{fbd}$. Function $t_{latest} : 2^U \to U$ yields the negative (or zero) upper bound that is not included in the time period $P(\subseteq U)$ given. This function is defined in Eq. (8).

$$t_{latest}(P) = \max((-\infty, 0) \cap P^c) \tag{8}$$

7 Duplication Within a Cycle

Each stub stretches and shrinks only once within one cycle in the schedule obtained Algorithm. 1 or Algorithm 1 plus Eq. (5). However, some edges can be morphed two or more times within a single cycle, without causing any crossings. In other words, focusing on certain edges may further reduce the average cycle length.

Hereafter, the start time of morphing with respect to an edge is treated as a set and is denoted by $T_{start}(e)$. Accordingly, the definition of $I_s(e)$ is changed to $I_s(e) = \{p \in I(e) | T_{start}(e/p) \neq \emptyset\}$. Function $P_{self}(e)$ is also extended.

Algorithm 2 presents an algorithm for scheduling multiple morphs within one cycle. The function $P^{(3)}_{fbd} : E \times \mathbb{R}_{\geq 0} \to 2^U$, defined by Eq. (9) yields the forbidden morphing start period of edge $e \in E$ when the cycle length is $c \in \mathbb{R}_{\geq 0}$.

$$P^{(3)}_{fbd}(e, c) = W(c, P^{(1)}_{crit}(e) \cup P^{(2)}_{self}(e)) \tag{9}$$

$$P^{(2)}_{self}(e) = \bigcup_{t \in T_{start}(e)} [t - \tau_{trip}(e), t + \tau_{trip}(e)), \tag{10}$$

where the function $W : \mathbb{R}_{\geq 0} \times 2^U \to 2^U$ is defined as $W(c, P) = P \cup (\bigcup_{[a,b) \subseteq P} [a + c, b + c))$. This adds one cycle length c to the time period P. If c is equal to the t_{total} obtained from Algorithm 1, the function W does not need to be applied. However, if c is shorter than t_{total}, there may be edges that morph across two cycles; therefore, the period is extended to two cycles to determine the forbidden morphing start period.

Algorithm 2. Scheduling multiple morphing within a single cycle

---

Input: E – Set of edges in the morphing group, the start time $t_{start}(e)$ of all $e \in E$,
$\quad$ t_{total} – total morphing time, t_{cycle} – morphing cycle length
Output: Set of start times $T_{start}(e)$ for all $e \in E$
1: **function** SCHEDULEDUPLICATION($E, t_{start}, t_{total}, t_{cycle}$)
2: $\quad$ **for** e in E **do**
3: $\quad\quad$ $T_{start}(e) \leftarrow \{t_{start}(e)\}$
4: $\quad$ **end for**
5: $\quad$ $E_1 \leftarrow E$
6: $\quad$ **while** $E_1 \neq \emptyset$ **do**
7: $\quad\quad$ $E_2 \leftarrow \emptyset$
8: $\quad\quad$ **for** e in $sort(E_1)$ **do**
9: $\quad\quad\quad$ $t_{start2} \leftarrow t_{earliest}(P_{fbd}(e, t_{cycle}))$
10: $\quad\quad\quad$ **if** $t_{start2} + \tau_{trip}(e) \leq t_{total}$ **then**
11: $\quad\quad\quad\quad$ $T_{start}(e) \leftarrow T_{start}(e) \cup \{t_{start2}\}$
12: $\quad\quad\quad\quad$ $E_2 \leftarrow E_2 \cup \{e\}$
13: $\quad\quad\quad$ **end if**
14: $\quad\quad$ **end for**
15: $\quad\quad$ $E_1 \leftarrow E_2$
16: $\quad$ **end while**
17: **end function**

---

8 Allowance of Crossings

We considered scheduling that allows for up to a certain number of crossings (*allowable crossing number*) n per edge. Thus far, we proceeded with the explanation assuming that there were no always-passing intersections. However, hereafter, we include always-passing intersections in our considerations.

Let the *controllable crossing number* be the allowable crossing number minus the number of always-crossing intersections. When the number of always-crossing intersections exceeded the allowable crossing number, let the controllable crossing number be zero. Because we cannot control the occurrence of crossings at the always-crossing intersections, we perform scheduling by ignoring these crossings based on the controllable crossing number.

As crossings at semi-avoidable intersections cannot be avoided, the number of crossings may exceed the controllable crossing number. Even in these cases, the number of crossings should be maintained as low as possible. For example, let us suppose two semi-avoidable intersections exist on an edge. Although crossings at these intersections cannot be avoided, it may be possible to schedule them such that no two crossings occur simultaneously.

Function $P_{fbd}^{(4)} : E \times \mathbb{R}_{\geq 0} \times \mathbb{N} \to 2^U$, which determines the forbidden morphing start period of edge e when the presence of always-passing intersections is allowed and when a certain number of crossings is allowed, can be expressed as in Eq. (11).

$$P_{fbd}^{(4)}(e, c, n) = W(c, P_{crit}^{(2)}(e, k_n(e)) \cup P_{self}^{(2)}(e) \cup P_{opst}(e, n)), \tag{11}$$

where $c \in \mathbb{R}_{\geq 0}$ represents the cycle length and is set to zero if undetermined. The role of W is the same as that described in Sect. 7. The allowable crossing number $n \in \mathbb{N}$ given in advance is a common condition for all the edges. However, the controllable crossing number differs from edge to edge because the number of always-crossing intersections differs accordingly. Therefore, let $k_n(e)$ denote the controllable crossing number for edge e. If a stub of edge e always passes through intersection p, let $C_p^e = U$, even if the morphing schedule of edge e remains undefined. $P_{opst} : E \times \mathbb{N} \to 2^U$ is a function used to find the critical time period of conditions for opposite edges. When no crossing is allowed, there is no need to consider these conditions, because when no crossing occurs for the target edge, the same condition applies to opposite edges as well. However, when allowing crossings and setting their upper limits, conditions for the target edge differ from those for the opposite edges. We explain the definition of the P_{opst} function in Sect. 8.2.

8.1   Satisfying Allowable Crossing Number for Target Edge

The critical period during which the controllable crossing number of edge e exceeds k is represented by $P_{crit}^{(2)}(e, k)$, as indicated in Eq. (12).

$$
P_{crit}^{(2)}(e, k) = \begin{cases} P_{cirt}^{(1)}(e) \cup \{\bigcup_{Q \in I_{sa}(e)\#2} P_{cSub}(e, Q)\} & \text{if } k = 0 \\ \bigcup_{Q \in I_{s'}(e)\#k+1} P_{cSub}(e, Q) & \text{otherwise} \end{cases} \tag{12}
$$

$$
P_{cSub}(e, Q) = \begin{cases} \emptyset & \text{if } O(e, Q) = U \\ \bigcap_{p \in Q} S_p^e(O(e, Q)) & \text{otherwise} \end{cases} \tag{13}
$$

$$
O(e, Q) = \bigcap_{p \in Q} C_p^{e/p}, \tag{14}
$$

where $I_a(e) = \{p \in I(e) | p \text{ is avoidable}\}$, $I_{sa}(e) = \{p \in I(e) | p \text{ is semi-avoidable}\}$, and $I_{s'}(e) = I_s(e) \cup \{p \in I_a(e) | e/p \text{ always passes } p\}$. We set Q in Eqs. (12) and (13) as the subset of intersections on e with two or $k+1$ elements. $O(e, Q)$ denotes the period during which the stubs of the opposite edges pass simultaneously at $\#(Q)$ points on edge e. If $O(e, Q) \neq \emptyset$, then crossings may occur simultaneously at all the intersection points in Q. To avoid this, e should not pass through these points during the time period. $P_{cSub}(e, Q)$, shown in Eq. (13), represents the critical time period when a stub of edge e starts and then passes through all the intersection points in that time period. However, in Eq. (13), $P_{cSub}(e, Q) = \emptyset$ when $O(e, Q) = U$. When the existence of always-passing intersections is allowed, unavoidable crossings may occur. Case $O(e, Q) = U$ represents an unavoidable situation. This indicates that all the opposite edges at the $\#(Q)$ intersection points in Q always pass and that simultaneous crossings with all of them are unavoidable. Therefore, for crossings at intersection points Q, the critical time period is $\emptyset$, and it does not affect the start time.

8.2   Satisfying Allowable Crossing Number for Opposite Edges

The critical time period in which the number of crossings of the opposite edges of edge e exceeds the allowable crossing number n is represented by $P_{opst}(e, n)$, as indicated by Eqs. (15) and (16), where $I_{a1}(e) = \{p \in I(e) | p$ is avoidable for $e\}$.

$$P_{opst}(e, n) = \bigcup_{p \in I_{a1}(e)} S_p^e(P_{oSub}(e/p, p, n)) \tag{15}$$

$$P_{oSub}(e', p, n) = \begin{cases} C_p^{e'} & \text{if } k_n(e') = 0 \wedge (p \text{ is avoidable for } e') \\ \bigcup_{q \in I_a(e')} X(q) & \text{if } k_n(e') = 0 \wedge (p \text{ is always-passing for } e') \\ \bigcup_{Q \in I_{s'}(e')\#k_n(e')} \bigcap_{q \in Q} X(q) & \text{otherwise,} \end{cases}$$

$$\tag{16}$$

where $X(q)$ denote the time period when a crossing occurs at point q. That is, $X(q) = C_q^{e_1} \cap C_q^{e_2}$ when e_1 and $e_2 = e_1/q$ cross at point q. $P_{oSub}(e', p, n)$ represents the critical time period at intersection p with respect to an opposite edge e' for the allowable crossing number n, as shown in Eq. (16). The right-hand side of Eq. (15) indicates the union of the forbidden morphing start periods when e passes through the intersection point with the opposite edge during this critical time period. The definition of P_{oSub} can be divided into three cases. (1) If $k_n(e') = 0$ and e' can avoid p, then the period $C_p^{e'}$ (at which e' passes intersection point p) is the critical time period. (2) If $k_n(e') = 0$ and e' always passes through p, then e should be allowed to pass through p, provided that all crossing time periods at semi-avoidable intersections that are avoidable for e' are avoided. This implies that the critical time period is the time period of crossing occurrence at the avoidable intersections for e'. (3) Otherwise, the critical time period is the time period in which more than $k_n(e')$ crossings occur simultaneously on e'.

8.3   Overlapping a Part of Each Cycle

The method proposed in Sect. 6 shortens the cycle length by determining the possible start time of the following cycle for each edge. The possible start time of each single edge was examined, however, the effect of shifting the start time of all the edges was not inspected. Therefore, if the allowable crossing number is greater than or equal to one, the method does not function properly.

Because we have not yet identified an efficient method to address this issue, we only present a simple countermeasure. The method involves affording a tentative shortened cycle length using the method described in Sect. 6, then checking the time period when the condition is violated, and extending the cycle length by the amount of time when the condition is violated.

9   Evaluation of Effectiveness

We implemented the scheduling algorithm described in Algorithm 1 and Algorithm 2 along with the functions in Java with JRE 16.0.2. The cycle length was defined as a real number (an element of $\mathbb{R}_{\geq 0}$) in the aforementioned explanations; however, in our implementation, it was defined as an int with ms as the unit.

Experiments were conducted to investigate the effects of each of the previously described factors: overlapping a part of each cycle, duplicating morphs in one cycle, and allowance of crossings. We prepared complete graphs with 7–13 nodes and laid out the nodes of each graph equally spaced around a circumference with a radius of 200 pixels. The speed of the tips of the stubs was 100 pixels/s and the tips were paused for 100 ms at the longest stub ratio.

The longest stub-ratio was set to $\eta = 50\%$ and the shortest stub-ratios δ were set to 4%, 9%, 16%, and 25%. Always-passing intersections were included in the case of $\delta = 25\%$ with seven nodes, $\delta \geq 16\%$ with 8–10 nodes, and $\delta \geq 9\%$ with 11–13 nodes. Furthermore, always-crossing intersections were included in the case of $\delta = 25\%$ with 10–13 nodes.

For each of these 28 combinations, morphing scheduling was performed with or without the application of overlapping, duplication within a cycle, and by changing the allowable crossing number from 0 to 10. In addition, considering the effects of *sort* in Algorithm 1, scheduling was performed with 100 different orders under the same conditions, including a descending order of the edge lengths and vice versa, as well as 98 randomly sorted orders.

9.1 Overlapping a Part of Each Cycle

Here, we examined the reduction rate of the cycle length and derived it as the ratio of the cycle length obtained by applying the proposed methods to the cycle length obtained by the scheduling algorithm proposed by Misue & Akasaka [6]. The number of samples was 2,400 for seven nodes, 3,600 for eight nodes, and 4,400 for each of the other cases. Figure 3(a) shows the quartiles of the reduction rates of cycle lengths. Although a certain effect is observed, it is found that this decreases as the number of nodes increases. The median value for the seven nodes is 0.771, but it increased to 0.910 for 13 nodes.

9.2 Duplication Within a Cycle

For example, if all the edges could morph twice within one cycle, the cycle length would be effectively halved. Hence, we considered the reduction rate as the number of edges to be morphed divided by the total number of morphs. The number of samples was the same as that used for the evaluation of the overlapping cycles. In all the cases, the overlapping a part of each cycle was not applied.

Figure 3(b) shows the quartiles of the reduction rates. It can be observed that the effectiveness improves as the number of nodes increases. Focusing on the median, when the number of nodes is seven, the median is one, and there is no reduction effect; however, when the number of nodes is 13, the median is 0.602. In other words, on an average, morphing can be performed nearly twice within one cycle.

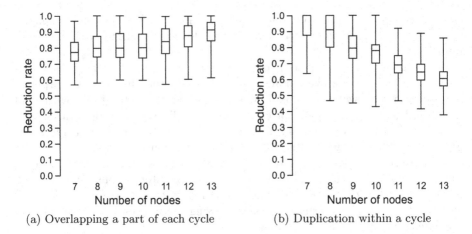

(a) Overlapping a part of each cycle   (b) Duplication within a cycle

Fig. 3. Effects of overlapping a part of each cycle and duplication within a cycle

9.3 Allowance of Crossings

We examined the reduction rate of the cycle length when crossings were allowed for each allowable crossing number. Figure 4 shows the quartiles of the reduction rate of cycle lengths. Figure 4(a) shows the case in which all types of intersections are included, and Fig. 4(b) shows the case in which only fully avoidable intersections are included. In both cases, the reduction effect improved as the allowable crossing number increased. However, in some cases, the cycle becomes longer around the allowable crossing numbers of 1 to 3. In the case of fully avoidable intersections, these cases are less frequent. In any case, the identification of the factors will be our focus in future studies.

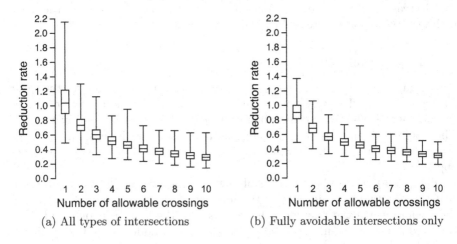

(a) All types of intersections   (b) Fully avoidable intersections only

Fig. 4. Effects of allowance of crossings

10   Conclusion

We developed three scheduling methods to shorten the morphing cycle in MED. The first method shortens a cycle by overlapping the end of the current cycle with the succeeding one. The second method shortens the average duration of a cycle by duplicating every morph by the allowable number of times in one cycle. The third method aims at shortening the cycle length by allowing a certain number of crossings at each edge. We incorporated these developed methods into a program and conducted evaluation experiments on complete graphs laid out on a circle to confirm the effectiveness of each method.

Acknowledgements. This work was supported by JSPS KAKENHI Grant Number JP21K11975.

References

1. Binucci, C., Liotta, G., Montecchiani, F., Tappini, A.: Partial edge drawing: homogeneity is more important than crossings and ink. In: 2016 7th International Conference on Information, Intelligence, Systems & Applications (IISA). pp. 1–6 (July 2016). https://doi.org/10.1109/IISA.2016.7785427
2. Bruckdorfer, T., Kaufmann, M.: Mad at edge crossings? Break the edges! In: Kranakis, E., Krizanc, D., Luccio, F. (eds.) FUN 2012. LNCS, vol. 7288, pp. 40–50. Springer, Heidelberg (2012). https://doi.org/10.1007/978-3-642-30347-0_7
3. Bruckdorfer, T., Kaufmann, M., Leibßle, S.: PED user study. In: Di Giacomo, E., Lubiw, A. (eds.) GD 2015. LNCS, vol. 9411, pp. 551–553. Springer, Cham (2015). https://doi.org/10.1007/978-3-319-27261-0_47
4. Burch, M.: A user study on judging the target node in partial link drawings. In: 2017 21st International Conference Information Visualisation (iV), pp. 199–204 (2017). https://doi.org/10.1109/iV.2017.43
5. Burch, M., Vehlow, C., Konevtsova, N., Weiskopf, D.: Evaluating partially drawn links for directed graph edges. In: van Kreveld, M., Speckmann, B. (eds.) GD 2011. LNCS, vol. 7034, pp. 226–237. Springer, Heidelberg (2012). https://doi.org/10.1007/978-3-642-25878-7_22
6. Misue, K., Akasaka, K.: Graph drawing with morphing partial edges. In: Archambault, D., Tóth, C.D. (eds.) Graph Drawing and Network Visualization, pp. 337–349. Springer, Cham (2019). https://doi.org/10.1007/978-3-030-35802-0
7. Nöllenburg, M.: Crossing layout in non-planar graph drawings. In: Hong, S.-H., Tokuyama, T. (eds.) Beyond Planar Graphs, pp. 187–209. Springer, Singapore (2020). https://doi.org/10.1007/978-981-15-6533-5_11
8. Purchase, H.: Which aesthetic has the greatest effect on human understanding? In: DiBattista, G. (ed.) GD 1997. LNCS, vol. 1353, pp. 248–261. Springer, Heidelberg (1997). https://doi.org/10.1007/3-540-63938-1_67
9. Purchase, H.C., Cohen, R.F., James, M.: Validating graph drawing aesthetics. In: Brandenburg, F.J. (ed.) GD 1995. LNCS, vol. 1027, pp. 435–446. Springer, Heidelberg (1996). https://doi.org/10.1007/BFb0021827
10. Schmauder, H., Burch, M., Weiskopf, D.: Visualizing dynamic weighted digraphs with partial links. In: Proceedings of 6th International Conference on Information Visualization Theory and Applications (IVAPP), pp. 123–130 (2015)

Linear Layouts

Queue Layouts of Two-Dimensional Posets

Sergey Pupyrev$^{(\boxtimes)}$ (iD)

Menlo Park, CA, USA
spupyrev@gmail.com

Abstract. The queue number of a poset is the queue number of its cover graph when the vertex order is a linear extension of the poset. Heath and Pemmaraju conjectured that every poset of width w has queue number at most w. The conjecture has been confirmed for posets of width $w = 2$ and for planar posets with 0 and 1. In contrast, the conjecture has been refused by a family of general (non-planar) posets of width $w > 2$.

In this paper, we study queue layouts of two-dimensional posets. First, we construct a two-dimensional poset of width $w > 2$ with queue number $2(w - 1)$, thereby disproving the conjecture for two-dimensional posets. Second, we show an upper bound of $w(w + 1)/2$ on the queue number of such posets, thus improving the previously best-known bound of $(w - 1)^2 + 1$ for every $w > 3$.

Keywords: Poset · Queue number · Width · Dimension · Linear extension

1 Introduction

Let G be a simple, undirected, finite graph with vertex set V and edge set E, and let σ be a total order of V. For a pair of distinct vertices u and v, we write $u <_\sigma v$ (or simply $u < v$), if u precedes v in σ. We also write $[v_1, v_2, \ldots, v_k]$ to denote that v_i precedes v_{i+1} for all $1 \leq i < k$; such a subsequence of σ is called a *pattern*. Two edges $(u, v) \in E$ and $(a, b) \in E$ *nest* if $u <_\sigma a <_\sigma b <_\sigma v$. A k-queue layout of G is a total order of V and a partition of E into subsets $E_1, E_2, \ldots, E_k$, called *queues*, such that no two edges in the same set E_i nest. The *queue number* of G, $\mathrm{qn}(G)$, is the minimum k such that G admits a k-queue layout. Equivalently, the queue number is the minimum k such that there exists an order σ containing no $(k + 1)$-*rainbow*, that is, a set of edges $\{(u_i, v_i); i = 1, 2, \ldots, k + 1\}$ forming pattern $[u_1, \ldots, u_{k+1}, v_{k+1}, \ldots, v_1]$ in σ.

Queue layouts can be studied for partially ordered sets (or simply *posets*). A poset over a finite set of elements X is a transitive and asymmetric binary relation $<$ on X. The main idea is that given a poset, one should lay it out respecting the relation. Two elements a, b of a poset, $P = (X, <)$, are called *comparable* if $a < b$ or $b < a$, and *incomparable*, denoted by $a \parallel b$, otherwise. Posets are visualized by their diagrams: Elements are placed as points in the plane and whenever $a < b$ in the poset and there is no element c with $a < c < b$,

P. Angelini and R. von Hanxleden (Eds.): GD 2022, LNCS 13764, pp. 353–360, 2023.
https://doi.org/10.1007/978-3-031-22203-0_25

there is a curve from a to b going upwards (that is y-monotone); see Fig. 1a. Such relations, denoted by $a \prec b$, are known as *cover relations*; they are essential in the sense that they are not implied by transitivity. The directed graph implicitly defined by such a diagram is the cover graph G_P of the poset P. Given a poset P, a linear extension L of P is a total order on the elements of P such that $a <_L b$, whenever $a <_P b$. Finally, the queue number of a poset P, denoted by $\mathrm{qn}(P)$, is the smallest k such that there exists a linear extension L of P for which the resulting layout of G_P contains no $(k+1)$-rainbow; see Fig. 1c.

Queue layouts of posets were first studied by Heath and Pemmaraju [5], who provided bounds on the queue number of posets in terms of their *width*, that is, the maximum number of pairwise incomparable elements. In particular, they observed that the size of a rainbow in a queue layout of a poset of width w cannot exceed w^2, and therefore, $\mathrm{qn}(P) \le w^2$ for every poset P. Furthermore, Heath and Pemmaraju conjectured that $\mathrm{qn}(P) \le w$ for a width-w poset P. The study of the conjecture received a notable attention in the recent years. Knauer, Micek, and Ueckerdt [6] confirmed the conjecture for posets of width $w = 2$ and for planar posets with 0 and 1. Later Alam et al. [1] constructed a poset of width $w \ge 3$ whose queue number is $w + 1$, thus refuting the conjecture for general non-planar posets. In the same paper Alam et al. improved the upper bound by showing that $\mathrm{qn}(P) \le (w-1)^2 + 1$ for all posets P of width w. Finally, Felsner, Ueckerdt, and Wille [4] strengthened the lower bound by presenting a poset of width $w > 3$ with $\mathrm{qn}(P) \ge w^2/8$.

In this short paper we refine our knowledge on queue layouts of posets by improving the known upper and lower bounds of the queue number of two-dimensional posets. Recall that the *dimension* of poset P is the least positive integer d for which there are d linear extensions (*realizers*) $L_1, \ldots, L_d$ of P so that $a < b$ in P if and only if $a < b$ in L_i for every $i \in \{1, \ldots, d\}$. *Two-dimensional* posets are described by realizers L_1 and L_2 and often represented by *dominance drawings* in which the coordinates of the elements are their positions in L_1 and L_2; see Fig. 1b. We emphasize that the existing lower bound constructions [1,4] are not two-dimensional. Thus, Felsner et al. [4] asked whether the conjecture of Heath and Pemmaraju holds for posets with dimension 2. Our first result answers the question negatively.

Theorem 1. *There exists a two-dimensional poset P of width $w > 1$ with $\mathrm{qn}(P) \ge 2(w-1)$.*

Observe that our construction and the proof of Theorem 1 for $w = 3$ is arguably much simpler than the one of Alam et al. [1], which is based on a tedious case analysis. Thus, it can be interesting on its own right.

Next we study the upper bound on the queue number of two-dimensional posets. Our result is the following theorem, which is an improvement over the known $(w-1)^2 + 1$ bound of Alam et al. [1] for every $w > 3$.

Theorem 2. *Let P be a two-dimensional poset with realizers L_1, L_2. Then there is a layout of P in at most $w(w+1)/2$ queues using either L_1 or L_2 as the vertex order.*

The paper is structured as follows. In Sect. 3 we prove Theorem 1 and in Sect. 2 we prove Theorem 2. Section 4 concludes the paper with interesting open questions.

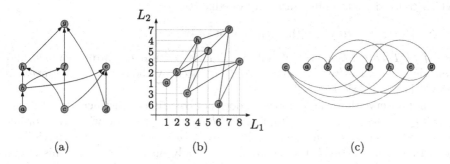

(a)   (b)   (c)

Fig. 1. A two-dimensional poset of width 3, its dominance drawing, and a 2-queue layout

2 An Upper Bound

Consider a two-dimensional poset $P = (X, <)$ of width $w \geq 1$ with realizers L_1 and L_2. In this section we study queue layouts of P using vertex orders L_1 or L_2, which we call *realizer-based*. It is well-known that the elements of P can be partitioned into w *chains*, that is, subsets of pairwise comparable elements. We fix such a partition and treat it as a function $\mathcal{C} : X \to \{1, \ldots, w\}$ such that if $\mathcal{C}(u) = \mathcal{C}(v)$ and $u \neq v$, then either $u < v$ or $v < u$.

We start with a property of a linear extension of a poset, whose proof follows directly from the absence of transitive edges in G_P. Recall that $\prec$ indicates cover relations of P, that is, edges of G_P.

Proposition 1. *A linear extension of a poset P with chain partition $\mathcal{C}$ does not contain pattern $[b_1, b_2, b_3]$, where $\mathcal{C}(b_1) = \mathcal{C}(b_2) = \mathcal{C}(b_3)$ and $b_1 \prec b_3$.*

The next observation, whose proof is immediate, provides a crucial property of realizer-based linear extensions of two-dimensional posets. In fact, a poset, P, admits a linear extension with such a property if and only if P has dimension 2; see for example [3] where such linear extensions are called *non-separating*.

Proposition 2. *Consider a two-dimensional poset P with realizers L_1, L_2 and chain partition $\mathcal{C}$. Let $[a_1, b, a_2]$ be a pattern in L_1 (or L_2) with $\mathcal{C}(a_1) = \mathcal{C}(a_2)$. Then either $a_1 < b$ or $b < a_2$.*

The next useful property in the section holds for realizer-based linear extensions of two-dimensional posets.

Proposition 3. *Consider a two-dimensional poset P with realizers L_1, L_2 and chain partition $\mathcal{C}$. Then L_1 (or L_2) does not contain pattern $[a_1, b_2, a, a_2, b_1]$, where $\mathcal{C}(a_1) = \mathcal{C}(a_2) = \mathcal{C}(a)$, $\mathcal{C}(b_1) = \mathcal{C}(b_2)$, and $a_1 \prec b_1$, $b_2 \prec a_2$.*

Proof. For the sake of contradiction, assume that $[a_1, b_2, a, a_2, b_1]$ is in L_1, with $\mathcal{C}(a_1) = \mathcal{C}(a_2) = \mathcal{C}(a)$, $\mathcal{C}(b_1) = \mathcal{C}(b_2)$, and $a_1 \prec b_1$, $b_2 \prec a_2$. Notice that $a_1 \parallel b_2$, as otherwise we have $a_1 < b_2 < b_1$ and the edge (a_1, b_1) is transitive. Hence by Proposition 2 applied to $[a_1, b_2, a]$, $b_2 < a$. Therefore, it holds that $b_2 < a < a_2$, which contradicts to non-transitivity of edge (b_2, a_2).

Now we ready to prove the main result of the section.

Proof of Theorem 2. Assume that poset P is partitioned into w chains, and consider a maximal rainbow, denoted T, induced by the order L_1. We need to prove that $|T| \le w(w + 1)/2$.

First observe that the rainbow, T, does not contain two distinct edges (a_1, b_1) and (a_2, b_2) with $\mathcal{C}(a_1) = \mathcal{C}(a_2)$ and $\mathcal{C}(b_1) = \mathcal{C}(b_2)$. Otherwise, the former edge nests the latter one and we have $a_1 < a_2 \prec b_2 < b_1$, which violates non-transitivity of (a_1, b_1). Therefore, we already have $|T| \le w^2$. (This is the argument of Heath and Pemmaraju for their original upper bound in [5])

Next we show two more configurations that are absent in T:

(i) For every pair of distinct chains, the rainbow does not contain edges (a_1, b_1), (b_2, a_2), and (a_3, a_4) with $\mathcal{C}(a_1) = \mathcal{C}(a_2) = \mathcal{C}(a_3) = \mathcal{C}(a_4)$ and $\mathcal{C}(b_1) = \mathcal{C}(b_2)$. For a contradiction, assume the rainbow contains the three edges. By Proposition 1, edge (a_3, a_4) cannot cover elements a_1 or a_2. Thus, L_1 contains pattern $[a_1, b_2, a_3, a_4, a_2, b_1]$ or $[b_2, a_1, a_3, a_4, b_1, a_2]$. Both patterns violate Proposition 3.

(ii) For every triple of distinct chains, the rainbow does not contain edges (a_1, b_1), (b_2, a_2), (a_3, c_3), and (c_4, a_4) with $\mathcal{C}(a_1) = \mathcal{C}(a_2) = \mathcal{C}(a_3) = \mathcal{C}(a_4)$, $\mathcal{C}(b_1) = \mathcal{C}(b_2)$, and $\mathcal{C}(c_3) = \mathcal{C}(c_4)$.
For a contradiction, assume T contains the four edges. Consider the innermost edge in the rainbow; without loss of generality, assume the edge is (a_1, b_1). Vertex a_1 is covered by two edges, (a_3, c_3) and (c_4, a_4), forming the pattern of Proposition 3; a contradiction.

Now observe that T may contain at most w *uni-colored* edges (that is, (u, v) such that $\mathcal{C}(u) = \mathcal{C}(v)$) and at most $w(w - 1)$ *bi-colored* edges (that is, (u, v) such that $\mathcal{C}(u) \ne \mathcal{C}(v)$).

On the one hand, if T contains exactly w uni-colored edges and $|T| > w(w + 1)/2$, then it must contain at least one pair of bi-colored edges (a_1, b_1), (b_2, a_2) with $\mathcal{C}(a_1) = \mathcal{C}(a_2)$, $\mathcal{C}(b_1) = \mathcal{C}(b_2)$. Together with the uni-colored edge from chain $\mathcal{C}(a_1)$, the triple forms the forbidden configuration (*i*).

On the other hand, if T contains at most $w - 1$ uni-colored edges and $|T| > w(w+1)/2$, then T contains two pairs of bi-colored edges, as in configuration (*ii*); a contradiction.

This completes the proof of the theorem.   □

Notice that the bound of Theorem 2 is worst-case optimal, as we show next.

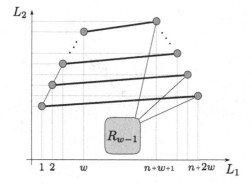

Fig. 2. A 2-dimensional poset of width $w \geq 1$, R_w, with a realizer-based order containing a $\big(w(w+1)/2\big)$-rainbow, which is comprised of w thick edges that nest all edges of R_{w-1}

Lemma 1. *There exists a two-dimensional poset of width $w \geq 1$, denoted R_w, with realizers L_1, L_2 such that its layout with vertex order L_1 contains a $\big(w(w+1)/2\big)$-rainbow.*

Proof. The poset R_w is built recursively. For $w = 1$, the poset consists of two comparable elements. For $w > 1$, we assume that R_{w-1} is constructed and described by realizers L_1^{w-1} and L_2^{w-1}. The poset R_w is constructed from R_{w-1} by adding $2w$ elements. Assume $|L_1^{w-1}| = n$ and the elements of R_{w-1} are indexed by $w + 1, \ldots, w + n$. We set L_1^w to the identity permutation and use

$$L_2^w = L_2^{w-1} \ \cup \ (1, n + 2w, 2, n + 2w - 1, \ldots, w, n + w + 1),$$

where $\cup$ denotes the concatenation of the two orders. Figure 2 illustrates the construction. It is easy to verify that the width of the new poset is exactly w. Observe that in the layout of R_w with order L_1^w, edges $(1, n+2w), \ldots, (w, n+w+1)$ form a w-rainbow and nest all edges of R_{w-1}. Therefore, the layout contains a $\big(w(w+1)/2\big)$-rainbow, as claimed.

We remark that Lemma 1 provides a poset whose queue layout with one of its realizers contains a $\big(w(w+1)/2\big)$-rainbow. It is straightforward to extend the construction (by concatenating R_w with its dual) so that both realizer-based vertex orders yield a rainbow of that size. However, the queue number of the poset (and the proposed extension) is at most w, which is achieved with a different, non-realizer-based, vertex order. Thus, a more delicate construction is needed to force a larger rainbow in every linear extension of a poset.

3 A Lower Bound

In this section we provide a new counter-example to the conjecture of Heath and Pemmaraju [5] by describing a two-dimensional poset of width $w \geq 3$ whose

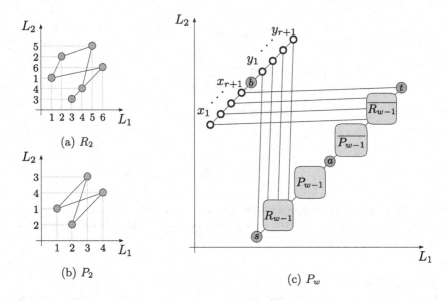

Fig. 3. A counter-example to the conjecture of Heath and Pemmaraju [5]: A two-dimensional poset, P_w, of width $w \geq 3$ with queue number exceeding w

queue number exceeds w. The poset, denoted P_w, is constructed recursively. The base case, P_2, is a four-element poset with $L_1 = (1, 2, 3, 4)$ and $L_2 = (2, 1, 4, 3)$; see Fig. 3b. The step of the construction is illustrated in Fig. 3c. Poset P_w consists of a copy of P_{w-1}, a copy of the poset R_{w-1} utilized in Lemma 1, the duals of the two posets, and a chain of additional elements. Recall that the *dual* of a poset, P, is the poset, $\overline{P}$, on the same set of elements such that $x < y$ in P if and only if $y < x$ in $\overline{P}$ for every pair of the elements x and y.

We now formally describe the construction. Denote by $L_1(P), L_2(P)$ the two realizers of a two-dimensional poset P. Let $\cup$ denote the concatenation of two sequences, and let $(x_1, x_2, \dots) \uplus (y_1, y_2, \dots)$ denote the *interleaving* of two equal-length sequences, that is, $(x_1, y_1, x_2, y_2, \dots)$. Assume that R_{w-1} contains r elements. Then we set

$$L_1(P_w) = (x_1, \dots, x_{r+1}) \cup b \cup s \cup y_1 \cup \left(L_1(R_{w-1}) \uplus (y_2, \dots, y_{r+1})\right) \cup$$
$$L_1(P_{w-1}) \cup a \cup L_1(\overline{P_{w-1}}) \cup L_1(\overline{R_{w-1}}) \cup t, \text{ and}$$
$$L_2(P_w) = s \cup L_2(R_{w-1}) \cup L_2(P_{w-1}) \cup a \cup L_2(\overline{P_{w-1}}) \cup$$
$$\left((x_1, \dots, x_r) \uplus L_2(\overline{R_{w-1}})\right) \cup x_{r+1} \cup t \cup b \cup (y_1, \dots, y_{r+1}).$$

We refer to Fig. 3 for the illustration of the construction. Now we prove that the constructed poset has queue number at least $2w - 2$.

Proof of Theorem 1. It is easy to verify that the constructed poset, P_w, is two-dimensional and has width exactly w. Furthermore, the poset is dual to itself, that is, $P_w = \overline{P_w}$ with a and b being the fixed points. Thus, we may assume that

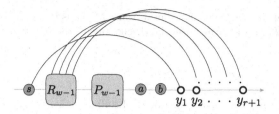

Fig. 4. A linear extension of poset P_w for the proof of Theorem 1 in which $a < b$

in the linear extension corresponding to the optimal queue layout of the poset, element a precedes b and we have $s < \cdots < a < b < y_1 < \cdots < y_{r+1}$. Next we consider the queue layout induced by the elements s, R_{w-1}, P_{w-1}, a, and $y_1, \ldots, y_{r+1}$; see Fig. 4.

We prove the theorem by induction. For $w = 2$, the claim holds trivially. For $w > 2$, we assume that $\mathrm{qn}(P_{w-1}) \geq 2(w-2)$ and distinguish two cases depending on the size of the maximum rainbow, T, formed by edges (s, y_1), $(v_1, y_2), \ldots, (v_r, y_{r+1})$, where $v_i, 1 \leq i \leq r$ are elements of R_{w-1}:

- if $|T| \geq 2$, then $\mathrm{qn}(P_w) \geq \mathrm{qn}(P_{w-1}) + |T| \geq 2(w-1)$, as all edges of P_{w-1} are nested by edges of T;
- if $|T| = 1$, then the elements of R_{w-1} must appear in the order induced by $L_1(R_{w-1})$, since otherwise at least two of the edges of T nest. By Lemma 1, the edges of R_{w-1} form a $\big(w(w-1)/2\big)$-rainbow. The rainbow is covered by edge (s, y_1), which yields $\mathrm{qn}(P_w) \geq \big(w(w-1)/2\big) + 1 \geq 2(w-1)$ for $w \geq 3$.

This completes the proof of Theorem 1. □

4 Conclusions

We disproved the conjecture of Heath and Pemmaraju for two-dimensional posets and answered a question posed by Felsner et al. [4]. A number of intriguing problems in the area remain unsolved.

- Is it possible to get a subquadratic upper bound on the queue number of two-dimensional posets of width w? A poset of Felsner et al. [4] that requires $w^2/8$ queues in every linear extension is not two-dimensional, which leaves a hope for an asymptotically stronger result than the one given by Theorem 2.
- What is the queue number of two-dimensional posets of width 3? By Theorem 1 and the result of Alam et al. [1], the value is either 4 or 5.
- Queue layouts of graphs are closely related to so-called *track layouts*, which are connected with the existence of low-volume three-dimensional graph drawings [2,7]. In particular, every t-track (undirected) graph has a $(t-1)$-queue layout, and every q-queue (undirected) graph has track number at most $4q \cdot 4q^{(2q-1)(4q-1)}$. We think it is interesting to study the relationship between the two concepts for directed graphs and posets.

Acknowledgments. We thank Jawaherul Alam, Michalis Bekos, Martin Gronemann, and Michael Kaufmann for fruitful initial discussions of the problem.

References

1. Alam, J.M., Bekos, M.A., Gronemann, M., Kaufmann, M., Pupyrev, S.: Lazy queue layouts of posets. In: GD 2020. LNCS, vol. 12590, pp. 55–68. Springer, Cham (2020). https://doi.org/10.1007/978-3-030-68766-3_5
2. Dujmović, V., Pór, A., Wood, D.R.: Track layouts of graphs. Discrete Math. Theor. Comput. Sci. **6**(2), 497–522 (2004). https://doi.org/10.46298/dmtcs.315
3. Dushnik, B., Miller, E.W.: Partially ordered sets. Am. J. Math. **63**(3), 600–610 (1941). https://doi.org/10.2307/2371374
4. Felsner, S., Ueckerdt, T., Wille, K.: On the queue-number of partial orders. In: Purchase, H.C., Rutter, I. (eds.) GD 2021. LNCS, vol. 12868, pp. 231–241. Springer, Cham (2021). https://doi.org/10.1007/978-3-030-92931-2_17
5. Heath, L.S., Pemmaraju, S.V.: Stack and queue layouts of posets. SIAM J. Discret. Math. **10**(4), 599–625 (1997). https://doi.org/10.1137/S0895480193252380
6. Knauer, K., Micek, P., Ueckerdt, T.: The queue-number of posets of bounded width or height. In: Biedl, T., Kerren, A. (eds.) GD 2018. LNCS, vol. 11282, pp. 200–212. Springer, Cham (2018). https://doi.org/10.1007/978-3-030-04414-5_14
7. Pupyrev, S.: Improved bounds for track numbers of planar graphs. J. Graph Algorithms Appl. **24**(3), 323–341 (2020). https://doi.org/10.7155/jgaa.00536

Recognizing DAGs with Page-Number 2
Is NP-complete

Michael A. Bekos[1] , Giordano Da Lozzo[2](✉) , Fabrizio Frati[2] ,
Martin Gronemann[3] , Tamara Mchedlidze[4] ,
and Chrysanthi N. Raftopoulou[5]

[1] Department of Mathematics, University of Ioannina, Ioannina, Greece
`bekos@uoi.gr`
[2] Department of Engineering, Roma Tre University, Rome, Italy
`{giordano.dalozzo,fabrizio.frati}@uniroma3.it`
[3] Algorithms and Complexity Group, TU Wien, Vienna, Austria
`mgronemann@ac.tuwien.ac.at`
[4] Department of Computer Science, Utrecht University, Utrecht, The Netherlands
`t.mtsentlintze@uu.nl`
[5] School of Applied Mathematics and Physical Sciences, NTUA, Athens, Greece
`crisraft@mail.ntua.gr`

Abstract. The page-number of a directed acyclic graph (a DAG, for short) is the minimum k for which the DAG has a topological order and a k-coloring of its edges such that no two edges of the same color cross, i.e., have alternating endpoints along the topological order. In 1999, Heath and Pemmaraju conjectured that the recognition of DAGs with page-number 2 is NP-complete and proved that recognizing DAGs with page-number 6 is NP-complete [*SIAM J. Computing*, 1999]. Binucci et al. recently strengthened this result by proving that recognizing DAGs with page-number k is NP-complete, for every $k \geq 3$ [*SoCG* 2019]. In this paper, we finally resolve Heath and Pemmaraju's conjecture in the affirmative. In particular, our NP-completeness result holds even for *st*-planar graphs and planar posets.

Keywords: Page-number · Directed acyclic graphs · Planar posets

1 Introduction

The problem of embedding graphs in books [27] has a long history of research with early results dating back to the 1970's. Such embeddings are specified by a linear order of the vertices along a line, called *spine*, and by a partition of the edges into sets, called *pages*, such that the edges in each page are drawn crossing-free in a half-plane delimited by the spine. The *page-number* of a graph is the minimum number of pages over all its book embeddings, while the page-number of a graph family is the maximum page-number over its members.

This research was partially supported by ΠΕΒΕ 2020, MIUR Project "AHeAD" under PRIN 20174LF3T8 and by H2020-MSCA-RISE project 734922 – "CONNECT".

P. Angelini and R. von Hanxleden (Eds.): GD 2022, LNCS 13764, pp. 361–370, 2023.
https://doi.org/10.1007/978-3-031-22203-0_26

An important branch of literature focuses on the page-number of planar graphs. An upper bound of 4 was known since 1986 [30], while a matching lower bound was only recently proposed [4,31]. Better bounds are known for several subfamilies [14,15]. A special attention has been devoted to the planar graphs with page-number 2 [3,8,11,13,19,21,25,28]. These have been characterized as the subgraphs of Hamiltonian planar graphs [17] and hence are called *subhamiltonian*. Recognizing subhamiltonian graphs turns out to be NP-complete [29].

If the input graph is directed and acyclic (a DAG, for short), then the linear vertex order of a book embedding is required to be a *topological order* [26]. Heath and Pemmaraju [16] showed that there exist planar DAGs whose page-number is linear in the input size. Certain subfamilies of planar DAGs, however, have bounded page-number [1,6,10,18], while recently it was shown that upward planar graphs have sublinear page-number [20], improving previous bounds [12]. From an algorithmic point of view, testing whether a DAG has page-number k is NP-complete for every fixed value of $k \geq 3$ [7], linear-time solvable for $k = 1$ [16], and fixed-parameter tractable with respect to the vertex cover number for every k [6] and with respect to the treewidth for st-graphs when $k = 2$ [7]. In contrast to the undirected setting, however, for $k = 2$ the complexity question has remained open since 1999, when Heath and Pemmaraju posed the following conjecture.

Conjecture 1 (Heath and Pemmaraju [16]). *Deciding whether a DAG has page-number 2 is NP-complete.*

Our Contribution. In this work, we settle Conjecture 1 in the positive. More precisely, we show that testing st-planar graphs for 2-page embeddability is NP-complete. In [2], we further show that the problem remains NP-complete for *planar posets*, i.e., upward-planar graphs with no transitive edges.

2   Preliminaries

A *plane embedding* of a connected graph is an equivalence class of planar drawings of the graph, where two drawings are equivalent if they define the same clockwise order of the incident edges at each vertex and the same clockwise order of the vertices along the outer face. The *flip* of a plane embedding produces a plane embedding in which the clockwise order of the incident edges at each vertex and the clockwise order of the vertices along the outer face is the reverse of the original one. A drawing of a DAG is *upward* if each edge is represented by a curve whose y-coordinates monotonically increase from the source to the sink, and it is *upward planar* if it is both upward and planar. An *upward planar embedding* is an equivalence class of upward planar drawings of a DAG, where two drawings are equivalent if they define the same plane embedding and the same left-to-right order of the outgoing (and incoming) edges at each vertex. A *plane DAG* is a DAG together with an upward planar embedding. A DAG is *st-planar* if it has a single source s and a single sink t, and admits a planar drawing with s and t on the outer face. It is known that every st-planar graph is upward planar [9,22]. An

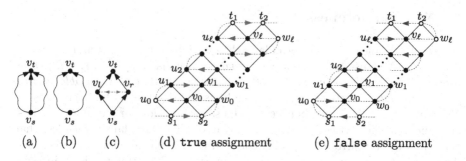

Fig. 1. Curly curves represent paths and straight-lines represent edges. Edges with no arrow are directed upward, also in subsequent figures. (a) Generalized diamond, (b) non-transitive face, (c) rhombus, and (d)-(e) the two subhamiltonian paths of a double ladder of even length ℓ.

st-*plane* graph is an st-planar graph together with an upward planar embedding in which s and t are incident to the outer face. As in the undirected case, a DAG G has page-number 2 if it is *subhamiltonian*, i.e., it is a spanning subgraph of an st-planar graph $\overline{G}$ that has a directed Hamiltonian st-path P [24]. In the previous definition, if G has a prescribed plane embedding, we additionally require that the plane embedding of $\overline{G}$ restricted to G coincides with the one of G. We say that P is a *subhamiltonian path* for G, and we refer to the edges of P that are not in G as *augmenting edges*. Further, $\overline{G}$ is an *HP-completion* of G.

A *generalized diamond* is an st-plane graph consisting of three directed paths from v_s to v_t, one of which is the edge $v_s v_t$ and appears between the other two paths in the upward planar embedding; see Fig. 1a. A face (by *face* of a plane DAG we always mean an *internal face*, unless otherwise specified) of a plane DAG whose boundary consists of two directed paths is an st-*face*. An st-face is *transitive* if one of these paths is an edge; *non-transitive*, otherwise (see Fig. 1b). A *rhombus* is a non-transitive st-face whose boundary paths have length 2; see Fig. 1c. From [24, Theorem 1], we obtain Property 1 which implies Property 2.

Property 1. *A Hamiltonian st-plane graph contains only transitive faces and no generalized diamond.*

Property 2. *Let G be a plane DAG and P be a subhamiltonian path for G. If G contains a rhombus (v_s, v_l, v_r, v_t) with source v_s and sink v_t, then P contains either the edge $v_l v_r$ or the edge $v_r v_l$, i.e., v_l and v_r are consecutive in P.*

The next property follows directly from Theorem 1 in [23] and Property 1. We provide a full proof in [2].

Property 3 ($\star$). *Let G be a plane DAG and P be a subhamiltonian path for G. If G contains a non-transitive face f with boundaries (v_s, w, v_t) and $(v_s, v_1, \ldots, v_r, v_t)$, then the augmenting edges of P inside f are either (i) the edge $w v_1$, or (ii) the edge $v_r w$, or (iii) edges $v_i w$ and $w v_{i+1}$ for some $1 \le i < r$.*

3   NP-completeness

Let ϕ be a Boolean 3-SAT formula with n variables $x_1, \ldots, x_n$ and m clauses $c_1, \ldots, c_m$. A clause of ϕ is *positive* (*negative*) if it has only positive (negative) literals. The *incidence graph* G_ϕ of ϕ is the graph that has *variable vertices* $x_1, \ldots, x_n$, *clause vertices* $c_1, \ldots, c_m$, and has an edge (c_j, x_i) for each clause c_j containing x_i or $\overline{x}_i$. Note that we use the same notation for variables (clauses) in ϕ and variable vertices (clause vertices) in G_ϕ. If ϕ has clauses with less than three literals, we introduce parallel edges in G_ϕ so that all clause vertices have degree 3 in G_ϕ; see, e.g., the dotted edge in Fig. 4. The formula ϕ is an instance of the NP-complete PLANAR MONOTONE 3-SAT problem [5], if each clause of ϕ is positive or negative, and G_ϕ has a plane embedding $\mathcal{E}_\phi$ to which the edges of a cycle $\mathcal{C}_\phi := x_1, \ldots, x_n$ can be added that separates positive and negative clause vertices. The problem asks whether ϕ is satisfiable. Next, we present our gadgets.

Double Ladder. A double ladder of even length ℓ is defined as follows. Its vertex set consists of two sources, s_1 and s_2, two sinks, t_1 and t_2, and vertices in $\cup_{i=0}^{\ell}\{u_i, v_i, w_i\}$. Its edge set consists of edges s_1u_0, s_1v_0, s_2v_0, s_2w_0, $u_\ell t_1$, $v_\ell t_1$, $v_\ell t_2$, $w_\ell t_2$, and $\cup_{i=0}^{\ell-1}\{u_iu_{i+1}, v_iu_{i+1}, v_iv_{i+1}, w_iv_{i+1}, w_iw_{i+1}\}$.

Property 4. *The double ladder has a unique upward planar embedding (up to a flip), shown in Figs. 1d and 1e.*

Proof. The embedding shown in Figs. 1d and 1e clearly is an upward planar embedding. The underlying graph of the double ladder has four combinatorial embeddings, which are obtained from the embedding in Figs. 1d and 1e, by possibly flipping the path $u_1u_0s_1v_0$ along u_1v_0 and the path $w_{\ell-1}w_\ell t_2v_\ell$ along $w_{\ell-1}v_\ell$. However, such flips respectively force s_1v_0 and $v_\ell t_2$ to point downward. Finally, since the outer face of the embedding in Figs. 1d and 1e is the only face containing at least one source and one sink, the claim follows. $\square$

Property 5. *Let G be a plane DAG with a subhamiltonian path P. If G contains a double ladder of length ℓ, then P contains the pattern $[\ldots u_iv_iw_i \ldots w_{i+1} v_{i+1}u_{i+1} \ldots]$ or $[\ldots w_iv_iu_i \ldots u_{i+1}v_{i+1}w_{i+1} \ldots]$ for $i = 0, \ldots, \ell - 1$.*

Proof. By Properties 2 and 4, u_i, v_i, w_i are consecutive along P, for $i = 0, \ldots, \ell$. The edge u_iu_{i+1} implies that u_i, v_i, w_i precede $u_{i+1}, v_{i+1}, w_{i+1}$. So, it remains to rule out patterns $[\ldots u_iv_iw_i \ldots u_{i+1}v_{i+1}w_{i+1} \ldots]$ and $[\ldots w_iv_iu_i \ldots w_{i+1}v_{i+1} u_{i+1} \ldots]$. If P contains one of them, then edges u_iu_{i+1}, v_iv_{i+1} and w_iw_{i+1} pairwise cross, implying that G has page-number at least 3; a contradiction. $\square$

Corollary 1. *There exist two subhamiltonian paths for the double ladder, shown in Figs. 1d and 1e.*

Variable Gadget: Let $x \in \{x_1, \ldots, x_n\}$. The variable gadget L_x for x is the double ladder of length $4d_x$, where d_x is the degree of x in G_ϕ. To distinguish between vertices of different variable gadgets, we denote the vertices of L_x with

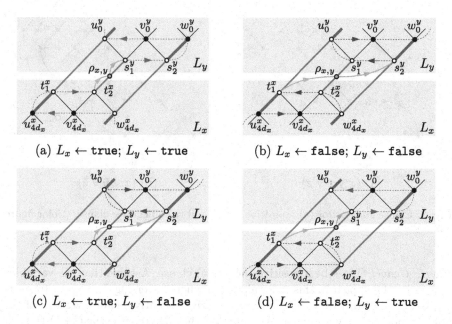

(a) $L_x \leftarrow$ **true**; $L_y \leftarrow$ **true**   (b) $L_x \leftarrow$ **false**; $L_y \leftarrow$ **false**

(c) $L_x \leftarrow$ **true**; $L_y \leftarrow$ **false**   (d) $L_x \leftarrow$ **false**; $L_y \leftarrow$ **true**

Fig. 2. The connector gadget for two variables having (a)-(b) the same truth assignment, and (c)-(d) the opposite truth assignment.

the superscript x, as in Fig. 2. Vertices s_1^x, s_2^x, u_0^x are the *bottom connectors* and $w_{4d_x}^x$, t_1^x, t_2^x are the *top connectors* of L_x. The two subhamiltonian paths of Corollary. 1 correspond to the truth assignments of x; Fig. 1d corresponds to **true**, while Fig. 1e to **false**. Also, we refer to the edges of L_x that are part of the subhamiltonian path of Fig. 1d (of Fig. 1e) as *true edges* (*false edges*, respectively). In particular, $u_{2j}^x u_{2j+1}^x$ and $w_{2j+1}^x w_{2j+2}^x$ are true edges of L_x, while $u_{2j+1}^x u_{2j+2}^x$ and $w_{2j}^x w_{2j+1}^x$ are false edges of L_x, for $j = 0, \ldots, 2d_x - 1$.

Connector Gadget: A connector gadget connects two variable gadgets L_x and L_y by means of three paths from the top connectors of L_x to the bottom connectors of L_y; see Fig. 2. These paths are: the edge $t_1^x u_0^y$, the length-2 path $t_2^x \rho_{x,y} s_1^y$, where $\rho_{x,y}$ is a newly introduced vertex, and the edge $w_{4d_x}^x s_2^y$.

Property 6. *Given subhamiltonian paths P_x for L_x and P_y for L_y, there is a subhamiltonian path P containing P_x and P_y for the graph obtained by adding a vertex $\rho_{x,y}$ and edges $t_1^x u_0^y, t_2^x \rho_{x,y}, \rho_{x,y} s_1^y, w_{4d_x}^x s_2^y$ to $L_x \cup L_y$.*

Proof. Each of P_x and P_y is one of the two subhamiltonian paths of Corollary 1; see Fig. 1. In particular, the last vertex of P_x is t_1^x or t_2^x, and the first vertex of P_y is s_1^y or s_2^y, depending on the truth assignments for x and y, respectively, as shown in Fig. 2. We obtain P by adding directed edges from the last vertex of P_x to $\rho_{x,y}$ and from $\rho_{x,y}$ to the first vertex of P_y. □

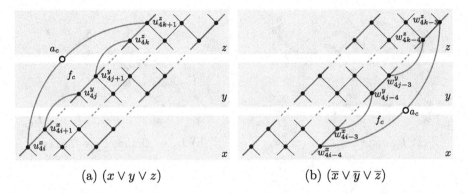

(a) $(x \vee y \vee z)$ (b) $(\overline{x} \vee \overline{y} \vee \overline{z})$

Fig. 3. Clause gadgets for (a) a positive clause and (b) a negative clause. (Color figure online)

Clause Gadget: Let c be a positive (negative) clause. Assume that the variables x, y and z of c appear in this order along $\mathcal{C}_\phi$, when traversing $\mathcal{C}_\phi$ from x_1 towards x_n. In $\mathcal{E}_\phi$, the edges between x and the positive (negative) clause vertices of G_ϕ appear consecutively around x. Assume that the edge (c, x) is the $(i+1)$-th such edge in a clockwise (counter-clockwise) traversal of the edges around x starting at the edge of $\mathcal{C}_\phi$ incoming x. Similarly, define indices j and k for y and z, respectively. Let L_x, L_y and L_z be the three variable gadgets for x, y and z.

The clause gadget C_c for c consists of an *anchor vertex* a_c, and four edges. If c is positive, these edges are $u^x_{4i}a_c$, $a_c u^z_{4k+1}$, $u^x_{4i+1}u^y_{4j}$ and $u^y_{4j+1}u^z_{4k}$ (green in Fig. 3a); otherwise, they are $w^x_{4i-4}a_c$, $a_c w^z_{4k-3}$, $w^x_{4i-3}w^y_{4j-4}$ and $w^y_{4j-3}w^z_{4k-4}$ (green in Fig. 3b). Note that C_c creates a non-transitive face f_c, called *anchor face*, whose boundary is delimited by the two newly-introduced edges incident to a_c and by a directed path whose edges alternate between three true edges (if c is positive) or three false edges (if c is negative) and the two newly-introduced edges not incident to a_c; see Fig. 3. The three true (or false) edges on the boundary of f_c stem from L_x, L_y, and L_z. The length of the double ladders ensures that, if $x = y$ (which implies that $j = i + 1$), then vertices u^x_{4i+1} and u^y_{4j} (w^x_{4i-3} and w^y_{4j-4}) are not adjacent in L_x and the edge $u^x_{4i+1}u^y_{4j}$ ($w^x_{4i-3}w^y_{4j-4}$) is well defined; this is the reason that we do not use vertices with indices $2, 3 \bmod 4$.

Theorem 1. *Recognizing whether a DAG has page-number 2 is NP-complete, even if the input is an st-planar graph.*

Proof. The problem clearly belongs to NP, as a non-deterministic Turing machine can guess an order of the vertices of an input graph and a partition of its edges into two pages, and check in polynomial time whether the order is a topological order and if so, whether any two edges in the same page cross.

Given an instance ϕ of Planar Monotone 3-SAT, we construct in polynomial time an st-planar graph H that has page-number 2 if and only if ϕ is satisfiable; see Fig. 4. We consider the variable gadgets $L_{x_1}, \ldots, L_{x_n}$, where $x_1, \ldots, x_n$ is the order of the variables along the cycle $\mathcal{C}_\phi$; for $i = 1, \ldots, n - 1$,

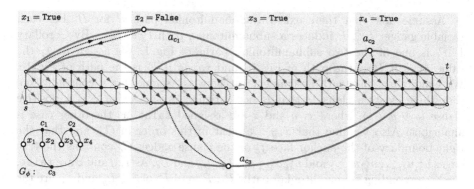

Fig. 4. The graph H obtained from $\phi = c_1 \wedge c_2 \wedge c_3$ with $c_1 = (x_1 \vee x_2 \vee x_3)$, $c_2 = (x_3 \vee x_4)$, and $c_3 = (\overline{x}_1 \vee \overline{x}_2 \vee \overline{x}_4)$. For space reasons, the variable gadgets have smaller length and the drawing is rotated by $45°$.

we connect L_{x_i} with $L_{x_{i+1}}$ using a connector gadget. For each positive (negative) clause c of ϕ, we add a clause gadget C_c using the true (false) edges of the variable gadgets. This yields a plane DAG with two sources $s_1^{x_1}$ and $s_2^{x_1}$ and two sinks $t_1^{x_n}$ and $t_2^{x_n}$. We add a source s connected with outgoing edges to $s_1^{x_1}$ and $s_2^{x_1}$, and a sink t connected with incoming edges to $t_1^{x_n}$ and $t_2^{x_n}$. The constructed graph H is st-planar. Since the underlying graph of H is a subdivision of a triconnected planar graph and since only one face of H contains s and t, it follows that H has a unique upward planar embedding. We next prove that H is subhamiltonian (and therefore has page-number 2) if and only if ϕ is satisfiable.

Assume first that ϕ is satisfiable. We show how to construct a subhamiltonian path P for H, by exploiting a satisfying truth assignment for ϕ. For $i = 1, \ldots, n$, we have that P contains the subhamiltonian path P_i for L_{x_i} shown in Fig. 1d if x_i is **true**, and the one shown in Fig. 1e otherwise. By Property 6, there is a subhamiltonian path P for the subgraph of H induced by the vertices of all variable and connector gadgets, containing $P_1, \ldots, P_n$ as subpaths. The path P starts from a source of L_{x_1} and ends at a sink of L_{x_n}; hence we can extend P to include s and t as its first and last vertices. We now extend P to a subhamiltonian path for H by including the anchor vertex of each clause gadget. Consider a positive clause $c = (x \vee y \vee z)$ with anchor vertex a_c; the case of a negative clause is similar. As ϕ is satisfied, at least one of x, y and z is **true**; assume w.l.o.g. that x is **true**. By construction, the anchor face f_c of C_c is non-transitive, with the anchor vertex a_c on its left boundary, and exactly one true edge of each of L_x, L_y, and L_z along its right boundary. Let $i \geq 0$ be such that $u_{4i}^x u_{4i+1}^x$ is the true edge of L_x on the right boundary of f_c. Since x is **true**, vertices u_{4i}^x and u_{4i+1}^x are consecutive in P. We extend P by visiting vertex a_c after u_{4i}^x and before u_{4i+1}^x. This corresponds to adding two augmenting edges $u_{4i}^x a_c$ and $a_c u_{4i+1}^x$ of P in the interior of f_c; see the black dashed edges of Fig. 4. At the end of this process, P is extended to a subhamiltonian path for H.

Assume now that there exists a subhamiltonian path P for H. For each variable gadget L_{x_i}, P induces a subhamiltonian path P_i for L_{x_i}. By Corollary 1, P_i is one of the two subhamiltonian paths of Fig. 1. We assign to x_i the value `true` if P_i is the path of Fig. 1d and `false` if P_i is the path of Fig. 1e. We claim that this truth assignment satisfies ϕ. Assume, for a contradiction, that there exists a clause c that is not satisfied. Assume that c is a positive clause $(x \vee y \vee z)$, where x, y and z are assigned `false`, as the other case is analogous. Also, assume that x, y, z appear in this order in C_ϕ, and that the right boundary of the anchor face f_c of the clause gadget C_c contains the true edges $u_{4i}^x u_{4i+1}^x$, $u_{4j}^y u_{4j+1}^y$ and $u_{4k}^z u_{4k+1}^z$ of L_x, L_y and L_z. As x, y and z are `false`, the corresponding subhamiltonian paths P_x, P_y and P_z of L_x, L_y and L_z are the ones of Fig. 1e. Hence, P contains the augmenting edges $u_{4i}^x v_{4i}^x$ and $v_{4i+1}^x u_{4i+1}^x$ of P_x, $u_{4j}^y v_{4j}^y$ and $v_{4j+1}^y u_{4j+1}^y$ of P_y and $u_{4k}^z v_{4k}^z$ and $v_{4k+1}^z u_{4k+1}^z$ of P_z. By Property 3 for the non-transitive face f_c, P contains either (i) the augmenting edge $a_c u_{4i+1}^x$, or (ii) the augmenting edge $u_{4k}^z a_c$, or (iii) for a pair of consecutive vertices, say u and u', along the right boundary of f_c, the augmenting edges $u a_c$ and $a_c u'$. Cases (i) and (ii) contradict the existence of augmenting edges $v_{4i+1}^x u_{4i+1}^x$ and $u_{4k}^z v_{4k}^z$ of P respectively. Similarly, in case (iii) the augmenting edges of P that belong to P_x, P_y, and P_z imply that $u \notin \{u_{4i}^x, u_{4j}^y, u_{4k}^z\}$ and $u' \notin \{u_{4i+1}^x, u_{4j+1}^y, u_{4k+1}^z\}$. Hence $u = u_{4i+1}^x$ and $u' = u_{4j}^y$ holds, or $u = u_{4j+1}^y$ and $u' = u_{4k}^z$. In both cases, the HP-completion of H contains a generalized diamond with $v_s = u$ and $v_t = u'$, violating Property 1. Hence at least one of variables x, y and z must be `true`, contradicting our assumption that c is not satisfied. □

We conclude by mentioning that our NP-completeness proof can be adjusted so that the constructed graph is a planar poset; refer to [2] for details.

References

1. Alzohairi, M., Rival, I.: Series-parallel planar ordered sets have pagenumber two. In: North, S. (ed.) GD 1996. LNCS, vol. 1190, pp. 11–24. Springer, Heidelberg (1997). https://doi.org/10.1007/3-540-62495-3_34

2. Bekos, M.A., Da Lozzo, G., Frati, F., Gronemann, M., Mchedlidze, T., Raftopoulou, C.N.: Recognizing DAGs with page-number 2 is NP-complete. CoRR abs/2208.13615 (2022). https://arxiv.org/abs/2208.13615

3. Bekos, M.A., Gronemann, M., Raftopoulou, C.N.: Two-page book embeddings of 4-planar graphs. Algorithmica 75(1), 158–185 (2016)

4. Bekos, M.A., Kaufmann, M., Klute, F., Pupyrev, S., Raftopoulou, C.N., Ueckerdt, T.: Four pages are indeed necessary for planar graphs. J. Comput. Geom. 11(1), 332–353 (2020)

5. de Berg, M., Khosravi, A.: Optimal binary space partitions for segments in the plane. Int. J. Comput. Geom. Appl. 22(3), 187–206 (2012)

6. Bhore, S., Da Lozzo, G., Montecchiani, F., Nöllenburg, M.: On the upward book thickness problem: combinatorial and complexity results. In: Purchase, H.C., Rutter, I. (eds.) GD 2021. LNCS, vol. 12868, pp. 242–256. Springer, Cham (2021). https://doi.org/10.1007/978-3-030-92931-2_18

7. Binucci, C., Da Lozzo, G., Di Giacomo, E., Didimo, W., Mchedlidze, T., Patrignani, M.: Upward book embeddings of ST-Graphs. In: Barequet, G., Wang, Y. (eds.) SoCG. LIPIcs, vol. 129, pp. 13:1–13:22. Schloss Dagstuhl-Leibniz-Zentrum fuer Informatik, Dagstuhl, Germany (2019)

8. Cornuéjols, G., Naddef, D., Pulleyblank, W.R.: Halin graphs and the travelling salesman problem. Math. Program. **26**(3), 287–294 (1983)

9. Di Battista, G., Tamassia, R.: Algorithms for plane representations of acyclic digraphs. Theor. Comput. Sci. **61**, 175–198 (1988)

10. Di Giacomo, E., Didimo, W., Liotta, G., Wismath, S.K.: Book embeddability of series-parallel digraphs. Algorithmica **45**(4), 531–547 (2006)

11. Ewald, G.: Hamiltonian circuits in simplicial complexes. Geom. Dedicata. **2**(1), 115–125 (1973)

12. Frati, F., Fulek, R., Ruiz-Vargas, A.J.: On the page number of upward planar directed acyclic graphs. J. Graph Algorithms Appl. **17**(3), 221–244 (2013)

13. de Fraysseix, H., de Mendez, P.O., Pach, J.: A left-first search algorithm for planar graphs. Discrete Comput. Geom. **13**, 459–468 (1995)

14. Guan, X., Yang, W.: Embedding planar 5-graphs in three pages. Dis. Appl. Math. **282**, 108–121 (2020)

15. Heath, L.: Embedding planar graphs in seven pages. In: FOCS, pp. 74–83. IEEE Comp. Soc. (1984)

16. Heath, L., Pemmaraju, S.V.: Stack and queue layouts of directed acyclic graphs: part II. SIAM J. Comput. **28**(5), 1588–1626 (1999)

17. Heath, L.S.: Embedding outerplanar graphs in small books. SIAM J. Algebraic Discrete Methods **8**(2), 198–218 (1987)

18. Heath, L.S., Pemmaraju, S.V., Trenk, A.N.: Stack and queue layouts of directed acyclic graphs: part I. SIAM J. Comput. **28**(4), 1510–1539 (1999)

19. Hoffmann, M., Klemz, B.: Triconnected planar graphs of maximum degree five are subhamiltonian. In: Bender, M.A., Svensson, O., Herman, G. (eds.) ESA. LIPIcs, vol. 144, pp. 58:1–58:14. Schloss Dagstuhl (2019)

20. Jungeblut, P., Merker, L., Ueckerdt, T.: A sublinear bound on the page number of upward planar graphs. In: Naor, J.S., Buchbinder, N. (eds.) SODA, pp. 963–978. SIAM (2022)

21. Kainen, P.C., Overbay, S.: Extension of a theorem of Whitney. Appl. Math. Lett. **20**(7), 835–837 (2007)

22. Kelly, D.: Fundamentals of planar ordered sets. Discret. Math. **63**(2–3), 197–216 (1987)

23. Mchedlidze, T., Symvonis, A.: Crossing-free acyclic Hamiltonian path completion for planar ST-digraphs. In: Dong, Y., Du, D.-Z., Ibarra, O. (eds.) ISAAC 2009. LNCS, vol. 5878, pp. 882–891. Springer, Heidelberg (2009). https://doi.org/10.1007/978-3-642-10631-6_89

24. Mchedlidze, T., Symvonis, A.: Crossing-optimal acyclic HP-completion for outerplanar ST-digraphs. J. Graph Algorithms Appl. **15**(3), 373–415 (2011)

25. Nishizeki, T., Chiba, N.: Planar Graphs: theory and algorithms, chap. 10. Hamiltonian Cycles, pp. 171–184. Dover Books on Mathematics, Courier Dover Publications (2008)

26. Nowakowski, R., Parker, A.: Ordered sets, pagenumbers and planarity. Order **6**(3), 209–218 (1989). https://doi.org/10.1007/BF00563521

27. Ollmann, T.: On the book thicknesses of various graphs. In: Hoffman, F., Levow, R., Thomas, R. (eds.) Southeastern Conference on Combinatorics, Graph Theory and Computing. Cong. Num., vol. 8, p. 459 (1973)

28. Rengarajan, S., Veni Madhavan, C.E.: Stack and queue number of 2-trees. In: Du, D.-Z., Li, M. (eds.) COCOON 1995. LNCS, vol. 959, pp. 203–212. Springer, Heidelberg (1995). https://doi.org/10.1007/BFb0030834
29. Wigderson, A.: The complexity of the Hamiltonian circuit problem for maximal planar graphs. Tech. Rep. TR-298, Princeton University (1982)
30. Yannakakis, M.: Embedding planar graphs in four pages. J. Comput. Syst. Sci. 38(1), 36–67 (1989)
31. Yannakakis, M.: Planar graphs that need four pages. J. Comb. Theory Ser. B 145, 241–263 (2020)

The Rique-Number of Graphs

Michael A. Bekos[1] (ID), Stefan Felsner[2] (ID), Philipp Kindermann[3](✉)(ID),
Stephen Kobourov[4] (ID), Jan Kratochvíl[5] (ID), and Ignaz Rutter[6] (ID)

[1] Department of Mathematics, University of Ioannina, Ioannina, Greece
bekos@uoi.gr
[2] Institut für Mathematik, Technische Universität Berlin, Berlin, Germany
felsner@math.tu-berlin.de
[3] Fachbereich IV - Informatikwissenschaften, Universität Trier, Trier, Germany
kindermann@uni-trier.de
[4] Department of Computer Science, University of Arizona, Tucson, AZ, USA
kobourov@cs.arizona.edu
[5] Department of Applied Mathematics, Charles University, Prague, Czech Republic
honza@kam.mff.cuni.cz
[6] Fakultät für Informatik und Mathematik, Universität Passau, Passau, Germany
rutter@fim.uni-passau.de

Abstract. We continue the study of linear layouts of graphs in relation
to known data structures. At a high level, given a data structure, the goal
is to find a linear order of the vertices of the graph and a partition of its
edges into pages, such that the edges in each page follow the restriction
of the given data structure in the underlying order. In this regard, the
most notable representatives are the stack and queue layouts, while there
exists some work also for deques.

In this paper, we study linear layouts of graphs that follow the restriction of a restricted-input queue (rique), in which insertions occur only at
the head, and removals occur both at the head and the tail. We characterize the graphs admitting rique layouts with a single page and we use the
characterization to derive a corresponding testing algorithm when the
input graph is maximal planar. We finally give bounds on the number of
needed pages (so-called rique-number) of complete graphs.

Keywords: Linear layout · Restricted-input queue · Rique-number

1 Introduction

Linear graph layouts form an important methodological tool, since they provide a
key-framework for defining different graph-parameters (including the well-known
cutwidth [1], bandwidth [14] and pathwidth [34]). As a result, the corresponding
literature is rather rich; see [35]. Such layouts typically consist of an order of the
vertices of a graph and an objective over its edges that one seeks to optimize.
In the closely-related area of permutations and arrangements, back in 1973,

---

This work was initiated at the Bertinoro Workshop on Graph Drawing 2022.

P. Angelini and R. von Hanxleden (Eds.): GD 2022, LNCS 13764, pp. 371–386, 2023.
https://doi.org/10.1007/978-3-031-22203-0_27

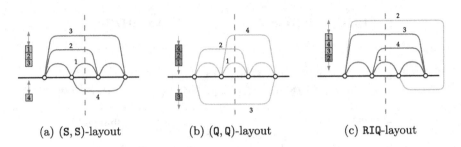

(a) (S, S)-layout   (b) (Q, Q)-layout   (c) RIQ-layout

Fig. 1. Different linear layouts of the complete graph K_4. The data structures are depicted in the states that corresponds to the dashed vertical line.

Pratt [32] introduced and studied several variants of linear layouts that one can derive by leveraging different data structures to capture the order of the vertices (e.g., stacks, queues and deques).

Formally, given k data structures $D_1, \ldots, D_k$, a graph G admits a $(D_1, \ldots, D_k)$-layout if there is a linear order $\prec$ of the vertices of G and a partition of the edges of G into k sets $E_1, \ldots, E_k$, called *pages*, such that for each page E_i in the partition, each edge (u, v) of E_i is processed by the data structure D_i by inserting (u, v) to D_i at u and removing it from D_i at v if $u \prec v$ in the linear layout. If the sequence of insertions and removals is feasible, then G is called a $(D_1, \ldots, D_k)$-graph. We denote the class of $(D_1, \ldots, D_k)$-graphs by $D_1 + \ldots + D_k$. For a certain data structure D, the D-number of a graph G is the smallest k such that G admits a $(D_1, \ldots, D_k)$-layout with $D = D_1 = \ldots = D_k$. This graph parameter has been the subject of intense research for certain data structures, as we discuss below.

1. If D is a *stack* (abbreviated by S), then insertions and removals only occur at the head of D; see Fig. 1a. It is known that a non-planar graph may have linear stack-number, e.g., the stack-number of K_n is $\lceil n/2 \rceil$ [13]. A central result here is by Yannakakis, who back in 1986 showed that the stack-number of planar graphs is at most 4 [36], a bound which was only recently shown to be tight [11]. Certain subclasses of planar graphs, however, allow for stack-layouts with fewer than four stacks, e.g., see [9,15,20,21,24,26,29–31,33].

2. If D is a *queue* (abbreviated by Q), then insertions only occur at the head and removals only at the tail of D; see Fig. 1b. In this context, a breakthrough by Dujmović et al. [19] states that the queue-number of planar graphs is at most 49, improving previous results [5,16–18]. Even though this bound was recently improved to 42 [10], the exact queue-number of planar graphs is not yet known; the current-best lower bound is 4 [10]. Again, several subclasses allow for layouts with significantly fewer than 42 queues, e.g., see [2,22,27,33].

3. If D is a *double-ended queue* or *deque* (abbreviated by DEQ), then insertions and removals can occur both at the head and the tail of D; we denote the deque-number of a graph G by $\deq(G)$. This definition implies that $S + S \subseteq DEQ \subseteq S + S + Q$. A characterization by Auer et al. [4] (stating that a graph has deque-number 1 if and only if it is a spanning subgraph of a planar graph with a Hamiltonian path) implies that the first containment is strict, because a maximal planar graph with a Hamiltonian path but not a Hamiltonian

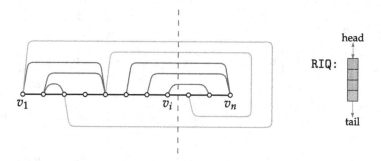

Fig. 2. A strongly 1-sided Hamiltonian path and the state of the RIQ that processes it right after processing the edges incident to v_i.

cycle (e.g., the Goldner-Harary graph [23]) admits a DEQ-layout, but not an (S, S)-layout. The second containment is also strict because (S, S, Q)-graphs can be non-planar (e.g., K_6 [3]). Hence, $S + S \subsetneq DEQ \subsetneq S + S + Q$ holds.

Our Contribution. In this work, we focus on the case where the data structure D is a *restricted-input queue* or *rique* (abbreviated by RIQ), in which insertions occur only at the head, and removals occur both at the head and the tail of D; see Fig. 1c. We first characterize the graphs with *rique*-number 1 as those admitting a planar embedding with a so-called *strongly 1-sided subhamiltonian path*, that is, a Hamiltonian path $v_1, \ldots, v_n$ in some plane extension of the embedding such that each edge (v_i, v_j) with $1 < i < j \leq n$ leaves v_i on the same side of the path; see Fig. 2. This characterization allows us to derive an inclusion relationship similar to the one above for deques (namely, $S, Q \subsetneq RIQ \subsetneq S + Q$; see Observation 2) and corresponding recognition algorithms for graphs with rique-number 1 under some assumptions (Theorem 3). Then, we focus on bounds on the rique-number of a graph G, which we denote by $\text{riq}(G)$. Our contribution is an edge-density bound for the graphs with rique-number k (Thoerem 5), and a lower and an upper bound on the rique-number of complete graphs (Therorem 6).

2 Preliminaries

We start with definitions that are central in Sect. 3. Given a rique-layout, we call an edge (u, v) a *head-edge* (*tail-edge*), if (u, v) is removed at v from the head (tail) of the RIQ. A *strongly 1-sided Hamiltonian path* of a plane graph is a Hamiltonian path $v_1, \ldots, v_n$ such that each edge (v_i, v_j) with $1 < i < j \leq n$ leaves v_i on the same side of the path, say w.l.o.g. the *left* one, i.e., between (v_{i-1}, v_i) and (v_i, v_{i+1}) in clockwise order around v_i (see Fig. 2). A plane graph is *strongly 1-sided Hamiltonian* if it contains a strongly 1-sided Hamiltonian path. A planar graph is *strongly 1-sided Hamiltonian* if it admits a planar embedding that contains a strongly 1-sided Hamiltonian path. A planar (plane) graph G is *strongly 1-sided subhamiltonian* if there exists a planar (plane) supergraph H of G that is strongly 1-sided Hamiltonian.

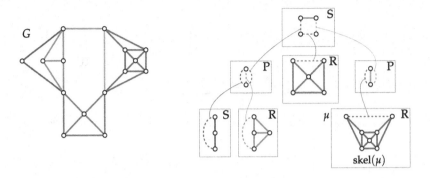

Fig. 3. An SPQR-tree, omitting the Q-nodes.

Another key-tool that we leverage in Sect. 4 is the SPQR-tree. This data structure, introduced by Di Battista and Tamassia [6,7], compactly represents all planar embeddings of a biconnected planar graph; see Fig. 3 for an example. It is unique and can be computed in linear time [25]. We assume familiarity with SPQR-trees; for a brief introduction refer to the full version of the paper [8].

3   Characterization of Graphs with Rique-Number 1

In this section, we discuss properties of graphs with rique-number 1. We first characterize these graphs in Lemma 1 in terms of the following forbidden pattern.

P.1 Three edges $\langle e_a, e_b, e_c \rangle$ with $e_a = (a, a')$, $e_b = (b, b')$ and $e_c = (c, c')$ form Pattern P.1 in a linear layout if and only if $a \prec b \prec c \prec b' \prec \{a', c'\}$; see Fig. 4.

Lemma 1. *A graph has rique-number 1 if and only if it admits a linear order avoiding Pattern P.1.*

Proof. Let G be a graph with rique-number 1 and assume for a contradiction that a linear order of it contains Pattern P.1. The edge e_a is inserted into data structure RIQ before the edge e_b is inserted, but removed after e_b is removed. Hence, e_b cannot be removed at the tail of RIQ, so it has to be removed at its head. However, the edge e_c is inserted after the edge e_b is inserted, but also removed after e_b is removed, so e_b also cannot be removed at the head; a contradiction.

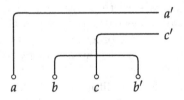

Fig. 4. Forbidden Pattern P.1

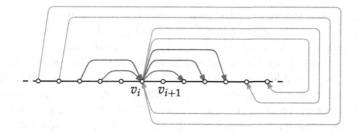

Fig. 5. Ordering of the edges around a vertex v_i.

For the other direction, assume that G has rique-number greater than 1. We will prove that every linear order of G contains Pattern P.1. Let $\prec$ be such an order. Since G has rique-number greater than 1 and all insertions into a RIQ happen on the same side, at some time b' there is an edge e_b to be removed that is neither at the head nor at the tail of RIQ. Since e_b is not at the head, there is some other edge e_c that was inserted into RIQ after e_b and is still there at time b'. Since e_b is not at the tail, there is some other edge e_a that was inserted into RIQ before e_b and still is there. Then, $\langle e_a, e_b, e_c \rangle$ form Pattern P.1.   □

We are now ready to completely characterize the graphs with rique-number 1.

Theorem 1. *A graph G has rique-number 1 if and only if G is planar strongly 1-sided subhamiltonian.*

Proof. First, assume that G can be embedded so that it contains a strongly 1-sided subhamiltonian path $v_1, \ldots, v_n$. For a contradiction, assume further that $\langle e_a = (a, a'), e_b = (b, b'), e_c = (c, c') \rangle$ form Pattern P.1 in the order $v_1, \ldots, v_n$. Note that e_a, e_b, and e_c leave a, b, and c on the left side, respectively. If e_b enters b' from the left, then e_b crosses e_c as $b \prec c \prec b' \prec c'$. So, e_b has to enter b' from the right. Then, however, e_b crosses e_a since $a \prec b \prec b' \prec a'$; a contradiction. So, by Lemma 1, G has rique-number 1.

Assume now that G has rique-number 1. By Lemma 1, G admits a linear order $v_1, \ldots, v_n$ avoiding Pattern P.1. W.l.o.g. we assume that G contains all edges in $\{(v_1, v_2), \ldots, (v_{n-1}, v_n)\}$ and prove that G is strongly 1-sided Hamiltonian.

Consider a vertex v_i. We order the edges around v_i counter-clockwise as follows; see Fig. 5. (i) The edge (v_i, v_{i+1}) (for $i < n$); (ii) the outgoing head-edges of v_i, ordered in increasing order by the index of the target vertex; (iii) the outgoing tail-edges of v_i, ordered in decreasing order by the index of the target vertex; (iv) the incoming head-edges of v_i, ordered in increasing order by the index of the source vertex; (v) the edge (v_{i-1}, v_i) (for $i > 1$); (vi) the incoming tail-edges of v_i, ordered in increasing order by the index of the source vertex. This ensures that all edges leave v_i on the correct side of the Hamiltonian path. It remains to be shown that this embedding is plane. To this end, assume that there are two edges (v_i, v_j) and (v_k, v_ℓ) that cross. W.l.o.g. we assume that $i < k$.

If (v_k, v_ℓ) is a head-edge, then it leaves and enters v_k and v_ℓ on the same side of the Hamiltonian path as (v_i, v_j) leaves v_i. Hence, (v_i, v_j) and (v_k, v_ℓ) cross only if (v_i, v_j) also enters v_j on the same side. So, (v_i, v_j) is also a head-edge with $i < k < j < \ell$. However, since (v_i, v_j) entered RIQ at the head before (v_k, v_ℓ), it cannot leave RIQ at the head before (v_k, v_ℓ); a contradiction.

If (v_k, v_ℓ) is a tail-edge, then (v_i, v_j) leaves v_i on the same side of the Hamiltonian path as (v_k, v_ℓ) leaves v_k, but (v_k, v_ℓ) enters v_ℓ on the other side. If (v_i, v_j) is a head-edge, then we must have $i < k < j$. However, since (v_i, v_j) entered RIQ at the head before (v_k, v_ℓ), it cannot leave RIQ at the head before (v_k, v_ℓ); a contradiction. Otherwise (v_i, v_j) is a tail-edge, and we must have $i < k < \ell < j$. However, since (v_i, v_j) entered RIQ at the head before (v_k, v_ℓ), it cannot leave RIQ at the tail after (v_k, v_ℓ); a contradiction.

It follows that no two edges cross, as desired. This concludes the proof.   □

The definition of a rique implies $\mathtt{X} \subseteq \mathtt{RIQ} \subseteq \mathtt{S} + \mathtt{Q}$, where $\mathtt{X} \in \{\mathtt{S}, \mathtt{Q}\}$. By Theorem 1, both inclusions are strict, as $K_4 \in \mathtt{RIQ}$ (see Fig. 1c) but it admits neither a stack-layout (since it is not outerplanar [13]) nor a queue-layout (since any linear order yields a 2-rainbow [28]), and K_6 admits an $(\mathtt{S}, \mathtt{Q})$-layout [3] but is not planar and therefore $K_6 \notin \mathtt{RIQ}$.

Observation 2. $\mathtt{X} \subsetneq \mathtt{RIQ} \subsetneq \mathtt{S} + \mathtt{Q}$, where $\mathtt{X} \in \{\mathtt{S}, \mathtt{Q}\}$

4   Recognition of Graphs with Rique-Number 1

With the characterization of Theorem 1 at hand, we now turn our focus to the recognition problem, where we present two algorithms: (i) the first one is simple and tests whether a plane graph is strongly 1-sided Hamiltonian, while (ii) the second one is more elaborate and tests whether a planar graph is strongly 1-sided Hamiltonian. Even though our algorithms do not solve the general case of testing whether a graph has rique-number 1 (or equivalently by Theorem 1 whether it is strongly 1-sided subhamiltonian), they can be leveraged for testing, e.g., whether a maximal planar graph or a 3-connected planar graph has rique-number 1.

Theorem 3. *Given a plane n-vertex graph G, there is an $O(n^2)$-time algorithm to test whether G is plane strongly 1-sided Hamiltonian.*

Proof. After guessing the first edge of the path, for which there are $O(n)$ choices, we assume that we have computed a subpath $\pi = v_1, \ldots, v_i$, $2 \leq i < n$ of a strongly 1-sided Hamiltonian path of G. We claim that the next vertex v_{i+1} is uniquely determined by π. Consider the edges of G incident to v_i in counterclockwise order, starting from the edge after (v_{i-1}, v_i). Let e be the first edge in this order, whose other endpoint does not lie on π. We choose this endpoint as v_{i+1}. This is correct, since choosing an endpoint of an edge preceding e visits a vertex twice, whereas choosing an endpoint of an edge succeeding e would imply that e leaves the resulting path on the wrong side. The above argument shows that, after guessing an initial edge, the remainder of the 1-sided Hamiltonian path is uniquely defined, if it exists. Since a single starting edge can be tested in $O(n)$ time, the overall time complexity of our algorithm is $O(n^2)$.   □

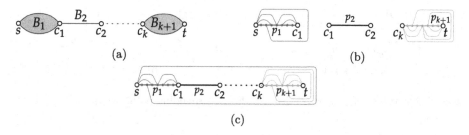

Fig. 6. (a) A block-cut tree; (b) a strongly 1-sided Hamiltonian embedding for each block; (c) a strongly 1-sided Hamiltonian embedding for the whole graph.

Corollary 1. *Given a maximal planar graph G with n vertices, there is an $O(n^2)$-time algorithm to test whether G has rique-number 1.*

Theorem 4. *Given a planar n-vertex graph G, there is an $O(n^4)$-time algorithm to test whether G is planar strongly 1-sided Hamiltonian.*

Proof. To prove the statement, we assume that the endpoints s, t of the Hamiltonian path are specified as part of the input and we show that testing whether G admits a planar embedding containing a strongly 1-sided Hamiltonian st-path can be done in $O(n^2)$ time. In the positive case, we say that G is *st-1-sided*.

If G is not biconnected, then for G to be st-1-sided its block-cut tree must be a path $B_1, c_1, B_2, \ldots, c_k, B_{k+1}$, such that $s \in B_1$ and $t \in B_{k+1}$ (or vice versa; here k denotes the number of cutvertices of G). We set $c_0 = s$ and $c_{k+1} = t$ and claim that G is st-1-sided if and only if each block B_i is $c_{i-1}c_i$-1-sided for $i = 1, \ldots, k+1$. The necessity is clear, we prove the sufficiency. Let $\mathcal{E}_i$ be a planar embedding of B_i containing a strongly 1-sided Hamiltonian $c_{i-1}c_i$-path p_i for $i = 1, \ldots, k+1$. We modify the embedding $\mathcal{E}_i$ such that the first edge of p_i lies on the outer face, and combine $\mathcal{E}_{i-1}$ and $\mathcal{E}_i$ in such a way that the first edge of p_i follows the last edge of p_{i-1} in counterclockwise order around c_i. Then the path p obtained by concatenating p_i, $i = 1, \ldots, k+1$ is a strongly 1-sided Hamiltonian path in the resulting embedding $\mathcal{E}$ of G; see Fig. 6.

Hence, we may assume that G consists of a single block. Since the case where G consists of a single edge can be handled trivially, we focus on the case where G is biconnected. To determine whether G is st-1-sided, we use a dynamic program based on an SPQR-tree $\mathcal{T}$ of G. We root $\mathcal{T}$ at an edge incident to t and for each node μ of $\mathcal{T}$ with poles u, v, we want to answer the following questions: If $s \notin \mathrm{pert}(\mu)$, we want to know for each of the two ordered pairs of poles $(x, y) \in \{(u, v), (v, u)\}$ whether $\mathrm{pert}(\mu)$ has an embedding with x, y on the outer face such that it contains a strongly 1-sided Hamiltonian path from x to y that starts with the edge that follows the parent edge counterclockwise around x; in the positive case. We define the set $L(\mu)$ as those ordered pairs (x, y) where this is the case. For a pair $(x, y) \in L(\mu)$, we denote by $\mathcal{E}_\mu(x, y)$ the corresponding embedding of $\mathrm{pert}(\mu)$ and by $P_\mu(x, y)$ the corresponding path. If $s \in \mathrm{pert}(\mu)$, then for each $x \in \{u, v\}$ and $Y \subseteq \{u, v\} \setminus \{x\}$ we want to know whether $\mathrm{pert}(\mu)$

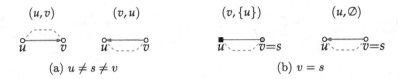

Fig. 7. The tuples of $L(\mu)$ for a Q-node; the corresponding paths are red. (Color figure online)

has an embedding $\mathcal{E}_\mu(x, Y)$ such that u, v are incident to the outer face and there is a strongly 1-sided path $P_\mu(x, Y)$ from s to x that visits all vertices of $\mathrm{pert}(\mu) - Y$. As above, for node μ, we define $L(\mu)$ as the set of all pairs (x, Y) where this is possible.

Consider the root r of $\mathcal{T}$ and let μ be its child with poles u, t. Then G is st-1-sided if and only if and only if $(t, \emptyset) \in L(\mu)$. The necessity is clear. For the sufficiency, observe that $P_\mu(t, \emptyset)$ is a strongly 1-sided st-path in the embedding of G obtained from $\mathcal{E}_\mu(t, \emptyset)$ by adding the edge ut in the outer face. We compute the set $L(\mu)$ for each node μ of $\mathcal{T}$ (together with corresponding embeddings of $\mathrm{pert}(\mu)$ and paths) by a bottom-up traversal of $\mathcal{T}$ as follows. Let μ be a node of $\mathcal{T}$ in this traversal with poles u, v. If μ is not a leaf in $\mathcal{T}$, we denote by $\mu_1, \dots, \mu_k$ its children, and we assume that $L(\mu_i)$ has already been computed for $i = 1, \dots, k$. We next distinguish cases based on the type of μ.

Case 1: μ is a Q-node. If $u \neq s \neq v$, then $L(\mu) = \{(u, v), (v, u)\}$. And for $(x, y) \in L(\mu)$ the embedding $\mathcal{E}_\mu(x, y)$ and the path $P_\mu(x, y)$ are trivial; see Fig. 7a. Otherwise, assume w.l.o.g. $s = v$. Then $L(\mu) = \{(v, \{u\}), (u, \emptyset)\}$. Again for $(x, Y) \in L(\mu)$, $\mathcal{E}_\mu(x, Y)$ and $P_\mu(x, Y)$ can be defined trivially; see Fig. 7b.

Case 2: μ is a P-node. Assume first that $s \notin \mathrm{pert}(\mu)$; see Fig. 8a. We show how to test whether $(v, u) \in L(\mu)$. The case of (u, v) is symmetric. First $(v, u) \in L(\mu)$ requires $k = 2$ and that only one of the children, say μ_1, is not a Q-node. If so, $(v, u) \in L(\mu)$ if and only if $(v, u) \in L(\mu_1)$. Also, $P_\mu(v, u) = P_{\mu_1}(v, u)$ and $\mathcal{E}_\mu(v, u)$ is obtained by embedding the edge represented by μ_2 to the left of $\mathcal{E}_{\mu_1}(v, u)$.

Now, consider the case that $s \in \mathrm{pert}(\mu)$. Assume first that s is a pole of μ; see Fig. 8a. Then, any 1-sided path of μ unavoidably visits the other pole. In fact, only a single child can be traversed, i.e., $k = 2$, and one child, say μ_2, is a Q-node. If this is not the case, $L(\mu) = \emptyset$. Otherwise, $L(\mu) = L(\mu_1)$. For $(x, Y) \in L(\mu_1)$, we set $p_\mu(x, Y) = p_{\mu_1}(x, Y)$ and we define $\mathcal{E}_\mu(x, Y)$ as the embedding obtained from $\mathcal{E}_{\mu_1}(x, Y)$ by putting the edge represented by μ_2 to its left parallel to it.

Assume now that s is not a pole and it lies, w.l.o.g., in $\mathrm{pert}(\mu_1)$. Let (x, Y) be a pair with $x \in \{u, v\}$, $Y \subseteq \{u, v\} \setminus \{x\}$. W.l.o.g. we assume $x = u$. The case $x = v$ is analogous. Then either $Y = \{v\}$ or $Y = \emptyset$. If $v \in Y$ (see Fig. 8b), then $(u, Y) \in L(\mu)$ if and only if $(u, Y) \in L(\mu_1)$ and $k = 2$ and μ_2 is a Q-node. In that case, we set $P_\mu(u, Y) = P_{\mu_1}(u, Y)$ and we define $\mathcal{E}_\mu(u, Y)$ as the embedding obtained from $\mathcal{E}_{\mu_1}(x, Y)$ by embedding the edge represented by μ_2 to its left. If $Y = \emptyset$ (see Fig. 8c), then we distinguish cases based on whether there is a second child, say μ_2, that is not a Q-node. If there is none, then μ_2 is a Q-node and then $(u, \emptyset) \in L(\mu)$ if and only if either $(u, \emptyset) \in L(\mu_1)$ or if $(v, \{u\}) \in L(\mu_1)$.

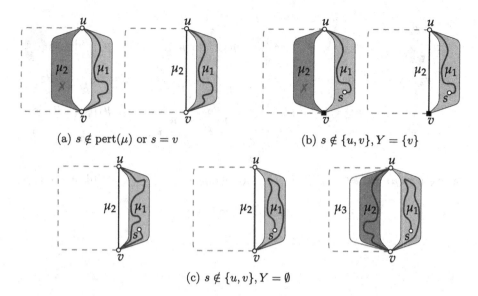

(a) $s \notin \mathrm{pert}(\mu)$ or $s = v$ (b) $s \notin \{u,v\}, Y = \{v\}$

(c) $s \notin \{u,v\}, Y = \emptyset$

Fig. 8. Paths $P(v,u)$, $P(u,Y)$ in a P-node. The vertices in Y are black.

In these cases, we set $P_\mu(u,\emptyset) = P_{\mu_1}(u,\emptyset)$ or $P_\mu(u,\emptyset) = P_{\mu_1}(v,\{u\}) \cdot (v,u)$. The embedding $\mathcal{E}_\mu(u,\emptyset)$ is obtained by embedding the edge represented by μ_2 on the left side of $\mathcal{E}_{\mu_1}(u,\emptyset)$ or $\mathcal{E}_{\mu_1}(v,\{u\})$, respectively. Otherwise μ_2 is not a Q-node. It is then necessary that $k \leq 3$ and if μ_3 exists, it must be a Q-node. Now, $(u,\emptyset) \in L(\mu)$ if and only if $(v,\{u\}) \in L(\mu_1)$ and $(v,u) \in L(\mu_2)$. In this case, we define $P_\mu(u,\emptyset) = P_{\mu_1}(v,\{u\}) \cdot P_{\mu_2}(v,u)$ and the embedding $\mathcal{E}_\mu(u,\emptyset)$ is obtained by embedding $\mathcal{E}_{\mu_2}(v,u)$ to the left of $\mathcal{E}_{\mu_1}(v,\{u\})$ and the edge represented by μ_3, if it exists, to the left of that.

Case 3: μ is an S-node. Let the children of μ be numbered so that v is a pole of μ_1. Further, we denote by v_i the pole shared by μ_i and μ_{i+1} for $i = 1, \ldots, k-1$. To ease the presentation, we also write $v_0 = v$ and $v_{k+1} = u$.

We start with the case that $s \notin \mathrm{pert}(\mu)$; see Fig. 9a. We show how to test whether $(v,u) \in L(\mu)$; the case of (u,v) is analogous. Then $(v,u) \in L(\mu)$ if and only if $(v_{i-1},v_i) \in L(\mu_i)$ for $i = 1, \ldots, k$. In that case, $P_\mu(v,u)$ is obtained by concatenating $P_{\mu_i}(v_{i-1},v_i)$ for $i = 1, \ldots, k$, while $\mathcal{E}_\mu(v,u)$ is obtained by merging $\mathcal{E}_{\mu_i}(v_{i-1},v_i)$ for $i = 1, \ldots, k$.

Now, consider the case that $s \in \mathrm{pert}(\mu)$. Consider a pair (x,Y) as above. We show the case $x = u$, the case $x = v$ can be handled analogously. If $s = v$, then we cannot avoid visiting s, and we proceed as in the case of (v,u) where s is not in $\mathrm{pert}(\mu)$. Now consider the case that s is not a pole; see Figs. 9b to 9d. Let i be the smallest index so that s belongs to $\mathrm{pert}(\mu_i)$ (observe that s belongs to more than one pertinent graphs if and only if it is a vertex of $\mathrm{skel}(\mu)$). If $i > 2$, then $L(\mu) = \emptyset$, i.e., there is no path from s to x that visits v_1; see Fig. 9b. Similarly, for $i = 2$ we have $(u,Y) \in L(\mu)$ if and only if μ_1 is a Q-node, $Y = \{v\}$, $(v_2,\emptyset) \in L(\mu_2)$, and $(v_{j-1},v_j) \in L(\mu_j)$ for $j = 3, \ldots, k$; see Fig. 9c. In this case, $P_\mu(x,Y)$

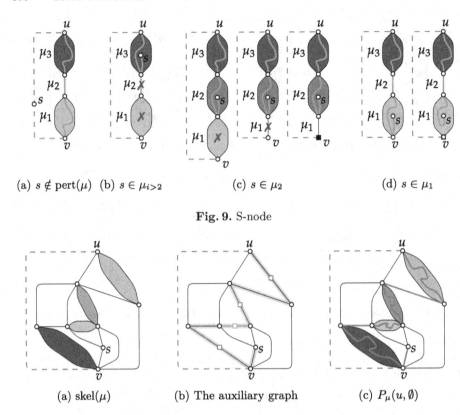

(a) $s \notin \operatorname{pert}(\mu)$ (b) $s \in \mu_{i>2}$ (c) $s \in \mu_2$ (d) $s \in \mu_1$

Fig. 9. S-node

(a) skel(μ)   (b) The auxiliary graph   (c) $P_\mu(u, \emptyset)$

Fig. 10. An R-node μ for the case that s is a vertex of skel(μ).

is composed by concatenating $P_{\mu_2}(v_2, \emptyset)$ with $P_{\mu_j}(v_{j-1}, v_j)$ for $j = 3, \ldots, k$, while the embedding $\mathcal{E}_\mu(x, Y)$ is obtained by merging the edge representing μ_1 with $\mathcal{E}_{\mu_2}(v_2, \emptyset)$ with the embeddings of $\mathcal{E}_{\mu_j}(v_{j-1}, v_j)$ for $j = 3, \ldots, k$. If $i = 1$, $(u, Y) \in L(\mu)$ if and only if $(v_1, Y) \in L(\mu_1)$ and $(v_{j-1}, v_j) \in L(\mu_j)$ for $j = 2, \ldots, k$; see Fig. 9d. In this case $P_\mu(x, Y)$ is composed by concatenating $P_{\mu_1}(v_1, Y)$ with $P_{\mu_j}(v_{j-1}, v_j)$ for $j = 2, \ldots, k$ and the embedding $\mathcal{E}(x, Y)$ is obtained by merging the embeddings $\mathcal{E}_{\mu_1}(v_1, Y)$ and $\mathcal{E}_{\mu_j}(v_{j-1}, v_j)$ for $j = 2, \ldots, k$ so that u and v lie on the outer face.

Case 4: μ **is an R-node.** If $s \notin \operatorname{pert}(\mu)$, then $P_\mu(v, u)$ must traverse every vertex in $\operatorname{pert}(\mu)$, starting with the edge e counterclockwise following the parent edge, with all other edges of $\operatorname{pert}(\mu)$ to the left of $P_\mu(v, u)$. Since v, u, and e lie on a common face, $P_\mu(v, u)$ follows only this face, so skel(μ) is outerplanar; a contradiction, as the skeleton of an R-node is triconnected.

Now, consider the case that $s \in \operatorname{pert}(\mu)$. We start with the case that s is a vertex of skel(μ); see Fig. 10. The path $P_\mu(u, Y)$ certainly must traverse the pertinent graphs of all children that are not Q-nodes and possibly also some of the Q-nodes. To model this, we consider the auxiliary plane graph obtained

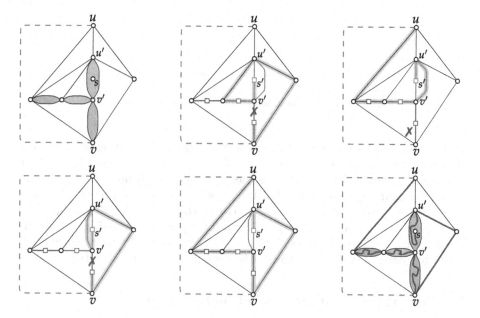

Fig. 11. An R-node μ for the case that s lies in a child ν.

from $\mathrm{skel}(\mu)$ by replacing each virtual edge that corresponds to a non-Q-node child by a path of length 2. We now employ the algorithm from Theorem 3 for both embeddings of the auxiliary graph. We try every edge incident to s as a possible starting edge and check when we arrive at u whether all vertices except the vertices in Y have been visited. If this is successful, let $v_1, \ldots, v_\ell$ be the corresponding path in $\mathrm{skel}(\mu)$ and let μ_i be the child corresponding to the virtual edge $\{v_i, v_{i+1}\}$ for $i = 1, \ldots, \ell - 1$. If further $(v_i, v_{i+1}) \in L(\mu_i)$ for $i = 1, \ldots, \ell - 1$, then $(v, u) \in L(\mu)$. In that case, $P_\mu(v, u)$ is obtained by concatenating $P_{\mu_i}(v_i, v_{i+1})$ for $i = 1, \ldots, \ell - 1$ and $\mathcal{E}_\mu(v, u)$ is obtained from the embedding of the auxiliary graph by replacing each path of length 2 that represents a non-Q-node child μ_i by $\mathcal{E}_{\mu_i}(v_i, v_{i+1})$. If this test is not successful we repeat the above steps with the flipped embedding of the auxiliary graph.

Otherwise s is contained in a child ν of μ with poles u', v'; see Fig. 11. We consider the same auxiliary graph H as above. Let s' be the vertex on the length-2 path between u' and v' in H. We add the edge (u', v') to H embedded either to the left or to the right of the path $\langle u', s', v' \rangle$; this way we obtain two different embeddings of the resulting graph. We now employ the algorithm from Theorem 3 for both embeddings. Again, we try both starting edges incident to s and for each of them, we check when we arrive at x whether all vertices except possibly the vertices in Y have been visited. This way, we obtain up to four solutions, depending on the starting edge and whether we use the edge (u', v') or not. Let $x' \in \{u', v'\}$ such that (s, x') is the starting edge of one such solution. If the path uses the edge (u', v'), then we have to check whether $(x', \emptyset) \in L(\nu)$; otherwise,

Fig. 12. Illustration of page i in the proof of Theorem 5.

we have to check whether $(x', \{u', v'\} \setminus \{x'\}) \in L(\nu)$. If the check is successful, then we compute the corresponding path $P_\mu(v, u)$ and embedding $\mathcal{E}_\mu(v, u)$ as in the case $s \notin \text{pert}(\mu)$. This finishes the description of the R-node.

We conclude by mentioning that the running time stems from the fact that in an R-node that contains s, we try $O(n)$ starting edges for the path, where each try takes $O(n)$ time. Therefore, for a fixed pair of endvertices s, t testing the existence of an embedding that is st-sided takes $O(n^2)$ time. Since there are $O(n^2)$ pairs of endvertices to try, the overall running time is $O(n^4)$. □

5   The Rique-number of Complete Graphs

In this section, we provide bounds on the density of graphs admitting k-page RIQ-layouts and on the *rique*-number of complete graphs.

Theorem 5. *Any graph G that admits a k-page RIQ-layout cannot have more than $(2n + 2)k - k^2 + (n - 3)$ edges.*

Proof. Let $v_1, \ldots, v_n$ be the linear order of the vertices and let $E_1, \ldots, E_k$ be the pages of a k-page RIQ-layout of G. Since, by Theorem 1, each page is a planar graph, it has at most $3n - 6$ edges. Since, however, the $n - 1$ so-called *spine edges* (v_i, v_{i+1}), $i = 1, \ldots, n - 1$ can be added as head-edges to every page, every page has at most $2n - 5$ non-spine edges. Next, we argue that there exists a k-page RIQ-layout $E'_1, \ldots, E'_k$ of G such that each vertex v_i, $1 \leq i \leq k$ contains edges only on pages $E'_1, \ldots, E'_i$. We start with $E'_i = E_i$, for each $1 \leq i \leq k$.

For $1 \leq i \leq k$, assume that the first $i - 1$ vertices $v_1, \ldots, v_{i-1}$ only have edges in $E'_1, \ldots, E'_{i-1}$ and consider the next vertex v_i (see Fig. 12). If v_i also only has edges in $E'_1, \ldots, E'_{i-1}$, then the claim follows. Otherwise, let (v_i, v_j), $i + 1 \leq j \leq n$ be the edge with j maximal that does not lie in $E'_1, \ldots, E'_{i-1}$ and assume w.l.o.g. that $(v_i, v_j) \in E'_i$. By our assumption, there is no edge that stems from $v_1, \ldots, v_{i-1}$. Further, the edge (v_i, v_j) blocks any possible tail-edge between two vertices in $v_{i+1}, \ldots, v_{j-1}$ in E'_i. Hence, all tail-edges that end in a vertex in $v_{i+1}, \ldots, v_{j-1}$ in E'_i stem from v_i. Thus, we can add all edges from v_i to $v_{i+1}, \ldots, v_{j-1}$ to E'_i as tail-edges. Since all edges from $v_1, \ldots, v_{i-1}$ to v_i lie in $E'_1, \ldots, E'_{i-1}$, by the choice of j, so do all edges from v_i to $v_{j+1}, \ldots, v_n$. Thus, $E'_{i+1}, \ldots, E'_k$ contain no edge of v_i. Since any page E'_i, $1 \leq i \leq k$ contains edges

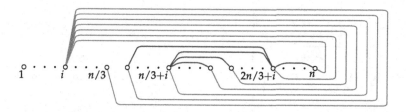

Fig. 13. Illustration of page i in the upper bound of Theorem 6.

Table 1. A summary of our results on the rique-number of K_n

n	4	5–7	8–11	12–14	15–17	18–21	22	23–24	25	26–28
riq(K_n)	1	2	3	4	5	6	6 or 7	7	7 or 8	8

of at most $n-i+1$ vertices, it has at most $2(n-i+1)-5 = 2n-2i-3$ non-spine edges. Hence, the number of edges in $E_1', \ldots, E_k'$ is at most

$$n - 1 + \sum_{i=1}^{k}(2n - 2i - 3) = (2n - 4)k - k^2 + (n - 1). \qquad \square$$

We are now ready to present our bounds on the rique-number of K_n.

Theorem 6. $0.2929(n - 2) \approx (1 - \frac{1}{\sqrt{2}})(n - 2) \leq \text{riq}(K_n) \leq \lceil n/3 \rceil \approx 0.3333n$

Proof. Let $k = \text{riq}(K_n)$. As K_n has $n(n - 1)/2$ edges, Theorem 5 implies:

$$(2n - 4)k - k^2 + (n - 1) \geq \frac{n(n - 1)}{2} \Leftrightarrow k^2 - (2n - 4)k + (\frac{n^2}{2} - \frac{3n}{2} + 1) \leq 0$$

The inequality above then gives the claimed lower bound as follows:

$$k \geq n - 2 - \frac{\sqrt{2}}{2}\sqrt{(n - 2)(n - 3)} \geq n - 2 - \frac{\sqrt{2}(n - 2)}{2} = (1 - \frac{1}{\sqrt{2}})(n - 2)$$

We now show how to compute a layout of K_n with $\lceil n/3 \rceil$ pages. Assume w.l.o.g. that n is divisible by 3. Take an arbitrary stack layout of the clique on vertices $v_{n/3+1}, \ldots, v_{v_n}$ on $n/3$ pages [13]. Then put on page i all edges of vertex v_i as tail-edges; see Fig. 13. $\qquad \square$

We conclude this section with a few more insights on the rique-number of complete graphs, which we derived by adjusting a formulation of the book embedding problem as a SAT instance [12]; for details see the full version of the paper [8]. This adjustment allowed us to obtain bounds on the rique-number of K_n for values of n in $[4, \ldots, 27]$; see Table 1 and Fig. 14 and 15 for page-minimal layouts of K_7 and K_{11}.

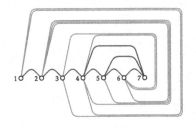
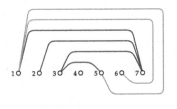

Fig. 14. A 2-page `RIQ`-layout of K_7

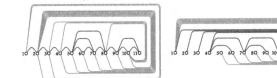
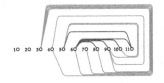

Fig. 15. A 3-page `RIQ`-layout of K_{11}

6   Conclusions and Open Problems

In this work, we continued the study of linear layouts of graphs in relation to known data structures, in particular, in relation to the restricted-input deque. Several problems are raised by our work: (i) the most important one is the complexity of the recognition of graphs with rique-number 1, (ii) another quite natural problem is to further narrow the gap between our lower and upper bounds on the rique-number of K_n; our experimental results indicate that there exist room for improvement in the upper bound, (iii) for complete bipartite graphs, we did not manage to obtain improved bounds (besides the obvious ones that one may derive from their stack- or queue-number), (iv) another interesting question regards the rique-number of planar graphs, which ranges between 2 and 4 (i.e., the upper bound by their stack-number); the same problem can be studied also for subclasses of planar graphs (e.g., planar 3-trees).

References

1. Adolphson, D., Hu, T.C.: Optimal linear ordering. SIAM J. Appl. Math. **25**(3), 403–423 (1973). https://doi.org/10.1137/0125042
2. Alam, J.M., Bekos, M.A., Gronemann, M., Kaufmann, M., Pupyrev, S.: Queue layouts of planar 3-trees. Algorithmica **82**(9), 2564–2585 (2020). https://doi.org/10.1007/s00453-020-00697-4
3. Alam, J.M., Bekos, M.A., Gronemann, M., Kaufmann, M., Pupyrev, S.: The mixed page number of graphs. Theor. Comput. Sci. **931**, 131–141 (2022). https://doi.org/10.1016/j.tcs.2022.07.036
4. Auer, C., Bachmaier, C., Brandenburg, F.J., Brunner, W., Gleißner, A.: Data structures and their planar graph layouts. J. Graph Algorithms Appl. **22**(2), 207–237 (2018). https://doi.org/10.7155/jgaa.00465

5. Bannister, M.J., Devanny, W.E., Dujmović, V., Eppstein, D., Wood, D.R.: Track layouts, layered path decompositions, and leveled planarity. Algorithmica 81(4), 1561–1583 (2018). https://doi.org/10.1007/s00453-018-0487-5
6. Battista, G.D., Tamassia, R.: Incremental planarity testing (extended abstract). In: Symposium on Foundations of Computer Science, pp. 436–441. IEEE Computer Society (1989). https://doi.org/10.1109/SFCS.1989.63515
7. Di Battista, G., Tamassia, R.: On-line graph algorithms with SPQR-trees. In: Paterson, M.S. (ed.) ICALP 1990. LNCS, vol. 443, pp. 598–611. Springer, Heidelberg (1990). https://doi.org/10.1007/BFb0032061
8. Bekos, M.A., Felsner, S., Kindermann, P., Kobourov, S., Kratovíl, J., Rutter, I.: The Rique-number of graphs (2022). https://doi.org/10.48550/ARXIV.2209.00424
9. Bekos, M.A., Gronemann, M., Raftopoulou, C.N.: Two-page book embeddings of 4-planar graphs. Algorithmica 75(1), 158–185 (2015). https://doi.org/10.1007/s00453-015-0016-8
10. Bekos, M.A., Gronemann, M., Raftopoulou, C.N.: On the Queue Number of Planar Graphs. In: Purchase, H.C., Rutter, I. (eds.) GD 2021. LNCS, vol. 12868, pp. 271–284. Springer, Cham (2021). https://doi.org/10.1007/978-3-030-92931-2_20
11. Bekos, M.A., Kaufmann, M., Klute, F., Pupyrev, S., Raftopoulou, C.N., Ueckerdt, T.: Four pages are indeed necessary for planar graphs. J. Comput. Geom. 11(1), 332–353 (2020)
12. Bekos, M.A., Kaufmann, M., Zielke, C.: The book embedding problem from a SAT-solving perspective. In: Di Giacomo, E., Lubiw, A. (eds.) GD 2015. LNCS, vol. 9411, pp. 125–138. Springer, Cham (2015). https://doi.org/10.1007/978-3-319-27261-0_11
13. Bernhart, F., Kainen, P.C.: The book thickness of a graph. J. Comb. Theory Ser. B 27(3), 320–331 (1979). https://doi.org/10.1016/0095-8956(79)90021-2
14. Chinn, P.Z., Chvatalova, J., Dewdney, A.K., Gibbs, N.E.: The bandwidth problem for graphs and matrices - a survey. J. Graph Theory 6(3), 223–254 (1982). https://doi.org/10.1002/jgt.3190060302
15. Cornuéjols, G., Naddef, D., Pulleyblank, W.R.: Halin graphs and the travelling salesman problem. Math. Program. 26(3), 287–294 (1983). https://doi.org/10.1007/BF02591867
16. Di Battista, G., Frati, F., Pach, J.: On the queue number of planar graphs. SIAM J. Comput. 42(6), 2243–2285 (2013). https://doi.org/10.1137/130908051
17. Dujmović, V.: Graph layouts via layered separators. J. Comb. Theory Ser. B 110, 79–89 (2015). https://doi.org/10.1016/j.jctb.2014.07.005
18. Dujmović, V., Frati, F.: Stack and queue layouts via layered separators. J. Graph Algorithms Appl. 22(1), 89–99 (2018). https://doi.org/10.7155/jgaa.00454
19. Dujmovic, V., Joret, G., Micek, P., Morin, P., Ueckerdt, T., Wood, D.R.: Planar graphs have bounded queue-number. J. ACM 67(4), 22:1-22:38 (2020). https://doi.org/10.1145/3385731
20. Ewald, G.: Hamiltonian circuits in simplicial complexes. Geom. Dedicata. 2(1), 115–125 (1973). https://doi.org/10.1007/BF00149287
21. de Fraysseix, H., de Mendez, P.O., Pach, J.: A left-first search algorithm for planar graphs. Discrete Comput. Geom. 13, 459–468 (1995). https://doi.org/10.1007/BF02574056
22. Ganley, J.L.: Stack and queue layouts of Halin graphs (1995)
23. Goldner, A., Harary, F.: Note on a smallest nonhamiltonian maximal planar graph. Bull. Malays. Math. Sci. Soc. 1(6), 41–42 (1975)
24. Guan, X., Yang, W.: Embedding planar 5-graphs in three pages. Discret. Appl. Math. (2019). https://doi.org/10.1016/j.dam.2019.11.020

25. Gutwenger, C., Mutzel, P.: A linear time implementation of SPQR-trees. In: Marks, J. (ed.) GD 2000. LNCS, vol. 1984, pp. 77–90. Springer, Heidelberg (2001). https://doi.org/10.1007/3-540-44541-2_8

26. Heath, L.S.: Embedding planar graphs in seven pages. In: Foundations of Computer Science, pp. 74–83. IEEE Computer Society (1984). https://doi.org/10.1109/SFCS.1984.715903

27. Heath, L.S., Leighton, F.T., Rosenberg, A.L.: Comparing queues and stacks as mechanisms for laying out graphs. SIAM J. Discrete Math. 5(3), 398–412 (1992). https://doi.org/10.1137/0405031

28. Heath, L.S., Rosenberg, A.L.: Laying out graphs using queues. SIAM J. Comput. 21(5), 927–958 (1992). https://doi.org/10.1137/0221055

29. Hoffmann, M., Klemz, B.: Triconnected planar graphs of maximum degree five are subhamiltonian. In: Bender, M.A., Svensson, O., Herman, G. (eds.) European Symposium on Algorithms. LIPIcs, vol. 144, pp. 58:1–58:14. Schloss Dagstuhl - Leibniz-Zentrum für Informatik (2019). https://doi.org/10.4230/LIPIcs.ESA.2019.58

30. Kainen, P.C., Overbay, S.: Extension of a theorem of Whitney. Appl. Math. Lett. 20(7), 835–837 (2007). https://doi.org/10.1016/j.aml.2006.08.019

31. Nishizeki, T., Chiba, N.: Planar Graphs: Theory and Algorithms, chap. 10. Hamiltonian Cycles, pp. 171–184. Dover Books on Mathematics, Courier Dover Publications (2008)

32. Pratt, V.R.: Computing permutations with double-ended queues, parallel stacks and parallel queues. In: Aho, A.V., Borodin, A., Constable, R.L., Floyd, R.W., Harrison, M.A., Karp, R.M., Strong, H.R. (eds.) ACM Symposium on Theory of Computing, pp. 268–277. ACM (1973). https://doi.org/10.1145/800125.804058

33. Rengarajan, S., Veni Madhavan, C.E.: Stack and queue number of 2-trees. In: Du, D.-Z., Li, M. (eds.) COCOON 1995. LNCS, vol. 959, pp. 203–212. Springer, Heidelberg (1995). https://doi.org/10.1007/BFb0030834

34. Robertson, N., Seymour, P.: Graph minors I excluding a forest. J. Comb. Theory Ser. B 35(1), 39–61 (1983). https://doi.org/10.1016/0095-8956(83)90079-5

35. Serna, M., Thilikos, D.: Parameterized complexity for graph layout problems. B. EATCS 86, 41–65 (2005)

36. Yannakakis, M.: Embedding planar graphs in four pages. J. Comput. Syst. Sci. 38(1), 36–67 (1989). https://doi.org/10.1016/0022-0000(89)90032-9

Contact and Visibility Graph Representations

Morphing Rectangular Duals

Steven Chaplick[1] , Philipp Kindermann[2] , Jonathan Klawitter[3,4(✉)] ,
Ignaz Rutter[5] , and Alexander Wolff[3]

[1] Maastricht University, Maastricht, The Netherlands
[2] Universität Trier, Trier, Germany
[3] Universität Würzburg, Würzburg, Germany
jo.klawitter@gmail.com
[4] University of Auckland, Auckland, New Zealand
[5] Universität Passau, Passau, Germany

Abstract. A rectangular dual of a plane graph G is a contact representations of G by interior-disjoint axis-aligned rectangles such that (i) no four rectangles share a point and (ii) the union of all rectangles is a rectangle. A rectangular dual gives rise to a regular edge labeling (REL), which captures the orientations of the rectangle contacts.

We study the problem of morphing between two rectangular duals of the same plane graph. If we require that, at any time throughout the morph, there is a rectangular dual, then a morph exists only if the two rectangular duals realize the same REL. Therefore, we allow intermediate contact representations of non-rectangular polygons of constant complexity. Given an n-vertex plane graph, we show how to compute in $\mathcal{O}(n^3)$ time a piecewise linear morph that consists of $\mathcal{O}(n^2)$ linear morphing steps.

Keywords: Morphing · Rectangular dual · Regular edge labeling · Lattice

1 Introduction

A *morph* between two representations (e.g., drawings) of the same graph G is a continuous transformation from one representation to the other. Preferably, a morph should preserve the user's "mental map", which means that, throughout the transformation, as little as necessary is changed to go from the source to target representation and that their properties are maintained [28]. For example, during a morph between two planar drawings, each intermediate drawing should also be planar. A *linear morph* moves each point along a straight-line segment at constant speed, where different points may have different speeds or may remain stationary. Note that a linear morph is fully defined by the source and target representation. A *piecewise linear morph* consists of a sequence of linear morphs, each of which is called a *step*.

Morphs are well studied for planar drawings. For example, it is known that piecewise linear planar morphs always exist between planar straight-line drawings [9] and that, for an n-vertex planar graph, $\mathcal{O}(n)$ steps suffice [1], which

P. Angelini and R. von Hanxleden (Eds.): GD 2022, LNCS 13764, pp. 389–403, 2023.
https://doi.org/10.1007/978-3-031-22203-0_28

is worst-case optimal. Further research on morphs includes, among others, the study of morphs of convex drawings [3,25], of orthogonal drawings [6,18], on different surfaces [10,23], and in higher dimensions [4].

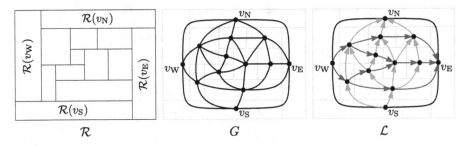

Fig. 1. A rectangular dual $\mathcal{R}$ for the graph G; the REL $\mathcal{L}$ induced by $\mathcal{R}$. (Color figures available online, though note the different arrow heads for red and blue edges.) (Color figure online)

Less attention has been given to morphs of alternative representations of graphs such as intersection and contact representations. A *geometric intersection representation* of a graph G is a mapping $\mathcal{R}$ that assigns to each vertex w of G a geometric object $\mathcal{R}(w)$ such that two vertices u and v are adjacent in G if and only if $\mathcal{R}(u)$ and $\mathcal{R}(v)$ intersect. In a *contact representation* we further require that, for any two vertices u and v, the objects $\mathcal{R}(u)$ and $\mathcal{R}(v)$ have disjoint interiors. Classic examples are interval graphs [7], where the objects are intervals of $\mathbb{R}$, or coin graphs [24], where the objects are interior-disjoint disks in the plane. Recently, Angelini et al. [2] studied morphs of right-triangle contact representations of planar graphs. They showed that one can test efficiently whether a morph exists (in which case a quadratic number of steps suffice). In this paper, we investigate morphs between contact representations of rectangles.

Rectangular Duals. A *rectangular dual* of a graph G is a contact representation $\mathcal{R}$ of G by axis-aligned rectangles such that (i) no four rectangles share a point and (ii) the union of all rectangles is a rectangle; see Fig. 1. Note that G may admit a rectangular dual only if it is planar and internally triangulated. Furthermore, a rectangular dual can always be augmented with four additional rectangles (one on each side) so that only these four rectangles touch the outer face of the representation. It is customary that the four corresponding vertices on the outer face of G are denoted by v_S, v_W, v_N, and v_E, and to require that $\mathcal{R}(v_S)$ is bottommost, $\mathcal{R}(v_W)$ is leftmost, $\mathcal{R}(v_N)$ is topmost, and $\mathcal{R}(v_E)$ is rightmost; see Fig. 1. The corresponding vertices are *outer*; the remaining ones are *inner*. Similarly, the four edges between the outer vertices are *outer*; the others are *inner*. A plane internally-triangulated graph has a representation with only four rectangles touching the outer face if and only if its outer face is a 4-cycle and it has no *separating triangle*, that is, a triangle whose removal disconnects the graph [25]. Such a graph is called a *properly-triangulated planar (PTP) graph*. For such a graph, a rectangular dual can be computed in linear time [22].

Historically, rectangular duals have been studied due to their applications in architecture [30], VLSI floor-planning [26,32], and cartography [17]. Morphs between rectangular duals are of interest, e.g., due to their relation to rectangular cartograms. Rectangular cartograms were introduced in 1934 [29] and combine statistical and geographical information in thematic maps, where geographic regions are represented as rectangles and scaled in proportion to some statistic. There has been a lot of work on efficiently computing rectangular cartograms [8, 20,31], see also the recent survey [27]. A morph between rectangular cartograms can visualize different data sets. Florisson et al. [13] implemented a method to construct rectangular cartograms by first extending the given map with "sea tiles" to obtain a rectangular dual, and then using a heuristic that moves maximal line segments until the area of the rectangles gets closer to the given data. They also used their heuristic to morph between two rectangular cartograms, but did not discuss when exactly this works and with what time complexity.

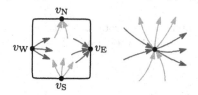

Fig. 2. Edge order at the four outer vertices and at an inner vertex in a REL.

Regular Edge Labelings. A combinatorial view of a rectangular dual of a graph G can be described by a coloring and orientation of the edges of G [22]. This is similar to how so-called Schnyder woods describe contact representations of planar graphs by triangles [14]. More precisely, a rectangular dual $\mathcal{R}$ gives rise to a 2-coloring and an orientation of the inner edges of G as follows. We color an edge $\{u, v\}$ blue if the contact segment between $\mathcal{R}(u)$ and $\mathcal{R}(v)$ is a horizontal line segment, and we color it red otherwise. We orient a blue (red) edge $\{u, v\}$ as uv if $\mathcal{R}(u)$ lies below (resp. left of) $\mathcal{R}(v)$; see Fig. 1. The resulting coloring and orientation has the following properties Fig. 2:

(1) For each outer vertex v_S, v_W, v_N, and v_E, the incident inner edges are blue outgoing, red outgoing, blue incoming, and red incoming, respectively.
(2) For each inner vertex, the incident edges form four clockwise (*cw*) ordered non-empty blocks: blue incoming, red incoming, blue outgoing, red outgoing.

A coloring and orientation with these properties is called a *regular edge labeling (REL)* or *transversal structure*. We let $\mathcal{L} = (L_1, L_2)$ denote a REL, where L_1 is the set of blue edges and L_2 is the set of red edges. Let $L_1(G)$ and $L_2(G)$ denote the two subgraphs of G induced by L_1 and L_2, respectively. Note that both $L_1(G)$ and $L_2(G)$ are st-graphs, that is, directed acyclic graphs with exactly one source and exactly one sink. Kant and He [22] introduced RELs as intermediate objects when constructing a rectangular dual of a PTP graph. It is well known that every PTP graph admits a REL and thus a rectangular dual [19,22]. A rectangular dual $\mathcal{R}$ *realizes* a REL $\mathcal{L}$ if the REL induced by $\mathcal{R}$ is $\mathcal{L}$.

We define the *interior* of a cycle to be the set of vertices and edges enclosed by, but not on the cycle. A 4-cycle is *separating* if there are other vertices both in its interior and in its exterior. A separating 4-cycle is *nontrivial* if its interior contains more than one vertex; otherwise it is *trivial*. We call non-separating 4-cycles also *empty 4-cycles*. (An empty 4-cycle contains exactly one edge.)

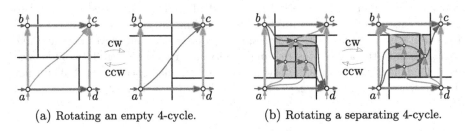

(a) Rotating an empty 4-cycle.   (b) Rotating a separating 4-cycle.

Fig. 3. Clockwise and counterclockwise rotations between RELs that recolor and reorient the edges inside an alternating 4-cycle $\langle a, b, c, d \rangle$ that is (a) empty or (b) separating.

If the edges of a cycle C alternate between red and blue, we say that C is *alternating*. We can move between different RELs of a PTP graph G by swapping the colors and reorienting the edges inside an alternating 4-cycle, see Fig. 3. This operation, which we call a *rotation* and which we define formally in the long version [11], connects all RELs of G. In fact, the RELs of G form a distributive lattice [15,16]. A 4-cycle C of G is called *rotatable* if it is alternating for at least one REL of G.

Important Related Work. Other combinatorial structures of graph representations also form lattices; see the work by Felsner and colleagues [12]. In the context of morphs, Barrera-Cruz et al. [5] exploited the lattice structure of Schnyder woods of a plane triangulation to obtain piecewise linear morphs between planar straight-line drawings. While their morphs require $\mathcal{O}(n^2)$ steps (compared to the optimum of $\mathcal{O}(n)$), they have the advantage that they are "visually pleasing" and that they maintain a quadratic-size drawing area between any two steps. To this end, Barrera-Cruz et al. showed that there is a path in the lattice of length $\mathcal{O}(n^2)$ between any two Schnyder woods. We show an analogous result for RELs. In order to morph between right-triangle contact representations, Angelini et al. [2] leveraged the lattice structure of Schnyder woods. They showed that if no separating triangle has to be flipped (a flip is a step in the lattice) between the source and the target Schnyder wood, then a morph with $\mathcal{O}(n^2)$ steps exists (else, no morph exists that uses right-triangle contact representations throughout).

Contribution. We consider piecewise linear morphs between two rectangular duals $\mathcal{R}$ and $\mathcal{R}'$ of the same PTP graph G. If $\mathcal{R}$ and $\mathcal{R}'$ realize the same REL, then a single step suffices, but if $\mathcal{R}$ and $\mathcal{R}'$ realize distinct RELs of G, then no rectangular-dual preserving morph exists (Sect. 2.1). Therefore, we propose a new type of morph where intermediate drawings are contact representations of G using convex polygons with up to five corners (Sect. 2.2). We show how to construct such a *relaxed* morph as a sequence of $\mathcal{O}(n^2)$ steps that implement moves in the lattice of RELs of G (Sect. 2.3). To this end, we make use of the following two results on paths in this lattice.

Proposition 1 ($\star$). *Given an n-vertex PTP graph G, the lattice of RELs of G has diameter $\mathcal{O}(n^2)$.*

Proposition 2 ($\star$). *Let G be an n-vertex PTP graph with RELs $\mathcal{L}$ and $\mathcal{L}'$. In the lattice of RELs of G, a shortest $\mathcal{L}$–$\mathcal{L}'$ path can be computed in $\mathcal{O}(n^3)$ time.*

We ensure that between any two morphing steps, our drawings remain on a quadratic-size section of the integer grid – like those of Barrera-Cruz et al. [5]. In order to evaluate the intermediate representations when our drawings are not on the integer grid, we use the measure *feature resolution* [21], that is, the ratio of the length of the longest segment over the shortest distance between two vertices or between a vertex and a non-incident segment. We show that the feature resolution in *any* intermediate drawing is bounded by $\mathcal{O}(n)$.

Finally, we investigate executing rotations in parallel; see Sect. 3. As a result, we can morph between any pair $(\mathcal{R}, \mathcal{R}')$ of rectangular duals of the given graph using $\mathcal{O}(1)$ times the minimum number of steps needed to get from $\mathcal{R}$ to $\mathcal{R}'$; however, our polygons have up to eight corners.

For statements marked with "$\star$", a proof is available in the full version [11].

2 Morphing Between Rectangular Duals

This section concerns morphs between two rectangular duals $\mathcal{R}$ and $\mathcal{R}'$ of a PTP graph G that realize the same REL, adjacent RELs, and finally any two RELs.

2.1 Morphing Between Rectangular Duals Realizing the Same REL

Theorem 3. *For a PTP graph G with rectangular duals $\mathcal{R}$ and $\mathcal{R}'$ (i) if $\mathcal{R}$ and $\mathcal{R}'$ realize the same REL, then there is a linear morph between them; (ii) otherwise, there is no morph between them (not even a non-linear one).*

Proof. Biedl et al. [6] studied morphs of orthogonal drawings. They showed that a single (planarity-preserving) linear morph suffices if all faces are rectangular and all edges are parallel in the two drawings, that is, any edge is either vertical in both drawings or horizontal in both drawings. We can apply this result to two rectangular duals $\mathcal{R}$ and $\mathcal{R}'$ precisely when they realize them same REL. A linear morph between them changes the x-coordinates of vertical line segments and the y-coordinates of horizontal line segments but does not change their relative order.

Now assume that $\mathcal{R}$ and $\mathcal{R}'$ realize different RELs $\mathcal{L}$ and $\mathcal{L}'$ of G, respectively. Then w.l.o.g. some contact segment s changes from being horizontal in $\mathcal{R}$ to being vertical in $\mathcal{R}'$. Since s must always be horizontal or vertical, it has to collapse to a point and then extend to a segment again. When s collapses, the intermediate representation is not a rectangular dual of G since four rectangles meet at a single point. Even worse, if a separating alternating cycle is rotated, then its interior contracts to a point, vanishes, and reappears rotated by 90°. □

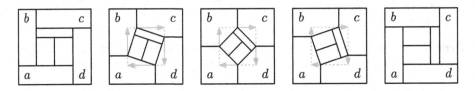

Fig. 4. A linear morph rotating the separating 4-cycle $\langle a, b, c, d \rangle$ emulates the rotation in the corresponding REL; see Fig. 3. The interior of the 4-cycle turns by 90° without changing its shape, while the outer contact segments move horizontally and vertically.

2.2   Morphing Between Rectangular Duals with Adjacent RELs

Let $\mathcal{R}$ and $\mathcal{R}'$ now realize different RELs $\mathcal{L}$ and $\mathcal{L}'$ of G, respectively. By Theorem 3, any continuous transformation between $\mathcal{R}$ and $\mathcal{R}'$ requires intermediate representations that are not rectangular duals of G, i.e., a morph in the traditional sense is not possible. We relax the conditions on a morph such that, in an intermediate contact representation of G, vertices can be represented by convex polygons of constant complexity – in this section, by 5-gons. However, we still require that these polygons form a tiling of the bounding rectangle of the representation. We call a transformation with this property a *relaxed morph*. (When we talk about linear morphing steps, we omit the adjective "relaxed".) The following statement describes relaxed morphs when $\mathcal{L}$ and $\mathcal{L}'$ are adjacent, that is, $(\mathcal{L}', \mathcal{L})$ is an edge in the lattice of RELs of G.

Proposition 4. *Let $\mathcal{R}$ and $\mathcal{R}'$ be two rectangular duals of an n-vertex PTP graph G realizing two adjacent RELs $\mathcal{L}$ and $\mathcal{L}'$ of G, respectively. Then, we can compute in $\mathcal{O}(n)$ time a 3-step relaxed morph between $\mathcal{R}$ and $\mathcal{R}'$. If $\mathcal{R}$ and $\mathcal{R}'$ have an area of at most $n \times n$ and feature resolution in $\mathcal{O}(n)$, then so has each representation throughout the morph.*

We assume w.l.o.g. that $\mathcal{L}'$ can be obtained from $\mathcal{L}$ by a cw rotation of an alternating 4-cycle C. The idea is to rotate the interior of the 4-cycle while simultaneously moving the contact segments that form the edges of C; see Figs. 4 and 5. To ensure that, except for the vertices of C all regions remain rectangles and that moving the contact segments of C does not change any adjacencies, the representation needs to satisfy certain requirements. Therefore, our relaxed morph from $\mathcal{R}$ to $\mathcal{R}'$ consists of three steps. First a *preparatory morph* from $\mathcal{R}$ to a rectangular dual $\mathcal{R}_1$ with REL $\mathcal{L}$ for which C satisfies conditions stated below; second a *main morph* which transforms $\mathcal{R}_1$ to a rectangular dual $\mathcal{R}_2$ whose REL is $\mathcal{L}'$, and third a *clean-up morph* that transforms $\mathcal{R}_2$ into $\mathcal{R}'$.

We first describe the main morph $\langle \mathcal{R}_1, \mathcal{R}_2 \rangle$ in detail, as this allows us to also infer the conditions under which it can be executed successfully. Then we describe the preparatory morph $\langle \mathcal{R}, \mathcal{R}_1 \rangle$, whose sole purpose is to ensure the conditions that are required for the main morph.

Main Morph $\langle \mathcal{R}_1, \mathcal{R}_2 \rangle$ *to Rotate* C. Let a, b, c, and d be the vertices of C in cw order where a is the vertex with an outgoing red and outgoing blue edge in C, i.e., it corresponds to the bottom-left rectangle of C.

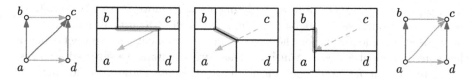

Fig. 5. A linear morph that rotates the inner contact segment of an empty 4-cycle.

Assume for now that C is separating. We have the following requirements for R_1, which become apparent shortly.

(P1) The rectangle I_C bounding the interior of C is a square.

Next, we consider the four maximal segments of $\mathcal{R}$ that contain one of the four borders of I_C, which we call *border segments*. Let s be the upper border segment of I_C and suppose its right endpoint lies on the left side of a rectangle $\mathcal{R}(x)$. Let S be the part in the horizontal strip defined by I_C that starts at I_C and ends at $\mathcal{R}(x)$.

(P2) The only horizontal segments that intersect S are border segments of I_C; see Fig. 6c.

We define (P2) for the other three border segments of I_C analogously. Next, assume that C is empty. Then the rectangle I_C degenerates to a segment s, and we assume w.l.o.g. that s is horizontal. Now I_C still has two vertical border segments, but the two horizontal border segments share the segment s. Let s have again its right endpoint on the left side of $\mathcal{R}(x)$. Let S be the rectangular area of height 1 directly below s that starts at b and ends at $\mathcal{R}(x)$. We have the following requirement if C is empty.

(P2') The only horizontal segment that intersects S is s; see Fig. 7c.

There is no requirement for the left side of s and the left vertical border segment of I_C. The requirements for the right vertical border segment is (P2).

We now describe the main morph for the case that C is a separating 4-cycle. In this case, the interior of C forms a square in $\mathcal{R}_1$ by (P1). Recall that the rotation of C from $\mathcal{L}$ to $\mathcal{L}'$ turns the interior of C by $90°$. During the morph, we move each corner of I_C to the coordinates of the corner that follows in cw order around I_C in $\mathcal{R}_1$; see Fig. 4. All other points in I_C are expressed as a convex combination of the corners of I_C and then move according to the movement of the corners. Furthermore, we move all points on the left border segment of I_C that are outside the boundary of I_C horizontally to the x-coordinate of the right side of I_C. We move the points on other border segments of I_C analogously.

This describes a single linear morph that results in a rectangular dual R_2 that realizes the REL $\mathcal{L}'$. Since I_C starts out as a square by (P1), throughout the morph, I_C remains a square, and by similarity all rectangles inside I_C remain rectangles. The rectangles a, b, c, and d become convex 5-gons. Furthermore, since outside I_C the horizontal border segments of I_C move an area that contains no other horizontal segments by (P2), no contact along a vertical segment arises or vanishes. Analogously, for the vertical border segments, no contact along a horizontal segment arises or vanishes. Hence, we maintain the same adjacencies.

Note that, if I_C would not be a square, then its corners would move at different speeds and I_C would deform to a rhombus where the inner angles are not $90°$, and so would all the rectangles inside I_C.

Next, consider the case that C is an empty 4-cycle. Recall that in this case, the rectangle I_C degenerates to a segment s and we assume that s is horizontal. Note that $\mathcal{R}'$ has a vertical contact between a and c, since we assume a cw rotation from $\mathcal{L}$ to $\mathcal{L}'$. We then we move the right endpoint of s vertically down by 1 and horizontally to the x-coordinate of the left endpoint of s; see Fig. 5. We also move all points on the border segments that contain the right endpoint of s accordingly. The rectangles a and c become convex 5-gons. Furthermore, since outside I_C only the horizontal border segments of I_C lie inside the area of height 1 below s by (P2'), no contact along a vertical segment arises or vanishes. Analogously, due to condition (P2) for the vertical border segments, no contact along a horizontal segment arises or vanishes. Hence, maintaining the same adjacencies.

To show that the feature resolution remains in $\mathcal{O}(n)$, note that both $\mathcal{R}_1$ and $\mathcal{R}_2$ are drawn on a $n \times n$ grid. Furthermore, the rectangles inside I_C are scaled during the morph, but since I_C is a square, the whole area inside I_C is scaled by at most $\sqrt{2}$. The distances outside I_C cannot become smaller than 1.

Lemma 5. *Let $\mathcal{R}_1$ and $\mathcal{R}_2$ be two rectangular duals of an n-vertex PTP graph G realizing two adjacent RELs $\mathcal{L}$ and $\mathcal{L}'$ of G, respectively, such that $\mathcal{R}_1$ satisfies (P1) and (P2) (or (P1) and (P2')). Then, we can compute in $\mathcal{O}(n)$ time a relaxed morph between $\mathcal{R}_1$ and $\mathcal{R}_2$. If $\mathcal{R}_1$ and $\mathcal{R}_2$ have an area of at most $n \times n$ and feature resolution in $\mathcal{O}(n)$, then so has each representation throughout the morph.*

Preparatory Morph $\langle \mathcal{R}, \mathcal{R}_1 \rangle$. We consider again the case where C is separating first. To obtain $\mathcal{R}_1$ from $\mathcal{R}$, we extend G to an auxiliary graph $\hat{G}$ that is almost a PTP graph but that contains empty chordless 4-cycles (which are represented by four rectangles touching in a single point). For $\hat{G}$, we compute an auxiliary

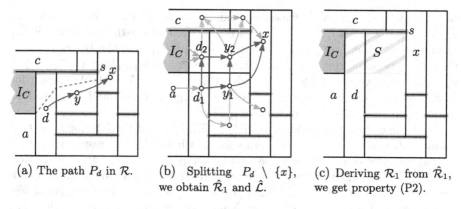

(a) The path P_d in $\mathcal{R}$.   (b) Splitting $P_d \setminus \{x\}$, we obtain $\hat{\mathcal{R}}_1$ and $\hat{\mathcal{L}}$.   (c) Deriving $\mathcal{R}_1$ from $\hat{\mathcal{R}}_1$, we get property (P2).

Fig. 6. We compute $\mathcal{R}_1$ via an auxiliary rectangular dual $\hat{\mathcal{R}}_1$ and an auxiliary REL $\hat{\mathcal{L}}$.

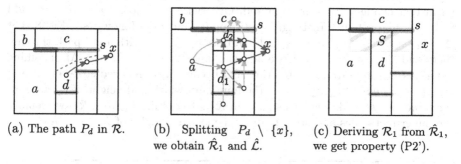

(a) The path P_d in $\mathcal{R}$.   (b) Splitting $P_d \setminus \{x\}$, we obtain $\hat{\mathcal{R}}_1$ and $\hat{\mathcal{L}}$.   (c) Deriving $\mathcal{R}_1$ from $\hat{\mathcal{R}}_1$, we get property (P2').

Fig. 7. Preparatory morph analogous Fig. 6 for the case when C is empty.

REL $\hat{\mathcal{L}}$ where the empty chordless 4-cycles of $\hat{G}$ are colored alternatingly. We then use the second step of the linear-time algorithm by Kant and He [22] to compute an (almost) rectangular dual $\hat{\mathcal{R}}_1$ of $\hat{G}$ that realizes $\hat{\mathcal{L}}$. By reversing the changes applied to G to obtain $\hat{G}$, we derive $\mathcal{R}_1$ from $\hat{\mathcal{R}}_1$. We explain the algorithm by Kant and He and why it also works for $\hat{G}$ in the full version [11].

We start with the changes to ensure (P2) for the upper border segment s of I_C; it works analogously for the other border segments. Let s end to the right again at $\mathcal{R}(x)$. Let P_d be the leftmost path in $L_2(G)$ from d to x. Let $y_1(R)$ and $y_2(R)$ denote the lower and upper y-coordinate of a rectangle R, respectively. Note that (P2) holds if for each vertex v on P_d we have $y_1(\mathcal{R}_1(v)) < y_2(\mathcal{R}_1(a))$. Therefore, from G to $\hat{G}$, we duplicate P_d by splitting each vertex v on $P_d \setminus \{x\}$ into two vertices v_1 and v_2; see Fig. 6. We then connect v_1 and v_2 with a blue edge. Let y be the successor of d on P_d. We assign the edges cw between (and including) dy and ad to d_1, and the edges cw between ad and dy to d_2. If $x = y$, the edge dy is assigned to both d_1 and d_2; otherwise we replace dy with d_1y_1 and d_2y_2. For all other vertices v on $P_d \setminus \{d, x\}$, let u be the predecessor and let w be the successor of v on P_d. We assign the edges cw between vw and uv

to v_1, and the edges cw between uv and vw to v_2; furthermore, we add the edges u_1v_1, v_1w_1, u_2v_2, and v_2w_2. As a result, there is a path from a to x in $\hat{\mathcal{R}}_1$ through the "upper" copies of the vertices in $P_d \setminus \{x\}$, and the bottom side of their corresponding rectangles are aligned. Hence, $y_1(\hat{\mathcal{R}}_1(v_1)) < y_1(\hat{\mathcal{R}}_1(v_2)) = y_2(\hat{\mathcal{R}}_1(a))$ for every $v \in P_d \setminus \{x\}$, and $y_1(\hat{\mathcal{R}}_1(x)) < y_2(\hat{\mathcal{R}}_1(a))$. We obtain for each v on $P_d \setminus \{x\}$ the rectangle $\mathcal{R}_1(v)$ by merging $\hat{\mathcal{R}}_1(v_1)$ and $\hat{\mathcal{R}}_1(v_2)$. This works analogously if C is empty; see Fig. 7.

Next, we describe how to ensure (P1) in $\mathcal{R}_1$, i.e., that the interior I_C of C is a square. Let w_C and h_C be the minimum width and height, respectively, of a rectangular dual of I_C. These values can be computed in $\mathcal{O}(I_C)$ time. Note that because of (P2), the algorithm by Kant and He [22] will draw I_C with minimum width and height in $\mathcal{R}_1$: the algorithm draws every horizontal line segment as low as possible, and because of (P2) there is no horizontal line segment to the right of I_C that forces the upper boundary segment of I_C to be higher; a symmetric argument applies to the left boundary segment of I_C. Hence, if $w_C = h_C$, then no further changes to $\hat{G}$ are required. Otherwise, if w.l.o.g. $w_C < h_C$, we add $h_C - w_C$ many buffer rectangles between I_C and d as follows; see Fig. 8. Let $\Delta = h_C - w_C$. From G to $\hat{G}$, we add vertices $v_1, \ldots, v_\Delta$ with a red path through them and, for $i \in \{1, \ldots, \Delta\}$, we add the blue edges av_i and v_ic. All incoming red edges of d in G from the interior of C become incoming red edges of v_1, and we add a red edge (v_Δ, d_2). In $\hat{G}$, the minimum width and height of I_C are now the same and I_C is drawn as a square in $\hat{\mathcal{R}}_1$. To obtain $\mathcal{R}_1$ from $\hat{\mathcal{R}}_1$, we remove the buffer rectangles and stretch all right-most rectangles of I_C to $y_1(\mathcal{R}_1(d))$.

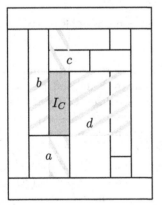
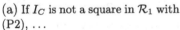

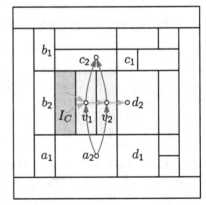

(a) If I_C is not a square in $\mathcal{R}_1$ with (P2), ...

(b) ...then we extend $\hat{\mathcal{R}}_1$ and $\hat{\mathcal{L}}$ with dummy vertices v_i inside $\langle a, b, c, d \rangle$.

Fig. 8. To ensure (P1), i.e., that I_C is a square in $\mathcal{R}_1$, we extend $\hat{\mathcal{R}}_1$ and $\hat{\mathcal{L}}$ further.

Concerning the running time, note that we can both find and split the paths for (P2) and add the extra vertices for (P1) in $\mathcal{O}(n)$ time. Since $\hat{G}$ and $\hat{\mathcal{L}}$ have a size in $\mathcal{O}(n)$, the algorithm by Kant and He [22] also runs in $\mathcal{O}(n)$ time.

Finally, we show that the area of $\mathcal{R}_1$ is bounded by $n \times n$. Observe that each triangle in G corresponds to a T-junction in $\mathcal{R}$ and thus to an endpoint of a maximal line segment. There are $2n - 4$ triangles in G and thus $n - 2$ inner maximal line segments besides the four outer ones. The algorithm by Kant and He [22] ensures that each x- and y-coordinate inside a rectangular dual contains a horizontal or vertical line segment, respectively. Note that $\hat{\mathcal{R}}_1$ contains exactly Δ more maximal line segments than $\mathcal{R}$. These were added inside C if in I_C the number of horizontal and vertical maximal line segments differed by at least Δ. Hence, $\mathcal{R}_1$ contains at most $n - 2$ vertical and at most $n - 2$ horizontal inner maximal line segments. Thus, the area of $\mathcal{R}_1$ is bounded by $n \times n$. Lastly, note that $\hat{\mathcal{R}}_1$ and $\mathcal{R}_1$ have the same size. Furthermore, we move points only away from each other, so the feature resolution remains in $\mathcal{O}(n)$.

Lemma 6. *Let $\mathcal{R}$ be a rectangular dual of an n-vertex PTP graph G realizing a REL $\mathcal{L}$ of G. Let C be an alternating separating 4-cycle in $\mathcal{L}$. Then, we can compute in $\mathcal{O}(n)$ time a rectangular dual $\mathcal{R}_1$ of G realizing $\mathcal{L}$ that satisfies the requirements (P1) and (P2). If $\mathcal{R}$ has an area of at most $n \times n$ and feature resolution $\mathcal{O}(n)$, then so has $\mathcal{R}_1$ and each representation throughout the morph.*

To prove Lemma 6, we do not use zig-zag moves, which were introduced for morphing orthogonal drawings [6, 18], since then we would not be able to bound the area by $n \times n$ throughout the morph. In order to keep a bound of $\mathcal{O}(n) \times \mathcal{O}(n)$, it seems that we would need a re-compactification step after each zig-zag move. Therefore, we keep the modifications in our morph as local as possible.

Let us now consider the morph $\langle \mathcal{R}_1, \mathcal{R}_2 \rangle$ again. Since $\mathcal{R}_1$ now satisfies (P1) and (P2), only the inside I_C of C, the four rectangles of C, and the border segments of I_C move. The target positions of these can be computed in $\mathcal{O}(n)$ time. The linear morph is then defined fully by the start and target positions. Furthermore, $\mathcal{R}_2$ and all intermediate representations have the same area as $\mathcal{R}_1$.

Proof (of Prop. 4). By Theorem 3 and Lemma 5, we can get from $\mathcal{R}$ via $\mathcal{R}_1$ and $\mathcal{R}_2$ to $\mathcal{R}'$ using three steps. The claims on the running time and the area follow from Theorem 3, Lemmas 5 and 6, and the observations above. □

2.3   Morphing Between Rectangular Duals

Combining results from the previous sections, we can now prove our main result.

Theorem 7. *Let G be an n-vertex PTP graph with rectangular duals $\mathcal{R}$ and $\mathcal{R}'$. We can find in $\mathcal{O}(n^3)$ time a relaxed morph between $\mathcal{R}$ and $\mathcal{R}'$ with $\mathcal{O}(n^2)$ steps that executes the minimum number of rotations. If $\mathcal{R}$ and $\mathcal{R}'$ have an area of at most $n \times n$ and feature resolution in $\mathcal{O}(n)$, then so does each representation throughout the morph.*

Proof. Let $\mathcal{L}$ and $\mathcal{L}'$ be the RELs realized by $\mathcal{R}$ and $\mathcal{R}'$, respectively. By Prop. 2 a shortest path between $\mathcal{L}$ and $\mathcal{L}'$ in the lattice of RELs of G can be computed in $\mathcal{O}(n^3)$ time, and its length is $\mathcal{O}(n^2)$ by Prop. 1. For each rotation along this path, we construct a relaxed morph with a constant number of steps in $\mathcal{O}(n)$ time by Prop. 4. The area and feature resolution also follow from Prop. 4.   □

3   Morphing with Parallel Rotations

We now show how to reduce the number of morphing steps by executing rotations in parallel. We assume that all separating 4-cycles in our PTP graph G are trivial.

Consider two cw rotatable separating 4-cycles C and C' that share a maximal horizontal line segment s as border segment; see Fig. 9. If C contains the left endpoint of s, a rotation of C would move s downwards while a rotation of C' would move s upwards. Therefore, such a morph skews angles such that they are not multiples of 90° even at vertices that are not incident to the interior of C or C'. To avoid such morphs, we say that C and C' are *conflicting*. For a set of cw rotatable separating 4-cycles $\mathcal{C}$ for $\mathcal{R}$, this gives rise to a *conflict graph* $K(\mathcal{C})$ with vertex set $\mathcal{C}$. Note that a separating 4-cycle can be in conflict with at most four other separating 4-cycles. Therefore, $K(\mathcal{C})$ has maximum degree four.

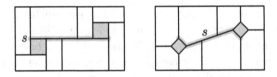

Fig. 9. Two conflicting separating 4-cycles that share the interior segment s.

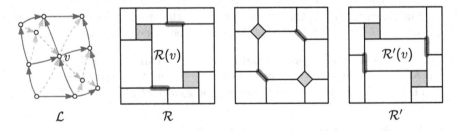

Fig. 10. When we rotate four cw alternating 4-cycles that share a vertex v in a single relaxed morph from $\mathcal{R}$ to $\mathcal{R}'$, then v is temporarily represented by a convex 8-gon.

Next, consider a separating 4-cycle C that shares a maximal horizontal line segment s with an empty 4-cycle C'; see Fig. 10. In this case, we can rotate and translate the inner contact segment of C' downwards, which allows us to simultaneously rotate C and C' without creating unnecessary skewed angles.

Also note that two cw rotatable empty 4-cycles may only overlap with one edge but may not contain an edge of the other. Hence, they are not conflicting.

To rotate a set $\mathcal{C}$ of alternating 4-cycles using $\mathcal{O}(1)$ steps, we divide $\mathcal{C}$ into color classes based on $K(\mathcal{C})$ and rotate one color class at a time.

Proposition 8 $(\star)$. *Let $\mathcal{R}$ be a rectangular dual of a PTP graph G with REL $\mathcal{L}$ whose separating 4-cycles are all trivial. Let $\mathcal{C}$ be a set of alternating 4-cycles of $\mathcal{R}$. Let $\mathcal{L}'$ be the REL obtained from $\mathcal{L}$ by executing all rotations in $\mathcal{C}$. There exists a relaxed morph with $\mathcal{O}(1)$ steps from $\mathcal{R}$ to a rectangular dual $\mathcal{R}'$ realizing $\mathcal{L}'$. The morph can be computed in linear time.*

Note that there exist rectangular duals with a linear number of alternating 4-cycles – extend Fig. 10 into a grid structure. Hence, parallelization can reduce the number of morphing steps by a linear factor. Even more, using Prop. 8, we obtain the following approximation result.

Theorem 9 $(\star)$. *Let G be a PTP graph whose separating 4-cycles are all trivial. Let $\mathcal{R}$ and $\mathcal{R}'$ be two rectangular duals of G, and let OPT be the minimum number of steps in any relaxed morph between $\mathcal{R}$ and $\mathcal{R}'$. Then we can construct in cubic time a relaxed morph consisting of $\mathcal{O}(\text{OPT})$ steps.*

4   Concluding Remarks

In the parallelization step, we considered only PTP graphs whose separating 4-cycles are trivial. It remains open how to parallelize rotations for RELs of PTP graphs with nontrivial separating 4-cycles, in particular, to construct morphs that execute rotations of nested 4-cycles in parallel. It would also be interesting to guarantee area bounds for morphs with parallel rotations.

During our relaxed morphs, we allow rectangles to temporarily turn into convex 5-gons (with four edges axis-aligned). Alternatively, one could insist that the intermediate objects remain ortho-polygons. This would require upt to six vertices per shape and would force not only the outer rectangles in Fig. 4 to change their shape, but also the rectangles in the interior. We find our approach more natural.

References

1. Alamdari, S., et al.: How to morph planar graph drawings. SIAM J. Comput. **46**(2), 824–852 (2017). https://doi.org/10.1137/16M1069171
2. Angelini, P., Chaplick, S., Cornelsen, S., Da Lozzo, G., Roselli, V.: Morphing contact representations of graphs. In: Barequet, G., Wang, Y. (eds.) Symp. Comput. Geom. (SoCG). LIPIcs, vol. 129, pp. 10:1–10:16. Schloss Dagstuhl - LZI (2019). https://doi.org/10.4230/LIPIcs.SoCG.2019.10
3. Angelini, P., Da Lozzo, G., Frati, F., Lubiw, A., Patrignani, M., Roselli, V.: Optimal morphs of convex drawings. In: Arge, L., Pach, J. (eds.) Symp. Comput. Geom. (SoCG). LIPIcs, vol. 34, pp. 126–140. Schloss Dagstuhl - LZI (2015). https://doi.org/10.4230/LIPIcs.SOCG.2015.126

4. Arseneva, E., et al.: Pole dancing: 3D morphs for tree drawings. J. Graph Alg. Appl. **23**(3), 579–602 (2019). https://doi.org/10.7155/jgaa.00503

5. Barrera-Cruz, F., Haxell, P., Lubiw, A.: Morphing schnyder drawings of Planar Triangulations. Discrete CMS. Geom. **61**(1), 161–184 (2018). https://doi.org/10.1007/s00454-018-0018-9

6. Biedl, T., Lubiw, A., Petrick, M., Spriggs, M.: Morphing orthogonal planar graph drawings. ACM Trans. Algorithms **9**(4) (2013). https://doi.org/10.1145/2500118

7. Booth, K.S., Lueker, G.S.: Testing for the consecutive ones property, interval graphs, and graph planarity using PQ-tree algorithms. J. Comput. Syst. Sci. **13**(3), 335–379 (1976). https://doi.org/10.1016/S0022-0000(76)80045-1

8. Buchin, K., Speckmann, B., Verdonschot, S.: Evolution strategies for optimizing rectangular cartograms. In: Xiao, N., Kwan, M.-P., Goodchild, M.F., Shekhar, S. (eds.) GIScience 2012. LNCS, vol. 7478, pp. 29–42. Springer, Heidelberg (2012). https://doi.org/10.1007/978-3-642-33024-7_3

9. Cairns, S.S.: Deformations of plane rectilinear complexes. Amer. Math. Monthly **51**(5), 247–252 (1944). https://doi.org/10.1080/00029890.1944.11999082

10. Chambers, E.W., Erickson, J., Lin, P., Parsa, S.: How to morph graphs on the torus. In: Marx, D. (ed.) Symp. Discrete Algorithms (SODA), pp. 2759–2778 (2021). https://doi.org/10.1137/1.9781611976465.164

11. Chaplick, S., Kindermann, P., Klawitter, J., Rutter, I., Wolff, A.: Morphing rectangular duals (2022). https://doi.org/10.48550/ARXIV.2112.03040

12. Felsner, S.: Rectangle and square representations of planar graphs. In: Pach, J. (ed.) Thirty Essays on Geometric Graph Theory, pp. 213–248. Springer, New York (2013). https://doi.org/10.1007/978-1-4614-0110-0_12

13. Florisson, S., van Kreveld, M.J., Speckmann, B.: Rectangular cartograms: construction & animation. In: Mitchell, J.S.B., Rote, G. (eds.) ACM Symp. Comput. Geom. (SoCG), pp. 372–373 (2005). https://doi.org/10.1145/1064092.1064152

14. de Fraysseix, H., de Mendez, P.O., Rosenstiehl, P.: On triangle contact graphs. Combin. Prob. Comput. **3**(2), 233–246 (1994). https://doi.org/10.1017/S0963548300001139

15. Fusy, É.: Transversal structures on triangulations, with application to straight-line drawing. In: Healy, P., Nikolov, N.S. (eds.) GD 2005. LNCS, vol. 3843, pp. 177–188. Springer, Heidelberg (2006). https://doi.org/10.1007/11618058_17

16. Fusy, É.: Transversal structures on triangulations: a combinatorial study and straight-line drawings. Discrete Math. **309**(7), 1870–1894 (2009). https://doi.org/10.1016/j.disc.2007.12.093

17. Gabriel, K.R., Sokal, R.R.: A new statistical approach to geographic variation analysis. Syst. Biol. **18**(3), 259–278 (1969). https://doi.org/10.2307/2412323

18. van Goethem, A., Speckmann, B., Verbeek, K.: Optimal morphs of planar orthogonal drawings II. In: Archambault, D., Tóth, C.D. (eds.) GD 2019. LNCS, vol. 11904, pp. 33–45. Springer, Cham (2019). https://doi.org/10.1007/978-3-030-35802-0_3

19. He, X.: On finding the rectangular duals of planar triangular graphs. SIAM J. Comput. **22**(6), 1218–1226 (1993). https://doi.org/10.1137/0222072

20. Heilmann, R., Keim, D.A., Panse, C., Sips, M.: Recmap: rectangular map approximations. In: Ward, M.O., Munzner, T. (eds.) IEEE Symp. Inform. Vis. (InfoVis), pp. 33–40 (2004). https://doi.org/10.1109/INFVIS.2004.57

21. Hoffmann, M., van Kreveld, M.J., Kusters, V., Rote, G.: Quality ratios of measures for graph drawing styles. In: Canadian Conf. Comput. Geom. (CCCG) (2014). http://www.cccg.ca/proceedings/2014/papers/paper05.pdf

22. Kant, G., He, X.: Regular edge labeling of 4-connected plane graphs and its applications in graph drawing problems. Theoret. Comput. Sci. **172**(1), 175–193 (1997). https://doi.org/10.1016/S0304-3975(95)00257-X

23. Kobourov, S.G., Landis, M.: Morphing planar graphs in spherical space. J. Graph Alg. Appl. **12**(1), 113–127 (2008). https://doi.org/10.7155/jgaa.00162

24. Koebe, P.: Kontaktprobleme der konformen Abbildung. Berichte über die Verhandlungen der Sächsischen Akademie der Wiss. zu Leipzig. Math.-Phys. Klasse **88**, pp. 141–164 (1936)

25. Koźmiński, K., Kinnen, E.: Rectangular duals of planar graphs. Netw. **15**(2), 145–157 (1985). https://doi.org/10.1002/net.3230150202

26. Leinwand, S.M., Lai, Y.: An algorithm for building rectangular floor-plans. In: 21st Design Automation Conference, pp. 663–664 (1984). https://doi.org/10.1109/DAC.1984.1585874

27. Nusrat, S., Kobourov, S.G.: The state of the art in cartograms. Comput. Graph. Forum **35**(3), 619–642 (2016). https://doi.org/10.1111/cgf.12932

28. Purchase, Helen C.., Hoggan, Eve, Görg, Carsten: How Important is the mental map an empirical investigation of a dynamic graph layout algorithm. In: Kaufmann, Michael, Wagner, Dorothea (eds.) GD 2006. LNCS, vol. 4372. Springer, Heidelberg (2007). https://doi.org/10.1007/978-3-540-70904-6_19

29. Raisz, E.: The rectangular statistical cartogram. Geogr. Review **24**(2), 292–296 (1934). https://doi.org/10.2307/208794. http://www.jstor.org/stable/208794

30. Steadman, P.: Graph theoretic representation of architectural arrangement. Archit. Res. Teach. **2**(3) 161–172 (1973)

31. van Kreveld, M.J., Speckmann, B.: On rectangular cartograms. Comput. Geom. Theory Appl. **37**(3), 175–187 (2007). https://doi.org/10.1016/j.comgeo.2006.06.002

32. Yeap, G.K.H., Sarrafzadeh, M.: Sliceable floorplanning by graph dualization. SIAM J. Discrete Math. **8**(2), 258–280 (1995). https://doi.org/10.1137/S0895480191266700

Visibility Representations of Toroidal and Klein-bottle Graphs

Therese Biedl[(✉)] [iD]

David R. Cheriton School of Computer Science,
University of Waterloo, Waterloo, ON N2L 3G1, Canada
biedl@uwaterloo.ca

Abstract. In this paper, we study visibility representations of graphs that are embedded on a torus or a Klein bottle. Mohar and Rosenstiehl showed that any toroidal graph has a visibility representation on a flat torus bounded by a parallelogram, but left open the question whether one can assume a rectangular flat torus, i.e., a flat torus bounded by a rectangle. Independently the same question was asked by Tamassia and Tollis. We answer this question in the positive. With the same technique, we can also show that any graph embedded on a Klein bottle has a visibility representation on the rectangular flat Klein bottle.

1 Introduction

Visibility representations are one of the oldest topics studied in graph drawing. Introduced as *horvert-drawings* by Otten and Van Wijk in 1978 [21], and independently as *S-representations* by Duchet, Hamidoune, Las Vergnas and Meyniel in 1983 [10], they consist of assigning disjoint horizontal segments to vertices and disjoint vertical segments to every edge such that for each edge the segment ends at the two vertex-segments of its endpoints and intersects no other vertex-segment. (Fig. 2(d) gives an example.) Later papers studied exactly which planar graphs have such visibility representations [23,24,27] and generalized them to the rolling cylinder [26], Möbius band [7], projective plane [16] or torus [20]. (There are numerous other generalizations, e.g. to higher dimensions [4], or permitting rectangles for vertices and horizontal and vertical edges [5], or permitting edges to go through a limited set of vertex-segments [8].)

The motivation for the current paper is the work by Mohar and Rosenstiehl [20], who showed that any *toroidal graph* (i.e., a graph that can be drawn on a torus without crossings) has a visibility representation on the *flat torus*, i.e., a parallelogram Q where opposite edges have been identified. They explicitly stated as open problem whether the same holds for a *rectangular flat torus*, i.e., where Q must be a rectangle—their method cannot be generalized to this case. (See also Fig. 5.) The same question was asked independently earlier by Tamassia and Tollis [26]. This paper answers this question in the positive.

Supported by NSERC. The author would like to thank Sam Barr for helpful input.

P. Angelini and R. von Hanxleden (Eds.): GD 2022, LNCS 13764, pp. 404–417, 2023.
https://doi.org/10.1007/978-3-031-22203-0_29

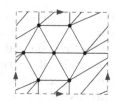 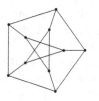 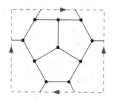

Fig. 1. The complete graph K_7 embedded on the rectangular flat torus and the Petersen-graph embedded on the rectangular flat Klein bottle.

Theorem 1. *Let G be a toroidal graph without loops. Then G has a visibility representation on the rectangular flat torus.*

There are quite a few graph drawing results for toroidal graphs; see Castelli Aleardi et al. [2] and the references therein for increasingly better results for straight-line drawings. Their approach is to convert the toroidal graph into a planar graph by deleting edges, then draw this planar graph, and then reinsert the edges. (Other papers [16,20] instead use a reduction approach, where the graph-size is reduced while staying in the same graph class until some small graph is reached, draw this graph, and then undo the reduction in the drawing.) We follow the first approach (i.e., delete edges to make the graph planar), but face a major challenge when wanting to reinsert an edge (v, w). For this, we need the segments of v and w to be visible across the horizontal boundary of the fundamental rectangle, and in particular, to share an x-coordinate. We achieve this by keeping two halves of each removed edge, connecting corresponding half-edges along paths, and then forcing these paths to be drawn along columns; the ability to do so may be of independent interest.

2 Background

We assume familiarity with graph theory and planar graphs, see for example Diestel's book [9]. Throughout, let $G = (V, E)$ be a connected graph without loops, with $|V| = n$ and $|E| = m$. A *map* M on a surface Σ is a 2-connected graph G together with an embedding of G in Σ such that every *face* (i.e., connected region of $\Sigma \setminus M$) is bounded by a simple cycle. Maps correspond naturally to rotation systems on the underlying graphs, up to homomorphisms among the embeddings [15]. Here a *rotation system* is a set of cyclic permutations ρ_v (for $v \in V$) where ρ_v corresponds to the clockwise cyclic order in which the edges incident to v emanate from v in the embedding. For ease of description we often assume that we have a map, though all algorithmic steps could be performed on the rotation system alone.

We study surfaces that have a *flat representation* consisting of a *fundamental parallelogram* Q in the plane with some sides identified. (We may assume that two sides of Q are horizontal, hence Q has a left/right/top/bottom side.) A *(standing) flat cylinder* is obtained by identifying the left and right side of Q

in the same direction (bottom-to-top). (We usually omit 'standing' since we will not discuss other kinds.) A *flat torus* is obtained from a flat cylinder by identifying the top and bottom side in the same direction (left-to-right), while a *flat Klein bottle* is obtained from a flat cylinder by identifying the top and bottom side in opposite direction. Figures 1, 5, 6 give some examples. A *rectangular flat torus* [*rectangular Klein bottle*] is a flat torus [flat Klein bottle] for which the fundamental parallelogram Q is required to be a rectangle.

Flat representations carry the local geometry of the plane; in particular when we speak of a *segment* or an *x-interval* then we specifically permit it to go across a side of the fundamental parallelogram Q. So for example in a flat cylinder $Q = [0, w] \times [0, h]$, an x-interval can have the form $[x', x'']$ for two x-coordinates $x' < x''$, but it can also have the form $[0, x''] \cup [x', w]$ for some $x'' < x'$. A *row/column* of Q is a horizontal/vertical line with integer coordinate that intersects the interior of Q.

A *visibility representation* of a graph G is a mapping of vertices into non-overlapping horizontal segments (called *vertex-segments*) and of edges of G into non-overlapping vertical segments (called *edge-segments*) such that for each edge (u, v), the associated edge-segment has its endpoints on the vertex-segments corresponding to u and v and it does not intersect any other vertex-segment.

3   Creating Visibility Representations

We first give an outline of our approach. Quite similar to what was was done for straight-line drawings of toroidal graphs [2], we remove a set of edges to convert the given graph into a planar graph. In contrast to the earlier work, we keep the edges but split each of them into two 'half-edges' that end at two new vertices s, t (Sect. 3.1). We will later need to re-connect these half-edges, and to this end, choose a 'path-system' that connects each pair of half-edges while keeping all the paths non-crossing and (after duplicating some edges) edge-disjoint (Sect. 3.2). Then we create a visibility representation on the flat cylinder for which these paths are drawn vertically. To be able to do so we first must argue that we can find an st-order that enumerates vertices of all paths in order (Sect. 3.3). Then we build the visibility representation (Sect. 3.4). Removing the segments of s and t and possibly inserting more columns gives the desired visibility representation. Figures 2 and 3, 4 illustrate the approach for K_7 and the Petersen-graph.

3.1   Making the Graph Planar

In this section we explain how to modify the input graph G to make it planar. We assume that G has no loop and comes embedded on a flat realization Q (either a torus or a Klein bottle). We first modify this embedding to achieve the following: (1) Every face is bounded by a simple cycle, so the embedding is a map. (2) No edge crosses the horizontal boundary of Q twice. (3) Parallelogram Q is a rectangle. (4) No vertex lies on the boundary of Q. (5) Edges intersect the boundary of Q in a finite set of points, and do not use a corner of Q. Conditions

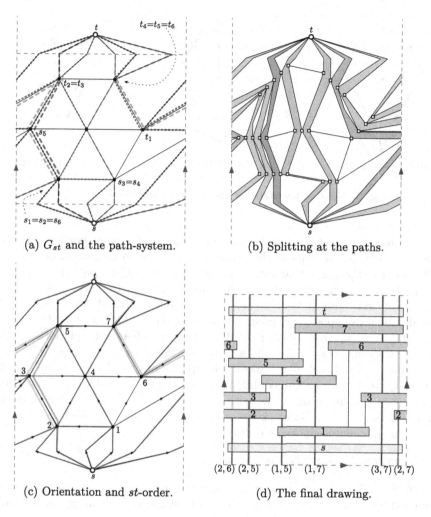

(a) G_{st} and the path-system.   (b) Splitting at the paths.

(c) Orientation and st-order.   (d) The final drawing.

Fig. 2. The construction for the complete graph K_7.

(1–5) can easily be achieved if arbitrary curves are allowed for edges as follows: (1) holds after adding sufficiently many edges (which can be deleted in the final visibility representation), (2) can be achieved by re-routing the horizontal boundary of Q along a so-called *tambourine* [2], (3) holds after a shear and (4–5) hold after locally re-routing.

Assume first that G is toroidal, so Q is a rectangular flat torus. Enumerate the edges that intersect the bottom side of Q as (s_i, t_i) (for $i = 1, \ldots, d$) from left to right, named such that part of the edge that goes upward from the bottom side ends at s_i for $i = 1, \ldots, d$. (This is feasible by condition (2) above.) Create a new graph G_{st} by removing edges (s_i, t_i) for $i = 1, \ldots, s$, adding a new vertex t incident to $t_1, \ldots, t_d$ and a new vertex s incident to $s_1, \ldots, s_d$. See Fig. 2(a).

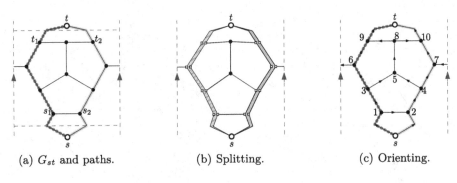

(a) G_{st} and paths.   (b) Splitting.   (c) Orienting.

Fig. 3. The first few steps for the Petersen-graph from Fig. 1(b).

Now assume that G is embedded on a rectangular flat Klein bottle Q instead. We construct G_{st} in almost the same way, but the enumeration of edges is different. Let the edges that cross the bottom side of Q be $(s_1, t_d), \ldots, (s_d, t_1)$ from left to right, named such that the part of the edge that goes upward from the bottom side ends at s_i for $i = 1, \ldots, d$. Since the top and bottom sides of Q are identified in opposite direction, the order of edges along the top side of Q is $(s_d, t_1), \ldots, (s_1, t_d)$ from left to right. Remove these edges and replace them by a vertex s incident to $s_1, \ldots, s_d$ and a vertex t incident to $t_1, \ldots, t_d$. See Fig. 3(a).

In both cases, by placing t above the top side of Q and s below the bottom side of Q, we obtain an embedding of G_{st} on the flat cylinder, so it is a *plane graph* (i.e., drawn on the plane with a fixed embedding). The edges incident to s lead to $s_1, \ldots, s_d$ (in clockwise order) and the edges incident to t lead to $t_1, \ldots, t_d$ (in counter-clockwise order).

Observation 1. *Graph G_{st} is 2-connected.*

Proof. Since G_{st} is a plane graph, 2-connectivity is equivalent to all faces being bounded by a simple cycle. This holds for all faces of G by assumption. The only faces of G_{st} that are not in G are those incident to s and t. These consist of part of the boundary of a face of G, plus two newly added edges that both end at s (or both end at t). So the boundary of these faces are simple cycles as well. □

3.2 Choosing Paths

We now show how to choose a set Π of paths in G_{st} that satisfy some properties. A path is called *simple* if no vertex repeats. Two simple edge-disjoint paths π, π' are *non-crossing* if at any vertex v that is interior to both the paths only touch, i.e., the edges of the paths appear in order π, π, π', π' in ρ_v.

Lemma 1. *There exists a planar graph $\hat{G}$ (obtained by duplicating edges of G_{st}) and a set of simple edge-disjoint non-crossing paths $\pi_1, \ldots, \pi_d$ in $\hat{G}$ such that path π_i begins with (s, s_i) and ends with (t_i, t) for $i = 1, \ldots, d$.*

Before giving the proof, we need to define the operation of *splitting* a map at a path π (also used in Figs. 2(b) and 3(b)). Temporarily direct π from one end to the other. Duplicate all interior vertices of π (say vertex v becomes v^ℓ and v^r) and duplicate all edges of π correspondingly. For any interior vertex v of π, and any edge e incident to v but not on π, we re-connect e to end at v^ℓ $[v^r]$ if e occurs before [after] the outgoing edge of π at v when enumerating ρ_v beginning with the incoming edge of π on v. Splitting at π creates a new face f_π bounded by the two copies of π.

Proof. Let π be a simple path that begins with (s, s_1) and ends with (t_1, t); this exists since G_{st} is 2-connected. Temporarily split graph G_{st} at π to obtain a planar graph $\tilde{G}$. The resulting new face f_π contains both s and t; for ease of description we assume that f_π is the outer-face of $\tilde{G}$.

Let $\tilde{G}^+$ be the graph obtained from $\tilde{G}$ by replacing any edge e that is not incident to s or t by a multi-edge that has $d+1$ copies of e. Any s-t-cut of $\tilde{G}^+$ either consists of the edges incident to s (then it has size $d+1$ since (s, s_1) exists twice in $\tilde{G}$) or of the edges incident to t (then it likewise has size $d+1$), or it contains some edge e not incident to either s or t and so has size at least $d+1$. By the max-flow-min-cut theorem therefore $\tilde{G}^+$ has a flow of value $d+1$ from s to t; equivalently, it has $d+1$ edge-disjoint paths $\pi_1, \ldots, \pi_{d+1}$ from s to t. Since s and t are both on the outer-face we can find these paths using *right-first search* [22]; this will automatically make them crossing-free.

Since the paths are crossing-free and use all edges incident to s, t, and since s and t are on the outer-face, there is no choice which pair of edges must be the first and last on each path. The clockwise order of edges at s (beginning after the outer-face) is $(s, s_1^r), \ldots, (s, s_d), (s, s_1^\ell)$. The counter-clockwise order of edges at t (beginning after the outer-face) is $(t, t_1^r), \ldots, (t, t_d), (t, t_1^\ell)$. Therefore path π_i begins with (s, s_i) and end with (t_i, t) for $i = 2, \ldots, d$, while π_1 and π_{d+1} use the copies of s_1 and t_1.

To obtain $\hat{G}$, re-combine any two vertices v^ℓ and v^r that resulted from splitting an interior vertex v of π, and keep all edges of $\tilde{G}^+$ except (s, s_1^ℓ) and (t_1^ℓ, t). Since these two edges were used by π_{d+1}, they were used by no other path in $\pi_1, \ldots, \pi_d$, and we have hence obtained our desired path-system. □

3.3   A Path-Constrained st-order

By Lemma 1, we can fix a supergraph $\hat{G}$ of G_{st} and a *path-system* Π, i.e., a set of simple edge-disjoint non-crossing paths from s to t. To draw $\hat{G}$, we add vertices one-by-one, and to draw the paths in Π vertically, we require a vertex-order with special properties.

We need some definitions. A *bipolar orientation* is an assignment of directions to the edges that is acyclic and has exactly one source and one sink. An *st-order* is a vertex order $v_1, \ldots, v_n$ such that orienting all edges from the lower-indexed to the higher-indexed vertex gives a bipolar orientation. Vice versa, for any bipolar orientation, enumerating the vertices in topological order gives an st-order. It is well-known that any 2-connected graph has a bipolar orientation, even if we fix a-priori which vertices should be the source and sink [17]; it can be found in linear time [11].

We say that a bipolar orientation *respects a path system* Π if every path in Π is directed from s to t in the bipolar orientation. We phrase the following result for an arbitrary graph H since it does not depend on the graph stemming from a toroidal or Klein-bottle graph and may be of independent interest.

Lemma 2. *Let H be a 2-connected plane graph with two vertices $s \neq t$. Let Π be a set of simple edge-disjoint crossing-free paths from s to t. Then H has a bipolar orientation that respects Π and has source s and sink t.*

Proof. Consider the graph $\hat{H}$ obtained from H by splitting H at each path in Π. See Figs. 2(b) and 3(b). Any face of $\hat{H}$ is either a face of H (then it is a simple cycle since H is 2-connected) or f is bounded by the two copies of some path $\pi \in \Pi$ (then it is a simple cycle since π is simple). So $\hat{H}$ is 2-connected and has a bipolar orientation $\hat{D}$ with source s and sink t.

It is well-known [24] that in $\hat{D}$ any face has a unique source and sink. In any face f_π bounded by two copies of some $\pi \in \Pi$, the unique source is s and the unique sink is t. Therefore both copies of π are directed from s to t and undoing the splitting gives the desired orientation. □

Note Added in Proof: This lemma is not correct, since the created orientation may not be acyclic. (This is unavoidable since the orientation of the paths already may induce a directed cycle.) With a much longer argument (not given here) it can be shown that with a suitable choice of paths (possibly after changing which edges of G are being cut to create G_{st}), we can get such a bipolar orientation for $\hat{G}$. So the rest of the paper appears to be correct.

3.4   Path-Constrained Visibility Representations

In this section, we give an easy construction of a visibility representation on the flat cylinder where a given path-system Π is drawn vertically. Formally, we say that a path π lies *on an exclusive column* ℓ (in a visibility representation Γ) if all edges of π are represented by segments on ℓ, and column ℓ intersects no vertex- or edge-segment except the ones that belong to vertices/edges of π.

Our approach to create visibility representations is quite different from prior constructions [16, 20, 21, 23, 24, 26, 27], which either read the coordinates for the segments directly from the orientation (using the length of the longest paths in the primal and dual graph), or reduced the graph (or its dual) by removing an edge somewhere in the graph, creating a representation recursively, and expanding. In contrast to this, we use here an incremental approach which resembles more the incremental approaches taken for straight-line drawings [2, 13] or orthogonal drawings [3]. This uses a vertex ordering and adds the vertices to the drawing one-by-one.

Theorem 2. *Let H be a 2-connected plane graph with two vertices s, t and let Π be a set of simple edge-disjoint non-crossing paths from s to t. Then H has a visibility representation on the flat cylinder such that each $\pi \in \Pi$ lies on an exclusive column.*

Proof. Fix a bipolar orientation using Lemma 2 and extract an st-order $v_1, \ldots, v_n$ from it; we know $v_1 = s$ and $v_n = t$ and the numbers along any path in Π increase from s to t. For $i = 1, \ldots, n$ let H_i be the subgraph induced by $v_1, \ldots, v_i$ and let the *cut* $E_{i:i+1}$ be the set of all edges (v_h, v_j) with $h \le i < j$. There is a natural cyclic order of the edges in $E_{i:i+1}$ implied by the embedding of H (specifically, if we contracted the vertices $v_1, \ldots, v_i$ into a supernode, then the order of $E_{i:i+1}$ would be the clockwise order of edges at this supernode). We will use induction on i to create a visibility representation of H_i on a flat cylinder that satisfies the following for $i < n$:

1. Every edge $e = (v_h, v_j)$, $h < j$ in cut $E_{i:i+1}$ is associated with a column that intersects v_h and that is empty above v_h.
2. The left-to-right order of columns associated with $E_{i:i+1}$ respects the cyclic order of edges in $E_{i:i+1}$.
3. For any path $\pi \in \Pi$, the sub-path of π in H_i lies on an exclusive column, and the same column is associated with the unique edge of π in $E_{i:i+1}$.

Figure 4(a-b) illustrates the following construction. For $i = 1$, we create the desired visibility representation simply by defining a horizontal line segment $s(v_1)$ for v_1 with y-coordinate 0 and width $|E_{1:2}|$, and assigning columns intersecting $s(v_1)$ to edges in $E_{1:2}$ in the correct order.

For $i > 1$, assume we have created a visibility representation of H_{i-1} already. Define edge-sets $E_i^- := \{(v_h, v_i) : h < i\}$ and $E_i^+ := \{(v_i, v_j) : i < j\}$; the former is non-empty by $i > 1$ since we have an st-order. It is well-known [17] that E_i^- is consecutive in the cyclic order of edges in $E_{i-1:i}$. By the invariant therefore there exists an x-interval X_i on the flat cylinder that intersects all columns associated with edges in E_i^- in its interior and intersects no other columns associated with $E_{i-1:i}$. Define the segment $s(v_i)$ of v_i to have x-range X_i and a y-coordinate that is higher than the one of all its neighbours in E_i^-. These edges can then be completed along their associated columns.

To associate columns with E_i^+, we insert new columns as needed. First consider any edge $e \in E_i^+$ in some path $\pi \in \Pi$. Since π begins at s and $i > 1$, and since indices increase along π, some edge $e' \in E_i^-$ also belongs to π. Associate the column of e' with e. Notice that this associates columns in the correct order, because if multiple paths $\pi_1, \ldots, \pi_k \in \Pi$ all went through v_i, then the counterclockwise order of their edges in E_i^- at v_i must be the same as the clockwise order of their edges in E_i^+ at v_i, otherwise two of these paths would cross at v_i.

Now consider any edge $e \in E_i^+$ that does not belong to a path in Π. Assign a ray upward from $s(v_i)$ to e, choosing rays such that all edges in E_i^+ use distinct rays/columns and their order reflects the order of edges at v_i. By stretching horizontal segments as needed, we can re-assign coordinates so that all inserted rays lie on integer coordinates, hence become new columns. This gives the desired visibility representation of H_i. □

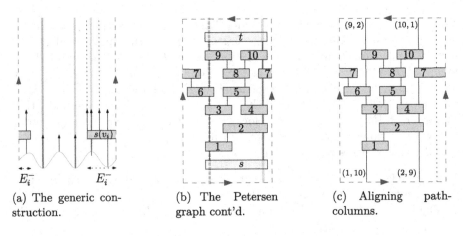

(a) The generic construction.

(b) The Petersen graph cont'd.

(c) Aligning path-columns.

Fig. 4. Creating visibility representations.

3.5 Putting it All Together

We now have all ingredients to prove our main result (Theorem 1): Any toroidal graph G without loops has a visibility representation on the rectangular flat torus. See Fig. 2 for the entire process.

Proof. Add edges to G until all its faces are simple cycles. As described in Subsects. 3.1-3.4, split G at edges (s_i, t_i) (for $i = 1, \ldots, d$) to obtain G_{st}, find a supergraph $\hat{G}$ with a path-system Π where path π_i begins with (s, s_i) and ends with (t_i, t), find an orientation that respects Π, and find a visibility representation Γ of $\hat{G}$ on the flat cylinder Q such that π_i is drawn along an exclusive column ℓ_i. Remove the segments that represent s and t and complete (s_i, t_i) along column ℓ_i. After re-interpreting Q as a rectangular flat torus this gives the desired visibility representation of G after deleting all added edges. □

With a bit more care when reconnecting edges, the same approach also works for Klein-bottle graphs.

Theorem 3. *Let G be a graph without loops embedded on the Klein bottle. Then G has a visibility representation on the rectangular flat Klein bottle.*

Proof. Exactly as in the previous proof, create a visibility representation Γ of $\hat{G}$ on the flat cylinder Q such that π_i is drawn along an exclusive column ℓ_i. Remove the segments that represent s and t and extend (s_i, s) and (t_i, t) along ℓ_i until they reach the horizontal boundary of Q.

We are not quite done yet, because we must ensure that column ℓ_i 'lines up' with column ℓ_{d+1-i} (for $i = 1, \ldots, \lfloor d/2 \rfloor$) so that edges (s_i, t_{d+1-i}) and (s_{d+1-1}, t_i) are connected correctly when interpreting Q as the flat Klein bottle. This is easily achieved by inserting columns. Namely, assume Q has x-range $[0, w]$ and let $x(\ell)$ denote the x-coordinate of column ℓ. For $i = 1, \ldots, \lfloor d/2 \rfloor$, while $x(\ell_i) < w - x(\ell_{d+1-i})$, insert an empty column to the left of ℓ_i, and while

$x(\ell_i) > w - x(\ell_{d+1-i})$, insert an empty column to the right of ℓ_{d+1-i} See Fig. 4(c). This maintains distances of $\ell_1, \ldots, \ell_{i-1}$ to the left boundary and distances of $\ell_{d+2-i}, \ldots, \ell_d$ to the right boundary of Q. So performing this for $i = 1, \ldots, \lfloor d/2 \rfloor$ gives the desired visibility representation on the flat Klein bottle.   □

We note here that our visibility representations exactly respect the given embedding. Under this restriction, the condition 'no loops' cannot be avoided. (This was essentially observed by Mohar and Rosenstiehl [20] already.) Namely, let M_0 be a graph with a single vertex v and two loops ℓ_1, ℓ_2 such that $\rho_v = \langle \ell_1, \ell_2, \ell_1, \ell_2 \rangle$. This is toroidal, but has no visibility representation on the rectangular flat torus that respects the embedding since the rotation scheme at v in such an embedding is necessarily $\ell_1, \ell_1, \ell_2, \ell_2$.

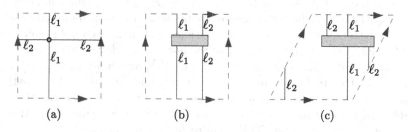

(a)   (b)   (c)

Fig. 5. (a) Graph M_0. (b) The only possible visibility representation on a rectangular flat torus. (c) An embedding-preserving visibility on the flat torus.

3.6 Grid-Size

We can give a bound on the grid-size of Theorem 1, assuming that the input is already a map (i.e., all faces are simple cycles). We say that a visibility representation *has grid-size* $w \times h$ if the fundamental rectangle Q intersects w columns and h rows, not counting the boundaries of Q. In our current approach, the visibility representation Γ_{st} of G_{st} uses significantly more area than it needs to since we may duplicate quite a few edges of G_{st} to obtain the path system (see also the discussion below). However, as for all visibility representations, one should apply compaction steps (similar as for VLSI design [18]) to reduce the size of the drawing. We claim that after doing this, the visibility representation Γ of a toroidal graph G has grid-size at most $(m - n) \times n$.

To see this, observe that we need at most n rows, since assigning row i to vertex v_i will certainly place it high enough and the rows for s, t can be deleted during compaction. As for the number of columns, each column must contain at least one edge, else it could have been deleted. Furthermore, we used a bipolar orientation of $\hat{G}$, which means that every vertex other than s and t has both an incoming and an outgoing edge. Since $\hat{G}$ is obtained from G_{st} by duplicating edges, the same holds in G_{st}. Vertices s and t are removed in the final visibility representation (but their incident edges remain and are re-combined). With the standard compaction steps, therefore at least one column at each vertex v is used for two edges incident to v. It follows that each vertex saves at least one column, hence the number of columns is $m - n$.

3.7   Run-time

Following the steps of our algorithm, it is very clear that our visibility representations can be found in polynomial time. In fact, the drawing in Theorem 2 can be found in linear time with standard-approaches: do not explicitly maintain the x-coordinates, but store the drawing implicitly by computing x-spans of vertex-segments and x-offsets of edge-segments from the left endpoints of their lower endpoints. The final drawing can then be computed with one pass over the entire graph after all vertices have been placed.

Unfortunately finding the drawings in Theorems 1 and 3 may take superlinear time since the supergraph $\hat{G}$ may have many extra edges. If G_{st} has $\Omega(n)$ disjoint edge-cuts that separate s and t, then each of the $|\Pi|$ paths must duplicate an edge in each edge-cut, leading to $\Omega(m + |\Pi|n)$ edges for $\hat{G}$. One can show that $|\Pi| \in O(\sqrt{n})$ can be achieved, because any toroidal graph has a non-contractible cycle of length $O(\sqrt{n})$ [1], and we can use such a cycle in the dual graph to find an embedding where $O(\sqrt{n})$ edges cross the horizontal side and hence necessitate a path in Π. With this choice we get $|\hat{G}| \in O(n^{1.5})$ and run-time $O(n^{1.5})$.

Reducing this to linear time seems not implausible: we need the paths in Π only to steer us towards placing edges in the visibility representation at a suitable place, and it may be possible to encode this in a smaller data structure that permits linear run-time. This remains for future work.

4   Other Drawing Styles

We close the paper by discussing how our results do (or do not) imply results in some other graph drawing styles that are closely related to visibility representations. The first drawing style that we consider are *orthogonal point-drawings*, where vertices are represented by points and every edge is a polygonal curve between its endpoints that uses only horizontal and vertical segments and does not intersect other edges or vertices. (These can only exist if the graph has maximum degree at most four.)

Theorem 4. *Every toroidal graph with maximum degree four has an orthogonal point-drawing on the rectangular flat torus. Every Klein-bottle graph with maximum degree four has an orthogonal point-drawing on the flat Klein bottle.*

Proof. Tamassia and Tollis [25] showed how to create orthogonal point-drawings by starting with a visibility representation and replacing vertex-segments locally by points and polygonal curves that connect to the edge-segments. The exact same transformations can be applied to any visibility representation that lies on a flat representation, so using it with Theorem 1 and Theorem 3 (after subdividing loops, if any) gives the desired orthogonal point-drawings.   □

Two other related drawing styles are grid contact and tessellation representations. A *bipartite graph* has a vertex-partition $V = W \cup B$ such that there are no edges within W or within B. In a *grid contact representation* of a bipartite graph, the vertices of W and B are assigned to horizontal and vertical segments, respectively, with all segments disjoint except that any segment of one kind may touch at both of its ends an interior point of a segment of the other kind, and such a common point occurs only if the two vertices are adjacent. See Fig. 6(b). It is well-known [12] that every planar bipartite graph has a grid contact representation in the plane, and Mohar and Rosenstiehl [20] showed that any toroidal bipartite graph has a grid contact representation on the flat (not necessarily rectangular) torus. A *tessellation representation* of a graph G is a grid contact representation of the bipartite graph whose vertices are the faces and vertices of G and whose edges are the incidences between them.[1] See Fig. 6(c).

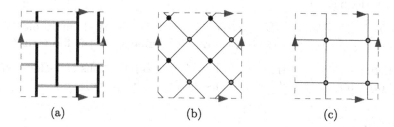

(a)   (b)   (c)

Fig. 6. (a) A set of segments that is a grid contact representation of $K_{4,4}$ (shown in (b)) or a tessellation representation of the graph in (c).

Mohar and Rosenstiehl constructed tessellation representations of toroidal graphs (on a flat torus), from which their results on grid contact representations and visibility representations follow easily. They must permit a non-rectangular flat torus because they reduce their graph to M_0 (or another single-vertex graph with loops), which cannot be represented on a rectangular flat torus. But does it help to have no loops?

Conjecture 1. Every toroidal graph without loops has a tessellation representation on the rectangular flat torus.

Conjecture 2. Every bipartite toroidal graph without loops has a grid contact representation on the rectangular flat torus.

At first sight one might think that Theorem 1 implies Conjecture 1, because Mohar and Rosenstiehl [20] show that a visibility representation can be converted to a tessellation representation. Alas, their definition of "visibility representation" uses the 'strong' model where *all* visibilities must lead to an edge, hence faces are triangles, and this is vital in their proof. On the positive side, their proof does not affect the shape of the flat representation, so using it one can show that Conjecture 1 holds for toroidal graphs where all faces are triangles.

---

[1] In contrast to earlier work [20], we use here *weak* models, where not all adjacencies that could be added must exist.

Finally we are interested in *segment intersection representations*, i.e., every vertex is assigned to a segment (of arbitrary slope) on the flat torus, with segments intersecting if and only if the vertices are adjacent. Such representations exist for all planar graphs [6], and one proof of this proceeds by representing a planar graph as the intersection-graph of L-shaped curves in the plane [14] and then converting the L-shaped curves into segments [19]. The corresponding questions on the flat torus appear to be open even if we drop 'rectangular':

Question 1. Does every simple toroidal graph have a segment intersection representation on the flat torus?

Question 2. Is every simple toroidal graph the intersection-graph of L-shaped curves on the flat torus?

Question 3. If a graph is the intersection-graph of L-shaped curves on the flat torus, then is it also the intersection-graph of segments on the flat torus?

Finally all these questions could be asked also for graphs embedded on the Klein bottle (or other surfaces, such as the projective plane).

References

1. Albertson, M., Hutchinson, J.: On the independence ratio of a graph. J. Graph Theor. **2**(1), 1–8 (1978). https://doi.org/10.1002/jgt.3190020102
2. Aleardi, L.C., Devillers, O., Fusy, É.: Canonical ordering for graphs on the cylinder, with applications to periodic straight-line drawings on the flat cyclinder and torus. J. Comput. Geom. 9(1), 391–429 (2018). https://doi.org/10.20382/jocg.v9i1a14
3. Biedl, T., Kant, G.: A better heuristic for orthogonal graph drawings. Comput. Geom.: Theor. Appl. **9**, 159–180 (1998). https://doi.org/10.1016/S0925-7721(97)00026-6
4. Bose, P., Everett, H., Fekete, S., Houle, M., Lubiw, A., Meijer, H., et al.: A visibility representation for graphs in three dimensions. J. Graph Algorithms Appl. **2**(3), 1–16 (1998). https://doi.org/10.7155/jgaa.00006
5. Bose, P., Dean, A., Hutchinson, J., Shermer, T.: On rectangle visibility graphs. In: North, S. (ed.) GD 1996. LNCS, vol. 1190, pp. 25–44. Springer, Heidelberg (1997). https://doi.org/10.1007/3-540-62495-3_35
6. Chalopin, J., Gonçalves, D.: Every planar graph is the intersection graph of segments in the plane. In: ACM Symposium on Theory of Computing (STOC 2009), pp. 631–638 (2009). https://doi.org/10.1145/1536414.1536500
7. Dean, A.M.: A layout algorithm for bar-visibility graphs on the Möbius band. In: Marks, J. (ed.) GD 2000. LNCS, vol. 1984, pp. 350–359. Springer, Heidelberg (2001). https://doi.org/10.1007/3-540-44541-2_33
8. Dean, A., Evans, W., Gethner, E., Laison, J., Safari, M.A., Trotter, W.: Bar k-visibility graphs. J. Graph Algorithms Appl. **11**(1), 45–59 (2007). https://doi.org/10.7155/jgaa.00136
9. Diestel, R.: Graph Theory. GTM, vol. 173. Springer, Heidelberg (2017). https://doi.org/10.1007/978-3-662-53622-3
10. Duchet, P., Hamidoune, Y., Vergnas, M.L., Meyniel, H.: Representing a planar graph by vertical lines joining different levels. Discret. Math. **46**(3), 319–321 (1983). https://doi.org/10.1016/0012-365X(83)90128-0

11. Even, S., Tarjan, R.E.: Computing an st-numbering. Theor. Comput. Sci. **2**, 436–441 (1976). https://doi.org/10.1016/0304-3975(76)90086-4
12. de Fraysseix, H., de Mendez, P.O., Pach, J.: Representation of planar graphs by segments. Intuitive Geom. **63**, 109–117 (1991)
13. de Fraysseix, H., Pach, J., Pollack, R.: How to draw a planar graph on a grid. Combinatorica **10**, 41–51 (1990). https://doi.org/10.1007/BF02122694
14. Gonçalves, D., Isenmann, L., Pennarun, C.: Planar graphs as L-intersection or L-contact graphs. In: SIAM Symposium on Discrete Algorithms (SODA 2018), pp. 172–184 (2018). https://doi.org/10.1137/1.9781611975031.12
15. Gross, J.L., Tucker, T.W.: Topological Graph Theory. John Wiley and Sons (1987)
16. Hutchinson, J.: A note on rectilinear and polar visibility graphs. Discret. Appl. Math. **148**(3), 263–272 (2005). https://doi.org/10.1016/j.dam.2004.12.004
17. Lempel, A., Even, S., Cederbaum, I.: An algorithm for planarity testing of graphs. In: Theory of Graphs, International Symposium Rome 1966, pp. 215–232. Gordon and Breach (1967)
18. Lengauer, T.: Combinatorial Algorithms for Integrated Circuit Layout. Teubner/Wiley & Sons, Stuttgart/Chicester (1990)
19. Middendorf, M., Pfeiffer, F.: The max clique problem in classes of string-graphs. Discret. Math. **108**(1–3), 365–372 (1992). https://doi.org/10.1016/0012-365X(92)90688-C
20. Mohar, B., Rosenstiehl, P.: Tessellation and visibility representations of maps on the torus. Discret. Comput. Geom. **19**(2), 249–263 (1998). https://doi.org/10.1007/PL00009344
21. Otten, R., van Wijk, J.: Graph representations in interactive layout design. In: IEEE International Symposium on Circuits and Systems, pp. 914–918. New York (1978)
22. Ripphausen-Lipa, H., Wagner, D., Weihe, K.: The vertex-disjoint Menger problem in planar graphs. SIAM J. Comput. **26**(2), 331–349 (1997). https://doi.org/10.1137/S0097539793253565
23. Rosenstiehl, P., Tarjan, R.E.: Rectilinear planar layouts and bipolar orientations of planar graphs. Discret. Comput. Geom. **1**(4), 343–353 (1986). https://doi.org/10.1007/BF02187706
24. Tamassia, R., Tollis, I.G.: A unified approach to visibility representations of planar graphs. Discret. Comput. Geom. **1**(4), 321–341 (1986). https://doi.org/10.1007/BF02187705
25. Tamassia, R., Tollis, I.: Planar grid embedding in linear time. IEEE Trans. Circuits Syst. **36**(9), 1230–1234 (1989)
26. Tamassia, R., Tollis, I.: Representations of graphs on a cylinder. SIAM J. Discret. Math. **4**(1), 139–149 (1991). https://doi.org/10.1137/0404014
27. Wismath, S.: Characterizing bar line-of-sight graphs. In: ACM Symposium on Computational Geometry (SoCG '85), pp. 147–152 (1985). https://doi.org/10.1145/323233.323253

Coloring Mixed and Directional Interval Graphs

Grzegorz Gutowski[1] , Florian Mittelstädt[2], Ignaz Rutter[3] ,
Joachim Spoerhase[4] , Alexander Wolff[2] , and Johannes Zink[2]([✉])

[1] Institute of Theoretical Computer Science, Faculty of Mathematics and Computer
Science, Jagiellonian University, Kraków, Poland
[2] Universität Würzburg, Würzburg, Germany
zink@informatik.uni-wuerzburg.de
[3] Universität Passau, Passau, Germany
[4] Max-Planck-Institut Saarbrücken, Saarbrücken, Germany

Abstract. A *mixed graph* has a set of vertices, a set of undirected edges,
and a set of directed arcs. A *proper coloring* of a mixed graph G is a
function c that assigns to each vertex in G a positive integer such that,
for each edge $\{u, v\}$ in G, $c(u) \neq c(v)$ and, for each arc (u, v) in G,
$c(u) < c(v)$. For a mixed graph G, the *chromatic number* $\chi(G)$ is the
smallest number of colors in any proper coloring of G. A *directional
interval graph* is a mixed graph whose vertices correspond to intervals on
the real line. Such a graph has an edge between every two intervals where
one is contained in the other and an arc between every two overlapping
intervals, directed towards the interval that starts and ends to the right.

Coloring such graphs has applications in routing edges in layered
orthogonal graph drawing according to the Sugiyama framework; the
colors correspond to the tracks for routing the edges. We show how to
recognize directional interval graphs, and how to compute their chromatic number efficiently. On the other hand, for *mixed interval graphs*,
i.e., graphs where two intersecting intervals can be connected by an edge
or by an arc in either direction arbitrarily, we prove that computing the
chromatic number is NP-hard.

Keywords: Mixed graphs · Mixed interval graphs · Directed interval
graphs · Recognition · Proper coloring

1 Introduction

A *mixed graph* is a graph that contains both undirected edges and directed arcs.
Formally, a mixed graph G is a tuple (V, E, A) where $V = V(G)$ is the set of

Work on this problem was initiated at the HOMONOLO Workshop 2021 in Nová
Louka, Czech Republic. G.G. is partially supported by the National Science Center of
Poland under grant no. 2019/35/B/ST6/02472.

P. Angelini and R. von Hanxleden (Eds.): GD 2022, LNCS 13764, pp. 418–431, 2023.
https://doi.org/10.1007/978-3-031-22203-0_30

vertices, $E = E(G)$ is the set of edges, and $A = A(G)$ is the set of arcs. We require that any two vertices are connected by at most one edge or arc. For a mixed graph G, let $U(G) = (V(G), E')$ denote the *underlying undirected graph*, where $E' = E(G) \cup \{\{u, v\}: (u, v) \in A(G)$ or $(v, u) \in A(G)\}$.

A *proper coloring* of a mixed graph G is a function c that assigns a positive integer to every vertex in G, satisfying $c(u) \neq c(v)$ for every edge $\{u, v\}$ in G, and $c(u) < c(v)$ for every arc (u, v) in G. It is easy to see that a mixed graph admits a proper coloring if and only if the arcs of G do not induce a directed circuit. For a mixed graph G with no directed circuit, we define the chromatic number $\chi(G)$ as the smallest number of colors in any proper coloring of G.

The concept of mixed graphs was introduced by Sotskov and Tanaev [17] and reintroduced by Hansen, Kuplinsky, and de Werra [9] in the context of proper colorings of mixed graphs. Coloring of mixed graphs was used to model problems in scheduling with precedence constraints [16]. It is NP-hard in general, and it was considered for some restricted graph classes, e.g., when the underlying graph is a tree, a series-parallel graph, a graph of bounded tree-width, or a bipartite graph [5,6,15]. Mixed graphs have also been studied in the context of (quasi-) upward planar drawings [2–4], and extensions of partial orientations [1,10].

Let $\mathcal{I}$ be a set of closed non-degenerate intervals on the real line. The *intersection graph* of $\mathcal{I}$ is the graph with vertex set $\mathcal{I}$ where two vertices are adjacent if the corresponding intervals intersect. An *interval graph* is a graph G that is isomorphic to the intersection graph of some set $\mathcal{I}$ of intervals. We call $\mathcal{I}$ an *interval representation* of G, and for a vertex v in G, we write $\mathcal{I}(v)$ to denote the interval that represents v. A *mixed interval graph* is a mixed graph G whose underlying graph $U(G)$ is an interval graph.

For a set $\mathcal{I}$ of closed non-degenerate intervals on the real line, the *directional intersection graph* of $\mathcal{I}$ is a mixed graph G with vertex set $\mathcal{I}$ where, for every two vertices $u = [l_u, r_u]$, $v = [l_v, r_v]$ with u starting to the left of v, i.e., $l_u \leqslant l_v$, exactly one of the following conditions holds:

$$u \text{ and } v \text{ are disjoint, i.e., } r_u < l_v \iff u \text{ and } v \text{ are independent in } G,$$
$$u \text{ and } v \text{ overlap, i.e., } l_u < l_v \leqslant r_u < r_v \iff \text{arc } (u, v) \text{ is in } G,$$
$$u \text{ contains } v, \text{ i.e., } r_v \leqslant r_u \iff \text{edge } \{u, v\} \text{ is in } G.$$

A *directional interval graph* is a mixed graph G that is isomorphic to the directional intersection graph of some set $\mathcal{I}$ of intervals. We call $\mathcal{I}$ a *directional representation* of G. Similarly to interval graphs, a directional interval graph may have several different directional representations. As there is no directed circuit in a directional interval graph G, $\chi(G)$ is well defined. Observe that the endpoints in any directional representation can be perturbed so that every endpoint is unique, and the modified intervals represent the same graph.

Further, we generalize directional interval graphs and directional representations to *bidirectional interval graphs* and *bidirectional representations*. There, we assume that we have two types of intervals, which we call *left-going* and *right-going*. For left-going intervals, the edges and arcs are defined as in directional intersection graphs. For right-going intervals, the symmetric definition applies,

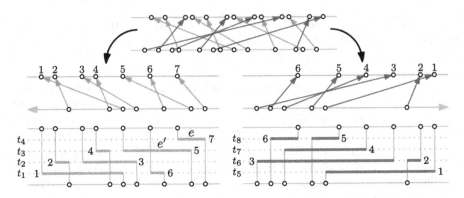

Fig. 1. Separate greedy assignment of left-going and right-going edges to tracks.

that is, we have an arc (u, v) if and only if $l_v < l_u \leqslant r_v < r_u$. Moreover, there is an edge for every pair of a left-going and a right-going interval that intersect.

Interval graphs are a classic subject of algorithmic graph theory whose applications range from scheduling problems to analysis of genomes [7]. Many problems that are NP-hard for general graphs can be solved efficiently for interval graphs. In particular, the chromatic number of (undirected) interval graphs [7] and directed acyclic graphs [9] can be computed in linear time.

In this paper we combine the research directions of coloring geometric intersection graphs and of coloring mixed graphs, by studying the coloring of mixed interval graphs. Our research is also motivated by the following application.

A subproblem that occurs when drawing layered graphs according to the Sugiyama framework [18] is the edge routing step. This step is applied to every pair of consecutive layers. Zink et al. [19] formalize this for orthogonal edges as follows. Given a set of points on two horizontal lines (corresponding to the vertices on two consecutive layers) and a perfect matching between the points on the lower and those on the upper line, connect the matched pairs of points by x- and y-monotone rectilinear paths. Since we can assume that no two points have the same x-coordinate, each pair of points can be connected by a path that consists of three axis-aligned line segments; a vertical, a horizontal, and another vertical one; see Fig. 1. We refer to the interval that corresponds to the vertical projection of an edge to the x-axis as the *span* of that edge. We direct all edges upward. This allows us to classify the edges into *left-* vs. *right-going*.

Now the task is to map the horizontal pieces to horizontal "tracks" between the two layers such that no two such pieces overlap and no two edges cross twice. This implies that any two edges whose spans intersect must be mapped to different tracks. If there is a left-going edge e whose span overlaps that of another left-going edge e' that lies further to the left (see Fig. 1), then e must be mapped to a higher track than e' to avoid crossings. The symmetric statement holds for pairs of right-going edges. The aim is to minimize the number of tracks in order to get a compact layered drawing of the original graph. This corresponds to minimizing the number of colors in a proper coloring of a bidirectional interval graph. Zink et al. solve this combinatorial problem heuristically. They greedily construct two colorings (of left-going edges and of right-going edges) and combine the colorings by assigning separate tracks to the two directions; see Fig. 1.

Our Contribution. We first show that the above-mentioned greedy algorithm of Zink et al. [19] colors directional interval graphs with the minimum number of colors; see Sect. 2. This yields a simple 2-approximation algorithm for the bidirectional case. Then, we prove that computing the chromatic number of a mixed interval graph is NP-hard; see Sect. 3. This result extends to proper interval graphs; see our full version [8]. Finally, we present an efficient algorithm that recognizes directional interval graphs; see Sect. 4. Our algorithm is based on PQ-trees and the recognition of two-dimensional posets. It can construct a directional interval representation of a yes-instance in quadratic time.

Proofs of statements with a "$\star$" are available in our full version [8] on arXiv.

2   Coloring Directional Interval Graphs

We prove that the greedy algorithm of Zink et al. [19] computes an optimal coloring for a given directional interval representation of G. If we are not given a representation (i.e., a set of intervals) but only the graph, we obtain a representation in quadratic time by Theorem 3. The greedy algorithm proceeds analogously to the classic greedy coloring algorithm for (undirected) interval graphs. Also our optimality proof follows, on a high level, the strategy of relating the coloring to a large clique. In our setting, however, the underlying geometry is more intricate, which makes the optimality proof as well as a fast implementation more involved. The algorithm works as follows; see Fig. 1 (left) for an example.

GREEDY ALGORITHM. Iterate over the given intervals in increasing order of their left endpoints. For each interval v, assign v the smallest available color $c(v)$. A color k is *available* for v if, for any interval u that has already been colored, $k \neq c(u)$ if u contains v and $k > c(u)$ if u overlaps v.

A naive implementation of the greedy algorithm runs in quadratic time. Using augmented binary search trees, we can speed it up to optimal $O(n \log n)$ time.

Lemma 1 ($\star$). *The greedy algorithm can be implemented to color n intervals in $O(n \log n)$ time, which is optimal assuming the comparison-based model.*

Next we show that the greedy algorithm computes an optimal proper coloring. This also yields a simple 2-approximation for the bidirectional case.

Theorem 1. *Given a directional representation of a directional interval graph G, the greedy algorithm computes a proper coloring of G with $\chi(G)$ many colors.*

Proof. The *transitive closure* G^+ of G is the graph that we obtain by exhaustively adding transitive arcs, i.e., if there are arcs (u, v) and (v, w), we add the arc (u, w) if absent. Clearly, no pair of adjacent intervals in the underlying undirected graph $U(G^+)$ of G^+ can have the same color in a proper coloring of G. Therefore, $\omega(U(G^+)) \leqslant \chi(G)$ where $\omega(U(G^+))$ denotes the size of a largest clique in $U(G^+)$. We show below that the greedy algorithm computes a coloring with at most $\omega(U(G^+))$ many colors, which must therefore be optimal. For $v \in V$ let $\mathcal{I}_{\mathrm{in}}(v)$ be the set of intervals having an arc to v in G.

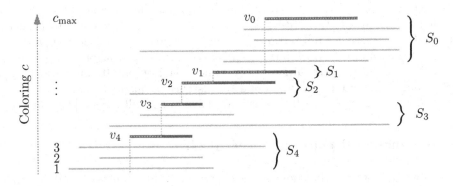

Fig. 2. A staircase and its intermediate intervals, which form a clique in $U(G^+)$.

Let c be the coloring computed by our greedy coloring algorithm. Since we always pick an available color, c is a proper coloring. To prove optimality of c, we show the existence of a clique in $U(G^+)$ of cardinality $c_{\max} = \max_{v \in V} c(v)$.

Consider an interval $v_0 = [l_0, r_0]$ of color $c_{\max}$. Among $\mathcal{I}_{\mathrm{in}}(v_0)$, let v_1 be the unique interval with the largest color (all intervals in $\mathcal{I}_{\mathrm{in}}(v_0)$ have different colors as they share the point l_0). We call v_1 the *step below* v_0. We repeat this argument to find the step v_2 below v_1 and so on. For some $t \geqslant 0$, there is a v_t without a step below it, namely where $\mathcal{I}_{\mathrm{in}}(v_t) = \emptyset$. We call the sequence $v_0, v_1, \ldots, v_t$ a *staircase* and each of its intervals a *step*; see Fig. 2. Clearly, (v_j, v_i) is an arc of G^+ for $0 \leqslant i < j \leqslant t$. In particular, the staircase is a clique of size $t+1$ in $U(G^+)$. Next we argue about the intervals with colors in-between the steps.

For a step $v_i = [l_i, r_i]$, $i \in \{0, \ldots, t\}$, let S_i denote the set of intervals that contain the point l_i and have a color $x \in \{c(v_{i+1}) + 1, c(v_{i+1}) + 2, \ldots, c(v_i)\}$; see Fig. 2. Note that $v_i \in S_i$ and, by the definition of steps, each interval in S_i contains v_i. Observe that $|S_i| = c(v_i) - c(v_{i+1})$, as otherwise the greedy algorithm would have assigned a smaller color to v_i. It follows that $c_{\max} = \sum_{i=0}^{t} |S_i|$.

We claim that $S = \bigcup_{i=0}^{t} S_i$ is a clique in $U(G^+)$. Let $u \in S_i$, $v \in S_l$ such that $u \cap v = \emptyset$ (otherwise they are clearly adjacent in $U(G^+)$). Assume without loss of generality that $i < l$. Let j, k be the largest and smallest index so that $v_j \cap u \neq \emptyset$ and $v_k \cap v \neq \emptyset$, respectively. Observe that $u \cap v = \emptyset$, $u \cap v_{i+1} \neq \emptyset$, and $v \cap v_{l-1} \neq \emptyset$ imply $i < j < l$ and $i < k < l$. Since u does not intersect v_{j+1}, it overlaps with v_j, i.e., G contains the arc (v_j, u) and likewise, since v does not intersect v_{k-1}, it overlaps with v_k, i.e., G contains the arc (v, v_k).

If $j < k$, then G^+ contains (v, v_k) and (v_k, v_j), and therefore (v, v_j). If $j \geqslant k$, then v_j is adjacent to both u and v, and since u, v are disjoint, v_j overlaps with u and v, i.e., G contains (v, v_j). In either case, the presence of (v, v_j) and (v_j, u) implies that G^+ contains (v, u). It follows that S forms a clique in $U(G^+)$.

Corollary 1 ($\star$). *There is an $O(n \log n)$-time algorithm that, given a bidirectional interval representation, computes a 2-approximation of an optimal proper coloring of the corresponding bidirectional interval graph.*

3   Coloring Mixed Interval Graphs

In this section, we show that computing the chromatic number of a mixed interval graph is NP-hard. Recall that the chromatic number can be computed efficiently for interval graphs [7], directed acyclic graphs [9], and directional interval graphs (Theorem 1). In other words, coloring interval graphs becomes NP-hard only if edges and arcs are combined in a non-directional way.

Theorem 2. *Given a mixed interval graph G and a number k, it is NP-complete to decide whether G admits a proper coloring with at most k colors.*

Proof. Containment in NP is clear since a specific coloring with k colors serves as a certificate of polynomial size. We prove NP-hardness by a polynomial-time reduction from 3-SAT. The high-level idea is as follows. We are given a 3-SAT formula Φ with variables $v_1, v_2, \ldots, v_n$, and clauses $c_1, c_2, \ldots, c_m$, where each clause contains at most three literals. A literal is a variable or a negated variable – we refer to them as a *positive* or a *negative* occurrence of that variable. From Φ, we construct in polynomial time a mixed interval graph G_Φ with the property that Φ is satisfiable if and only if G_Φ admits a proper coloring with $6n$ colors.

To prove that G_Φ is a mixed interval graph, we present an interval representation of $U(G_\Phi)$ and specify which pairs of intersecting intervals are connected by a directed arc, assuming that all other pairs of intersecting intervals are connected by an edge. The graph G_Φ has the property that the color of many of the intervals is fixed in every proper coloring with $6n$ colors. In our figures, the x-dimension corresponds to the real line that contains the interval, whereas we indicate its color by its position in the y-dimension – thus, we also refer to a color as a *layer*. In this model, our reduction has the property that Φ is satisfiable if and only if the intervals of G_Φ admit a drawing that fits into $6n$ layers.

Our construction consists of a *frame* and n *variable gadgets* and m *clause gadgets*. Each variable gadget is contained in a horizontal strip of height 6 that spans the whole construction, and each clause gadget is contained in a vertical strip of width 4 and height $6n$. The strips of the variable gadgets are pairwise disjoint, and likewise the strips of the clause gadgets are pairwise disjoint.

Frame. See Fig. 3c. The frame consists of six intervals $f_i^1, f_i^2, \ldots, f_i^6$ for each of the variables v_i, $i = 1, \ldots, n$. All of these intervals start at position 0 and extend from the left into the construction. The intervals f_i^2, f_i^4, f_i^6 end at position 1. The intervals f_i^1 and f_i^5 extend to the very right of the construction. Interval f_i^3 ends at position 3. Further, there are arcs (f_i^j, f_i^{j+1}) for $j = 1, \ldots, 5$ and (f_i^6, f_{i+1}^1) for $i = 1, \ldots, n-1$. This structure guarantees that any proper coloring with colors $\{1, 2, \ldots, 6n\}$ assigns color $6(i-1) + j$ to interval f_i^j.

Variable Gadget. See Figs. 3a and 3b. For each variable v_i, $i = 1, \ldots, n$, we have two intervals v_i^{false} and v_i^{true}, which start at position 2 and extend to the very right of the construction. Moreover, they both have an incoming arc from f_i^1 and an outgoing arc to f_i^5. This guarantees that they are drawn in the layers of f_i^2 and f_i^4, however their ordering can be chosen freely. We say that v_i is set to true if v_i^{true} is below v_i^{false}, and v_i is set to false otherwise.

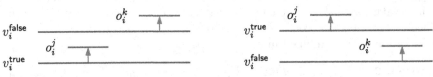

(a) v_i is true and appears positively in c_j, and negatively in c_k.

(b) v_i is false and appears positively in c_j, and negatively in c_k.

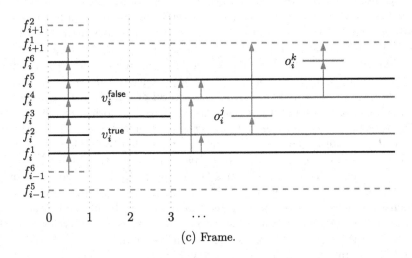

(c) Frame.

Fig. 3. A variable gadget for a variable v_i.

For each occurrence of v_i in a clause c_j, $j = 1, \ldots, m$, we create an interval o_i^j within the clause gadget of c_j. There is an arc $(v_i^{\mathsf{true}}, o_i^j)$ for a positive occurrence and an arc $(v_i^{\mathsf{false}}, o_i^j)$ for a negative occurrence as well as an arc (o_i^j, f_{i+1}^1) if $i < n$. This structure guarantees that o_i^j is drawn either in the same layer as f_i^3 or as f_i^6. However, drawing o_i^j in the layer of f_i^3 (which lies between v_i^{true} and v_i^{false}) is possible if and only if the chosen truth assignment of v_i satisfies c_j.

Clause Gadget. See Fig. 4. Our clause gadget starts at position $4j$, relative to which we describe the following positions. Consider a fixed clause c_j that contains variables v_i, v_k, v_ℓ. We create an interval s_j of length 3 starting at position 1. The key idea is that s_j can be drawn in the layer of f_i^6, f_k^6 or f_ℓ^6, but only if o_i^j, o_k^j or o_ℓ^j, each of which has length 1 and starts at position 3, is not drawn there. This is possible iff the corresponding variable satisfies the clause.

To ensure that s_j does not occupy any other layer, we block all the other layers. More precisely, for each variable v_z with $z \notin \{i, k, \ell\}$, we create

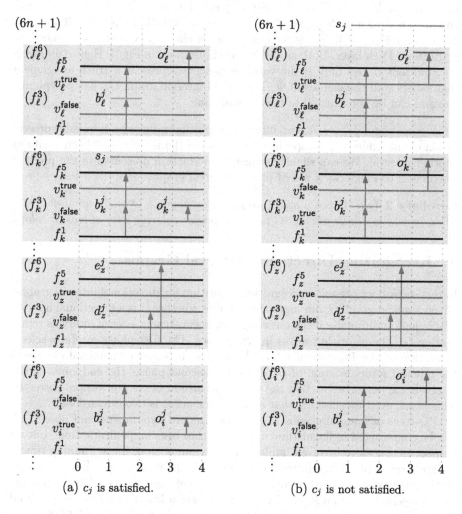

Fig. 4. A clause gadget for a clause $c_j = v_i \vee \neg v_k \vee v_\ell$, where $z \notin \{i, k, \ell\}$.

dummy intervals d_z^j, e_z^j of length 3 starting at position 1 that have arcs from f_z^1 and to f_{z+1}^1. These arcs force d_z^j, e_z^j to be drawn in the layers of f_z^3 and f_z^6, thereby ensuring that s_j is not placed in any layer associated with the variable z.

Similarly, for each $z \in \{i, k, \ell\}$, we create a blocker b_z^j of length 1 starting at position 1 that has arcs from f_z^1 and to f_z^5. This fixes b_z^j to the layer of f_z^3 (since the layers of f_z^2 and f_z^4 are occupied by v_z^{true} and v_z^{false}), thereby ensuring that, among all layers associated with v_z, s_j can only be drawn in the layer of f_z^6.

Correctness. Consider for each clause c_j with variables v_i, v_k, and v_ℓ the corresponding clause gadget. To achieve a total height of at most $6n$, s_j needs to be drawn in the same layer as some interval of the frame. Due to the presence of the dummy intervals, the only available layers are the ones of f_z^6 for $z \in \{i, k, \ell\}$. However, the layer of f_z^6 is only free if o_z^j is not there, which is the case if and only if o_z^j is drawn in the layer of f_z^3. By construction, this is possible if and only if the variable v_z is in the state that satisfies clause j. Otherwise we need an extra $(6n+1)$-th layer. Both situations are illustrated in Fig. 4. Hence, $6n$ layers are sufficient if and only if the variable gadgets represent a truth assignment that satisfies all the clauses of Φ. The mixed interval graph G_Φ has polynomial size and can be constructed in polynomial time.

A *proper interval graph* is an interval graph that admits an interval representation of the underlying graph in which none of the intervals properly contains another interval. We can slightly adjust the reduction presented in the proof of Theorem 2 to make G_Φ a *mixed* proper interval graph.

Corollary 2 ($\star$). *Given a mixed proper interval graph G and a number k, it is NP-complete to decide whether G admits a proper coloring with at most k colors.*

4   Recognizing Directional Interval Graphs

In this section we present a recognition algorithm for directional interval graphs. Given a mixed graph G, our algorithm decides whether G is a directional interval graph, and additionally if the answer is yes, it constructs a set of intervals representing G. The algorithm works in two phases. The first phase carefully selects a rotation of the PQ-tree of $U(G)$. This fixes the order of maximal cliques in the interval representation of $U(G)$. In the second phase, the endpoints of the intervals are perturbed so that the edges and arcs in G are represented correctly. This is achieved by checking that an auxiliary poset is two-dimensional.

PQ-trees of interval graphs [13] and realizers of two-dimensional posets [14] can be constructed in linear time. Our algorithm runs in quadratic time, but we suspect that a more involved implementation can achieve linear running time.

For a set of pairwise intersecting intervals on the real line, let the *clique point* be the leftmost point on the real line that lies in all the intervals. Given an interval representation of an interval graph G, we get a linear order of the maximal cliques of G by their clique points from left to right. Booth and Lueker [13] showed that a graph G is an interval graph if and only if the maximal cliques of G admit a *consecutive arrangement*, i.e., a linear order such that, for each vertex v, all the maximal cliques containing v occur consecutively in the order. They have also introduced a data structure called PQ-tree that encodes all possible consecutive arrangements of G. We present our algorithm in terms of modified PQ-trees (MPQ-trees, for short) as described by Korte and Möhring [11,12]. We briefly describe MPQ-trees in the next few paragraphs; see [12] for a proper introduction.

An *MPQ-tree* T of an interval graph G is a rooted, ordered tree with two types of nodes: P-nodes and Q-nodes, joined by links. Each node can have any number of children and a set of consecutive links joining a Q-node x with children is called a *segment* of x. Further, each vertex v in G is assigned either to one of the P-nodes, or to a segment of some Q-node. Based on this assignment, we *store* v in the links of T. If v is assigned to a P-node x, we store v in the link just above x in T (adding a dummy link above the root of T). If v is assigned to a segment of a Q-node x, we store v in each link of the segment. For a link $\{x, y\}$, let S_{xy} denote the set of vertices stored in $\{x, y\}$. We say that v is *above* (*below*, resp.) a node x if v is stored in any of the links on the upward path (in any of the links on some downward path, resp.) from x in T. We write A_x^T (B_x^T, resp.) for the set of all vertices in G that are above (below, resp.) node x.

The *frontier* of T is the sequence of the sets A_x^T, where x goes through all leaves in T in the order of T. Given an MPQ-tree T, one can obtain another MPQ-tree, which is called a *rotation* of T, by arbitrarily permuting the order of the children of P-nodes and by reversing the orders of the children of some Q-nodes. The defining property of the MPQ-tree T of a graph G is that each leaf x of T corresponds to a maximal clique A_x^T of G and the frontiers of rotations of T correspond bijectively to the consecutive arrangements of G. Observe that any two vertices adjacent in G are stored in links that are connected by an upward path in T. We say that T *agrees* with an interval representation $\mathcal{I}$ of G if the order of the maximal cliques of G given by their clique points in $\mathcal{I}$ from left to right is the same as in the frontier of T. We assume the following properties of the MPQ-tree (see [12], Lemma 2.2):

- For a P-node x with children $y_1, \ldots, y_k$, for every $i = 1, \ldots, k$, there is at least one vertex stored in link $\{x, y_i\}$ or below y_i, i.e., $S_{xy_i} \cup B_{y_i}^T \neq \emptyset$.
- For a Q-node x with children $y_1, \ldots, y_k$, we have $k \geqslant 3$. Further, for $S_i = S_{xy_i}$, we have:
  - $S_1 \cap S_k = \emptyset$, $B_{y_1}^T \neq \emptyset$, $B_{y_k}^T \neq \emptyset$, $S_1 \subsetneq S_2$, $S_k \subsetneq S_{k-1}$,
  - $(S_i \cap S_{i+1}) \setminus S_1 \neq \emptyset$, $(S_{i-1} \cap S_i) \setminus S_k \neq \emptyset$, for $i = 2, \ldots, k-1$.

A *partially ordered set*, or a *poset* for short, is a transitive directed acyclic graph. A poset P is *total* if, for every pair of vertices u and v, there is either an arc (u, v) or an arc (v, u) in P. We can conveniently represent a total poset P by a linear order of its vertices $v_1 < v_2 < \cdots < v_n$ meaning that there is an arc (v_i, v_j) for each $1 \leqslant i < j \leqslant n$. A poset P is *two-dimensional* if the arc set of P is the intersection of the arc sets of two total posets on the same set of vertices as P. McConnell and Spinrad [14] gave a linear-time algorithm that, given a directed graph D as input, decides whether D is a two-dimensional poset. If the answer is yes, the algorithm also constructs a *realizer*, that is, (in this case) two linear orders (R_1, R_2) on the vertex set of D such that

$$\text{arc } (u, v) \text{ is in } D \iff [(u < v \text{ in } R_1) \wedge (u < v \text{ in } R_2)].$$

The main result of this section is the following theorem.

Theorem 3 ($\star$). *There is an algorithm that, given a mixed graph G, decides whether G is a directional interval graph. The algorithm runs in $O\big(|V(G)|^2\big)$ time and produces a directional representation of G if G admits one.*

The algorithm runs in two phases that we introduce in separate lemmas.

Lemma 2 (Rotating PQ-trees). *There is an algorithm that, given a directional interval graph G, constructs an MPQ-tree T that agrees with some directional representation of G.*

Proof. Given a mixed graph G, if G is a directional interval graph, then clearly $U(G)$ is an interval graph and we can construct an MPQ-tree T of $U(G)$ in linear time using the algorithm by Korte and Möhring [12]. We call a rotation of T *directional* if it agrees with some directional representation of G. As we assume G to be a directional interval graph, there is at least one directional rotation $\tilde{T}$ of T, and our goal is to find some directional rotation of T. Our algorithm decides the rotation of each node in T independently.

Rotating Q-nodes. Let $y_1, \ldots, y_k$ be the children of a Q-node x in T. We are to decide whether to reverse the order of the children of x. Let $S_i = S_{xy_i}$, let $\ell = \max\{i : S_1 \cap S_i \neq \emptyset\}$, and let $u \in S_1 \cap S_\ell$. We have $\ell < k$, and there is some vertex $v \in (S_\ell \cap S_{\ell+1}) \setminus S_1$. This implies that u and v are assigned to overlapping segments of x. Thus, the intervals representing u and v overlap in every interval representation of $U(G)$. Hence, u and v are connected by an arc in G, and the direction of this arc determines the only possible rotation of x in any directional rotation of T, e.g., if (u, v) is an arc in G and the segment of u is to the right of the segment of v, then reverse the order of the children of x.

Rotating P-nodes. Let $y_1, \ldots, y_k$ be the children of a P-node x in T. For each $i = 1, \ldots, k$, let $B_i = S_{xy_i} \cup B_{y_i}^T$, and let $B = \bigcup_{i=1}^{k} B_i$. The properties of the MPQ-tree give us that (i) every vertex in A_x^T is adjacent in $U(G)$ to every vertex in B, (ii) none of the B_i is empty, and (iii) for any two vertices $b_i \in B_i$, $b_j \in B_j$ with $i \neq j$, we have that b_i and b_j are independent in G.

Assume that there is an arc (b_i, a) directed from some $b_i \in B_i$ to some $a \in A_x^T$. We claim that any rotation T' of T that does not put y_i as the first child of x is not directional. Assume the contrary. Let y_j, $j \neq i$ be the first child of x in T', let $\mathcal{I}$ be a directional representation that agrees with T', and let b_j be some vertex in B_j. The left endpoint of $\mathcal{I}(a)$ is to the right of the left endpoint of $\mathcal{I}(b_i)$ as (b_i, a) is an arc. The right endpoint of $\mathcal{I}(b_j)$ is to the left of the left endpoint of $\mathcal{I}(b_i)$ as T' puts y_j before y_i. Thus, $\mathcal{I}(b_j)$ and $\mathcal{I}(a)$ are disjoint, a contradiction.

Similarly, there are directed arcs from A_x^T to at most one set of type B_i. If there are any, the corresponding child y_i is in the last position in every directional rotation of T. Our algorithm rotates the child y_i (y_j) with an arc from B_i to A_x^T (from A_x^T to B_j) to the first (last) position, should such children exist, and leaves the other children as they are in T. It remains to show that the resulting rotation of T is directional; see the appendix of our full version [8].

Lemma 3 (Perturbing Endpoints). *There is an algorithm that, given an MPQ-tree T that agrees with some directional representation of a graph G, constructs a directional representation $\mathcal{I}$ of G such that T agrees with $\mathcal{I}$.*

Proof. The frontier of T yields a fixed order of maximal cliques $C_1, \ldots, C_k$ of G. Given this order, we construct the following auxiliary poset D. First, we add two independent chains of length $k+1$ each: vertices $a_1, \ldots, a_{k+1}$ with arcs (a_i, a_j) for $1 \leqslant i < j \leqslant k+1$, and vertices $b_1, \ldots, b_{k+1}$ with arcs (b_i, b_j) for $1 \leqslant i < j \leqslant k+1$. Then, for each vertex v in G, let $\mathrm{lc}(v)$ and $\mathrm{rc}(v)$ denote the indices of the leftmost and of the rightmost clique in which v is present, respectively. Now we add to D vertex v plus, for $1 \leqslant i \leqslant \mathrm{lc}(v)$, the arc (a_i, v) and, for $1 \leqslant i \leqslant \mathrm{rc}(v)$, the arc (b_i, v). Further, for each arc (u, v) in G, we add (u, v) to D. Lastly, for any two vertices u and v that are independent in G and that fulfill $\mathrm{rc}(u) < \mathrm{lc}(v)$, we add an arc (u, v) to D. We claim that G is a directional interval graph if and only if D is a two-dimensional poset.

First assume that G is a directional interval graph and fix a directional interval representation of G whose intervals all have distinct endpoints. For $i = 1, \ldots, k$, let L_i be the sequence of all the vertices v in G for which $\mathrm{lc}(v) = i$, in the order of their left endpoints. Similarly, let R_i be the sequence of all the vertices v in G for which $\mathrm{rc}(v) = i$, in the order of their right endpoints. The following two linear orders L and R of the vertices of D yield a realizer of D:

$$L = b_1 < b_2 < \ldots < b_k < a_1 < L_1 < a_2 < L_2 < \ldots < a_k < L_k < a_{k+1},$$
$$R = a_1 < a_2 < \ldots < a_k < b_1 < R_1 < b_2 < L_2 < \ldots < b_k < R_k < b_{k+1}.$$

Now, for the other direction, assume that we have a two-dimensional realizer of D. As b_{k+1} and a_1 are independent in D, we have that $b_{k+1} < a_1$ in exactly one of the orders in the realizer. We call this order L, and the other one R. As a_{k+1} and b_1 are independent in D and $b_1 < b_{k+1} < a_1 < a_{k+1}$ in L, we have that $a_{k+1} < b_1$ in R. For each $i = 1, \ldots, k$, define L_i as the sequence of vertices in G appearing between a_i and a_{i+1} in the order L. Similarly, let R_i be the sequence of vertices in G appearing between b_i and b_{i+1} in the order R. Observe that, for every vertex v, we have that $a_{\mathrm{lc}(v)} < v$ in D and that $a_{\mathrm{lc}(v)+1}$ and v are independent in D. As $a_{\mathrm{lc}(v)+1} \leqslant a_{k+1} < b_1 \leqslant b_{\mathrm{rc}(v)} < v$ in R, we have $v < a_{\mathrm{lc}(v)+1}$ in L. Thus, v is in $L_{\mathrm{lc}(v)}$ and, by a similar argument, v is in $R_{\mathrm{rc}(v)}$.

Now we are ready to construct a directional interval representation $\mathcal{I}$ of G. For each $i = 1, \ldots, k$, we select $|L_i|$ different real points in $(i - \frac{1}{2}, i)$ and $|R_i|$ different real points in $(i, i + \frac{1}{2})$. For a vertex v that appears on the i-th position in $L_{\mathrm{lc}(v)}$ and on the j-th position in $R_{\mathrm{rc}(v)}$, we choose the i-th point in $(\mathrm{lc}(v) - \frac{1}{2}, \mathrm{lc}(v))$ as the left endpoint, and the j-th point in $(\mathrm{rc}(v), \mathrm{rc}(v) + \frac{1}{2})$ as the right endpoint. Such a set of intervals is a directional interval representation of G. First, observe that any two intervals intersect if and only if they have a common clique. Next, if there is an arc (u, v) in G, then the arc (u, v) is also in D, $u < v$ holds both in L and in R, the corresponding intervals overlap, and $\mathcal{I}(u)$ starts and ends to the left of $\mathcal{I}(v)$. Last, if there is an edge $\{u, v\}$ in G, then u and v are independent in D, $u < v$ in one of the orders in the realizer, and $v < u$ in the other. Thus, one of the intervals $\mathcal{I}(u)$ and $\mathcal{I}(v)$ must contain the other.

Theorem 3 follows easily from Lemmas 2 and 3; see in our full version [8].

5 Open Problems

Can we recognize directional interval graphs in linear time? Can we recognize bidirectional interval graphs in polynomial time? Can we color bidirectional interval graphs optimally, or at least find α-approximate solutions with $\alpha < 2$?

References

1. Bang-Jensen, J., Huang, J., Zhu, X.: Completing orientations of partially oriented graphs. J. Graph Theory **87**(3), 285–304 (2018). https://doi.org/10.1002/jgt.22157
2. Binucci, C., Didimo, W.: Computing quasi-upward planar drawings of mixed graphs. Comput. J. **59**(1), 133–150 (2016). https://doi.org/10.1093/comjnl/bxv082
3. Binucci, C., Didimo, W., Patrignani, M.: Upward and quasi-upward planarity testing of embedded mixed graphs. Theor. Comput. Sci. **526**, 75–89 (2014). https://doi.org/10.1016/j.tcs.2014.01.015
4. Frati, F., Kaufmann, M., Pach, J., Tóth, C.D., Wood, D.R.: On the upward planarity of mixed plane graphs. J. Graph Algorithms Appl. **18**(2), 253–279 (2014). https://doi.org/10.7155/jgaa.00322
5. Furmańczyk, H., Kosowski, A., Żyliński, P.: A note on mixed tree coloring. Inf. Process. Lett. **106**(4), 133–135 (2008). https://doi.org/10.1016/j.ipl.2007.11.003
6. Furmańczyk, H., Kosowski, A., Żyliński, P.: Scheduling with precedence constraints: mixed graph coloring in series-parallel graphs. In: Proceedings of PPAM 2007, pp. 1001–1008 (2008). https://doi.org/10.1007/978-3-540-68111-3_106
7. Golumbic, M.C.: Algorithmic Graph Theory and Perfect Graphs. Academic Press, New York (1980). https://doi.org/10.1002/net.3230130214
8. Gutowski, G., Mittelstädt, F., Rutter, I., Spoerhase, J., Wolff, A., Zink, J.: Coloring mixed and directional interval graphs. arXiv report (2022). http://arxiv.org/abs/2208.14250
9. Hansen, P., Kuplinsky, J., de Werra, D.: Mixed graph colorings. Math. Methods Oper. Res. **45**, 145–160 (1997). https://doi.org/10.1007/BF01194253
10. Klavík, P., et al.: Extending partial representations of proper and unit interval graphs. Algorithmica **77**(4), 1071–1104 (2016). https://doi.org/10.1007/s00453-016-0133-z
11. Korte, N., Möhring, R.H.: Transitive orientation of graphs with side constraints. In: Proceedings of WG 1985, pp. 143–160 (1985)
12. Korte, N., Möhring, R.H.: An incremental linear-time algorithm for recognizing interval graphs. SIAM J. Comput. **18**(1), 68–81 (1989). https://doi.org/10.1137/0218005

13. Lueker, G.S., Booth, K.S.: A linear time algorithm for deciding interval graph isomorphism. J. ACM **26**(2), 183–195 (1979). https://doi.org/10.1145/322123.322125
14. McConnell, R.M., Spinrad, J.P.: Modular decomposition and transitive orientation. Discrete Math. **201**(1), 189–241 (1999). https://doi.org/10.1016/S0012-365X(98)00319-7
15. Ries, B., de Werra, D.: On two coloring problems in mixed graphs. Eur. J. Comb. **29**(3), 712–725 (2008). https://doi.org/10.1016/j.ejc.2007.03.006
16. Sotskov, Y.N.: Mixed graph colorings: a historical review. Mathematics **8**(3), 385 (2020). https://doi.org/10.3390/math8030385
17. Sotskov, Y.N., Tanaev, V.S.: Chromatic polynomial of a mixed graph. Vestsi Akademii Navuk BSSR. Seryya Fizika-Matematychnykh Navuk **6**, 20–23 (1976)
18. Sugiyama, K., Tagawa, S., Toda, M.: Methods for visual understanding of hierarchical system structures. IEEE Trans. Syst. Man Cybern. **11**(2), 109–125 (1981). https://doi.org/10.1109/TSMC.1981.4308636
19. Zink, J., Walter, J., Baumeister, J., Wolff, A.: Layered drawing of undirected graphs with generalized port constraints. Comput. Geom. **105–106**(101886), 1–29 (2022). https://doi.org/10.1016/j.comgeo.2022.101886

Outside-Obstacle Representations with All Vertices on the Outer Face

Oksana Firman[1]([✉])[iD], Philipp Kindermann[2][iD], Jonathan Klawitter[3][iD], Boris Klemz[1][iD], Felix Klesen[1][iD], and Alexander Wolff[1][iD]

[1] Institut für Informatik, Universität Würzburg, Würzburg, Germany
{oksana.firman,jonathan.klawitter,boris.klemz,
felix.klesen}@uni-wuerzburg.de
[2] Informatikwissenschaften, Universität Trier, Trier, Germany
kindermann@uni-trier.de
[3] University of Auckland, Auckland, New Zealand
jo.klawitter@gmail.com

Abstract. An *obstacle representation* of a graph G consists of a set of polygonal obstacles and a drawing of G as a *visibility graph* with respect to the obstacles: vertices are mapped to points and edges to straight-line segments such that each edge avoids all obstacles whereas each non-edge intersects at least one obstacle. Obstacle representations have been investigated quite intensely over the last few years. Here we focus on *outside-obstacle representations* (OORs) that use only one obstacle in the outer face of the drawing. It is known that every outerplanar graph admits such a representation [Alpert, Koch, Laison; DCG 2010].

We strengthen this result by showing that every (partial) 2-tree has an OOR. We also consider restricted versions of OORs where the vertices of the graph lie on a convex polygon or a regular polygon. We characterize when the complement of a tree and when a complete graph minus a simple cycle admits a convex OOR. We construct regular OORs for all (partial) outerpaths, cactus graphs, and grids.

Keywords: Obstacle representation · Visibility graph · Outside obstacle

1 Introduction

Recognizing graphs that have a certain type of geometric representation is a well-established field of research dealing with, e.g., geometric intersection graphs, visibility graphs, and graphs admitting certain contact representations. Given a set $\mathcal{C}$ of *obstacles* (here, simple polygons without holes) and a set P of points in the plane, the *visibility graph* $G_{\mathcal{C}}(P)$ has a vertex for each point in P and an edge pq for any two points p and q in P that can *see* each other, that is, the line segment $\overline{pq}$ connecting p and q does not intersect any obstacle in $\mathcal{C}$. An *obstacle representation* of a graph G consists of a set $\mathcal{C}$ of obstacles in the plane and a mapping of the vertices of G to a set P of points such that $G = G_{\mathcal{C}}(P)$. The

© The Author(s), under exclusive license to Springer Nature Switzerland AG 2023
P. Angelini and R. von Hanxleden (Eds.): GD 2022, LNCS 13764, pp. 432–440, 2023.
https://doi.org/10.1007/978-3-031-22203-0_31

mapping defines a straight-line drawing Γ of $G_{\mathcal{C}}(P)$. We planarize Γ by replacing all intersection points by dummy vertices. The outer face of the resulting planar drawing is a closed polygonal chain Π_Γ where vertices and edges can occur several times. We call the complement of the closure of Π_Γ the *outer face* of Γ. We differentiate between two types of obstacles: *outside* obstacles lie in the outer face of the drawing, and *inside* obstacles lie in the complement of the outer face; see Fig. 1a.

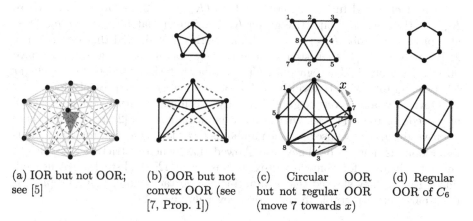

(a) IOR but not OOR; see [5]

(b) OOR but not convex OOR (see [7, Prop. 1])

(c) Circular OOR but not regular OOR (move 7 towards x)

(d) Regular OOR of C_6

Fig. 1. Inside- and outside-obstacle representations (IORs and OORs). (Color figures are available online.)

Every graph trivially admits an obstacle representation: take an arbitrary straight-line drawing without collinear vertices and "fill" each face with an obstacle. This, however, can lead to a large number of obstacles, which motivates the optimization problem of finding an obstacle representation with the minimum number of obstacles. For a graph G, the *obstacle number* obs(G) is the smallest number of obstacles that suffice to represent G as a visibility graph.

In this paper, we focus on *outside* obstacle representations (OORs), that is, obstacle representations with a single outside obstacle and without any inside obstacles. For such a representation, it suffices to specify the positions of the vertices; the outside obstacle is simply the whole outer face of the representation. In an OOR every non-edge must thus intersect the outer face. We also consider three special types: In a *convex* OOR, the vertices must be in convex position; in a *circular* OOR, the vertices must lie on a circle; and in a *regular* OOR, the vertices must form a regular n-gon.

In general, the class of graphs representable by outside obstacles is not closed under taking subgraphs, but the situation is different for graphs admitting a *reducible* OOR, meaning that all of its edges are incident to the outer face:

Observation 1. *If a graph G admits a reducible OOR, then every subgraph of G also admits such a representation.*

Previous Work. Alpert et al. [1] introduced the notion of the obstacle number of a graph in 2010. They also introduced *inside* obstacle representations, i.e., representations without an outside obstacle. They characterized the class of graphs that have an inside obstacle representation with a single convex obstacle and showed that every outerplanar graph has an OOR. Chaplick et al. [5] proved that the class of graphs with an inside obstacle representation is incomparable with the class of graphs with an OOR. They showed that any graph with at most seven vertices has an OOR, which does not hold for a specific 8-vertex graph.

Alpert et al. [1] further showed that $\text{obs}(K_{m,n}^*) \leq 2$ for any $m \leq n$, where $K_{m,n}^*$ is the complete bipartite graph $K_{m,n}$ minus a matching of size m. They also proved that $\text{obs}(K_{5,7}^*) = 2$. Pach and Sarıöz [10] showed that $\text{obs}(K_{5,5}^*) = 2$. Berman et al. [4] suggested some necessary conditions for a graph to have obstacle number 1. They gave a SAT formula that they used to find a *planar* 10-vertex graph (with treewidth 4) that has no 1-obstacle representation.

Obviously, any n-vertex graph has obstacle number $\mathcal{O}(n^2)$. Balko et al. [3] improved this to $\mathcal{O}(n \log n)$. On the other hand, Balko et al. [2] showed that there are n-vertex graphs whose obstacle number is $\Omega(n/\log\log n)$, improving previous lower bounds, e.g., [1,6]. They also showed that, when restricting obstacles to *convex* polygons, for some n-vertex graphs, even $\Omega(n)$ obstacles are needed. Furthermore, they showed that computing the obstacle number of a graph G is fixed-parameter tractable in the vertex cover number of G.

Our Contribution. We first strengthen the result of Alpert et al. [1] regarding OORs of outerplanar graphs by showing that every (partial) 2-tree admits a reducible OOR with all vertices on the outer face; see Sect. 2. Equivalently, every graph of treewidth at most two, which includes outerplanar and series-parallel graphs, admits such a representation. Then we establish two combinatorial conditions for convex OORs (see Sect. 3). In particular, we introduce a necessary condition that can be used to show that a given graph does *not* admit a convex OOR as, e.g., the graph in Fig. 1b. We apply these conditions to characterize when the complement of a tree and when a complete graph minus a simple cycle admits a convex OOR. We construct *regular* reducible OORs for all outerpaths, grids, and cacti; see Sect. 4. The result for grids strengthens an observation by Dujmović and Morin [6, Fig. 1], who showed that grids have (outside) obstacle number 1.

The complete proofs of our claims are in the full version [7].

Notation. For a graph G, let $V(G)$ be the vertex set of G, and let $E(G)$ be the edge set of G. Arranging the vertices of G in circular order $\sigma = \langle v_1, \ldots, v_n \rangle$, we write, for $i \neq j$, $[v_i, v_j)$ to refer to the sequence $\langle v_i, v_{i+1}, \ldots, v_{j-1} \rangle$, where indices are interpreted modulo n. Sequences (v_i, v_j) and $[v_i, v_j]$ are defined analogously.

2   Outside-Obstacle Representations for Partial 2-Trees

The graph class of *2-trees* is recursively defined as follows: K_3 is a 2-tree. Further, any graph is a 2-tree if it is obtained from a 2-tree G by introducing a new vertex x and making x adjacent to the endpoints of some edge uv in G. We say that x is *stacked* on uv. The edges xu and xv are called the *parent edges* of x.

Theorem 1. *Every 2-tree admits a reducible OOR with all vertices on the outer face.*

Proof sketch. Every 2-tree T can be constructed through the following iterative procedure: (1) Start with one edge, called the *base* edge and mark its vertices as *inactive*. Stack any number of vertices onto the base edge and mark them as *active*. During the entire procedure, every present vertex is marked either as active or inactive. Moreover, once a vertex is inactive, it remains inactive for the remainder of the construction. (2) Pick one active vertex v and stack any number of vertices onto each of its two parent edges. All the new vertices are marked as active and v as inactive. (3) If there are active vertices remaining, repeat step (2). We construct a drawing of T by geometrically implementing this iterative procedure, so that after every step of the algorithm the present part of the graph is realized as a straight-line drawing satisfying the following invariants:

(i) Each vertex v not incident to the base edge is associated with an open circular arc C_v that lies completely in the outer face and whose endpoints belong to the two parent edges of v. Moreover, v is located at the center of C_v and the parent edges of v are below v.

(ii) Each non-edge intersects the circular arc of at least one of its incident vertices.

(iii) For each active vertex v, the region R_v enclosed by C_v and the two parent edges of v is *empty*, meaning that R_v is not intersected by any edges, vertices, or circular arcs.

(iv) Every vertex is incident to the outer face.

It is easy to see that once the procedure terminates with a drawing that satisfies invariants (i)–(iv), we obtain the desired representation (in particular, invariants (i) and (ii) together imply that each non-edge intersects the outer face).

Construction. To carry out step (1), we draw the base edge horizontally and place the stacked vertices on a common horizontal line above the base edge, see Fig. 2a. Circular arcs that satisfy the invariants are now easy to define. Suppose we have obtained a drawing Γ of the graph obtained after step (1) and some number of iterations of step (2) such that Γ is equipped with a set of circular arcs satisfying the invariants (i)–(iv). We describe how to carry out another iteration of step (2) while maintaining the invariants. Let v be an active vertex. By invariant (i), both parent edges of v are below v. Let e_ℓ and e_r be the left and right parent edge, respectively. Let $\ell_1, \ell_2, \ldots, \ell_i$ and $r_1, r_2, \ldots, r_j$ be the vertices stacked onto e_ℓ and e_r, respectively. We refer to $\ell_1, \ell_2, \ldots, \ell_i$ and $r_1, r_2, \ldots, r_j$ as the *new* vertices; the vertices of Γ are called *old*. We place all the new vertices on a common horizontal line h that intersects R_v above v, see Fig. 2b. The vertices $\ell_1, \ell_2, \ldots, \ell_i$ are placed inside R_v, to the right of the line $\overline{e_\ell}$ extending e_ℓ. Symmetrically, $r_1, r_2, \ldots, r_j$ are placed inside R_v, to the left of the line $\overline{e_r}$ extending e_r.

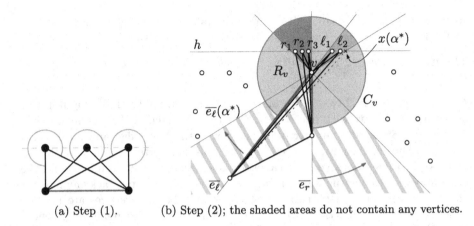

(a) Step (1).   (b) Step (2); the shaded areas do not contain any vertices.

Fig. 2. Construction steps in the proof of Theorem 1.

We place $\ell_1, \ell_2, \ldots, \ell_i$ close enough to e_ℓ and $r_1, r_2, \ldots, r_j$ close enough to e_r such that the following properties are satisfied: (a) None of the parent edges of the new vertices intersect C_v. (b) For each new vertex, the unbounded open cone obtained by extending its parent edges to the bottom does not contain any vertices.

Each of the old vertices retains its circular arc from Γ. By invariants (i) and (iii) for Γ, it is easy to define circular arcs for the new vertices that satisfy invariant (i). Using invariants (i)–(iv) for Γ and properties (a) and (b), it can be shown that all invariants are satisfied.   □

3   Convex Outside Obstacle Representations

We start with a sufficient condition. Suppose that we have a convex OOR Γ of a graph G. Let σ be the clockwise circular order of the vertices of G along the convex hull. If all neighbors of a vertex v of G are consecutive in σ, we say that v has the *consecutive-neighbors property*, which implies that all non-edges incident to v are consecutive around v and trivially intersect the outer face in the immediate vicinity of v; see Fig. 3a.

Lemma 1 (Consecutive-neighbors property). *A graph G admits a convex OOR with circular vertex order σ if there is a subset V' of $V(G)$ that covers all non-edges of G and each vertex of V' has the consecutive-neighbors property with respect to σ.*

Next, we derive a necessary condition. For any two consecutive vertices v and v' in σ that are not adjacent in G, we say that the line segment $g = \overline{vv'}$ is a *gap*. Then the *gap region* of g is the inner face of $\Gamma + vv'$ incident to g; see the gray region in Fig. 3b. We consider the gap region to be open, but add to it the relative interior of the line segment $\overline{vv'}$, so that the non-edge vv' intersects

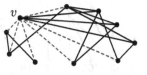
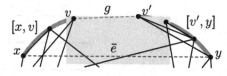

| (a) Vertex v has the CNP | (b) Gap g is a candidate gap for the non-edge $\bar{e}$. |

Fig. 3. Examples for the consecutive-neighbors property (CNP) and a candidate gap.

its own gap region. Observe that each non-edge $\bar{e} = xy$ that intersects the outer face has to intersect some gap region in an OOR. Suppose that g lies between x and y with respect to σ, that is, $[v, v'] \subseteq [x, y]$. We say that g is a *candidate gap* for $\bar{e}$ if there is no edge that connects a vertex in $[x, v]$ and a vertex in $[v', y]$. Note that $\bar{e}$ can only intersect gap regions of candidate gaps.

Lemma 2 (Gap condition). *A graph G admits a convex OOR with circular vertex order σ only if there exists a candidate gap with respect to σ for each non-edge of G.*

It remains an open problem whether the gap condition is also sufficient. Nonetheless, we can use the gap condition for no-certificates. To this end, we derived a SAT formula from the following expression, which checks the gap condition for every non-edge of a graph G:

$$\bigwedge_{xy \notin E(G)} \left[\bigvee_{v \in [x,y]} \left(\bigwedge_{u \in [x,v], w \in (v,y)} uw \notin E(G) \right) \vee \bigvee_{v \in [y,x]} \left(\bigwedge_{u \in [y,v], w \in (v,x)} uw \notin E(G) \right) \right]$$

We have used this formula to test whether all connected cubic graphs with up to 16 vertices admit convex OORs. The only counterexample we found was the Petersen graph. The so-called Blanuša snarks, the Pappus graph, the dodecahedron, and the generalized Peterson graph $G(11, 2)$ satisfy the gap condition. The latter three graphs do admit convex OORs [8].

The smallest graph (and the only 6-vertex graph) that does not satisfy the gap condition is the wheel graph W_6 [7, Prop. 1]. Hence, W_6 does not admit a *convex* OOR, but it does admit a (non-convex) OOR; see Fig. 1b.

In the following, we consider "dense" graphs, namely the complements of trees. For any graph G, let $\bar{G} = (V(G), \bar{E}(G))$ with $\bar{E}(G) = \{uv \mid uv \notin E(G)\}$ be the complement of G. A *caterpillar* is a tree where all vertices are within distance at most 1 of a central path.

Theorem 2. *For any tree T, the graph $\bar{T}$ has a convex OOR if and only if T is a caterpillar.*

Proof sketch. First, we show that for every caterpillar C, the graph $\bar{C}$ has a circular OOR. To this end, we arrange the vertices of the central path P on a circle in the order given by P. Then, for each vertex of P, we insert its leaves as

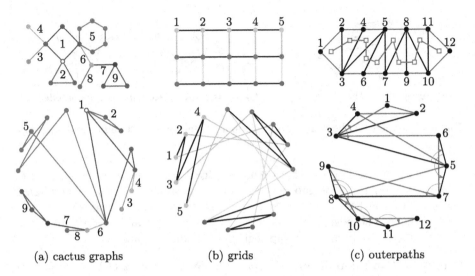

(a) cactus graphs   (b) grids   (c) outerpaths

Fig. 4. Graph classes that admit reducible regular OORs (see Theorem 4).

an interval next to it. The result is a circular OOR since every non-edge of $\bar{C}$ intersects the outer face in the vicinity of the incident path vertex (or vertices). Second, we show that if T is a tree that is not a caterpillar, then for any circular vertex order, there exists at least one non-edge of $\bar{T}$ that is a diagonal of a quadrilateral formed by edges of $\bar{T}$. □

Another class of dense graphs consists of complete graphs from which we remove the edge set of a simple (not necessarily Hamiltonian) cycle. Using Lemma 2, we can prove the following theorem similarly as Theorem 2.

Theorem 3. *Let $3 \le k \le n$. Then the graph $G_{n,k} = K_n - E(C_k)$, where C_k is a simple k-cycle, admits a convex OOR if and only if $k \in \{3, 4, n\}$.*

4   Regular Outside Obstacle Representations

This section deals with regular OORs. A *cactus* is a connected graph where every edge is contained in at most one simple cycle. An *outerpath* is a graph that admits an *outerpath* drawing, i.e., an outerplanar drawing whose weak dual is a path. A *grid* is the Cartesian product $P_k \square P_\ell$ of two simple paths P_k, P_ℓ.

Theorem 4. *The following graphs have reducible regular OORs:*
1. every cactus; 2. every grid; 3. every outerpath.

Proof sketch. For cacti, we use a decomposition into *blocks* (i.e., maximal 2-connected subgraphs or bridges). We start with an arbitrary block and insert its child blocks as intervals next to the corresponding cut vertices etc.; see Fig. 4a.

For a grid, we lay out each horizontal path in a separate arc, in a zig-zag manner. Then we add the vertical edges accordingly; see Fig. 4b. Our strategy for (maximal) outerpaths relies on a specific stacking order. We start with a triangle. Then we always place the next inner edge (black in Fig. 4c) such that it avoids the empty arc that corresponds to the previous inner edge.   □

Every graph with up to six vertices – except for the graph in Fig. 1b – and every outerplanar graph with up to seven vertices admits a regular OOR (see [7, Prop. 1] and [9], respectively). The 8-vertex outerplanar graph in Fig. 1c (and only it [9]), however, does not admit any regular OOR [7, Prop. 2].

Our representations for cacti, outerpaths, and complements of caterpillars depend only on the vertex order. Hence, given such a graph with n vertices, every cocircular point set of size n is *universal*, i.e., can be used for an OOR.

5   Open Problems

(1) What is the complexity of deciding whether a given graph admits an OOR? (2) Is the gap condition sufficient, i.e., does every graph with a circular vertex order satisfying the gap condition admit a convex OOR? (3) Does every graph that admits a *convex* OOR also admit a *circular* OOR? (4) Does every outerplanar graph admit a (reducible) convex OOR? (5) Does every connected cubic graph *except the Peterson graph* admit a convex OOR?

References

1. Alpert, H., Koch, C., Laison, J.D.: Obstacle numbers of Graphs. Discrete Comput. Geom. **44**(1), 223–244 (2009). https://doi.org/10.1007/s00454-009-9233-8
2. Balko, M., Chaplick, S., Gupta, S., Hoffmann, M., Valtr, P., Wolff, A.: Bounding and computing obstacle numbers of graphs. In: Chechik, S., Navarro, G., Rotenberg, E., Herman, G. (eds.) Proceedings of 30th Annual European Symposium on Algorithms (ESA 2022). LIPIcs, vol. 244, pp. 11:1–11:13. Schloss Dagstuhl – Leibniz-Zentrum für Informatik (2022). https://doi.org/10.4230/LIPIcs.ESA.2022.11
3. Balko, M., Cibulka, J., Valtr, P.: Drawing graphs using a small number of obstacles. Discrete Comput. Geom. **59**(1), 143–164 (2018)
4. Berman, L.W., Chappell, G.G., Faudree, J.R., Gimbel, J., Hartman, C., Williams, G.I.: Graphs with obstacle number greater than one. J. Graph Algorithms Appl. **21**(6), 1107–1119 (2017). https://doi.org/10.7155/jgaa.00452
5. Chaplick, S., Lipp, F., Park, J., Wolff, A.: Obstructing visibilities with one obstacle. In: Hu, Y., Nöllenburg, M. (eds.) GD 2016. LNCS, vol. 9801, pp. 295–308. Springer, Cham (2016). https://doi.org/10.1007/978-3-319-50106-2_23
6. Dujmović, V., Morin, P.: On obstacle numbers. Electr. J. Combin. **22**(3), 3.1–7 (2015). https://doi.org/10.37236/4373
7. Firman, O., Kindermann, P., Klawitter, J., Klemz, B., Klesen, F., Wolff, A.: Outside-obstacle representations with all vertices on the outer face. Arxiv report (2022). http://arxiv.org/abs/2202.13015

8. Goldschmied, C.: 1-Hindernis-Sichtbarkeitsgraphen von kubischen Graphen. Bachelor's thesis, Institut für Informatik, Univ. Würzburg (2021). http://www1.pub.informatik.uni-wuerzburg.de/theses/2021-goldschmied-bachelor.pdf

9. Lang, L.: Regelmäßige Außenhindernisrepräsentation von kleinen planaren Graphen. Bachelor's thesis, Inst. für Informatik, Univ. Würzburg (2022). http://www1.pub.informatik.uni-wuerzburg.de/theses/2022-lang-bachelor.pdf

10. Pach, J., Sarıöz, D.: On the structure of graphs with low obstacle number. Graphs Combin. **27**(3), 465–473 (2011). https://doi.org/10.1007/s00373-011-1027-0

Arrangements of Pseudocircles: On Digons and Triangles

Stefan Felsner[✉][iD], Sandro Roch[✉][iD], and Manfred Scheucher[✉][iD]

Institut für Mathematik, Technische Universität Berlin, Berlin, Germany
{felsner,roch,scheucher}@math.tu-berlin.de

Abstract. In this article, we study the cell-structure of simple arrangements of pairwise intersecting pseudocircles. The focus will be on two problems from Grünbaum's monograph from the 1970's.

First, we discuss the maximum number of digons or touching points. Grünbaum conjectured that there are at most $2n-2$ digon cells or equivalently at most $2n-2$ touchings. Agarwal et al. (2004) verified the conjecture for cylindrical arrangements. We show that the conjecture holds for any arrangement which contains three pseudocircles that pairwise form a touching. The proof makes use of the result for cylindrical arrangements. Moreover, we construct non-cylindrical arrangements which attain the maximum of $2n - 2$ touchings and have no triple of pairwise touching pseudocircles.

Second, we discuss the minimum number of triangular cells (triangles) in arrangements without digons and touchings. Felsner and Scheucher (2017) showed that there exist arrangements with only $\lceil \frac{16}{11} n \rceil$ triangles, which disproved a conjecture of Grünbaum. Here we provide a construction with only $\lceil \frac{4}{3} n \rceil$ triangles. A corresponding lower bound was obtained by Snoeyink and Hershberger (1991).

Keywords: Arrangement of pseudocircles · Touching · Empty lense · Cylindrical arrangement · Arrangement of pseudoparabolas · Grünbaum's conjecture

1 Introduction

An *arrangement* $\mathcal{A}$ *of pairwise intersecting pseudocircles* is a collection of $n(\mathcal{A})$ simple closed curves on the sphere or plane such that any two of the curves either touch in a single point or intersect in exactly two points where they cross. Throughout this article, we consider all arrangements to be *simple*, that is, no three pseudocircles meet in a common point. An arrangement $\mathcal{A}$ partitions the

A part of this work was initiated at a workshop of the collaborative DACH project *Arrangements and Drawings* in Gathertown. We thank the organizers and all the participants for the inspiring atmosphere. S. Roch was funded by the DFG-Research-Training-Group 'Facets of Complexity' (DFG-GRK 2434). M. Scheucher was supported by the DFG Grant SCHE 2214/1-1.

P. Angelini and R. von Hanxleden (Eds.): GD 2022, LNCS 13764, pp. 441–455, 2023.
https://doi.org/10.1007/978-3-031-22203-0_32

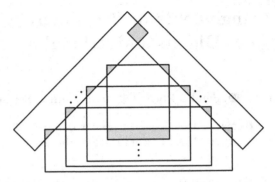

Fig. 1. An illustration of the construction by Grünbaum [6, Figure 3.28]: an arrangement of $n \geq 4$ pairwise intersecting pseudocircles with exactly $2n - 2$ digons. Digons are highlighted gray. (Color figure online)

plane into cells. A cell with exactly k crossings on its boundary is a k-*cell*, 2-cells are also called *digons* and 3-cells are *triangles*. The number of k-cells of an arrangement $\mathcal{A}$ is denoted as $p_k(\mathcal{A})$.

The study of cells in arrangements started about 100 years ago when Levi [7] showed that, in an arrangement of at least three pseudolines in the projective plane, every pseudoline is incident to at least three triangles. In the 1970's, Grünbaum [6] intensively investigated arrangements of pseudolines and initiated the study of arrangements of pseudocircles.

1.1 Digons and Touchings

Concerning digons in arrangements of pairwise intersecting pseudocircles, Grünbaum [6] presented a construction with $2n - 2$ digons (depicted in Fig. 1) and conjectured that these arrangements have the maximum number of digons[1].

Conjecture 1 (Grünbaum's digon conjecture [6, Conjecture 3.6]). Every simple arrangement $\mathcal{A}$ of n pairwise intersecting pseudocircles has at most $2n-2$ digons, i.e., $p_2 \leq 2n - 2$.

It was shown by Agarwal et al. [1, Corollary 2.12] that Conjecture 1 holds for simple cylindrical arrangements. An intersecting arrangement of pseudocircles is *cylindrical* if there is a pair of cells which are separated by each pseudocircle of the arrangement. More specifically, they showed that the number of touchings in an intersecting arrangement of n pseudo-parabolas is at most $2n - 4$ [1, Theorem 2.4]. An *intersecting arrangement of pseudoparabolas* is a collection of infinite x-monotone curves, called *pseudoparabolas*, where each pair of them either have a single touching or intersect in exactly two points where they cross. Every cylindrical arrangement of pseudocircles can be represented as an arrangement

---

[1] Originally the conjecture was stated as to include non-simple arrangements which are *non-trivial*, i.e., non-simple arrangements with at least 3 crossing points.

Fig. 2. Contracting some of the digons to touchings.

of pseudoparabolas and vice versa. From an arrangement of pseudoparabolas one can directly obtain a drawing of an arrangement of pseudocircles on the lateral surface of a cylinder so that the pseudocircles wrap around the cylinder. The two separating cells correspond to the top and the bottom of the cylinder.

Agarwal et al. [1, Theorem 2.13] showed for intersecting arrangements of pseudocircles that the number of digons is at most linear in n. The proof is based on the fact that every arrangement of intersecting pseudocircles can be stabbed by constantly many points. That is, there exists an absolute constant k, called the *stabbing number*[2], such that for every arrangement of n pseudocircles in the plane there exists a set of k points with the property that each pseudocircle contains at least one of the points in its interior [1, Corollary 2.8]. Therefore, the arrangement can be decomposed into constantly many cylindrical subarrangements. The linear upper bound then follows from the fact that each pair of subarrangements contributes at most linearly many digons. In [5] we verified Grünbaum's digon conjecture for up to 7 pseudocircles.

Here we show that Grünbaum's digon conjecture (Conjecture 1) holds for arrangements which contain three pseudocircles that pairwise form a digon. Before we state the result as a theorem, let us introduce some notation. For an arrangement $\mathcal{A}$ of pseudocircles and any selection of its digons, we can perform a perturbation so that the selected digons become touching points. Figure 2 gives an illustration. It is therefore sufficient to find an upper bound on the number of touchings to prove Grünbaum's digon conjecture. We define the *touching graph* $T(\mathcal{A})$ to have the pseudocircles of $\mathcal{A}$ as vertices, and two vertices form an edge if the two corresponding pseudocircles touch.

Theorem 1. *Let $\mathcal{A}$ be a simple arrangement of n pairwise intersecting pseudocircles. If the touching graph $T(\mathcal{A})$ contains a triangle, then there are at most $2n - 2$ touchings, i.e., $p_2 \leq 2n - 2$.*

Theorem 1 in particular shows that Grünbaum's construction with $2n - 2$ touchings is maximal for arrangements with triangles in the touching graph. However, the maximum number of touchings in general arrangements remains unknown. In Sect. 3 we construct a family of arrangements of n pseudocircles

---

[2] In the literature, the stabbing number is also referred to as *piercing number* or *transversal number*.

which have exactly $2n - 2$ touchings and a triangle free touching graph. This family witnesses that the conjectured upper bound (Conjecture 1) can also be achieved in the cases not covered by Theorem 1.

Proposition 1. *For $n \in \{11, 14, 15\}$ and $n \geq 17$ there exists a simple arrangement $\mathcal{A}_n$ of n pairwise intersecting pseudocircles with no triangle in the touching graph $T(\mathcal{A}_n)$ and with exactly $p_2(\mathcal{A}_n) = 2n - 2$ touchings.*

1.2 Triangles in Digon-Free Arrangements

In this context we assume that all arrangements are digon- and touching-free. It was shown by Levi [7] that every arrangement of n pseudolines in the projective plane contains at least n triangles. Since arrangements of pseudolines are in correspondence with arrangements of *great-pseudocircles* (see e.g. [4, Section 4]), it follows directly that an arrangement of n great-pseudocircles contains at least $2n$ triangles, i.e., $p_3 \geq 2n$.

Grünbaum conjectured that every digon-free intersecting arrangement on n pseudocircles contains at least $2n - 4$ triangles [6, Conjecture 3.7]. Snoeyink and Hershberger [10] proved a sweeping lemma for arrangements of pseudocircles. Using this powerful tool, they concluded that in every digon-free intersecting arrangement every pseudocircle has two triangles on each of its two sides (interior and exterior). This immediately implies the lower bound $p_3(\mathcal{A}) \geq 4n/3$; see Sect. 4.2 in [10].

In [5] we constructed an infinite family of digon-free arrangements with $p_3 < \frac{16}{11}n$ which shows that Grünbaum's conjecture is wrong and verified that the lower bound $p_3 \geq 4n/3$ by Snoeyink and Hershberger is tight for $6 \leq n \leq 14$. Here we show that their bound is tight for all $n \geq 6$:

Theorem 2. *For every $n \geq 6$, there exists a simple digon-free arrangement $\mathcal{A}_n$ of n pairwise intersecting pseudocircles with $p_3(\mathcal{A}_n) = \lceil \frac{4}{3}n \rceil$ triangles. Moreover, these arrangements are cylindrical.*

All arrangements constructed in Sect. 4 contain a specific arrangement $\mathcal{A}_6$ (depicted on the left of Fig. 11) as a subarrangement. This remarkable arrangement has been studied as the arrangement $\mathcal{N}_6^\Delta$ in [4] where it was shown that $\mathcal{N}_6^\Delta$ is *non-circularizable*, i.e., $\mathcal{N}_6^\Delta$ cannot be represented by an arrangement of proper circles. As a consequence, all arrangements constructed in Sect. 4 are as well non-circularizable. In fact, all known counter-examples to Grünbaum's triangle conjecture contain $\mathcal{N}_6^\Delta$ and are therefore non-circularizable. Hence, Grünbaum's conjecture may still be true when restricted to arrangements of proper circles.

Conjecture 2 (Weak Grünbaum triangle conjecture, [5, Conjecture 2.2]). Every simple digon-free arrangement $\mathcal{A}$ of n pairwise intersecting circles has at least $2n - 4$ triangles.

1.3   Related Work and Discussion

In the proof of Theorem 1 we make use of a triangle (K_3) in the touching graph to bound the number of digons in the arrangement. It would be interesting whether other subgraphs like C_4 or $K_{3,3}$ can also be used to bound the number of digons.

The focus of this article is on arrangements of pairwise intersecting pseudocircles. For the setting of arrangements, where pseudocircles do not necessarily pairwise intersect, a classical construction of Erdős [3] gives arrangements of n unit circles with $\Omega(n^{1+c/\log\log n})$ touchings. An upper bound of $O(n^{3/2+\epsilon})$ on the number of digons in *circle* arrangements was shown by Aronov and Sharir [2]. The precise asymptotics, however, remain unknown. Moreover, we are not aware of an upper bound for *pseudocircles*.

Problem 1. Determine the maximum number of touchings among all simple arrangements of n circles and pseudocircles, respectively.

It is also worth noting that, for the very restrictive setting of arrangements of n pairwise intersecting unit-circles, Pinchasi showed an upper bound of $p_2 \leq n+3$ [8, Lemma 3.4 and Corollary 3.10].

Concerning arrangements with digons, the number of triangles behaves different than in digon-free arrangements. While our best lower bound so far is $p_3 \geq 2n/3$, we managed to verify that $p_3 \geq n-1$ is a tight lower bound for $3 \leq n \leq 7$ using a computer-assisted exhaustive enumeration [5]. It remains open, whether $p_3 \geq n-1$ is a tight lower bound for every $n \geq 3$.

Conjecture 3 ([5, Conjecture 2.10]). Every simple arrangement of $n \geq 3$ pairwise intersecting pseudocircles has at least $n-1$ triangles, i.e., $p_3 \geq n-1$.

Concerning the maximum number of triangles in intersecting arrangements, in [5] we have shown an upper bound $p_3 \leq \frac{4}{3}\binom{n}{2} + O(n)$ which is optimal up to a linear error term. In fact, while $\frac{4}{3}\binom{n}{2}$ is an upper bound for arrangements of great-pseudocircles, we managed to find an intersecting arrangement with no digons, no touchings, and $\frac{4}{3}\binom{n}{2} + 1$ triangles. However, since we are not aware of an infinite family of such arrangements, it remains an interesting question to determine the exact maximum number of triangles.

Problem 2. Determine the maximum number of triangles among all simple arrangements of n pairwise intersecting pseudocircles.

2   Proof of Theorem 1

Since the touching graph $T(\mathcal{A})$ contains a triangle, there are three pseudocircles in $\mathcal{A}$ that pairwise touch. Let $\mathcal{K}$ be the subarrangement induced by these three pseudocircles and let $\triangle$ and $\triangle'$ denote the two open triangle cells in $\mathcal{K}$. We label the three touching points, which are also the corners of $\triangle$ and $\triangle'$, as a, b, c. Furthermore, we label the three boundary arcs of $\triangle$ (resp. $\triangle'$) as α, β, γ (resp. α', β', γ'), as shown in Fig. 3(a).

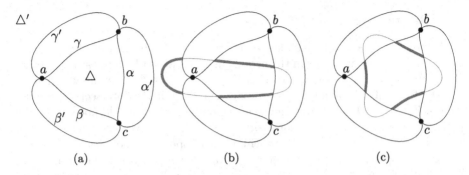

Fig. 3. (a) An illustration of the subarrangement $\mathcal{K}$. (b) and (c), respectively, illustrate an additional pseudocircle C (red). The pc-arcs inside $\triangle$ and $\triangle'$, respectively, are highlighted. (Color figure online)

Assume that all digons in $\mathcal{A}$ are contracted to touchings. The intersection of a pseudocircle $C \in \mathcal{A} \setminus \mathcal{K}$ with $\triangle \cup \triangle'$ results in three connected segments, which we denote as the three *pc-arcs* of C, see Figs. 3(b) and 3(c). Note that two of the pc-arcs induced by C may share an endpoint if C forms a touching with one of the pseudocircles from $\mathcal{K}$; Fig. 5 shows such a touching.

Each pc-arc in $\triangle$ connects two of α, β or γ while a pc-arc in $\triangle'$ connects two of α', β' and γ'. Depending on the boundary arcs on which they start and end, they belong to one of the types $\alpha\beta$, $\beta\gamma$, $\alpha\gamma$, $\alpha'\beta'$, $\beta'\gamma'$ or $\alpha'\gamma'$.

Claim 1. *If two pc-arcs inside $\triangle$ (resp. $\triangle'$) have a touching or cross twice, then they are of the same type.*

Proof. We prove the claim for $\triangle$; the argument for $\triangle'$ is the same. Suppose towards a contradiction that two distinct pseudocircles C, C' from $\mathcal{A} \setminus \mathcal{K}$ contain pc-arcs $A \subset C \cap \triangle$ and $A' \subset C' \cap \triangle$ of different types that have a touching or cross twice. For simplicity, consider only the arrangement induced by the five pseudocircles $\mathcal{K} \cup \{C, C'\}$. By symmetry we may assume that A is of type $\alpha\gamma$ and A' is of type $\alpha\beta$. We may further assume that A and A' have a touching, since otherwise, if they cross twice, they form a digon and we can contract it. This allows us to distinguish four cases which are depicted in Fig. 4 (up to further possible contractions of digons formed between C and the pseudocircles of $\mathcal{K}$).

Case 1: C separates a from b and c.
Case 2: C separates b from a and c.
Case 3: C separates c from a and b.
Case 4: C does not separate a, b, c.

In the next paragraph we show that in neither case, it is possible to extend the arc A' to a pseudocircle C' intersecting the three pseudocircles of $\mathcal{K}$. This is a contradiction.

Extend A' starting from its endpoint on α. The only way to reach γ or γ', avoiding an invalid, additional intersection with C, is via the

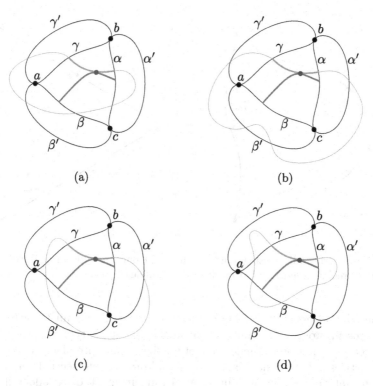

Fig. 4. (a)–(d) illustrate Cases 1–4 from the proof of Claim 1. The pseudocircles C and C' are highlighted blue and red, respectively. The pc-arcs A and A' are emphasized. (Color figure online)

pseudocircle $\beta \cup \beta'$. But the other endpoint of A' already lies on β, so either the pseudocircle extending A' has at least three intersections with $\beta \cup \beta'$ or it misses $\gamma \cup \gamma'$. Both are prohibited in an intersecting arrangement extending $\mathcal{K}$. This completes the proof of Claim 1. □

Next we transform $\mathcal{A}$ into another intersecting arrangement $\mathcal{A}'$ by redrawing the pc-arcs within $\triangle$ and $\triangle'$ such that the pairwise intersections and touchings are preserved and all crossings and touchings of each arc type are concentrated in a narrow region as depicted in Fig. 5. First we apply an appropriate homeomorphism on the drawing so that $\triangle$ becomes a proper triangle ($\triangle'$ will be treated in an analogous manner). For the arc type $\alpha\beta$ we place a small rectangular region $R_{\alpha\beta}$ within $\triangle$ that lies close to the vertex c. We now redraw all pc-arcs of type $\alpha\beta$ so that

- all crossings and touchings between pc-arcs of type $\alpha\beta$ lie inside $R_{\alpha\beta}$,
- every pc-arc of type $\alpha\beta$ intersects $R_{\alpha\beta}$ on opposite sites, and
- for every pc-arc of type $\alpha\beta$, the removal of $R_{\alpha\beta}$ leaves two straight line segments which connect $R_{\alpha\beta}$ to α and β (i.e., the boundary segments of $\triangle$).

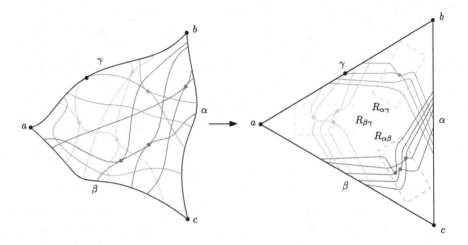

Fig. 5. Concentrate all crossings and touchings of one arc type in a narrow region. The narrow regions are indicated by dashed rectangles.

We proceed analogously for the arc types $\alpha\gamma$ and $\beta\gamma$. By Claim 1 touchings and double crossings only occur between pc-arcs of the same type and therefore lie in the rectangular regions. Since the rectangular regions are placed close enough to the vertices a, b, c of the triangle $\triangle$, no additional intersections or touching points are introduced and we obtain an arrangement $\mathcal{A}'$ of pseudocircles with the same intersections and touchings as $\mathcal{A}$. The combinatorics of the resulting arrangement $\mathcal{A}'$ may however differ from $\mathcal{A}$ since the transformation typically changes the intersection orders of the pseudocircles. We conclude:

Observation. *The transformation preserves the incidence relation between any pair of pc-arcs, that is, two pc-arcs in $\mathcal{A}$ are disjoint/cross in one point/cross in two points/touch if and only if the two corresponding pc-arcs in $\mathcal{A}'$ are disjoint/cross in one point/cross in two points/touch.*

This implies that $\mathcal{A}'$ is indeed again an arrangement of $n(\mathcal{A}') = n(\mathcal{A})$ pairwise intersecting pseudocircles with identical touching graph $T(\mathcal{A}') = T(\mathcal{A})$. In particular, the number of touchings is preserved.

Claim 2. *The arrangement induced by $\mathcal{A}' \setminus \mathcal{K}$ is cylindrical.*

Proof. For each pseudocircle $C \in \mathcal{A}' \setminus \mathcal{K}$, the intersection

$$C \cap (\triangle \cup \triangle') = (C \cap \triangle) \cup (C \cap \triangle')$$

consists of three pc-arcs, and each of these three pc-arcs is of a different type. The first arc is of type $\alpha\beta$ or $\alpha'\beta'$ (depending on whether it is inside $\triangle$ or $\triangle'$), the second is of type $\beta\gamma$ or $\beta'\gamma'$, and the third is of type $\alpha\gamma$ or $\alpha'\gamma'$.

Now we redraw $\mathcal{A}'$ on a cylinder as illustrated in Fig. 6. Since all crossings and touchings of the arc type are within a small region, all pseudocircles from

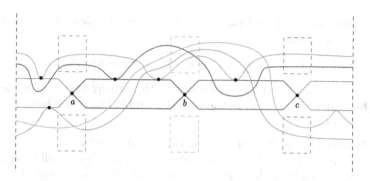

Fig. 6. A cylindrical drawing of $\mathcal{A}' \setminus \mathcal{K}$.

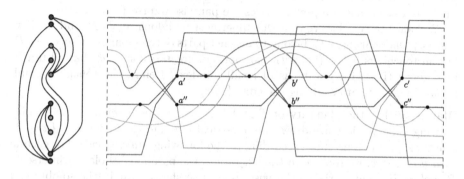

Fig. 7. Replace each of the three pseudocircles of $\mathcal{K}$ by two new pseudocircles so that the entire arrangement is now cylindrical. The green (resp. red and blue) pseudocircle from Fig. 6 is replaced by a new green and a new darkgreen (resp. red and darkred, and blue and darkblue) pseudocircle. On the left: the touching graph $T(\mathcal{A}'')$ of the arrangement. (Color figure online)

$\mathcal{A}' \setminus \mathcal{K}$ wrap around the cylinder, and hence the arrangement induced by $\mathcal{A}' \setminus \mathcal{K}$ is cylindrical. This completes the proof of Claim 2. $\qquad\square$

Next we replace the three pseudocircles of $\mathcal{K}$ by six pseudocircles as illustrated in Fig. 7, so that the resulting arrangement $\mathcal{A}''$ is cylindrical. Each of the three touching points a, b, c in $\mathcal{K}$ is replaced by two new touching points and altogether we obtain touchings $a', a'', b', b'', c', c''$. Hence, when transforming $\mathcal{A}$ into $\mathcal{A}''$, the number of pseudocircles is increased by 3 and the number of touchings is also increased by 3.

Agarwal et al. [1] proved the $p_2 \leq 2n - 2$ upper bound on the number of touchings in cylindrical arrangements of n pairwise intersecting pseudocircles by bounding the number of touchings in an arrangement of pairwise intersecting pseudoparabolas. They show that their touching graph is planar and bipartite [1, Theorem 2.4]. In fact, the drawing of $\mathcal{A}''$ in Fig. 7 can be seen as an intersecting arrangement of pseudoparabolas. We review the ideas of their proof to verify the following claim.

Claim 3. $T(\mathcal{A}'')$ *is planar, bipartite, and has at most* $2n - 5$ *edges.*

Proof. Label the pseudoparabolas with starting segments sorted from top to bottom as $P_1, \ldots, P_n$. In the touching graph $T(\mathcal{A}'')$, we label the corresponding vertices as $1, \ldots, n$.

Bipartiteness: The bipartition comes from the fact that the digons incident to a fixed pseudoparabola P_j are either all from below or all from above. Suppose that a pseudoparabola P_j has a touching from above with P_i and from below with P_k. It follows that P_i is above P_j everywhere and P_k is below P_j everywhere. Hence, P_i and P_k are separated by P_j and cannot intersect – this contradicts the assumption that the pseudocircles are pairwise intersecting.

We now further observe that the uppermost pseudoparabola P_1 and the lowermost pseudoparabola P_n belong to distinct parts of the bipartition, because P_1 has all touchings below (i.e. with parabolas of greater index); P_n has all touchings above (i.e. with parabolas of smaller index). Hence, the touching graph remains bipartite after adding the edge $\{1, n\}$.

Planarity: For the planarity of $T(\mathcal{A}'')$, Agarwal et al. [1] create a particular drawing: The vertices are drawn on a vertical line and each edge $e = \{u, v\}$ is drawn as y-monotone curve according to the following *drawing rule*: For each w with $u < w < v$, we route e to the left of w if the pseudoparabola P_w intersects P_u before P_v, and to right otherwise. It is then shown that in the so-obtained drawing $\mathcal{D}$, each pair of independent edges has an even number of intersections. Hence, the Hanani–Tutte theorem (cf. Sect. 3 in [9]) implies that $T(\mathcal{A}'')$ is planar.

Notice that $\{1, n\}$ is not an edge in $T(\mathcal{A}'')$, since by construction, the lowermost and uppermost pseudocircles do not touch. We further observe that, since all edges in $\mathcal{D}$ are drawn as y-monotone curves, the entire drawing lies in a box which is bounded from above by vertex 1 and from below by vertex n. Hence, we can draw an additional edge from 1 to n which is routed entirely outside of the box and does not intersect any other edge. Again, by the Hanani–Tutte theorem, we have planarity. Since any planar bipartite graph on n vertices has at most $2n - 4$ edges, we conclude that $T(\mathcal{A}'')$ has at most $2n - 5$ edges. This completes the proof of Claim 3.   □

We are now ready to finalize the proof of Theorem 1. From Claim 3 we obtain that $p_2(\mathcal{A}) + 3 = p_2(\mathcal{A}'') \leq 2(n + 3) - 5$, and therefore $p_2(\mathcal{A}) \leq 2n - 2$. This completes the argument.

3   Proof of Proposition 1

The proof of Proposition 1 is based on the *blossom operation*, which allows to dissolve certain triangles in the touching graph. We will apply the blossom operation to arrangements whose touching graphs are wheel graphs to obtain arrangements with the desired properties.

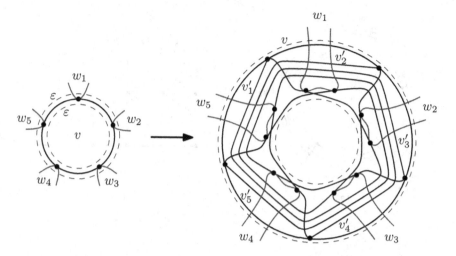

Fig. 8. An illustration of the blossom operation applied on the pseudocircle v of an arrangement.

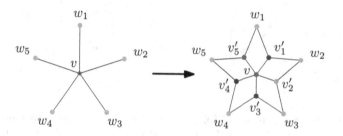

Fig. 9. Blossom operation applied on v: Modification of the touching graph.

The Blossom Operation. Let $\mathcal{A}$ be an arrangement of pairwise intersecting pseudocircles, let v be a pseudocircle in $\mathcal{A}$, and let $w_1, \ldots, w_d$ be the pseudocircles in $\mathcal{A}$ which form touchings with v in this particular circular order along v. As illustrated in Fig. 8, the blossom operation relaxes the touchings between v and $w_1, \ldots, w_d$ to digons and inserts d new pseudocircles $v'_1, \ldots, v'_d$ inside and very close to v so that

- $v'_1, \ldots, v'_d$ form a cylindrical arrangement,
- v touches $v'_1, \ldots, v'_d$, and
- w_i touches v'_{i-1} and v'_i (indices modulo d).

Since the new pseudocircles $v'_1, \ldots, v'_d$ are added in an ε-small area close to v, it is ensured that each v'_i intersects all other pseudocircles. Hence, the obtained arrangement is again an arrangement of pairwise intersecting pseudocircles.

Figure 9 shows the effect of the blossom operation on the touching graph. Note that in these graph drawings the circular orders of the edges incident to

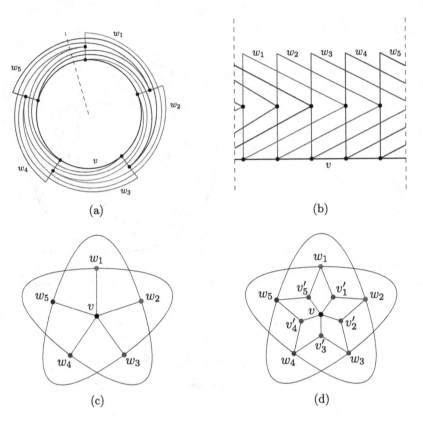

Fig. 10. (a) An arrangement $\mathcal{A}$ of 6 pseudocircles, (b) its cylindrical representation, (c) its touching graph $T(\mathcal{A})$, and (d) the touching graph $T(\mathcal{A}')$ after applying the blossom operation to v.

a vertex coincide with the orders in which the touchings appear on the corresponding pseudocircle.

The blossom operation increases the number of pseudocircles $n(\mathcal{A})$ by d while it increases the number of touchings $p_2(\mathcal{A})$ by $2d$. Hence, when applied to an arrangement $\mathcal{A}$ with exactly $p_2(\mathcal{A}) = 2n(\mathcal{A}) - 2$ touchings, the blossom operation yields again an arrangement $\mathcal{A}'$ with $p_2(\mathcal{A}') = 2n(\mathcal{A}') - 2$ touchings.

Moreover, the blossom operation can be used to eliminate certain triangles in the touching graph. Assume w_i and w_j have a common touching, so v, w_i, w_j form a triangle in the touching graph. Then the blossom operation on v destroys this triangle without creating a new one if and only if, along the pseudocircle v, the two touchings with w_i and w_j are not consecutive. In Fig. 9 a triangle $\{v, w_1, w_2\}$ would result in the new triangle $\{v'_1, w_1, w_2\}$, while a triangle $\{v, w_1, w_3\}$ would not yield a new triangle.

Using the blossom operation, we are now able to prove Proposition 1.

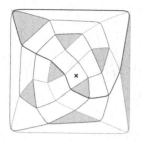 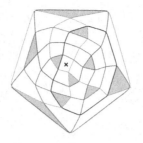 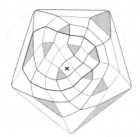

Fig. 11. Digon- and touching-free intersecting arrangements of $n = 6, 7, 8$ pseudocircles with 8, 10, 11 triangles, respectively. Each of the three arrangements is cylindrical, the common interior is marked with a cross. Triangular cells are highlighted gray. [5, Fig. 2] (Color figure online)

Proof (of Proposition 1). Let $n' \geq 11$ be an integer with $n' \equiv 3$ (mod 4). Then $n = \frac{n'+1}{2}$ is an even integer with $n \geq 6$. As illustrated in Fig. 10(a) and Fig. 10(b), we can construct an arrangement $\mathcal{A}$ of n pseudocircles with $p_2 = 2n - 2$ touchings such that the touching graph $T(\mathcal{A})$ is the wheel graph W_n.

In this construction the *central* pseudocircle v has a touching with each of the pseudocircles $w_1, \ldots, w_{n-1}$ and each w_i touches v, $w_{i+n/2}$, and $w_{i-n/2}$ (indices modulo $n - 1$); see Fig. 10(c).

All triangles in $T(\mathcal{A})$ contain the central vertex v and for each such triangle $\{v, w_i, w_j\}$, the touchings of the pseudocircles w_i and w_j with the pseudocircle v are not consecutive on v. Therefore, applying the blossom operation to v eliminates all triangles and the resulting arrangement $\mathcal{A}'$ of $n' = 2n - 1$ pairwise intersecting pseudocircles has $p_2(\mathcal{A}') = 2n' - 2$ touchings and a triangle-free touching graph $T(\mathcal{A}')$; see Fig. 10(d). This completes the argument for $n' \geq 11$ with $n \equiv 3$ (mod 4).

To give a construction for $n'' = 14$ and for all integers $n'' \geq 17$, note that the blossom operation can be applied to pseudocircles with exactly three touchings. The constructed examples with $n \equiv 3$ (mod 4) have pseudocircles with three touchings and the blossom operation applied to such a pseudocircle preserves the property.

Since $n'' = 14$ and every integer $n'' \geq 17$ can be written as $n' + 3k$ with $n' \in \{11, 15, 19\}$ and $k \in \mathbb{N} \cup \{0\}$ we obtain arrangements $\mathcal{A}''$ of n'' pseudocircles with $p_2(\mathcal{A}'') = 2n'' - 2$ touchings. This completes the proof of Proposition 1. $\square$

4   Proof of Theorem 2

We denote by $\mathcal{A}_6$, $\mathcal{A}_7$, and $\mathcal{A}_8$ the three arrangements shown in Fig. 11. These three arrangements on 6, 7, and 8 pseudocircles, respectively, are digon- and touching-free and contain 8, 10, and 11 triangles, respectively. In each of the three arrangements, there is a pseudocircle C and four incident triangles which are alternatingly inside and outside of C in the cyclic order around C. In fact, this *alternation property* holds for each pseudocircle of these three arrangements.

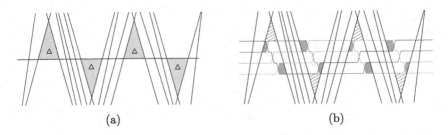

Fig. 12. Replacing one pseudocircle with the alternation property (i.e., four triangles on alternating sides) by a particular arrangement of four pseudocircles.

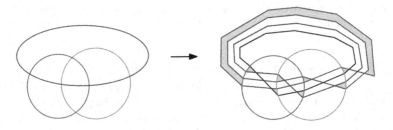

Fig. 13. Extending the Krupp arrangement (left) to the arrangement $\mathcal{A}_6$ (right).

To recursively construct $\mathcal{A}_n$ for $n \geq 9$, we replace a pseudocircle C with the alternation property from $\mathcal{A}_{n-3}$ by a particular arrangement of four pseudocircles as depicted in Fig. 12.

With this replacement we destroy 4 triangles incident to C in the original arrangement, and in total the four new pseudocircles are incident to eight new triangles. Hence, we have $p_3(\mathcal{A}_n) = p_3(\mathcal{A}_{n-3}) + 4 = \lceil \frac{4}{3}(n-3) \rceil + 4 = \lceil \frac{4}{3}n \rceil$.

Moreover, the so-obtained arrangement is cylindrical as the cell marked with the cross lies inside each pseudocircle, and for each of the four new pseudocircles, there are four new triangles (among the eight new triangles) that lie on alternating sides. This allow us to recurse by using one of the four new pseudocircles in the role of C for the next iteration. This completes the proof.

It is worth noting that $\mathcal{A}_6$ can be created with the same construction as illustrated in Fig. 13 by extending the Krupp arrangement of three pseudocircles, in which all cells are triangles.

References

1. Agarwal, P.K., Nevo, E., Pach, J., Pinchasi, R., Sharir, M., Smorodinsky, S.: Lenses in arrangements of pseudo-circles and their applications. J. ACM **51**(2), 139–186 (2004). https://doi.org/10.1145/972639.972641
2. Aronov, B., Sharir, M.: Cutting circles into pseudo-segments and improved bounds for incidences. Discrete Comput. Geom. **28**(4), 475–490 (2002). https://doi.org/10.1007/s00454-001-0084-1

3. Erdős, P.: On sets of distances of n points. Am. Math. Mon. **53**(5), 248–250 (1946). https://doi.org/10.2307/2305092
4. Felsner, S., Scheucher, M.: Arrangements of pseudocircles: on circularizability. Discrete Comput. Geom. Ricky Pollack Memorial Issue **64**, 776–813 (2020). https://doi.org/10.1007/s00454-019-00077-y
5. Felsner, S., Scheucher, M.: Arrangements of pseudocircles: triangles and drawings. Discrete Comput. Geom. **65**, 261–278 (2021). https://doi.org/10.1007/s00454-020-00173-4
6. Grünbaum, B.: Arrangements and spreads. In: CBMS Regional Conference Series in Mathematics, vol. 10. AMS (1972). https://doi.org/10.1090/cbms/010
7. Levi, F.: Die Teilung der projektiven Ebene durch Gerade oder Pseudogerade. Berichte über die Verhandlungen der Sächsischen Akademie der Wissenschaften zu Leipzig, Mathematisch-Physische Klasse **78**, 256–267 (1926)
8. Pinchasi, R.: Gallai-sylvester theorem for pairwise intersecting unit circles. Discrete Comput. Geom. **28**(4), 607–624 (2002). https://doi.org/10.1007/s00454-002-2892-3
9. Schaefer, M.: Toward a theory of planarity: Hanani-Tutte and planarity variants. J. Graph Algorithms Appl. **17**(4), 367–440 (2013). https://doi.org/10.7155/jgaa.00298
10. Snoeyink, J., Hershberger, J.: Sweeping arrangements of curves. In: Discrete and Computational Geometry: Papers from the DIMACS Special Year, DIMACS, vol. 6, pp. 309–349. AMS (1991). https://doi.org/10.1090/dimacs/006/21

GD Contest Report

Graph Drawing Contest Report

Philipp Kindermann[1], Fabian Klute[2], Tamara Mchedlidze[2],
and Wouter Meulemans[3(✉)]

[1] Universität Trier, Trier, Germany
`kindermann@uni-trier.de`
[2] Universiteit Utrecht, Utrecht, The Netherlands
`{f.n.klute,t.mtsentlintze}@uu.nl`
[3] TU Eindhoven, Eindhoven, The Netherlands
`w.meulemans@tue.nl`

Abstract. This report describes the 29th Annual Graph Drawing Contest, held in conjunction with the 30th International Symposium on Graph Drawing and Network Visualization (GD'22) in Tokyo, Japan. Due to the continuing global COVID-19 pandemic, the conference and thus also the contest was held in a hybrid format, with both on-site and online participants. The mission of the Graph Drawing Contest is to monitor and challenge the current state of the art in graph-drawing technology.

1 Introduction

Following the tradition of the past years, the Graph Drawing Contest was divided into two parts: the *creative topics* and the *live challenge*.

Creative topics were comprised by two data sets. The first data set was the *Opera Network*: The data represent a collection of opera performances that took place across Europe between 1775 and 1833. The second data set showed a an *Aesthetic Experience Network*: The data set represents 8 networks that model an aesthetic experience of the viewers when observing artworks. The data sets were published about half a year in advance, and contestants submitted their visualizations before the conference started.

The live challenge took place during the conference in a format similar to a typical programming contest. Teams were presented with a collection of *challenge graphs* and had one hour to submit their highest scoring drawings. This year's topic was similar to last year's: minimize edge-length ratio in a planar polyline drawing graph with vertex locations restricted to a grid and a maximum number of bends per edge allowed.

Overall, we received 26 submissions: 9 submissions for the creative topics and 17 submissions for the live challenge (10 manual and 7 automatic).

2 Creative Topics

The general goal of the creative topics was to model each data set as a graph and visualize it with complete artistic freedom, and with the aim of communicating

P. Angelini and R. von Hanxleden (Eds.): GD 2022, LNCS 13764, pp. 459–470, 2023.
https://doi.org/10.1007/978-3-031-22203-0_33

as much information as possible from the provided data in the most readable and clear way.

We received 8 submissions for the first topic, and 1 for the second. Submissions were evaluated according to four criteria:

(i) Readability and clarity of the visualization,
(ii) aesthetic quality,
(iii) novelty of the visualization concept, and
(iv) design quality.

We noticed overall that it is a complex combination of several aspects that make a submission stand out. These aspects include but are not limited to the understanding of the structure of the data, investigation of the additional data sources, applying intuitive and powerful data visual metaphors, careful design choices, combining automatically created visualizations with post-processing by hand, as well as keeping the visualization, especially the text labels, readable. For each topic, we selected the top five submissions before the conference, which were printed on large poster boards and presented at the Graph Drawing Symposium. We also made all the submissions available on the contest website in the form of a virtual poster exhibition. During the conference, we presented these submissions and announced the winners. For a complete list of submissions, refer to http://www.graphdrawing.org/gdcontest/contest2022/results.html. Eight of the submissions were accompanied by an online tool, which are linked on the web page.

2.1   Opera Networks

The data represents a collection of opera performances that took place across Europe between 1775 and 1833.

Each row corresponds to a performance and contains the following information:

- The performance title (`title`)
- The librettist's name (`libertist`)
- The composer's name (`composer`)
- The performance year (`performance_year`)
- The city in which the performance tool place (`placename`)
- `rism_id` - unique identifier corresponding to the performance that gives a possibility to extract more information about the performance from RISM database

The data was extracted from the RISM database[1] and was offered by Frans Wiering[2] – professor of Utrecht University studying Musicology.

---

[1] https://opac.rism.info/main-menu-/kachelmenu/help.
[2] https://www.uu.nl/medewerkers/FWiering.

There are several possibilities on how a network can be extracted from this data. We left it to the participants to decide how and whether to model this data set as a network. The possible research questions that can drive this modeling were pointed by Frans Wiering and are as follows:

- How performances travelled geographically and in time?
- How Italian/Viennese operas travelled to Europe?
- Which operas stayed at same place and which went over Europe?
- Are there patterns in collaborations among composers and libertists, also over time?

3rd Place: Joshua Rutschmann, Marc Seelmann, Patrizia Lenhart, Tim Scholl, Mike Fu, Vincent Lafragola, and Sarah Altenkrüger (Universität Tübingen). The contest committee likes this layout for its simplicity and easy readability of the data captured by the visualization. Representing the composer to librettist relations via a small graph is a choice that nicely inserts this information into the visualization without adding a lot of visual complexity. Also, the choice of laying out the visualization in the style of an opera seating arrangement leads to a pleasingly looking picture that invites exploration of the data. Clustering the geographic information by countries is a good choice, though the colors do not necessarily support the easy identification of geographic areas. The provided online tool adds the missing information like opera names as easy-to-read hover items.

Tool: http://operanetwork.cs.uni-tuebingen.de/

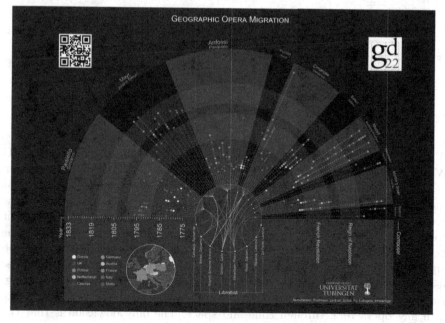

2nd Place: Richard Brath (Uncharted Software). The committee finds this visualization to be not only pleasant to look at, but also provoking to explore the data. It allows for an easy exploration and analysis of many aspects of the data. Most notably, the south-to-north pattern of the operas over time is very clearly visible, achieved via the color-scheme and the choice of layout. Also, the choice of repeating librettists at both sides of the visualization supports well the tracking of composer-librettist and opera-librettist relations which otherwise might have been hard to follow. However, finding all occurrences of a single location is somewhat difficult: the lines connecting them are hard to follow due to the majority being near vertical and the color scheme is a bit too subtle for this purpose.

Tool: https://codepen.io/Rbrath/full/ZEoYepb

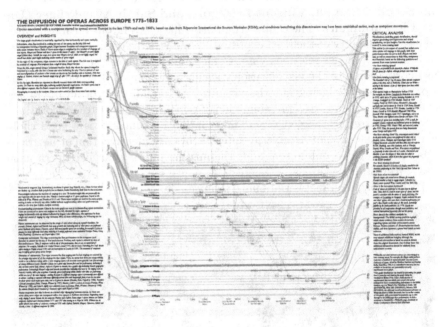

Winner: Thomas Depian, Michael Huber, and Wilhelm Wanecek (TU Wien). The committee finds this visualization to be mesmerizing and beautiful to look at. At the same time it also well supports analyzing and answering most questions posted with the challenge. The well thought-out space-central view makes locating the opera-city relation straightforward and the metro-style layout provides a familiar way of tracing the movement of an opera over time. The legend explains the visualization in a good fashion and the small bundled graph supports well the identification of composer-librettist relations. The committee also appreciates that various algorithmic tools were used to create this drawing. Finally, the online version of this visualization adds the ability to highlight the path any opera took through time and space. The only downside is that the temporal information is more difficult to assess and compare between cities.

Tool: https://opera-network.netlify.app

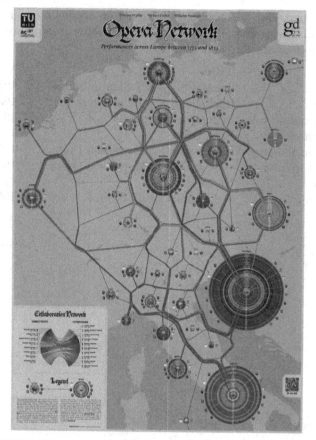

❝ In our Opera Network, we grouped performances by their operas, which we identified by the triple title, composer, and librettist. On a map of Europe, cities comprised of concentric rings, each representing a performance, were positioned close to their actual location using a force-based layout. Then, performances within the same group were chronologically connected on directed paths running along the edges of a generalized Voronoi diagram. For this, the crossing-optimal path bundling algorithm by Pupyrev et al. (2016) was used, followed by optimizations to make the paths more homogeneous. Finally, we assigned colors by mixing a base color for each composer with a shade of grey for each librettist, visualized with the composer-librettist collaborations in a chord diagram. On top, to make the data exploration easier, our interactive version allows highlighting the paths of performances individually or by composer/librettist.
Thomas Depian ❞

2.2 Aesthetic Experience Network

The data set represents 8 networks that model an aesthetic experience of the viewers when observing artworks. The analyzed artworks are 8 paintings by Klee, Kandinsky, Mortensen, Miro and Winter:

Artist	Title	Year
Paul Klee	Zeichen in Gelb / Sign in Yellow	1937
Paul Klee	Blick aus Rot / Be aware of Red	1937
Wassily Kandinsky	Regungen / Impulses	1928
Wassily Kandinsky	Untitled	1934
Richard Mortensen	velsesstykker / Mortensen Pink	1922
Richard Mortensen	velsesstykker / Mortensen Orange	1922
Joan Mirò	Untitled	1961
Fritz Winter	Siebdruck 6 / Silkscreen 6	1950

Each of the 14 nodes represents one of the two polarities of an aesthetic effect: (i) positive – negative; (ii) active – passive; (iii) still – lively; (iv) sad – happy; (v) peaceful – aggressive; (vi) hard – soft; (vii) cold – warm; (viii) light – heavy; (ix) rough – smooth; (x) spiritual – bodily; (xi) feminine – masculine; (xii) cautious – intrusive; (xiii) like – dislike; (xiv) interesting – uninteresting.

The edges are weighted by conditional dependence relations among aesthetic effects: If two aesthetic effects are connected in the resulting graph, they are dependent after controlling for all other symptoms. Thus, a negative dependency between A and B indicates a positive dependency between A and the opposite of B. This data is a result of the research presented in the paper *Associating With Art: A Network Model of Aesthetic Effects* by Specker et al. [1] and the full set of collected data is available online[3]. When sharing the data for the challenge, the authors of the paper said they are curious "how to visualize this data set for an art historical audience or other audience that does not know about network theory."

---

[3] https://osf.io/zqxbm/.

Winner: Axel Kuckuk, Henry Förster, and Sarah Gester (Universität Tübingen). The contest committee liked that the layout is easy to read and clearly displays the individual as well as the aggregated data. Taking a rather minimalistic approach with well-separated sub-figures, the authors create a visualization that conveys well the overall data at a glance for each piece of art. The committee also liked the meta-level of representing this particular data set about art again as a piece or as pieces of art hanging in an exhibition.

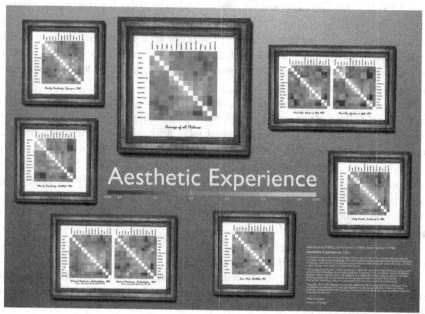

❝ The latest work by the artist trio Kuckuk, Förster and Gerster, who are renowned for creatively exploring how humans influence data and vice-versa, was also unveiled at the conference. The series of paintings titled "Aesthetic Experiences" showcases correlations of different aesthetic effects experienced by the audience of classic abstract art pieces in the shape of heatmaps. "In the design, we strived to present the data in a clear and minimalistic way, omitting numbers for the most part to encourage a playful and courious interaction with the data, searching for differences and similarities between the networks.", Kuckuk said. Acclaimed experts appeared ecstatic following the unveiling of this marvellous piece of art. As graph drawing expert Kindermann put it: "Readability is fine". His further statement "How they obtained the ordering isn't immediately clear." is courtesy to the unusually creative thought process of the artists. ❞

Henry Förster

3   Live Challenge

The live challenge took place during the conference and lasted exactly one hour. During this hour, local participants of the conference could take part in the manual category (in which they could attempt to draw the graphs using a supplied tool: http://graphdrawing.org/gdcontest/tool/), or in the automatic category (in which they could use their own software to draw the graphs). Because of the global COVID-19 pandemic, we allowed everybody in both categories to participate remotely. To coordinate the contest, give a brief introduction, answering questions, and giving participants the possibility to form teams, we were kindly provided with both a room in the conference building, and a the Zoom stream for the conference; furthermore, participants could also meet and follow the contest via a dedicated room in gather.town.

The challenge focused on minimizing the planar polyline edge-length ratio on a fixed grid. The *planar edge-length ratio* of a straight-line drawing is defined as the ratio between the length of longest edge and minimal Euclidean distance between two neighboring vertices. This slightly changed from last year to allow for better scores to more correspond to nicer drawings. There has been recent attention to this topic with several publications. The *planar polyline edge-length ratio* is a generalization of the planar edge-length ratio where edges do not have to be straight-line segments, but can be polylines with a maximum number of bends per edge defined by the input.

The input graphs were planar undirected graphs. For the manual category, each graph came already with a planar drawing.

The results were judged solely with respect to the edge-length ratio; other aesthetic criteria were not taken into account. This allows an objective way to evaluate each drawing.

3.1   The Graphs

In the manual category, participants were presented with six graphs. These were arranged from small to large and chosen to contain different types of graph structures. In the automatic category, participants had to draw the same six graphs as in the manual category, and in addition another seven larger graphs. Again, the graphs were constructed to have different structure.

For illustration, we include the third graph, which was given a seemingly random graph with initial ratio 22, but that can be drawn with uniform edge lengths, except for one edge. The best manual solution we received (by team *kuneri nashi*), and the best automatic solution we received (by team *OMEGA*) are given below.

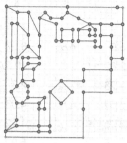 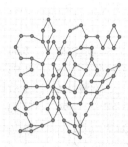 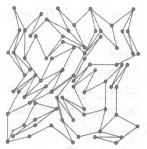

Provided drawing

edge-length ratio 22

Best manual solution
kuneri nashi
edge-length ratio 1.12

Best automatic solution
OMEGA
edge-length ratio 1.21

For the complete set of graphs and submissions, refer to the contest website at http://www.graphdrawing.org/gdcontest/contest2022/results.html. The graphs are still available for exploration and solving Graph Drawing Contest Submission System: https://graphdrawingcontest.appspot.com.

Similarly to the past years, the committee observed that manual (human) drawings of graphs often display a deeper understanding of the underlying graph structure than automatic and therefore gain in readability. However, excepting the instance above, the automatic techniques by *OMEGA* managed to outperform the manual solutions when measured purely on edge-length ratio. For the larger graphs, we gave ample space to ensure that finding some embedding would be feasible in the given time. This allowed for most techniques to solve most instances. However, the fourth instance of the larger graphs was still restricted in grid size, though it was given an initial embedding. Nonetheless, only *OMEGA* managed to roughly halve the initial edge-length ratio, suggesting that working in such a confined space is still challenging, even with a given embedding.

3.2 Results: Manual Category

Below we present the full list of scores for all teams. The numbers listed are the edge-length ratios of the drawings; the horizontal bars visualize the corresponding scores.

graph	1	2	3	4	5	6
Final Exam	1.78	2.23	8.54	5.0	4.85	8.27
Minimal is criminal	1.2	1.96	6.08	2.82	2.77	10.57
M. Gronemann Memorial T.	1.08	1.66	2.51	2.23	2.45	2.3
OneLayoutToRuleThemAll	2.23	2.28	4.03	5.09	4.8	4.03
kuneri nashi	1.07	1.71	1.11	3.16	5.0	1.92
Yoshio Okamoto	1.41	2.23	10.0	10.52	9.26	5.99
Good luck	1.5	2.21	2.76	4.0	4.25	2.4
New keyboard, who dis?	1.41	2.82	2.69	3.16	2.16	2.38
Greedy Unicorns	1.2	1.8	7.0	6.32	5.39	4.46

Third place: **New keyboard, who dis?**, consisting of Anaís Villedieu, Jules Wulms, and Soeren Nickel.

Second place: **kuneri nashi**, consisting of Felix Klesen and Johannes Zink

Winner: **Martin Gronemann Memorial Team**, consisting of Fouli Argyriou and Henry Förster.

> ❝ Since the formation of our contest team in 2019, it was the first time we had to compete without Martin Gronemann. So with heavy heart only the two of us kept going with the contest as Martin has untimely passed on in life after leaving academia. As we know he would be proud of us, we followed first of all the basic rule that is independent of the problem to be solved as Martin taught us: "So at the beginning submit all instances right away with their initial layout to ensure that at the end we have a feasible entry for every instance (also, it puts pressure on the competitors right away)." For the particular problem of this year's contest, we realized that bends were completely useless. So, when drawing one of the graphs, as a first step we deleted all bends. Then, we reintroduced planarity before shortening the longest edge (or very rarely elongating the shortest one). As a result, we achieved an edge length ratio of less than 3 in all contest graphs, or as Martin put it "Ihr habt aber auch nen geilen Job gemacht wenn ich mir die Punkte anschaue". ❞
>
> *Fouli Argyriou and Henry Förster*

3.3   Results: Automatic Category

In the following we present the full list of scores for all teams that participated in the automatic category. The numbers listed are the edge-length ratios of the drawings; the horizontal bars visualize the corresponding scores.

graph	1	2	3	4	5	6	7	8	9	10	11	12	13
el_ratio	1.58	2.0	5.61	4.24	5.14	3.65	10.6	67.13	69.0		58.48		480.0
HopingFor2DigitRatios	1.41	1.68	9.06	3.0	5.0	4.64	15.22	208.72	94.9	158.47	58.47	146.12	687.25
Golden Ratio	1.58	1.89	6.01	7.3	4.19	5.09	46.09	68.0	123.0	158.47	42.48	234.05	932.0
OMEGA	1.02	1.41	1.21	2.0	1.26	1.04	4.46	36.77	8.91	70.0	105.76	3.98	97.66
TUW-ELR1	1.11	1.68	3.16	2.82	2.17	3.25	44.57	152.58	219.61	158.47	129.87	122.56	879.75
Wo ECTS?	1.11	3.16	4.47	3.16	8.24	4.16	34.01	695.08	845.0		177.49	290.0	1760.0

Third place: **Golden Ratio**, consisting of Andreas Krystallidis, Leonid Darovskikh, and Manuel Bacher.

Second place: **TUW-ELR1**, consisting of lexander Dobler, Oliver Pilizar, and Sebastian Uhl.

Winner: **OMEGA**, consisting of Laurent Moalic, Dominique Schmitt, and Julien Bianchetti.

❝ Our team comes from the University of Haute-Alsace in Mulhouse (France). Julien is a Master's student in Computer Science, Laurent a researcher in Combinatorial Optimization, and Dominique a researcher in Computational Geometry. Our algorithm uses the OGDF library to first construct a planar straight-line drawing of the graph, whose external face is its largest face. The graph is then shrinked (if needed) to fit in the given grid. This is done by a simulated annealing algorithm, which attracts the nodes towards the centroid of the graph. Finally, the nodes are moved in random order with the aim to shorten the edges whose length is closer to the longest edge than to the shortest one, and to lengthen the other edges. Bends are added randomly and treated as nodes, except that they are allowed to overlap their two neighbors, in which case they are removed. Again, a simulated annealing approach is used to avoid falling in a local optimum. The algorithm was run simultaneously on all thirteen graphs of the competition on a laptop computer with a 12-core 2.7GHz processor. It worked well with all graphs except graphs 10 and 11. For graph 10, it did not succeed in fitting the generated drawing in the grid. Now, since graph 10 was given with coordinates, we had just to run the ratio-optimization phase on the given graph. Graph 11 was given with a grid that was too large for our algorithm. So we treated graph 11 with a slower version of the algorithm, which still achieved a ratio of 8 in about 10 minutes. Unfortunately, we submitted the wrong solution at the competition (the initial drawing rather than the improved one). Our algorithm also achieves good results on last year's competition by a straightforward adaptation of the edge-length computation. ❞

Laurent Moalic, Dominique Schmitt, and Julien Bianchetti

Acknowledgments. The contest committee would like to thank the organizing and program committee of the conference; the organizers who provided us with a room with hardware for the live challenge and monetary prizes; the generous sponsors of the symposium; and all the contestants for their participation. Special thanks goes to Franz Wiering for providing the data for the Opera Network and to Specker, Fried, Rosenberg, and Leder for sharing the data for the Aesthetic Experiences Network. Further details including all submitted drawings and challenge graphs can be found at the contest website:

http://www.graphdrawing.org/gdcontest/contest2022/results.html

Reference

1. Specker, E., Fried, E.I., Rosenberg, R., Leder, H.: Associating with art: a network model of aesthetic effects. Collabra Psychol. **7**(1), 24085 (2021)

Posters

Visualizing Node-Specific Hierarchies in Directed Networks

Mykyta Shvets, Ehsan Moradi, and Debajyoti Mondal[(✉)]

Department of Computer Science, University of Saskatchewan, Saskatoon, Canada
{mdy678,ehm486}@mail.usask.ca, dmondal@cs.usask.ca

Introduction. Force-directed layout algorithms are commonly used to compute aesthetically-pleasing visualizations of networks and reveal communities in the associated networks. For undirected networks, the forces on the nodes are often modeled with springs or electric charges [1–3, 8]. For directed networks, one can apply this algorithm ignoring the directions of the edges. However, edge directions are crucial to understand the hierarchical structure or directed paths in a directed network. Therefore, Sugiyama and Misue [7] proposed a magnetic spring model that in addition to having the spring forces, would contain magnetic forces that act at the two ends of an edge (Fig. 1(a)). Given a magnetic field over the plane (e.g., a linear field or a polar field), the magnetic forces attempt to rotate the edge to align its orientation to the direction of the magnetic field line (Fig. 1(b)–(c)) to reveal the directionality in the network.

Our Contribution. We propose a modification to the magnetic-spring idea by removing the fixed magnetic field and adding polar fields that are anchored at some user-specified nodes. In other words, the chosen nodes act as polar fields that can move along with the nodes (Fig. 1(d)) unless the users choose to pin them at a fixed location. The goal for exploring such a model is to visualize the hierarchical structure around the selected nodes in a directed network. For example, consider selecting some classes in a class dependency network of a software system as poles. Then one can expect our visualization to reveal the hierarchy of dependencies around each selected class and to visualize the influences of the selected classes over the whole network.

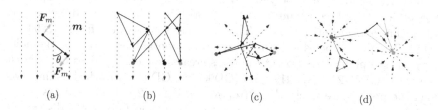

(a)   (b)   (c)   (d)

Fig. 1. (a) Magnetic forces acting on an edge. (b)–(c) A network drawn with a linear and a polar magnetic field, respectively. (d) Our approach, where two nodes are selected as poles (orange and green disks). An edge $\overrightarrow{(v, w)}$ is colored by the color of a pole p if the vertex w is closer to p than any other poles. (Color figure online)

The Work Is Supported in Part by NSERC.

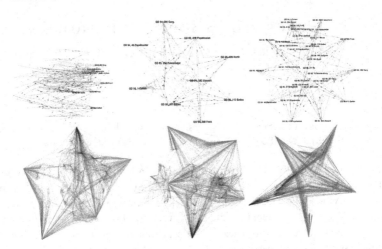

Fig. 2. (top) Drawing GD'01 using linear field, with our approach, and with fixed poles. (bottom) Dependency network for software Lucene. (left) Force layout with pole separation—relevant subnetworks are colored. (mid) Changes after adding magnetic force. (right) Revealed hierarchies after adding pole gravity.

Our experimental results with real-life datasets show the effectiveness of the proposed method in visualizing node-specific hierarchies in directed networks.

Our Methods and Experiments. We augment the force layout approach with pole separation forces that create space around the selected nodes, pole gravity forces that pull the colored subnetwork closer to the pole, and magnetic forces that align the directed edges towards the poles. Figure 2(top-left) depicts a visualization generated by Sugiyama and Misue's [6] approach with a linear field for the GD'01 dataset (i.e., citation network among GD papers), where 10 nodes are selected for investigation. Figure 2(top-mid) shows the visualization of our method with both pole separation and pole gravity forces activated. Figure 2(top-right) shows the visualization for the same technique but with the poles pinned at the corners of a regular polygon. While Sugiyama and Misue's [6] approach shows the overall directionality of the network, the influence of individual poles is less cluttered in our approach. The poles are some papers from GD'94 and GD'95. The pinned poles version reveals the temporal hierarchy near the poles. However, the pinned position of the poles creates some long edges between the pole pairs (GD95-254, GD-419) and (GD96-101, GD96-113), which is improved when the poles move freely (Fig. 2(top-mid)). Figure 2(bottom) depicts a software dependency network [4, 5], where the separation force helps discern the poles easily but it is not enough to separate different subnetworks as they may be densely connected with each other. With pole gravity and magnetic forces we can better see the hierarchy around the poles.

Although our work shows potential, we envision to make our system interactive and to investigate edge bundling techniques to better visualize the hierarchical structure around the nodes.

References

1. Eades, P., McKay, B.D.: An algorithm for generating subsets of fixed size with a strong minimal change property. Inf. Process. Lett. **19**(3), 131–133 (1984). https://doi.org/10.1016/0020-0190(84)90091-7
2. Kamada, T.: On visualization of abstract objects and relations. Ph. D. Dissertation in University of Tokyo **86** (1988)
3. Moradi, E., Mondal, D.: BigGraphVis: leveraging streaming algorithms and GPU acceleration for visualizing big graphs. CoRR abs/2108.00529 (2021)
4. Subelj, L.: Software dependency. http://wwwlovre.appspot.com/support.jsp
5. Subelj, L., Bajec, M.: Software systems through complex networks science: review, analysis and applications. In: Proceedings of the First International KDD Workshop on Software Mining, Software Mining, pp. 9–16. ACM (2012). https://doi.org/10.1145/2384416.2384418
6. Sugiyama, K., Misue, K.: Graph drawing by the magnetic spring model. J. Vis. Lang. Comput. **6**(3), 217–231 (1995). https://doi.org/10.1006/jvlc.1995.1013
7. Sugiyama, K., Misue, K.: A simple and unified method for drawing graphs: magnetic-spring algorithm. In: Tamassia, R., Tollis, I.G. (eds.) GD 1994. LNCS, vol. 894, pp. 364–375. Springer, Heidelberg (1995). https://doi.org/10.1007/3-540-58950-3_391
8. Tamassia, R. (ed.): Handbook of Graph Drawing and Visualization. CRC Press, Boca Raton, Florida, USA (2013). Chapman and Hall/CRC, ISBN 978-1-5848-8412-5

Can an NN Model Plainly Learn Planar Layouts?

Simon van Wageningen[(✉)] and Tamara Mchedlidze

Utrecht University, Utrecht, The Netherlands
{s.vanwageningen,t.mtsentlintze}@uu.nl

Introduction. Planar graph drawings tend to be aesthetically pleasing [1]. Planar graphs and their drawings have been extensively studied in graph drawing literature [2] and can be generated efficiently [2, 3]. However, there are no practical layout algorithms for the graphs that are nearly planar. Thus, force directed algorithms [4] often fail at detecting the planar substructure in such graphs. The attempts to formalize near-planarity (1-planar [5], RAC [6] and quasi-planar [7]) lead to NP-hard recognition problems [8, 9].

Due to the fact that the formalization of near-planarity immediately leads to NP-hard problems, we turn our attention towards Neural Networks (NNs). NNs have already been used for graph layout evaluation [10, 11], discrimination [12] and more relevantly for graph layout generation [13–15].

Our far-reaching goal is to investigate whether NNs are capable of producing drawings of nearly planar graphs that clearly depict large planar substructures. Such NNs are expected to be able to produce near-planar drawings of planar graphs. Therefore, as a first step towards our goal, we investigate whether NNs are successful in producing planar drawings of planar graphs. Additionally, we briefly explore the effectiveness of the model in generalizing beyond planarity.

Method. We refer the reader to the full version [16] for the details of the experiments. We reuse the LSTM model and Procrustes Statistic[1] (PS) loss function (LF) of Wang et al. [15, 17], who showed the model to be successful in producing planar drawings of grids and stars. Note that the PS LF ensures that coordinate-based patterns can be learned.

Since drawings with less stress are shown to correlate with positive preferences [18], additional experiments are also conducted using a supervised stress[2] (SuS) LF. We expect the SuS LF to be more capable than the PS LF when randomness is introduced to node coordinates.

We train 8 models on 8 different graph classes and layouts: Grids, Grids with all diagonals ($Grids_d$), Grids with random diagonals ($Grids_{rd}$), Delaunay Triangulations, 2-star caterpillar ($Caterp2$), 3-star caterpillar ($Caterp3$), randomized radial trees ($RRTrees$) and randomized Stress Majorization trees ($RSMTrees$). The number of graphs in a dataset (72–1000) and the graph sizes (18–625 nodes)

---

[1] PS LF calculates the differences between original and predicted node coordinates, after a series of transformations.

[2] Differences between original and predicted pairwise stress values, for each node pair, are computed.

© The Author(s), under exclusive license to Springer Nature Switzerland AG 2023
P. Angelini and R. von Hanxleden (Eds.): GD 2022, LNCS 13764, pp. 476–479, 2023.
https://doi.org/10.1007/978-3-031-22203-0

Table 1. Averaged performance of conventional techniques and model with SuS LF on multiple instances of different graph classes. Stress s is in 1e+7, bolded entries indicate interesting differences.

Graph class		LF	QM	QM	QM	FD QM	QM	QM	SM QM	QM	QM
Train	Test	SuS	nc	s	ar	nc	s	ar	nc	s	ar
Grids	Grids	1.54	**6.30**	4.62	0.36	**15.90**	6.18	0.50	**0.61**	4.88	0.93
Grids$_d$	Grids$_d$	3.27	**303**	9.76	0.23	**417**	12.40	0.19	**426**	9.55	0.73
Grids$_{rd}$	Grids$_{rd}$	2.60	3.31	4.24	0.21	27.5	6.10	0.26	3.04	4.82	0.57
Grids	Grids$_{rd}$	12.80	194	4.65	**0.0028**	27.5	6.10	**0.26**	3.04	4.82	**0.57**
Grids$_d$	Grids$_{rd}$	55.40	489	3.02	0.0046	27.5	6.10	0.26	3.04	4.82	0.57
Delaunay	Delaunay	21.80	200	3.25	0.0086	59.40	3.57	0.026	90	3.10	0.038
Caterp2	Caterp2	39.80	0	4.64	0.051	0	4.24	0.090	0	5.39	0.15
Caterp3	Caterp3	34.50	0.14	3.48	**0.089**	0.39	3.75	**0.061**	0	3.98	**0.067**
RR Trees	RR Trees	50.00	51.30	6.41	0.0054	3.87	5.13	0.052	30.90	5.63	0.13
RSM Trees	RSM Trees	27.30	55.90	5.63	0.0036	3.68	5.28	0.048	32.80	5.66	0.13

vary, depending on the graph class. Moreover, the testing datasets are comprised of multiple instances of similar sized graphs, as to make valid averaged-out comparisons. We evaluate the performance by visually inspecting the layouts and computing three quality metrics: the number of crossings (nc), the stress (s) and the angular resolution [19] (ar). The quality metrics values are compared with two conventional layout techniques: ForceAtlas2 [20, 21] (FD) and Stress Majorization [22] (SM).

Results. Table 1 showcases the results of the experiments with the SuS LF. On average, the models trained with the SuS LF outperform the models trained with the PS LF. Additionally, the model trained on Grids with the SuS LF outperforms the FD algorithm, in terms of number of crossings (nc) and stress (s). The model trained on $Caterp3$ shows a better angular resolution (ar) and stress than the FD and SM layouts. On average, the models trained with SuS show better stress scores than the conventional FD and SM techniques. However, w.r.t. the ar and the nc the results tend to worsen. Moreover, when some randomness is introduced to the training data ($RRTrees$ & $RSMTrees$), the models have difficulties generalizing, produce sub-optimal layouts and have unfavorable QM results. When it comes to generalizing beyond planarity, a model trained on Grids and tested on Grids$_{rd}$ shows poor results.

To conclude, our results indicate that planar graph classes can be learned by a Neural Network, and the produced planar drawings can score better than those produced by conventional techniques. We note that the loss function and the presence of randomness in graph data can have major effects on the model's learning capabilities. In the future, the combination of multiple loss functions should be explored as well as different Neural Network architectures.

References

1. Purchase, H.: Which aesthetic has the greatest effect on human understanding? In: DiBattista, G. (ed.) GD 1997. LNCS, vol. 1353, pp. 248–261. Springer, Heidelberg (1997). https://doi.org/10.1007/3-540-63938-1_67
2. Hopcroft, J., Tarjan, R.: Efficient planarity testing. J. ACM **21**(4), 549–568 (1974)
3. Tamassia, R.: Planar straight line drawing algorithms. In: Tamassia, R. (eds.) Handbook on Graph Drawing and Visualization, pp. 193–222. Chapman and Hall/CRC (2013)
4. Kobourov, S. G.: Force directed drawing algorithms. In: Tamassia, R. (eds.) Handbook on Graph Drawing and Visualization, pp. 383–408. Chapman and Hall/CRC (2013)
5. Borodin, O.V.: Solution of the Ringel problem on vertex-face coloring of planar graphs and coloring of 1-planar graphs. Metody Diskret. Analiz **41**(12), 108 (1984)
6. Didimo, W., Eades, P., Liotta, G.: Drawing graphs with right angle crossings. Theor. Comput. Sci. **412**(39), 5156–5166 (2011)
7. Agarwal, P.K., Aronov, B., Pach, J., Pollack, R., Sharir, M.: Quasi planar graphs have a linear number of edges. Combinatorica **17**(1), 1–9 (1997). https://doi.org/10.1007/BF01196127
8. Argyriou, E.N., Bekos, M.A., Symvonis, A.: The straight-line RAC drawing problem is NP-hard. In: Černá, I., et al. (eds.) SOFSEM 2011. LNCS, vol. 6543, pp. 74–85. Springer, Heidelberg (2011). https://doi.org/10.1007/978-3-642-18381-2_6
9. Korzhik, V.P., Mohar, B.: Minimal obstructions for 1-immersions and hardness of 1-planarity. J. Graph Theory **72**(1), 30–71 (2013)
10. Haleem, H., Wang, Y., Puri, A., Wadhwa, Sahil., Qu, H.: Evaluating the readability of force directed graph layouts: a deep learning approach. Comput. Graph. Appl. (IEEE) **39**(4), 40–53 (2019)
11. Giovannangeli, L., Bourqui, R., Giot, R., Auber, D.: Toward automatic comparison of visualization techniques: Application to graph visualization. CoRR (2019)
12. Klammler, M., Mchedlidze, T., Pak, A.: Aesthetic discrimination of graph layouts. In: Biedl, T., Kerren, A. (eds.) GD 2018. LNCS, vol. 11282, pp. 169–184. Springer, Cham (2018). https://doi.org/10.1007/978-3-030-04414-5_12
13. Kwon, O.H., Ma, K.L.: A deep generative model for graph layout. IEEE Trans. Vis. Comput. Graph. **26**(1), 665–675 (2019)
14. Giovannangeli, L., Lalanne, F., Auber, D., Giot, R., Bourqui, R.: Deep neural network for DrawiNg networks, $(DNN)^2$. In: Purchase, H.C., Rutter, I. (eds.) GD 2021. LNCS, vol. 12868, pp. 375–390. Springer, Cham (2021). https://doi.org/10.1007/978-3-030-92931-2_27
15. Wang, Y., Jin, Z., Wang, Q., Cui, W., Ma, T., Qu, H.: DeepDrawing: a deep learning approach to graph drawing. IEEE Trans. Vis. Comput. Graph. (2019). https://doi.org/10.48550/ARXIV.1907.11040
16. van Wageningen, S., Mchedlidze, T.: Can an NN Model plainly learn Planar Layouts? ADD REST ARXIV
17. DeepDrawing Python. https://github.com/jiayouwyhit/deepdrawing. Accessed 31 Aug 2022
18. Chimani, M., et al.: People prefer less stress and fewer crossings. In: GD 2014. LNCS, vol. 8871, pp. 523–524. Springer, Germany (2014). ISBN 978-3-662-45802-0
19. Garg, A., Tamassia, R.: Planar drawings and angular resolution: algorithms and bounds. In: van Leeuwen, J. (ed.) ESA 1994. LNCS, vol. 855, pp. 12–23. Springer, Heidelberg (1994). https://doi.org/10.1007/BFb0049393

20. Jacomy, M., Venturini, T., Heymann, S., Bastian, M.: ForceAtlas2, a continuous graph layout algorithm for handy network visualization designed for the gephi software. PLOS ONE **9**(6). https://doi.org/10.1371/journal.pone.0098679
21. ForceAtlas2 Python. https://github.com/bhargavchippada/forceatlas2. Accessed 10 Jul 2022
22. Gansner, E.R., Koren, Y., North, S.: Graph drawing by stress majorization. In: Pach, J. (ed.) GD 2004. LNCS, vol. 3383, pp. 239–250. Springer, Heidelberg (2005). https://doi.org/10.1007/978-3-540-31843-9_25

Edge Bundling by Density-Based Pathfinding Approach

Ryosuke Saga[✉][ID], Tomoki Yoshikawa, and Tomoharu Nakashima[ID]

Osaka Metropolitan University, 1-1 Gakuencho, Nakaku, Sakai, Japan
{r.saga,tomoharu.nakashima}@omu.ac.jp

1 Introduction

Edge bundling facilitates the understanding of the main flow of edges by transforming them into a bundled state based on certain rules. Various approaches exist, including methods based on the hierarchical structure of the graph [3], geometry [1], and force-directed models [4, 12] (see [17]). Some studies that have been published in recent years consider edge bundling itself mathematically, such as the proposal of faithfulness, which is a graph information fidelity metric [8], and metrics for quantitatively evaluating edge bundling from the viewpoint of structural aesthetics [9, 10]. There is also a tendency to regard edge bundling as an optimization problem. For example, it can be solved as an optimization problem for the combination of bundled edges [2] and sometimes as a control point placement problem for edges [11].

In this paper, we propose a new edge bundling method based on the idea of viewing edge bundling as the minimization of edge drawing cost. Here, we consider that the edge drawing cost consists of the density of the surrounding edges. A higher density improves the quality of edge bundling; thus, we consider that drawing regions with low edge density incurs a drawing cost. In other words, the edge path that passes through the highest-density region possible, i.e., the edge path with the lowest edge drawing cost, is considered to naturally form a bundle. This edge drawing cost is created from a density map that consists of a group of regions divided by grid of 2D/3D graph layouts.

One of the methods similar to this proposed approach is Winding Roads [6]. This method is based on grid, such as the density map used in this proposal, and summarizes the routes on that grid. In that case, the paths on the grids are considered based on the shortest path information between each node. However, unlike this study, the problem does not consider edge density. Another prominent model that considers edge density is kernel density estimation-based edge bundling (KDEEB) [5]. KDEEB deforms edges by repeatedly generating a density map by using kernel density estimation and moving the edge component points according to the gradient of the density map. In contrast, the proposed method creates a real-valued density map with the number of edges calculated for each fixed region of the graph and performs edge bundling by using pathfinding methods.

© The Author(s), under exclusive license to Springer Nature Switzerland AG 2023
P. Angelini and R. von Hanxleden (Eds.): GD 2022, LNCS 13764, pp. 480–482, 2023.
https://doi.org/10.1007/978-3-031-22203-0

Fig. 1. Comparison result (FDEB, KDEEB, and proposed method for US Airline

2   Implementation and Case Study

The pipeline simply consists of the following steps;

1. Input the graph structure and graph layout
2. Create an edge drawing cost map
3. Run pathfinding to minimize the cumulative edge drawing cost for each edge
4. Post-processing

The density map is created from the edge density on the cells of a grid from the graph layout information divided based on grid size s. Then, to express the edge drawing cost, the density within each cell is subtracted from the maximum density on all cells. Considering that the cost increases with each pass through each cell, a path is found such that the search cost from the node source to the target is low so that the final cumulative cost is low. Although multiple edge bundling results may be obtained due to the possibility of finding multiple final paths, the edge bundling candidates are prioritized based on quantitative indicators and other factors. Finally, post-processing such as smoothing and visual encoding is performed.

One feature of this pipeline is that it is relatively easy to implement because the pathfinding can be solved as an edge bundling, and the computational cost can be easily estimated based on the number of cells of a grid. In addition, because the pathfinding method and edge drawing cost can be defined independently, it can be applied to directed and undirected graphs by changing the edge drawing cost map, for example, and a faster pathfinding algorithm can be selected. In addition, unlike image-based techniques such as KDEEB, where the result depends on the layout size, the bundling granularity can be adjusted by adjusting the grid size according to the size of the layout.

We compared and verified the proposed method with force-directed edge bundling (FDEB) [4], KDEEB, which are representative edge bundling methods, and the proposed method with Dijkstra's and A* algorithm (Fig. 1). From the results, we could find the differences among methods. The graphs generated by the proposed method tend to be more linear than other methods, although it depends on the post-processing. Also, if only one candidate is produced, then, outputting at high speed is possible even without a GPU. Comparisons with different grid sizes show that the bundled thickness varies with the size of the grid, and that too large a grid size results in a large collapse. Moreover, there is room for improvement as the proposed method may not work well for graphs with homogeneous densities.

References

1. Cui, W., Zhou, H., Qu, H., Wong, P.C., Li, X.: Geometry-based edge clustering for graph visualization. IEEE Trans. Vis. Comput. Graph. **14**(6), 1277–1284 (2008)
2. Ferreira, J.D.M., Do Nascimento, H.A., Foulds, L.R.: An evolutionary algorithm for an optimization model of edge bundling. Information **9**(7), 154 (2018)
3. Holten, D.: Hierarchical edge bundles: visualization of adjacency relations in hierarchical data. IEEE Trans. Vis. Comput. Graph. **12**(5), 741–748 (2006)
4. Holten, D., Van Wijk, J.J.: Force-directed edge bundling for graph visualization. Comput. Graph. Forum **28**(3), 983–990 (2009). https://doi.org/10.1111/j.1467-8659.2009.01450.x
5. Hurter, C., Ersoy, O., Telea, A.: Graph bundling by kernel density estimation. Comput. Graph. Forum **31**(3pt1), 865–874 (2012). https://doi.org/10.1111/j.1467-8659.2012.03079.x
6. Lambert, A., Bourqui, R., Auber, D.: Winding roads: routing edges into bundles. Comput. Graph. Forum **29**(3), 853–862 (2010)
7. Lhuillier, A., Hurter, C., Telea, A.: State of the art in edge and trail bundling techniques. Comput. Graph. Forum **36**(3), 619–645 (2017). https://doi.org/10.1111/cgf.13213
8. Nguyen, Q., Eades, P., Hong, S.H.: On the faithfulness of graph visualizations. In: 2013 IEEE Pacific Visualization Symposium (PacificVis), pp. 209–216 (2013). https://doi.org/10.1109/PacificVis.2013.6596147
9. Saga, R.: Quantitative evaluation for edge bundling based on structural aesthetics. In: 18th Eurographics Conference on Visualization, EuroVis 2016 - Posters, pp. 17–19. Eurographics Association (2016). https://doi.org/10.2312/eurp.20161131
10. Saga, R.: Validation of quantitative measures for edge bundling by comparing with human feeling. In: 20th Eurographics Conference on Visualization, EuroVis 2018 - Posters, pp. 25–27. Eurographics Association (2018). https://doi.org/10.2312/eurp.20181121
11. Saga, R., Yoshikawa, T., Wakita, K., Sakamoto, K., Schaefer, G., Nakashima, T.: A genetic algorithm optimising control point placement for edge bundling. In: Proceedings of the 15th International Joint Conference on Computer Vision, Imaging and Computer Graphics Theory and Applications, pp. 217–222 (2020)
12. Yamashita, T., Saga, R.: Edge bundling in multi-attributed graphs. In: Yamamoto, S. (ed.) HIMI 2015, Part I. LNCS, vol. 9172, pp. 138–147. Springer, Cham (2015). https://doi.org/10.1007/978-3-319-20612-7_14

The Witness Unit Disk Representability Problem

Giuseppe Liotta[1] , Maarten Löffler[2], Fabrizio Montecchiani[1] ,
Alessandra Tappini[1] , and Soeren Terziadis[3]([⊠])

[1] University of Perugia, Via G. Duranti 93, 06125 Perugia, Italy
{giuseppe.liotta,fabrizio.montecchiani,alessandra.tappini}@unipg.it
[2] Utrecht University, Princetonplein 5, Kamer BBL411, 3584 CC Utrecht,
The Netherlands
m.loffler@uu.nl
[3] Technische Universität Wien, Favoritenstraße 9-11, E192-01, 1040 Vienna, Austria
sterziadis@ac.tuwien.ac.at

1 Introduction

Intersection representations of graphs map vertices to copies of geometric objects
in the plane. Two such objects intersect if and only if their vertices are connected
with an edge. We introduce and study a representation model, which uses an
underlying intersection representation (in our case a unit-disk intersection representation D) and an additional set of objects, which prevent edges, which in
our case is a set W of points (called *witnesses*). The graph G corresponding to
a *witness unit disk representation* (wUDR) $R = (D, W)$ is the graph implied by
D without all edges between two vertices u, v, if the intersection of the disks
corresponding to u and v contains a witness $w \in W$ (see example figure below).

Definition 1. *The unit disk witness number $w(G)$ of a graph G is the minimum
number of witnesses k, s.t., there exists a wUDR $R = (D, W)$, s.t., $|W| = k$. For
a graph class $\mathcal{G}$ we define $w(\mathcal{G})$ as the maximum unit disk witness number over
all graphs $G \in \mathcal{G}$.*

Intersection representations using rectangles [8], L-shapes [9] and curves [7]
have, among others, all been the focus of previous research. Intersection representations with disks have received special attention. It is well-known that the
graphs admitting an intersection representation with pairwise interior-disjoint
disks (also called a *contact representation*) are exactly the planar graphs [11],

---

This project was started during a research visit at the university of Perugia.

© The Author(s), under exclusive license to Springer Nature Switzerland AG 2023
P. Angelini and R. von Hanxleden (Eds.): GD 2022, LNCS 13764, pp. 483–486, 2023.
https://doi.org/10.1007/978-3-031-22203-0

while recognition of graphs that can be represented with overlapping disks is
∃ℝ-hard [13]. Recognizing graphs that can be represented by unit-disk intersection representations is NP-hard for planar graphs [6] and even outerplanar
graphs [5]. wUDR are also related to relative neighborhood graphs [10] and proximity drawings [14], where other vertices can be considered to "interfere" with
edges. Witness proximity drawings that adopt proximity regions different from
unit disks have been studied in [1–4, 12].

2   Contributions

Computing the Graph. Given a wUDR with n disks, w witnesses, we can use
a left to right sweepline algorithm, which uses the left- and rightmost points
of every disk, as well as the intersections between disks (of which there can be
$\Theta(n^2)$ many) as events and keeps track of all disks containing regions, which are
intersected by the sweepline.

Theorem 1. *A graph G can be computed from its wUDR in time $O(n^2 \log n)$.*

Stars, Unit Interval and Complete k-partite Graphs. We present upper/lower
bounds on the unit disk witness numbers of the class of stars, unit interval
bigraphs[1] and complete k-partite graphs.

Theorem 2. *For a star S, $w(S) = 0$ if $n < 7$, and $w(G) = 1$ otherwise.*

Proof (Sketch). If $n < 7$, S is a unit disk graph, otherwise we place the disks
of vertices of degree 1 s.t. there is a region contained in all disks, but not in the
disk of the center vertex. In this region we can place a single witness.

Theorem 3. *For a unit interval bigraph B, $w(B) \leq 2$.*

Proof. Given B (together with its unit interval bigraph representation I), we
can scale I to a width of 2, offset all intervals of one color vertically s.t. intervals
which only shared one endpoint have a distance of 2 between their centers, place
all disks of vertices at the centers of their intervals and remove all overlap of
disks of vertices of one color with one witness each.

Theorem 4. *There are complete k-partite graphs, which require k witnesses and
k witnesses are always sufficient for any complete k-partite graphs.*

Proof (Sketch). Any graph whose partitions have more than 6 vertices require
one witness per partition. It can be shown via construction that this is always
sufficient.

---

[1] Graphs, that admit a representation where every vertex is represented by a red or
blue unit interval on the real line and two vertices are connected if their intervals
intersect and are of different color.

Trees. Finally we provide a construction method for a wUDR for a given tree T with height h, which uses $h - 1$ witnesses.

Theorem 5. *For a tree T, $w(T) \leq h - 1$ and a wUDR with $h - 1$ witnesses can be constructed in polynomial time.*

Proof (Sketch). Let $\mathcal{T}$ be a tree of height h, in which every vertex (except the ones on level $h + 1$) has n children. The proof uses the fact that any tree T on n vertices is an induced sub-graph of $\mathcal{T}$ and that the class of graphs that have a wUDR with c witnesses for any constant c is hereditary. We provide a construction method for $\mathcal{T}$, placing all disks of the same layer almost on top of each other all containing a single witness. These layers are then stacked to create the correct inter-layer connections. Finally we only actually place the disks corresponding to vertices of T.

References

1. Aronov, B., Dulieu, M., Hurtado, F.: Witness (Delaunay) graphs. Comput. Geom. **44**(6–7), 329–344 (2011)
2. Aronov, B., Dulieu, M., Hurtado, F.: Witness Gabriel graphs. Comput. Geom. **46**(7), 894–908 (2013)
3. Aronov, B., Dulieu, M., Hurtado, F.: Mutual witness proximity graphs. Inf. Process. Lett. **114**(10), 519–523 (2014)
4. Aronov, B., Dulieu, M., Hurtado, F.: Witness rectangle graphs. Graphs Comb. **30**(4), 827–846 (2014). https://doi.org/10.1007/s00373-013-1316-x
5. Bhore, S., Löffler, M., Nickel, S., Nöllenburg, M.: Unit disk representations of embedded trees, outerplanar and multi-legged graphs. In: Purchase, H.C., Rutter, I. (eds.) GD 2021. LNCS, vol. 12868, pp. 304–317. Springer, Cham (2021). https://doi.org/10.1007/978-3-030-92931-2_22
6. Breu, H., Kirkpatrick, D.G.: Unit disk graph recognition is NP-hard. Comput. Geom. Theory Appl. **9**(1–2), 3–24 (1998). https://doi.org/10.1016/S0925-7721(97)00014-X
7. Chalopin, J., Gonçalves, D., Ochem, P.: Planar graphs have 1-string representations. Discrete Comput. Geom. **43**(3), 626–647 (2009). https://doi.org/10.1007/s00454-009-9196-9
8. Felsner, S.: Rectangle and square representations of planar graphs. In: Pach, J. (eds.) Thirty Essays on Geometric Graph Theory, pp. 213–248. Springer, New York (2013). https://doi.org/10.1007/978-1-4614-0110-0_12
9. Gonçalves, D., Isenmann, L., Pennarun, C.: Planar graphs as L-intersection or L-contact graphs. In: Discrete Algorithms (SODA 2018), pp. 172–184. SIAM (2018). https://doi.org/10.1137/1.9781611975031.12
10. Jaromczyk, J., Toussaint, G.: Relative neighborhood graphs and their relatives. Proc. IEEE **80**(9), 1502–1517 (1992). https://doi.org/10.1109/5.163414
11. Koebe, P.: Kontaktprobleme der konformen Abbildung. Ber. Sächs. Akad. Wiss. Leipzig, Math.-Phys. Klasse **88**, 141–164 (1936)
12. Lenhart, W., Liotta, G.: Mutual witness Gabriel drawings of complete bipartite graphs. In: Graph Drawing and Network Visualization - 30th International Symposium, GD 2022. LNCS. Springer (2022). https://doi.org/10.1007/978-3-031-22203-0_3

13. McDiarmid, C., Müller, T.: Integer realizations of disk and segment graphs. J. Comb. Theory Ser. B **103**(1), 114–143 (2013). https://doi.org/10.1016/j.jctb.2012. 09.004
14. Tamassia, R.: Handbook of Graph Drawing and Visualization. CRC Press, Boca Raton, Florida, USA (2013). Chapman and Hall/CRC, ISBN 978-1-5848-8412-5

Aggregating Hypergraphs by Node Attributes

David Trye[✉] ⓘ, Mark Apperley ⓘ, and David Bainbridge ⓘ

University of Waikato, Hamilton, New Zealand
dgt12@students.waikato.ac.nz, {mapperle,davidb}@waikato.ac.nz

PAOHVis [2, 5] displays hypergraphs [1, 3] in a matrix where rows represent nodes (dots) and columns represent hyperedges (vertical lines). We propose extensions to PAOHVis for leveraging repeated hyperedges in non-simple hypergraphs, and displaying multiple node attributes. This is accomplished through two aggregation functions: *count-based*, which targets low-level detail, and *binary*, for high-level overview. In doing so, we introduce a domain-agnostic framework for consolidating hypergraphs by one or more categorical node attributes.

Preliminary results indicate that these enhancements provide a clearer picture of overall patterns and distributions of hypergraph data. Consider Fig. 1, which illustrates the different aggregation levels applied to a fictional co-authorship dataset. There are 12 nodes (people) and 17 hyperedges (papers), 13 of which are distinct. Nodes are coloured and subsequently consolidated by the gender of the author. The legend summarises node/category frequencies for the respective hypergraph. Additional categorical node attributes (e.g. affiliation, position and field) can be displayed and aggregated at the same time, provided these are mapped to different visual channels (e.g. shape, outline and texture). Unless they are strongly correlated, aggregating a larger number of attributes greatly reduces the number of identical hyperedges, resulting in a less compact visualisation. Thus, it may be more fruitful to aggregate hypergraphs by each attribute in turn, rather than attempting to visualise all attributes at once.

Count-Based Aggregation. This kind of aggregation shows, for each hyperedge, the exact number of nodes per category. Hyperedges with the corresponding number of nodes in each category (e.g. all papers authored by exactly two men and one woman) are combined. The original size of each hyperedge is preserved and nodes are stacked as tightly as possible, from the top row downwards, in descending order of overall category frequency. This layout facilitates comparisons of hyperedge size, which can be difficult to assess in non-aggregated hypergraphs. The original nodes (people) can no longer be reliably identified, since the same node in a repeated hyperedge may represent a different person across separate instances.

Count-based aggregation is useful for tasks relating to category frequency and overall set size. The middle panel of Fig. 1. shows that all papers have between two and four authors, which was not so apparent in the non-aggregated chart (left panel), due to the different line lengths. It is also easier to see that papers tend to have more male than female authors, but that the paper with the most authors of the same gender is written by four women (and no men).

© The Author(s), under exclusive license to Springer Nature Switzerland AG 2023
P. Angelini and R. von Hanxleden (Eds.): GD 2022, LNCS 13764, pp. 487–489, 2023.
https://doi.org/10.1007/978-3-031-22203-0

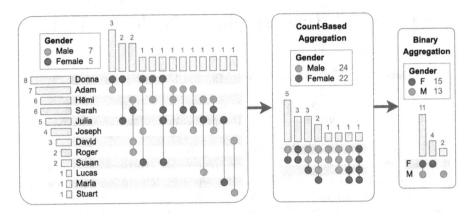

Fig. 1. Different levels of aggregation for a single node attribute (gender).

Binary Aggregation. Hypergraphs can be further aggregated by collapsing each category with multiple occurrences in a hyperedge into a single node. The bar chart then shows the number of hyperedges that contain at least one node from *precisely* the corresponding categories. While this has been partially implemented in PAOHVis, it is not currently possible to consolidate identical hyperedges, which is essential for obtaining a quick overview of hypergraphs that are very dense, especially since (certain) hyperedges are likely to elicit higher counts, given the smaller number of possible category combinations. If an attribute has more than two categories, the data can be aggregated even further, so that all hyperedges are *flattened* into pairwise combinations.

Binary aggregation helps analysts to see how many distinct categories tend to occur in a hyperedge (e.g. do all categories occur together or only some?) and whether particular combinations of categories are dominant. The right-most panel of Fig. 1 shows that, while papers tend to have more male authors, there are more papers authored solely by women (four) than by men (two).

In conclusion, building on PAOHVis, we advocate the consolidation of any repeated hyperedges and the encoding of their frequency in an aligned bar chart above each hyperedge. The result is visually similar to *UpSet* [4] but functionally different, with bar height denoting hyperedge multiplicity rather than set intersection size. This economises horizontal space, while also drawing attention to the distribution of recurrent hyperedges, especially when sorted by frequency.

Aggregation by node attributes is useful in situations where it is less important to know precisely which entities occur in relationships and more important to understand what *kinds* of entities they tend to be (e.g. to investigate a possible gender bias or to see how many papers have female-only or male-only authors). As the level of aggregation increases, more information about the original nodes and hyperedges is lost, in order to reveal more general patterns. It may be beneficial to view all levels of aggregation in conjunction, rather than in isolation.

References

1. Berge, C.: Graphs and hypergraphs (1973)
2. Buono, P., Valdivia, P.: Applications of dynamic hypergraph visualization. In: Proceedings of the 2022 International Conference on Advanced Visual Interfaces. AVI 2022. Association for Computing Machinery, New York 2022). https://doi.org/10.1145/3531073.3534495
3. Fischer, M.T., Frings, A., Keim, D.A., Seebacher, D.: Towards a survey on static and dynamic hypergraph visualizations. IEEE Trans. Vis. Comput. Graph. **2**(3), 81–85 (2021). https://doi.org/10.1109/VIS49827.2021.9623305
4. Lex, A., Gehlenborg, N., Strobelt, H., Vuillemot, R., Pfister, H.: Upset: visualization of intersecting sets. IEEE Trans. Vis. Comput. Graph. **20**(12), 1983–1992 (2014). https://doi.org/10.1109/TVCG.2014.2346248
5. Valdivia, P., Buono, P., Plaisant, C., Dufournaud, N., Fekete, J.D.: Analyzing dynamic hypergraphs with parallel aggregated ordered hypergraph visualization. IEEE Trans. Vis. Comput. Graph. **27**(1), 1–13 (2021). https://doi.org/10.1109/TVCG.2019.2933196

Author Index

Ahmed, Reyan 261, 319
Aichholzer, Oswin 16, 49
Alegría, Carlos 127
Apperley, Mark 487
Arleo, Alessio 271

Bainbridge, David 487
Bekos, Michael A. 144, 361, 371
Biedl, Therese 404
Binucci, Carla 201, 304
Bögl, Markus 271
Bourqui, Romain 61

Chaplick, Steven 175, 389
Cheong, Otfried 247
Cornelsen, Sabine 111

Da Lozzo, Giordano 127, 361
Di Battista, Giuseppe 127, 289
Di Giacomo, Emilio 175, 304
Diatzko, Gregor 111
Didimo, Walter 157, 201, 289

Eades, Peter 93
El Maalouly, Nicolas 16

Felsner, Stefan 371, 441
Filipov, Velitchko 271
Firman, Oksana 432
Fox, Jacob 219
Frati, Fabrizio 127, 175, 361

Ganian, Robert 175
García, Alfredo 40, 49
Geladaris, Vasileios 188
Giot, Romain 61
Giovannangeli, Loann 61
Gray, Kathryn 319
Grilli, Luca 289
Gronemann, Martin 144, 361
Grosso, Fabrizio 127, 289
Gutowski, Grzegorz 418

Hong, Seok-Hee 93
Huroyan, Vahan 77

Kaufmann, Michael 157
Kindermann, Philipp 371, 389, 432, 459
Klawitter, Jonathan 389, 432
Klemz, Boris 432
Klesen, Felix 432
Klute, Fabian 459
Knorr, Kristin 16
Kobourov, Stephen 77, 261, 319, 371
Kratochvíl, Jan 371
Kryven, Myroslav 261

Lalanne, Frederic 61
Lenhart, William J. 25, 304
Li, Mingwei 319
Lionakis, Panagiotis 188
Liotta, Giuseppe 25, 157, 304, 483
Loffler, Maarten 483

M. Reddy, Meghana 16
Mchedlidze, Tamara 361, 459, 476
Meidiana, Amyra 93
Meulemans, Wouter 459
Miksch, Silvia 271
Miller, Jacob 77
Misue, Kazuo 336
Mittelstädt, Florian 418
Mondal, Debajyoti 473
Montecchiani, Fabrizio 144, 304, 483
Moradi, Ehsan 473
Mulzer, Wolfgang 16

Nakashima, Tomoharu 480
Nöllenburg, Martin 232, 304

Obenaus, Johannes 16
Ortali, Giacomo 157, 289

Pach, János 219
Parada, Irene 49

Patrignani, Maurizio   127, 201, 289
Paul, Rosna   16
Pfister, Maximilian   247
Pupyrev, Sergey   353

Raftopoulou, Chrysanthi N.   175, 361
Roch, Sandro   441
Rutter, Ignaz   371, 389, 418

Saga, Ryosuke   480
Scheucher, Manfred   441
Schlipf, Lena   247
Shvets, Mykyta   473
Simonov, Kirill   175
Sorge, Manuel   232
Spoerhase, Joachim   418
Suk, Andrew   3, 219
Symvonis, Antonios   144, 304

Tappini, Alessandra   289, 483
Tejel, Javier   40
Terziadis, Soeren   232, 483
Tollis, Ioannis G.   188
Trye, David   487

van Wageningen, Simon   476
Villedieu, Anaïs   232
Vogtenhuber, Birgit   16, 40, 49

Weinberger, Alexandra   16, 40, 49
Wolff, Alexander   389, 418, 432
Wu, Hsiang-Yun   232
Wulms, Jules   232

Yoshikawa, Tomoki   480

Zeng, Ji   3
Zink, Johannes   418

Printed in the United States
by Baker & Taylor Publisher Services